Design of Prestressed Concrete to AS3600-2009

Second Edition

Design of Prestressed Concrete to AS3600-2009

Second Edition

Raymond Ian Gilbert
Neil Colin Mickleborough
Gianluca Ranzi

CRC Press
Taylor & Francis Group
Boca Raton   London   New York

CRC Press is an imprint of the
Taylor & Francis Group, an **informa** business

A SPON PRESS BOOK

CRC Press
Taylor & Francis Group
6000 Broken Sound Parkway NW, Suite 300
Boca Raton, FL 33487-2742

Printed on acid-free paper
Version Date: 20150507

International Standard Book Number-13: 978-1-4665-7269-0 (Paperback)

Library of Congress Cataloging-in-Publication Data

Gilbert, R. I., 1950-
   [Design of prestressed concrete]
   Design of prestressed concrete to AS3600-2009 / Raymond Ian Gilbert, Neil Colin Mickleborough, and Gianluca Ranzi. -- Second edition.
      pages cm
   Revised editon of: Design of prestressed concrete / R.I. Gilbert & N.C. Mickleborough. 1990.
   Includes bibliographical references and index.
   ISBN 978-1-4665-7269-0 (acid-free paper) 1. Prestressed concrete construction. 2. Structural design. I. Mickleborough, N. C. (Neil C.) II. Ranzi, Gianluca, 1972- III. Title.

TA683.9.G52 2016
624.1'83412--dc23   2015014596

Visit the Taylor & Francis Web site at
http://www.taylorandfrancis.com

and the CRC Press Web site at
http://www.crcpress.com

Contents

Preface   *xvii*
Acknowledgements   *xxi*
Notation and sign convention   *xxiii*

1 Basic concepts   **1**

  1.1   *Introduction 1*
  1.2   *Methods of prestressing 4*
     1.2.1   *Pretensioned concrete 4*
     1.2.2   *Post-tensioned concrete 5*
     1.2.3   *Other methods of prestressing 6*
  1.3   *Transverse forces induced by draped tendons 6*
  1.4   *Calculation of elastic stresses 9*
     1.4.1   *Combined load approach 10*
     1.4.2   *Internal couple concept 11*
     1.4.3   *Load balancing approach 12*
     1.4.4   *Introductory example 13*
  1.5   *Introduction to structural behaviour –*
     Initial to ultimate load 16
  Reference 19

2 Design procedures and applied actions   **21**

  2.1   *Limit states design philosophy 21*
  2.2   *Structural modelling and analysis 22*
     2.2.1   *Structural modelling 22*
     2.2.2   *Structural analysis 24*
  2.3   *Actions and combinations of actions 26*
     2.3.1   *General 26*
     2.3.2   *Combinations of actions*
          for the strength limit states 29

2.3.3   Combinations of actions
        for the stability limit states 30
2.3.4   Combinations of actions
        for the serviceability limit states 30
2.4 Design for the strength limit states 31
2.4.1   General 31
2.4.2   Strength checks for use with linear elastic
        analysis, simplified methods of analysis
        and for statically determinate structures 32
2.4.3   Strength checks for use with other
        methods of structural analysis 32
2.5 Design for the serviceability limit states 35
2.5.1   General 35
2.5.2   Deflection limits 36
2.5.3   Vibration control 39
2.5.4   Crack width limits 39
2.6 Design for durability 40
2.7 Design for fire resistance 43
2.8 Design for robustness 43
References 46

3 Prestressing systems   49

3.1 Introduction 49
3.2 Types of prestressing steel 49
3.3 Pretensioning 51
3.4 Post-tensioning 53
3.5 Bonded and unbonded post-tensioned construction 59
3.6 Circular prestressing 61
3.7 External prestressing 62
Reference 63

4 Material properties   65

4.1 Introduction 65
4.2 Concrete 65
4.2.1   Composition of concrete 66
4.2.2   Strength of concrete 66
4.2.3   Strength specifications in AS3600-2009 70
        4.2.3.1   Characteristic compressive strength 70
        4.2.3.2   Mean in situ compressive strength 71
        4.2.3.3   Tensile strength 71

                4.2.3.4   Stress–strain curves for
                            concrete in compression 72
        4.2.4   Deformation of concrete 73
                4.2.4.1   Discussion 73
                4.2.4.2   Instantaneous strain 75
                4.2.4.3   Creep strain 77
                4.2.4.4   Shrinkage strain 82
        4.2.5   Deformational characteristics
                specified in AS3600-2009 84
                4.2.5.1   Introduction 84
                4.2.5.2   Elastic modulus 84
                4.2.5.3   Creep coefficient 85
                4.2.5.4   Shrinkage strain 88
                4.2.5.5   Thermal expansion 92
    4.3 Steel reinforcement 92
        4.3.1   General 92
        4.3.2   Specification in AS3600-2009 94
                4.3.2.1   Strength and ductility 94
                4.3.2.2   Elastic modulus 95
                4.3.2.3   Stress–strain curves 95
                4.3.2.4   Coefficient of thermal expansion 97
    4.4 Steel used for prestressing 97
        4.4.1   General 97
        4.4.2   Specification in AS3600-2009 99
                4.4.2.1   Strength and ductility 99
                4.4.2.2   Elastic modulus 100
                4.4.2.3   Stress–strain curve 100
                4.4.2.4   Steel relaxation 101
    References 104

5 Design for serviceability   107

    5.1 Introduction 107
    5.2 Concrete stresses both at transfer
         and under full service loads 108
    5.3 Maximum jacking forces 111
    5.4 Determination of prestress and
         eccentricity in flexural members 112
        5.4.1   Satisfaction of stress limits 112
        5.4.2   Load balancing 120
    5.5 Cable profiles 122

5.6   Short-term analysis of uncracked cross-sections 123
      5.6.1   General 123
      5.6.2   Short-term cross-sectional analysis 126
5.7   Time-dependent analysis of uncracked cross-sections 142
      5.7.1   Introduction 142
      5.7.2   The age-adjusted effective modulus method 142
      5.7.3   Long-term analysis of an uncracked
              cross-section subjected to combined axial
              force and bending using AEMM 144
      5.7.4   Discussion 161
5.8   Short-term analysis of cracked cross-sections 164
      5.8.1   General 164
      5.8.2   Assumptions 166
      5.8.3   Analysis 166
5.9   Time-dependent analysis of cracked cross-sections 176
      5.9.1   Simplifying Assumption 176
      5.9.2   Long-term analysis of a cracked cross-
              section subjected to combined axial
              force and bending using the AEMM 177
5.10 Losses of prestress 181
      5.10.1 Definitions 181
      5.10.2 Immediate losses 182
              5.10.2.1 Elastic deformation losses 182
              5.10.2.2 Friction in the jack and anchorage 183
              5.10.2.3 Friction along the tendon 183
              5.10.2.4 Anchorage losses 186
              5.10.2.5 Other causes of
                        immediate losses 187
      5.10.3 Time-dependent losses of prestress 187
              5.10.3.1 Discussion 187
              5.10.3.2 Shrinkage losses 188
              5.10.3.3 Creep losses 190
              5.10.3.4 Relaxation losses 190
5.11 Deflection calculations 193
      5.11.1 General 193
      5.11.2 Short-term moment curvature
              relationship and tension stiffening 196
      5.11.3 Short-term deflection 201
      5.11.4 Long-term deflection 207
              5.11.4.1 Creep-induced curvature 208
              5.11.4.2 Shrinkage-induced curvature 209

5.12 Crack control 215
    5.12.1 Flexural crack control 215
    5.12.2 Crack control for restrained shrinkage
         and temperature effects 217
    5.12.3 Crack control at openings and discontinuities 219
References 219

6 Ultimate flexural strength   221

6.1 Introduction 221
6.2 Flexural behaviour at overloads 221
6.3 Ultimate flexural strength 224
    6.3.1 Assumptions 224
    6.3.2 Idealised rectangular compressive
         stress block for concrete 224
    6.3.3 Prestressed steel strain components
         (for bonded tendons) 227
    6.3.4 Determination of M_{uo} for a singly
         reinforced section with bonded tendons 229
    6.3.5 Determination of M_{uo} for sections
         containing non-prestressed
         reinforcement and bonded tendons 233
6.4 Approximate procedure in AS3600-2009 241
    6.4.1 Bonded tendons 241
    6.4.2 Unbonded tendons 245
6.5 Design calculations 247
    6.5.1 Discussion 247
    6.5.2 Calculation of additional non-prestressed
         tensile reinforcement 247
    6.5.3 Design of a doubly reinforced cross-section 251
6.6 Flanged sections 254
6.7 Ductility and robustness of prestressed concrete beams 260
    6.7.1 Introductory remarks 260
    6.7.2 Calculation of hinge rotations 262
    6.7.3 Quantifying ductility and
         robustness of beams and slabs 263
References 265

7 Ultimate strength in shear and torsion   267

7.1 Introduction 267
7.2 Shear in beams 267

7.2.1   *Inclined cracking 267*

7.2.2   *Effect of prestress 268*

7.2.3   *Web reinforcement 270*

7.2.4   *Shear strength 273*

    7.2.4.1   *Flexure-shear cracking 276*

    7.2.4.2   *Web-shear cracking 277*

7.2.5   *Summary of design requirements for shear 279*

    7.2.5.1   *Design equation 280*

7.3   *Torsion in beams 288*

7.3.1   *Compatibility torsion and equilibrium torsion 288*

7.3.2   *Effects of torsion 289*

7.3.3   *Design provisions for torsion 291*

    7.3.3.1   *Compatibility torsion 291*

    7.3.3.2   *Equilibrium torsion 292*

7.4   *Shear in slabs and footings 299*

7.4.1   *Punching shear 299*

7.4.2   *Design for punching shear 301*

    7.4.2.1   *Introduction and definitions 301*

    7.4.2.2   *Shear strength with no moment transfer 302*

    7.4.2.3   *Shear strength with moment transfer 303*

References 311

8   Anchorage zones   **313**

8.1   *Introduction 313*

8.2   *Pretensioned concrete – Force transfer by bond 314*

8.3   *Post-tensioned concrete anchorage zones 318*

8.3.1   *Introduction 318*

8.3.2   *Methods of analysis 322*

    8.3.2.1   *Single central anchorage 325*

    8.3.2.2   *Two symmetrically placed anchorages 326*

8.3.3   *Reinforcement requirements 329*

8.3.4   *Bearing stresses behind anchorages 330*

8.4   *Strut-and-tie modelling 346*

8.4.1   *Introduction 346*

8.4.2   *Concrete struts 347*

    8.4.2.1   *Types of struts 347*

    8.4.2.2   *Strength of struts 349*

    8.4.2.3   *Bursting reinforcement in bottle-shaped struts 349*

8.4.3   Steel ties 351
8.4.4   Nodes 351
References 352

9 Composite members   355

9.1   Types and advantages of composite construction 355
9.2   Behaviour of composite members 356
9.3   Stages of loading 358
9.4   Determination of prestress 361
9.5   Methods of analysis at service loads 363
       9.5.1   Introductory remarks 363
       9.5.2   Short-term analysis 364
       9.5.3   Time-dependent analysis 366
9.6   Ultimate flexural strength 392
9.7   Horizontal shear transfer 393
       9.7.1   Discussion 393
       9.7.2   Provisions for horizontal shear 395
9.8   Ultimate shear strength 398
       9.8.1   Introductory remarks 398
       9.8.2   Web-shear cracking 399
       9.8.3   Flexure-shear cracking 400
References 403

10 Design procedures for determinate beams   405

10.1   Introduction 405
10.2   Types of section 405
10.3   Initial trial section 407
       10.3.1   Based on serviceability requirements 407
       10.3.2   Based on strength requirements 408
10.4   Design procedures: fully prestressed beams 410
       10.4.1   Beams with varying eccentricity 411
       10.4.2   Beams with constant eccentricity 427
10.5   Design procedures: partially-prestressed beams 437
       10.5.1   Discussion 437
Reference 445

11 Statically indeterminate members   447

11.1   Introduction 447
11.2   Tendon profiles 449

11.3   Continuous beams 452
    11.3.1   Effects of prestress 452
    11.3.2   Determination of secondary
        effects using virtual work 453
    11.3.3   Linear transformation of a tendon profile 459
    11.3.4   Analysis using equivalent loads 461
        11.3.4.1   Moment distribution 462
    11.3.5   Practical tendon profiles 471
    11.3.6   Members with varying cross-
        sectional properties 475
    11.3.7   Effects of creep 476
11.4   Statically indeterminate frames 480
11.5   Design of continuous beams 484
    11.5.1   General 484
    11.5.2   Service load range – Before cracking 484
    11.5.3   Service load range – After cracking 487
    11.5.4   Overload range and ultimate
        strength in bending 488
        11.5.4.1   Behaviour 488
        11.5.4.2   Permissible moment
            redistribution at ultimate 490
        11.5.4.3   Secondary effects at ultimate 491
    11.5.5   Steps in design 492
References 505

12   Two-way slabs: Behaviour and design   507

12.1   Introduction 507
12.2   Effects of prestress 510
12.3   Balanced load stage 513
12.4   Initial sizing of slabs 515
    12.4.1   Existing guidelines 515
    12.4.2   Serviceability approach for the
        calculation of slab thickness 516
        12.4.2.1 The slab system factor, K 519
    12.4.3   Discussion 520
12.5   Other serviceability considerations 522
    12.5.1   Cracking and crack control in prestressed slabs 522
    12.5.2   Long-term deflections 523
12.6   Design approach – General 525
12.7   One-way slabs 525

12.8   Two-way edge-supported slabs 526
    12.8.1   Load balancing 526
    12.8.2   Methods of analysis 528
12.9   Flat plate slabs 540
    12.9.1   Load balancing 540
    12.9.2   Behaviour under unbalanced load 542
    12.9.3   Frame analysis 544
    12.9.4   Direct design method 546
    12.9.5   Shear strength 547
    12.9.6   Deflection calculations 548
    12.9.7   Yield line analysis of flat plates 561
12.10 Flat slabs with drop panels 566
12.11 Band-beam and slab systems 567
References 568

13   Compression and tension members   571

13.1   Types of compression members 571
13.2   Classification and behaviour of compression members 572
13.3   Cross-sectional analysis: Compression and bending 573
    13.3.1   The strength interaction diagram 573
    13.3.2   Ultimate strength analysis 575
    13.3.3   Design interaction curves 586
    13.3.4   Biaxial bending and compression 587
13.4   Slenderness effects 588
    13.4.1   Background 588
    13.4.2   Moment magnification method 591
13.5   Reinforcement requirements for
    compression members 597
13.6   Transmission of axial force through a floor system 597
13.7   Tension members 599
    13.7.1   Advantages and applications 599
    13.7.2   Behaviour 599
References 605

14   Detailing: Members and connections   607

14.1   Introduction 607
14.2   Principles of detailing 608
    14.2.1   When is steel reinforcement required? 608
    14.2.2   Objectives of detailing 609

14.2.3   *Sources of tension 610*
    14.2.3.1   *Tension caused by bending*
        (and axial tension) 610
    14.2.3.2   *Tension caused by load reversals 610*
    14.2.3.3   *Tension caused by shear and torsion 611*
    14.2.3.4   *Tension near the supports of beams 611*
    14.2.3.5   *Tension within the supports*
        of beams or slabs 612
    14.2.3.6   *Tension within connections 613*
    14.2.3.7   *Tension at concentrated loads 613*
    14.2.3.8   *Tension caused by directional*
        changes of internal forces 613
    14.2.3.9   *Other common sources of tension 616*
14.3   *Anchorage of deformed bars in tension 616*
  14.3.1   *Introductory remarks 616*
  14.3.2   *Development length for*
    deformed bars in tension 618
    14.3.2.1   *Basic development length 618*
    14.3.2.2   *Refined development length 620*
    14.3.2.3   *Length required to develop a*
        stress lower than the yield stress 622
    14.3.2.4   *Development length of a deformed bar*
        in tension with a standard hook or cog 622
    14.3.2.5   *Development length of*
        plain bars in tension 623
  14.3.3   *Lapped splices for bars in tension 625*
14.4   *Anchorage of deformed bars in compression 625*
  14.4.1   *Introductory remarks 625*
  14.4.2   *Development length of deformed*
    bars in compression 626
    14.4.2.1   *Basic development length 626*
    14.4.2.2   *Refined development length 626*
    14.4.2.3   *Development length to develop*
        a stress lower than the yield stress 627
    14.4.2.4   *Development length of a deformed bar*
        in compression with a hook or cog 627
    14.4.2.5   *Development length of plain*
        bars in compression 627
  14.4.3   *Lapped splices for bars in compression 628*
14.5   *Stress development and coupling of tendons 628*

14.6 Detailing of beams 628
    14.6.1 Anchorage of longitudinal
         reinforcement: General 628
    14.6.2 Anchorage of stirrups 632
    14.6.3 Detailing of support and loading points 636
14.7 Detailing of columns 640
    14.7.1 General requirements 640
    14.7.2 Requirements of AS3600-2009 642
14.8 Detailing of beam-column connections 644
    14.8.1 Introduction 644
    14.8.2 Knee connections (or two-member connections) 645
        14.8.2.1 Opening moments 645
        14.8.2.2 Closing moments 647
    14.8.3 Exterior three-member connections 649
    14.8.4 Interior four-member connections 650
14.9 Detailing of corbels 651
    14.9.1 Introduction 651
    14.9.2 Design procedure 651
14.10 Joints in structures 656
    14.10.1 Introduction 656
    14.10.2 Construction joints 656
    14.10.3 Control joints (contraction joints) 657
    14.10.4 Shrinkage strips 660
    14.10.5 Expansion joints 660
    14.10.6 Structural joints 661
References 662

Index   665

Preface

For the design of prestressed concrete structures, a sound understanding of structural behaviour at all stages of loading is essential. Also essential is a thorough knowledge of the design criteria specified in the relevant design standard, including the rules and requirements and their background. The aim of this book is to present a detailed description and explanation of the behaviour of prestressed concrete members and structures both at service loads and at ultimate loads and, in doing so, provide a comprehensive guide to structural design. Much of the text is based on first principles and relies only on the principles of mechanics and the properties of concrete and steel, with numerous worked examples. However, where the design requirements are code specific, this book refers to the provisions of the *Australian Standard for Concrete Structures* AS3600-2009 and, where possible, the notation is the same as in AS3600-2009. A companion edition in accordance with the requirements of Eurocode 2 is also available, with the same notation as in the European standard.

The first edition of the book was published almost 25 years ago, so a comprehensive update and revision is long overdue. This edition contains the most up-to-date and recent advances in the design of modern prestressed concrete structures, as well as the fundamental aspects of prestressed concrete behaviour and design that were so well received in the first edition. The text is written for senior undergraduate and postgraduate students of civil and structural engineering and also for practising structural engineers. It retains the clear and concise explanations and the easy-to-read style of the first edition.

Between them, the authors have almost 100 years of experience in the teaching, research and design of prestressed concrete structures, and this book reflects this wealth of experience. The work has also gained much from the membership of Professor Gilbert on the committees of Standards Australia and the American Concrete Institute and his involvement in the development of AS3600-2009 over the past 35 years.

The scope of the work ranges from an introduction to the fundamentals of prestressed concrete to in-depth treatments of the more advanced topics in modern prestressed concrete structures. The basic concepts of prestressed concrete are introduced in Chapter 1, and the limit states design philosophies used in Australian practice are outlined in Chapter 2. The hardware required to pre-tension and post-tension concrete structures is introduced in Chapter 3, including some construction considerations. Material properties relevant to design are presented and discussed in Chapter 4. A comprehensive treatment of the design of prestressed concrete beams for serviceability is presented in Chapter 5. The instantaneous and time-dependent behaviour of cross-sections under service loads is discussed in considerable detail, and methods for the analysis of both uncracked and cracked cross-sections are considered. Techniques for determining the section size, the magnitude and eccentricity of prestress, the losses of prestress and the deflection of members are outlined. Each aspect of design is illustrated by numerical examples.

Chapters 6 and 7 deal with the design of members for strength in bending, shear and torsion, and Chapter 8 covers the design of the anchorage zones in both pre-tensioned and post-tensioned members. A guide to the design of composite prestressed concrete beams is provided in Chapter 9 and includes a detailed worked example of the analysis of a composite trough girder footbridge. Chapter 10 discusses design procedures for statically determinate beams. Comprehensive self-contained design examples are provided for fully prestressed and partially prestressed, post-tensioned and pre-tensioned concrete members.

Chapter 11 covers the analysis and design of statically indeterminate beams and frames, and provides guidance on the treatment of secondary effects at all stages of loading. Chapter 12 provides a detailed discussion of the analysis and design of two-way slab systems, including aspects related to both strength and serviceability. Complete design examples are provided for panels of an edge-supported slab and a flat slab. The behaviour of axially loaded members is dealt with in Chapter 13. Compression members, members subjected to combined bending and compression, and prestressed concrete tension members are discussed and design aspects are illustrated by examples. Guidelines for successful detailing of the structural elements and connections in prestressed concrete structures are outlined in Chapter 14.

As in the first edition, the book provides a unique focus on the treatment of serviceability aspects of design. Concrete structures are prestressed to improve behaviour at service loads and thereby increase the economical range of concrete as a construction material. In conventional prestressed structures, the level of prestress and the position of the tendons are usually based on the considerations of serviceability. Practical methods for accounting for the nonlinear and time-dependent effects of cracking, creep, shrinkage and relaxation are presented in a clear and easy-to-follow format.

The authors hope that *Design of Prestressed Concrete to AS3600-2009* will be a valuable source of information and a useful guide for students and practitioners of structural design.

Raymond Ian Gilbert
Neil Colin Mickleborough
Gianluca Ranzi
Sydney, New South Wales, Australia

Acknowledgements

The authors acknowledge the support given by their respective institutions and by the following individuals and organisations for supplying the photographs contained herein:

Brian Lim (VSL International Limited)
Brett Gibbons (VSL Australia)

Notation and sign convention

All symbols are also defined in the text where they first appear. Throughout the book, we have assumed that tension is positive and compression is negative and that positive bending about a horizontal axis causes tension in the bottom fibres of a cross-section.

A	cross-sectional area
A_b	cross-sectional area of a reinforcing bar
A_c	area of the concrete part of the cross-section; *or* smallest cross-sectional area of a concrete strut measured normal to the line of action of the strut
$\underline{A_g}$	gross cross-sectional area
$\overline{A_k}$	area of the age-adjusted transformed section at time τ_k
A_m	an area enclosed by the median lines of the walls of a single cell of a box section
A_{min}	minimum required area of a cross-section
A_p	cross-sectional area of prestressed steel
$A_{p(i)}$	cross-sectional area of the prestressed steel at the i-th level
A_{pc}	area of the precast element of a composite cross-section
A_{pt}	cross-sectional area of the tendons in the zone that will be tensile under ultimate load conditions
A_s	cross-sectional area of the non-prestressed steel
$A_{s(i)}$	cross-sectional areas of non-prestressed steel reinforcement at the i-th level
A_{sb}	cross-sectional area of the transverse reinforcement required for bursting in an anchorage zone
A_{sc}	cross-sectional area of the non-prestressed compressive reinforcement
A_{sc}^+	additional area of the longitudinal reinforcement that should be provided to resist the torsion-induced tensile forces in the flexural compressive zone
A_{sf}	cross-sectional area of fully anchored reinforcement crossing the interface in a composite member

$A_{s(min)}$	minimum cross-sectional area of non-prestressed reinforcement
A_{ss}	cross-sectional area of the transverse reinforcement required for spalling in an anchorage zone
A_{st}	cross-sectional areas of the non-prestressed tensile reinforcement
A_{st}^{+}	additional area of the longitudinal reinforcement that should be provided to resist the torsion-induced tensile forces in the flexural tensile zone
A_{sv}	cross-sectional area of shear reinforcement at each stirrup location
$A_{sv.min}$	cross-sectional area of minimum shear reinforcement
A_{sw}	cross-sectional area of bar used for closed torsional stirrup
$A_{sw.\,min}$	minimum cross-sectional area of the closed stirrup (Equation 7.41)
$(A_s)_{min}$	minimum area of non-prestressed reinforcement required for crack control (Equations 5.184 through 5.186)
A_t	area of the polygon with vertices at the centres of the longitudinal bars at the corners of a closed stirrup
A_{tr}	cross-sectional area of a transverse reinforcing bar
A'	area of concrete under the idealised rectangular compressive stress block
A_0	the area of the transformed section at time τ_0
A_1	bearing area
A_2	largest area of the concrete surface geometrically similar to A_1
a	shear span, equal to the distance between the lines of action of an applied load and the adjacent support reaction; *or* horizontal projection of a strut
a	width of the torsion strip (Figure 7.14), distance along a beam between the points of zero bending moment, *or* distance from the face of the column to the applied load on a corbel
a_m	average axis distance
a_s	axis distance
a_{sup}	length of a support in the direction of the span (see Figure 2.1)
a_v	distance from the section at which shear is being considered to the face of the nearest support
B	first moment of the area of a cross-section about the reference axis
B_c	first moment of the concrete part of the cross-section about the reference axis

$\bar{B}_k$	first moment of the age-adjusted transformed section at time τ_k
B_0	the first moment of the area of the transformed section about the reference axis at first loading (i.e. at time τ_0)
b	width of the compressive zone of the cross-section of a beam or slab *or* smaller cross-sectional dimension of a rectangular column or the diameter of a circular column
b_{ef}	effective width of the flange of a flanged cross-section
b_f	width of the contact surface between the precast and in situ parts of a composite cross-section
b_o	width of an opening adjacent to the critical shear perimeter (Figure 7.13)
b_{tr}	width of the transformed flange of a composite cross-section (Equation 9.1)
b_v	effective width of the web for shear calculations (see Equation 7.6)
b_w	width of the web of a flanged cross-section
C	resultant compressive force, carry-over factor, or Celsius
$C(t, \tau)$	specific creep at time t produced by a sustained unit stress first applied at τ
C_b	transverse compressive force behind an anchorage plate caused by bursting
C_c	compressive force carried by the concrete
C_s	compressive force in the non-prestressed steel
$c\ (c_1)$	concrete cover
c_d	the smaller of the concrete covers to the deformed bar or half the clear distance to the next parallel bar developing stress
c_1, c_2	side dimensions of a column
D	overall depth of a cross-section
D_b	overall depth of the beam in a beam and slab system
D_e	depth of the symmetrical prism within an anchorage zone
D_{min}	minimum overall depth
D_s	depth of a slab
$(D_s)_{min}$	minimum effective slab thickness
$\mathbf{D}_k$	matrix of cross-sectional rigidities at time τ_k (Equation 5.84)
$\mathbf{D}_0$	matrix of cross-sectional rigidities at time τ_0 (Equation 5.40)
d	effective depth from the extreme compressive fibre to the resultant tensile force at the ultimate strength in bending
d_b	bar diameter
d_c	width of the idealised strut

$d_{c(1)}$, $d_{c(2)}$	depth from top fibre to the centroid of the precast and in situ elements, respectively (see Figure 9.3)
d_d	diameter of a prestressing duct
d_{id}	minimum diameter of bend in a reinforcing bar
d_n	depth from the extreme compressive fibre to the neutral axis
d_{n1}	depth from the extreme compressive fibre to the neutral axis for a section containing only prestressing steel (see Figure 6.10a)
d_o	depth to the bottom layer of tensile reinforcement
d_{om}	mean value of d_o averaged around the critical shear perimeter
d_p	depth to the prestressed steel
d_{pc}	depth to the plastic centroid of the cross-section (Figure 13.3)
$d_{p(i)}$	depth to the i-th level of prestressed steel
d_{ref}	depth of reference axis below the top fibre of the cross-section
d_s	depth to the non-prestressed steel
$d_{s(i)}$	depth to the i-th level of non-prestressed steel
d_{sc}	depth to the non-prestressed compressive steel
E	earthquake action
E_c	elastic modulus of concrete
E_{cp}	elastic modulus of concrete at transfer
E_{c1}, E_{c2}	elastic moduli of concrete in precast and in-situ elements, respectively, of a composite cross-section
E_d	the design action effect
E_e, $\bar{E}_e$	effective modulus of concrete (Equation 4.16) and age-adjusted effective modulus of concrete (Equation 4.18), respectively
$E_e(\tau_k,\tau_0)$	effective modulus of concrete at time τ_k for concrete first loaded at τ_0 (Equation 5.54)
$\bar{E}_e(\tau_k,\tau_0)$	age-adjusted effective modulus of concrete at time τ_k for concrete first loaded at time τ_0 (Equation 5.55)
E_p	elastic modulus of prestressing steel
$E_{p(i)}$	elastic moduli of the i-th level of prestressed steel
E_s	elastic modulus of non-prestressed steel reinforcing bars
$E_{s(i)}$	elastic moduli of the i-th level of non-prestressed steel
e	eccentricity of prestressing force or load in a compression member
e	base of Napierian logarithms
e^*	eccentricity of the pressure line from the centroidal axis
e_{AB}	axial deformation of member AB
e_{max}	maximum possible eccentricity of prestress
e_{min}	minimum acceptable eccentricity of prestress

e_o	initial eccentricity of load in a slender column
e_{pc}	eccentricity of prestress measured from the centroidal axis of the precast part of a composite cross-section
$F(x), \overline{F}(x)$	functions of x (Equation 11.5)
F_b	design strength of concrete in bearing
F_c	compressive stress limits for concrete under full load
F_c^*	absolute value of the design force in the compressive zone due to flexure
F_{cp}	compressive stress limits for concrete immediately after transfer
$F_{d.ef}$	effective design service load per unit length or area, used in serviceability design
F_e	action caused by earth pressure
F_{lp}	action caused by liquid pressure
F_{sn}	action caused by snow loads
F_t	tensile stress limits for concrete under full load
F_{tp}	tensile stress limits for concrete immediately after transfer
$\overline{F}_{e,0}$	age-adjusted creep factor (Equation 5.58)
$\mathbf{F}_k$	matrix relating applied actions to strain at time τ_k (Equation 5.90)
$\mathbf{F}_0$	matrix relating applied actions to strain at time τ_0 (Equation 5.44)
f_B	flexibility coefficient associated with a release at point B
f_b	average ultimate bond stress
f_{cm}	mean compressive concrete cylinder strength
f_{cmi}	mean compressive in-situ strength of concrete
f_{cp}	mean compressive strength of concrete at transfer
f_{ct}	uniaxial tensile strength of concrete
$f_{ct.f}$	flexural tensile strength of concrete
$f_{ct.sp}$	splitting tensile strength of concrete
f_{cu}	compressive strength of concrete from cube tests
f_{cv}	limiting concrete shear stress of a cross-section or the critical shear perimeter
f_{pb}	characteristic minimum breaking strength of prestressing steel
f_{bpt}	bond stress in the transmission length (Equation 8.1)
f_{py}	yield strength (0.1% proof stress) for prestressing steel
f_{su}	characteristic tensile strength (peak stress) of reinforcement
f_{sy}	characteristic yield strength of reinforcement
$f_{sy.f}$	characteristic yield strength of reinforcement used as fitments
$\mathbf{f}_{cr,k}$	vector of actions at time τ_k that accounts for creep during previous time period (Equation 5.85)

$f_{cs,k}$	vector of actions at time τ_k that accounts for shrinkage during previous time period (Equation 5.86)
$f_{p,init}$	vector of initial prestressing forces (Equation 5.42)
$f_{p.rel,k}$	vector of relaxation forces at time τ_k (Equation 5.88)
$f_c', f_c'(28)$	characteristic compressive strength of concrete at 28 days
f_{cp}'	characteristic compressive strength of concrete at the time of transfer
f_{ct}'	characteristic uniaxial tensile strength of concrete
$f_{ct.f}'$	characteristic flexural tensile strength of concrete
G	permanent action (dead load)
g	permanent action per unit length or per unit area
g_p	permanent distributed load normal to shear interface per unit length (N/mm)
h	dimension of anchorage plate *or* drape of tendon
h_o	height of a primary crack
h_x, h_y	drape of the tendons running in the x- and y-directions, respectively
I	second moment of area (moment of inertia) about centroidal axis of a cross-section
I_{av}	average second moment of area after cracking
I_c	second moment of the area of the concrete part of the cross-section
I_{c1}, I_{c2}	second moments of the area of precast and in-situ elements on a composite cross-section
I_{cr}	second moment of the area of a cracked cross-section
I_{ef}	effective second moment of area after cracking
I_g	second moment of the area of the gross cross-section
$\overline{I_k}$	second moment of the area of the age-adjusted transformed section at time τ_k
I_{uncr}	second moment of the area of the uncracked cross-section
I_0	second moment of the area of the transformed section about the reference axis at first loading (i.e. at time τ_0)
i, j, k	integers
$J(t,\tau)$	the creep function at time t due to a stress first applied at τ
J_t	torsional constant
K	a factor that accounts for the position of the bars being anchored with respect to the transverse reinforcement *or* slab system factor
k	decay factor for the post-peak response of concrete in compression *or* effective length factor (Figure 13.7)
kl_u	effective length of a column
k_{AB}	stiffness coefficient for member AB
k_{co}	cohesion coefficient given in Table 9.1

k_m	factor used to determine the moment magnifier for a slender column (Equation 13.30)
k_r, k_{r1}, k_{r2}	shrinkage curvature coefficients (Equations 5.180 to 5.183)
k_u	the ratio of the depth to the neutral axis from the extreme compressive fibre to the effective depth (i.e. d_n/d) at ultimate strength under any combination of bending and compression
k_{uo}	the ratio of the depth to the neutral axis from the extreme compressive fibre to the effective depth (i.e. d_n/d) at ultimate strength in bending without axial force
k_1, k_2	factors used for the determination of stress in the prestressing steel at ultimate (Equation 6.19)
k_1, k_2, k_3, k_4, k_5	material multiplication constants for creep and shrinkage (Equations 4.21 and 4.29) *or* factors affecting the development length in tension (Equation 14.3)
k_6	factor affecting the development length in compression (Equation 14.9)
L	centre to centre distance between supports (span)
L_a	length of anchorage zone measured from the loaded face (Figure 8.3b)
L_b, L_c	lengths of a beam and a column, respectively
L_{di}	length of tendon associated with draw-in losses (Equation 5.138)
L_e	effective length of a column
L_{ef}	effective span of a beam or slab, that is the lesser of the centre to centre distance between the supports and the clear span plus depth (i.e. $L_n + D$)
L_o	distance between points of zero bending moment in a beam or L minus 0.7 times the sum of the values of a_{sup} at each end of the span (see Figure 2.1)
L_n	clear span (i.e. the distance between the faces of the supports)
L_p	development length for a pre-tensioned tendon (Equation 8.3)
L_{pa}	length of the tendon from the jacking end to a point at a distance 'a' from that end
L_{pt}	transmission length for a pre-tensioned tendon (Table 8.1)
L_t	transverse span or width of design strip
L_{sc}	development length of a bar for a compressive stress less than the yield stress
L_{st}	development length of a bar for a tensile stress less than the yield stress
$L_{sy.c}$	development length of a reinforcing bar to develop the characteristic yield strength in compression

$L_{sy.t}$	development length of a reinforcing bar to develop the characteristic yield strength in tension
$L_{sy.t.lap}$	the tensile lap length for either contact or non-contact splices
L_u	unsupported length of a column
L_x, L_y	shorter and longer orthogonal span lengths, respectively, in two-way slabs
l, ℓ	internal level arm
l_b	lever arms associated with bursting moment or the length of the bursting zone parallel to the axis of a strut
l_c	distance of compressive force in the concrete above the non-prestressed tensile steel (Figure 6.6)
l_h	length of a plastic hinge in the direction of the member axis
l_p	distance of force in the prestressed tendon above the non-prestressed tensile steel (Figure 6.6)
l_s	lever arm associated with spalling moment or distance of compressive force in the steel above the non-prestressed tensile steel (Figure 6.6)
ln	natural logarithm
M	bending moment
$\bar{M}$	virtual bending moment
M_b	bursting moment; or moment transferred to front face of a column
M_{cr}	cracking moment
$M_{cr,cs}$	cracking moment accounting for tension caused by early shrinkage
M_{ext}	externally applied moment about reference axis
M_{FEM}	fixed-end moment
M_G	moment caused by the permanent actions
M_{int}	internal moment about reference axis
$M_{int,k}$	internal moment about reference axis at time τ_k
$M_{int,0}$	internal moment about reference axis at time τ_0
M_o	moment at a cross-section at transfer, total static moment in a two-way flat slab, or decompression moment
M_{pc}	moment in the post-cracking range (see Figure 6.1)
M_{ps}	secondary moment due to prestress in a continuous member
M_{pt}	total moment due to prestress in a continuous member
M_Q	moment caused by the imposed actions
$M_{R,0}$	sum of the external moment and the resultant moment about the centroidal axis caused by the compressive prestress at time τ_0 (Equation 5.166)
M_s	spalling moment or moment transferred to the side face of a column

M_{sus}	moment caused by the sustained loads
M_{sw}	moment caused by self-weight
M_T	moment caused by total service loads
M_u	ultimate strength in bending at a cross-section of an eccentrically loaded compressive member
M_{ub}	the balanced moment at the ultimate limit state (when $k_{uo} = 0.003/[0.003 + f_{sy}/E_s]$)
M_{uo}	ultimate flexural strength of a cross-section in bending without axial force
M_{uo1}	ultimate flexural strength of a cross-section containing only prestressing steel in bending without axial force (Equation 6.25)
$(M_{uo})_{min}$	minimum required strength in bending at a critical cross-section
M_{unbal}	unbalanced moment
M_{ux}, M_{uy}	ultimate strength in bending about the major and minor axes, respectively, of a column under the design axial force N^*
M^*	factored design bending moment at a cross-section for the strength limit state
M_s^*	maximum bending moment at the section based on the short-term serviceability load or construction load
M_v^*	design moment transferred from a slab to a column through the critical shear perimeter
M_x^*, M_y^*	design bending moment in a column about the major and minor axes, respectively; or positive design bending moment at mid-span in a slab in the x- and y-directions, respectively
M_1, M_3	moments applied to the precast member at transfer (stage 1) and immediately prior to composite action (stage 3), respectively
M_4	additional moment applied to the composite cross-section in load stage 4
m_p	number of layers of prestressed steel
m_s	number of layers of non-prestressed reinforcement
m_u, m_u'	ultimate moment of resistance per unit length along a positive and a negative yield line, respectively
N	axial force
$\bar{N}$	virtual axial force
N_c	critical buckling load (Equation 13.24)
$N_{c,k}$	the axial forces resisted by the concrete at time τ_k
$N_{c,0}$	the axial forces resisted by the concrete at time τ_0
N_{cr}	tensile axial force at cracking
N_{ext}	externally applied axial force
N_{int}	internal axial force

$N_{int,k}$	initial axial force at time τ_k
$N_{int,0}$	initial axial force at time τ_0
$N_{p,k}$	the axial forces resisted by the prestressing steel at time τ_k
$N_{p,0}$	the axial forces resisted by the prestressing steel at time τ_0
$N_{R,0}$	sum of the external axial force (if any) and the resultant compressive force exerted on the cross-section by the tendons at time τ_0 (Equation 5.165)
$N_{s,k}$	the axial forces resisted by the non-prestressed reinforcement at time τ_k
$N_{s,0}$	the axial forces resisted by the non-prestressed reinforcement at time τ_0
N_u	ultimate axial strength
N_{ub}	ultimate axial force at the balanced failure point
N_{uo}	ultimate strength in the compression of an axially loaded cross-section without bending
N^*	factored design axial force on a cross-section
N_f^*	design axial load in a fire situation
n_c	modular ratio of in-situ concrete in a composite section (E_{c2}/E_{c1})
n_p	modular ratio for prestressed steel (E_p/E_c)
$n_{p(i),0}$	modular ratio for the i-th level of prestressed steel at time τ_0
n_s	modular ratio for non-prestressed steel (E_s/E_c)
$n_{s(i),0}$	modular ratio for the i-th level of non-prestressed steel at time τ_0
$\bar{n}_{es,k}$	age-adjusted modular ratio for non-prestressed steel $(E_s/\bar{E}_{e,k})$
$\bar{n}_{ep,k}$	age-adjusted modular ratio for prestressed steel $(E_p/\bar{E}_{e,k})$
P	prestressing force or applied load
P_e	effective prestressing force after time-dependent losses
P_i	prestressing force immediately after transfer
$P_{init(i)}$	for post-tensioning, the prestressing force immediately after stressing the tendon for a pre-tensioned tendon, the prestressing force immediately before transfer
P_j	prestressing force at the jack before transfer
P_{pb}	the breaking force in a prestressing tendon $(= f_{pb} A_p)$
P_v	vertical component of prestressing force
P_x, P_y	prestressing forces in a slab in the x- and y-directions, respectively
p	reinforcement ratio
p_{cw}	web reinforcement ratio for compressive reinforcement (see Equation 5.171)
p_w	web reinforcement ratio for tensile reinforcement (see Equation 5.171)
Q	imposed action (live load) or first moment of an area about the centroidal axis

q	imposed action per unit length or per unit area
q_p	prestressing steel index (Equation 10.9)
q_s	reinforcing steel index (Equation 10.10)
q_t	transverse tension per unit length (Equation 14.1)
R	design relaxation in the prestressing steel (in percent), reaction force, *or* radius of curvature
$R_{A,0}, R_{B,0}, R_{I,0}$	cross-sectional rigidities at time τ_0 (Equations 5.33, 5.34 and 5.37)
$R_{A,k}, R_{B,k}, R_{I,k}$	cross-sectional rigidities at time τ_k (Equations 5.77, 5.78 and 5.80)
$R_{A,p}, R_{B,p}, R_{I,p}$	contribution to section rigidities provided by the bonded tendons (Equations 5.111 to 5.113)
$R_{A,s}, R_{B,s}, R_{I,s}$	contribution to section rigidities provided by the steel reinforcement (Equations 5.108 to 5.110)
R_b	basic relaxation of a tendon
$R_{cyl/cu}$	ratio of cube and cylinder strengths
R_d	design strength
R_u	ultimate strength
$R_{u.sys}$	capacity of the entire structure
R_{1000}	relaxation of prestressing steel (in percent) after 1000 hours
r	radius of gyration
$\mathbf{r}_{ext,k}$	vector of applied actions at time τ_k (Equation 5.70)
$\mathbf{r}_{ext,0}$	vector of applied actions at time τ_0 (Equation 5.39)
$\mathbf{r}_{int,k}$	vector of internal actions at time τ_k (Equation 5.71)
S_u	ultimate value of various actions
s	spacing between fitments
s_b	clear distance between bars of the non-contact lapped splice
s_t, s_v	spacing between stirrups required for torsion and shear, respectively, along the longitudinal axis of the member
T	resultant tensile force, twisting moment (torque); *or* temperature
T_b	transverse tension resulting from bursting in an anchorage zone or in a bottle-shaped strut
$T_{b.cr}$	bursting force required to cause first cracking
T_b^*	bursting tension caused by the factored design loads at the strength limit state
$T_{b.s}^*$	bursting tension caused by the design loads at the serviceability limit state
T_{cr}	twisting moment at first cracking
T_p	tension in the prestressed steel
T_s	tension in the non-prestressed steel; *or* twisting moment transferred to the side face of column

T_u	ultimate strength in torsion
T_{uc}	torsional strength of a beam without torsional reinforcement (Equation 7.23)
T_{us}	torsional strength of a beam containing torsional reinforcement (Equation 7.28)
$T_{u.max}$	maximum torsional strength of a cross-section (Equation 7.33)
T^*	factored design torsion for the strength limit state
$\Delta T_{c.s}$	restraining force at the level of non-prestressed steel
$\Delta T_{c.p}$	restraining force at the level of prestressed steel
t	time; or flange or wall thickness
t_h	hypothetical thickness of a member (taken as $2A_g/u_e$)
U	internal work
u	perimeter of the critical section for punching shear
u_e	portion of section perimeter exposed to the atmosphere plus half the total perimeter of any voids contained within the section
u_t	perimeter of the polygon with vertices at the centres of the longitudinal bars at the corners of a closed stirrup
V	shear force
V_b	shear transferred to the front face of a column
V_{cr}	shear force acting with torque T_{cr} at first cracking
V_{loads}	shear force caused by external loads
V_o	shear force corresponding to the decompression moment M_o
V_s	shear transferred to the side face of a column
V_t	shear force required to produce a web shear crack
V_u	ultimate shear strength
V_{uc}	contribution of concrete to the shear strength
V_{uo}	shear strength of the critical shear perimeter with no moment transferred
V_{us}	contribution of transverse steel to the shear strength
$V_{u.max}$	maximum shear strength of a beam (Equation 7.8)
$V_{u.min}$	shear strength of a beam containing the minimum shear reinforcement (Equation 7.6)
V^*	factored design shear force for the strength limit state
v	deflection
v_C	deflection at mid-span
v_{cc}	deflection due to creep
v_{cx}, v_{mx}	deflection of the column strip and the middle strip in the x-direction
v_{cy}, v_{my}	deflection of the column strip and the middle strip in the y-direction
v_i	deflection immediately after transfer
v_{max}	maximum permissible total deflection or maximum deflection of a flat plate

v_{cs}	deflection due to shrinkage
v_{sus}	short-term deflection caused by the sustained loads
v_{tot}	total deflection
v_{var}	deflection due to variable loads
W	actions arising due to wind *or* external work
W_1	elastic energy (see Figure 6.18)
W_2	plastic energy (see Figure 6.18)
w	uniformly distributed load *or* horizontal projection of a strut-and-tie node
w_b	the balanced load
w_G	uniformly distributed permanent loads
w_p	distributed transverse load exerted on a member by a parabolic tendon profile
w_{px}, w_{py}	transverse loads exerted by tendons in the x- and y-directions, respectively
w_Q	uniformly distributed imposed loads
w_s	maximum superimposed service load
w_{sw}	self-weight
w_{tot}	total equivalent long-term load (Equation 10.4)
w_u	collapse load
w_{ub}	unbalanced load
$w_{ub.sus}$	sustained part of the unbalanced load
w_v	variable or transient part of the uniform load
w^*	factored design load for the strength limit state
w^*_{cr}	maximum final crack width (Table 2.5)
x	direction of member axis
x, y	shorter and longer overall dimensions of the rectangular parts of a solid section
x_1, y_1	shorter and larger dimension of a closed rectangular tie
y	direction perpendicular to the member axis; *or* distance above reference level
y_b	distance from reference axis to the bottom fibre
$y_{n,0}$	distance from the reference axis to the neutral axis at time τ_0
$y_{p(i)}$	y-coordinates of i-th level of prestressed steel
$y_{s(i)}$	y-coordinates of i-th level of non-prestressed reinforcement
y_t	distance from reference axis to the top fibre
Z	section modulus of uncracked cross-section
Z_b	bottom fibre section modulus (I/y_b)
$(Z_b)_{min}$	minimum bottom fibre section modulus (Equation 5.5)
$Z_{b.pc}$	bottom fibre section modulus of the precast part of a composite cross-section
$Z_{b.comp}$	bottom fibre section modulus of a composite cross-section
Z_t	top fibre section modulus (I/y_t)
$(Z_t)_{min}$	minimum top fibre section modulus (Equation 5.6)

$Z_{t.pc}$	bottom fibre section modulus of the precast part of a composite cross-section
z	coordinate axis, vertical projection of a strut; *or* internal moment lever arm
$\bar{z}$	distance from extreme fibre to centroidal axis
α	a parameter to account for the effect of cracking and reinforcement quantity on the restraint to creep (Equation 5.178); the slope of the prestress line (Figure 5.25); constant associated with the development of concrete strength with time (Equation 4.2); factor that depends on the support conditions of a two-way edge-supported slab; factor to account for the restraint provided by the slab around the critical shear perimeter; angle of divergence between bottled shape compression fields and idealised parallel sided strut (Equations 8.18 and 8.19, Figure 8.26); *or* a moment coefficient (Figure 11.25)
α_b, α_t	section properties relating to bottom and top fibres (A/Z_b and A/Z_t, respectively)
$\alpha_{b.pc}, \alpha_{t.pc}$	section properties relating to bottom and top fibres of a precast section ($A_{pc}/Z_{b.pc}$ and $A_{pc}/Z_{t.pc}$, respectively)
α_n	factor to define the shape of a biaxial bending contour (Equation 13.22)
α_{tot}	is the sum in radians of the absolute values of successive angular deviations of the tendon over the length L_{pa}
α_v	angle between the inclined shear reinforcement and the longitudinal tensile reinforcement
α_1, α_2	constants associated with the development of shrinkage strain and the creep coefficient, respectively; *or* fractions of the span L shown in Figure 11.14
α_2	stress intensity factor for idealised rectangular stress block (see Figure 6.2)
β	an effective compression strength factor (Equation 2.9); deflection coefficient; tension stiffening constant to account for load duration (Equation 5.174); constant associated with the development of concrete strength with time (Equation 4.2); *or* the ratio of the compressive force in the in situ slab and the total compressive force on the cross-section, i.e. C_{slab}/C
β_h	ratio of the longest overall dimension Y of the effective loaded area to the overall dimension X measured in the perpendicular direction (Figure 7.13)
β_n	factor that accounts for the high level of strain incompatibility between the ties and struts entering a node (Equation 8.25)

β_p	angular deviation (in radians/m) due to wobble effects in the straight or curved parts of the tendon
β_s	strut efficiency factor (Equation 8.17)
β_x, β_y	moment coefficients in a two-way slab (Equations 12.19 and 12.20)
$\beta_1, \beta_2, \beta_3$	factors affecting the contribution of the concrete to the shear strength (Equation 7.9)
$\chi(t, \tau)$	ageing coefficient at time t for concrete first loaded at τ
χ^*	final ageing coefficient at time infinity
Δ	an increment or a change; lateral displacement (sway) at the top of a column slip; frame displacement due to prestress (see Figure 11.19); or draw-in (in mm) at the anchorage of a tendon
Δ_{slip}	slip of the tendon at an anchorage (Equation 5.137)
Δ_u	deflection at collapse after full plastic deformation (see Figure 6.20)
Δ_y	deflection at first yield (see Figure 6.20)
$\Delta\sigma_c(\tau_k)$	change in stress between τ_0 and τ_k (see Figure 5.11)
ΔF_p	change in force in the prestressing steel with time
$\Delta F_{s(i)}$	change in force in the i-th layer of non-prestressed reinforcement with time
ΔM	increment of moment
$\Delta M(\tau_k)$	increment of moment gradually applied about the reference axis in the time interval $(\tau_k - \tau_0)$
ΔN	increment of axial force
$\Delta N(\tau_k)$	the increment of axial force gradually applied in the time interval $(\tau_k - \tau_0)$
$\Delta T_{c.p(i)}$	restraining force at the i-th level of prestressing steel gradually applied in the time interval $(\tau_k - \tau_0)$
$\Delta T_{c.s(i)}$	restraining force at the i-th level of non-prestressed reinforcement gradually applied in the time interval $(\tau_k - \tau_0)$,
Δt_k	the time interval $(\tau_k - \tau_0)$
$\Delta\varepsilon_{r,k}$	change in strain at the level of the reference axis at time τ_k
$\Delta\varepsilon_{p,0}$	change in strain in the tendon immediately after transfer
$\Delta\sigma_p$	change in stress in the tendon
$\Delta\sigma_{p.cc}$	change in stress in the tendon due to creep (Equation 5.147)
$\Delta\sigma_{p.cs}$	change in stress in the tendon due to shrinkage (Equation 5.144)
$\Delta\sigma_{p, rel}$	change in stress in the tendon due to relaxation (Equation 5.148)
$\Delta\sigma_{p, time}$	time-dependent change in stress in the tendon
$\Delta\delta$	time-dependent lateral displacement of a slender column

δ	a factor that depends on the support conditions (Equation 12.14); lateral deflection of a column; *or* moment magnifier
δ_b	moment magnification factor for a braced column (Equation 13.29)
δ_i	initial lateral displacement of a slender column
δ_s	moment magnification factor for a sway column (Equation 13.32)
ε	strain
$\varepsilon_a(z)$	axial strain at the centroid of a member
$\varepsilon_{aA}, \varepsilon_{aB}$	axial strain at ends A and B of a span, respectively
ε_{aC}	axial strain at mid-span
ε_c	concrete strain
$\varepsilon_c(t)$	total concrete strain at time t
ε_{cc}	creep strain of concrete
$\varepsilon_{cc}(t, \tau)$	creep strain at time t due to a stress first applied at τ
$\varepsilon_{cc}^*(\tau)$	final creep strain at time infinity
ε_{cu}	extreme concrete fibre compressive strain at the ultimate limit state in combined bending and axial load
ε'_{cu}	extreme concrete fibre compressive strain at the ultimate limit state under concentric axial load
ε_{ce}	instantaneous strain in the concrete at the level of the pre-stressed steel due to the effective prestress (Equation 6.8)
$\varepsilon_{ce}(t)$	instantaneous or elastic component of concrete strain at time t
ε_{cmi}	strain corresponding to the mean peak in situ stress f_{cmi}
$\varepsilon_{cp,0}$	instantaneous strain in the concrete at the tendon level
ε_{cs}	shrinkage strain
$\varepsilon_{cs}(t)$	shrinkage strain at time t
ε_{csd}	drying shrinkage strain
$\varepsilon_{csd.b}$	final drying basic shrinkage strain
$\varepsilon_{csd.b}^*$	basic drying shrinkage strain that depends on the quality of the local aggregates
ε_{cse}	autogenous shrinkage strain
ε_{cs}^*	final design shrinkage strain at time infinity
ε_{cse}^*	final autogenous shrinkage strain
$\varepsilon_{cs}(28)$	shrinkage strain after 28 days of drying
ε_{cu}	extreme fibre compressive strain at the ultimate limit state
ε_k	strain in the concrete at time τ_k
$\boldsymbol{\varepsilon}_k$	vector of strain at time τ_k (Equations 5.83 and 5.89)
ε_p	strain in the prestressed steel
$\varepsilon_{p,0}$	strain in the prestressed steel at time τ_0
ε_{pe}	strain in the prestressed steel due to the effective pre-stress (Equation 6.10)

ε_{pi}	strain in the prestressed steel immediately after transfer
$\varepsilon_{p(i),0}$	strain in the i-th layer of prestressed steel at time τ_0
$\varepsilon_{p(i),k}$	strain in the i-th layer of prestressed steel at time τ_k
$\varepsilon_{p(i),\,init}$	initial strain in the i-th layer of prestressing steel before transfer in a pre-tensioned member and before grouting in a post-tensioned member
$\varepsilon_{p.rel(i),k}$	the tensile creep strain in the i-th prestressing tendon at time τ_k
ε_{pt}	tensile strain at the level of the prestressed steel in the post-cracking range and at the ultimate moment (see Figure 6.1 and Equation 6.11)
ε_{pu}	strain at the maximum stress of a prestressing tendon
ε_{py}	yield strain in the prestressed steel
ε_r	strain at the level of the reference axis
$\varepsilon_{r,k}$	strain at the level of the reference axis at time τ_k
$\varepsilon_{r,0}$	strain at the level of the reference axis at time τ_0
ε_s	strain in the non-prestressed steel
$\varepsilon_{sc}, \varepsilon_{st}$	strain in the non-prestressed compressive and tensile steel, respectively
ε_{su}	uniform strain at maximum stress (uniform elongation) corresponding to the onset of necking
ε_{sy}	yield strain in the non-prestressed steel
$\varepsilon_T(t)$	temperature strain at time t
ε_0	strain in the concrete at time τ_0
$\boldsymbol{\varepsilon}_0$	vector of strain at time τ_0 (Equations 5.41 and 5.43)
$\phi, \phi_s, \phi_{st}, \phi_{sys}$	strength reduction factors
γ	ratio of the depth of the idealised rectangular compressive stress block to the depth of the neutral axis at the ultimate strength in bending or combined bending and compression (see Figure 6.2); or ratio of initial stress to tensile strength of prestressing steel ($\sigma_{p,init}/f_{pb}$)
γ_1, γ_2	angles between the direction of the orthogonal reinforcement and the axis of the strut
η	the ratio of the concrete strain and the strain corresponding to the mean peak in-situ stress (i.e. $\varepsilon_c/\varepsilon_{cmi}$)
φ_{cc}	creep coefficient
$\varphi_{cc}(t, \tau)$	creep coefficient at time t due to a stress first applied at τ
$\varphi_{cc.b}$	basic creep coefficient (Table 4.4)
φ_{cc}^*	final creep coefficient at time infinity
$\varphi_p(t, \sigma_{p,init})$	creep coefficient in the prestressing steel (Equation 4.31)
$\varphi_{p(i)}$	creep coefficient for the prestressing steel at time τ_k
κ	curvature
κ_A, κ_B	curvature at ends A and B of a span, respectively
κ_C	curvature at mid-span

$\kappa_{cc}(t)$	creep-induced curvature at time t
κ_{cr}	curvature on the cracked cross-section
$\kappa_{cs}(t)$	shrinkage-induced curvature at time t
$\kappa_{cs,0}$	initial curvature induced by shrinkage prior to loading
κ_k	long-term curvature at time τ_k
κ_p	curvature of prestressing tendon
κ_{cs}	curvature induced by shrinkage
κ_{sus}	curvature caused by the sustained loads
$\kappa_{sus,0}$	curvature caused by the sustained loads at time τ_0
κ_u	curvature at ultimate
$(\kappa_u)_{min}$	minimum curvature at ultimate for a ductile cross-section (Equation 6.18)
κ_{uncr}	curvature on the uncracked cross-section
κ_y	curvature at first yield
κ_0	initial curvature at time τ_0
λ	long-term deflection multiplication factor; *or* a factor accounting for the amount of transverse steel across the splitting plane at a bar anchorage
λ_m	maximum percentage decrease or increase in the moment at an interior support of a continuous beam due to redistribution (see Figure 11.24)
λ_r	factor that accounts for the effect on the relaxation of the shortening of the concrete due to creep and shrinkage
μ	friction curvature coefficient (Equation 5.136); *or* a coefficient of friction given in Table 9.1
μ_{fi}	ratio of the design axial load in a fire to the design resistance of a column (i.e. $N_f^*/\phi N_u$)
ν	Poisson's ratio for concrete
θ	slope, angle measured between the axis of the strut and the axis of a tie passing through a common node; *or* angle of dispersion (Figure 12.6)
θ_A, θ_B	slope at A and B, respectively
θ_h	rotation at a plastic hinge
θ_p	angle of inclination of prestressing tendon
θ_t	angle between a torsion crack and the beam axis
θ_v	angle between the axis of the concrete compression strut and the longitudinal axis of the member
$(\theta_v)_{min}$	minimum value of θ_v (Equation 7.5)
ρ	density of concrete (in kg/m^3)
ρ_p	transverse compressive pressure (in MPa) at the ultimate limit state along the development length perpendicular to the plane of splitting
Σ	sum of
σ	stress
σ_b	stress in the bottom fibre of a cross-section

σ_c	stress in the concrete
$\sigma_c(t)$	stress in the concrete at time t
$\sigma_{c,0}$, $\sigma_c(\tau_0)$	stress in the concrete at time τ_0
$\sigma_{c,k}$, $\sigma_c(\tau_k)$	stress in the concrete at time τ_k
σ_{cp}	stress in the concrete at the level of the prestressing steel; or the average prestress P/A
$\sigma_{cp,0}$	stress in the concrete at the tendon level at time τ_0
σ_{cs}	maximum shrinkage-induced tensile stress on the uncracked section (Equation 5.170)
σ_{cu}	maximum (peak) concrete stress in uniaxial compression
σ_p	stress in the prestressed steel
$\sigma_{p(i),0}$	stress in the i-th layer of prestressed steel at time τ_0
$\sigma_{p(i),k}$	stress in the i-th layer of prestressed steel at time τ_k
σ_{pe}	stress in the prestressed steel due to the effective prestressing force (Equation 6.9)
σ_{pi}	stress in the prestressed steel immediately after transfer
σ_{pj}	stress in the tendon at the jacking end (Equation 5.136)
σ_{pu}	stress in the prestressed steel at ultimate
σ_s	stress in the non-prestressed steel or permissible steel stress
$\sigma_{s(i),0}$	stress in the i-th layer of non-prestressed steel at time τ_0
$\sigma_{s(i),k}$	stress in the i-th layer of non-prestressed steel at time τ_k
σ_{sb}	permissible steel stress in the transverse bursting reinforcement required for crack control
σ_{sc}, σ_{st}	stress in the non-prestressed compressive and tensile steel, respectively
σ_t	stress in the top fibre of a cross-section
σ_x, σ_y	longitudinal and transverse stress in the anchorage zone; or average stress imposed by the longitudinal prestress in each direction of a two-way slab
σ_1, σ_2	principal stresses in the concrete
τ	a time instant or a shear stress
τ_d	age of concrete immediately after the concrete sets or at the end of moist curing when drying shrinkage begins
τ_h	horizontal shear stress at the interface between elements acting compositely
τ_k	age of concrete after a period of sustained load and shrinkage
τ_0	age of concrete at first loading
ζ	a distribution coefficient that accounts for the moment level and the degree of cracking on the effective moment of inertia (Equation 5.173)
Ω	a factor that depends on the time-dependent loss of prestress in the concrete; or vertical dimension of a node in a strut-and-tie model
ψ_c	a combination factor for imposed actions (live loads) used in assessing the design load for strength

ψ_E a combination factor for earthquake actions used in assessing the design load for strength

ψ_s short-term imposed action (live load) factor used in assessing the design load for serviceability

ψ_ℓ long-term imposed action (live load) factor used in assessing the design load for both strength and serviceability

Chapter 1

Basic concepts

1.1 INTRODUCTION

For the construction of mankind's infrastructure, reinforced concrete is the most widely used structural material. It has maintained this position since the end of the nineteenth century and will continue to do so for the foreseeable future. Because the tensile strength of concrete is low, steel bars are embedded in the concrete to carry the internal tensile forces. Tensile forces may be caused by loads or deformations, or by load-independent effects such as temperature changes and shrinkage.

Consider the simple reinforced concrete beam shown in Figure 1.1a, where the external loads cause tension in the bottom of the beam leading to cracking. Practical reinforced concrete beams are usually cracked under day-to-day service loads. On a cracked section, the applied bending moment M is resisted by compression in the concrete above the crack and tension in the bonded reinforcing steel crossing the crack (Figure 1.1b).

Although the steel reinforcement provides the cracked beam with flexural strength, it does not prevent cracking, and it does not prevent a loss of stiffness when cracking occurs. Crack widths are approximately proportional to the strain, and hence stress, in the reinforcement. Steel stresses must therefore be limited to some appropriately low value under in-service conditions in order to avoid excessively wide cracks. In addition, large steel strain in a beam is the result of large curvature, which in turn is associated with large deflection. There is little benefit to be gained, therefore, by using higher strength steel or concrete, since in order to satisfy serviceability requirements, the increased capacity afforded by higher strength steel cannot be utilised.

Prestressed concrete is a particular form of reinforced concrete. Prestressing involves the application of an initial compressive load to the structure to reduce or eliminate the internal tensile forces and, thereby, control or eliminate cracking. The initial compressive load is imposed and sustained by highly tensioned steel reinforcement (tendons) reacting on the concrete. With cracking reduced or eliminated, a prestressed concrete section is considerably stiffer than the equivalent (usually cracked) reinforced

1

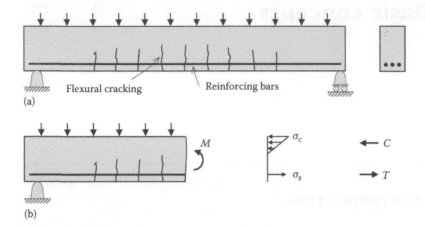

Figure 1.1 A cracked reinforced concrete beam. (a) Elevation and section. (b) Free-body diagram, stress distribution and resultant forces *C* and *T*.

concrete section. Prestressing may also impose internal forces that are of opposite sign to the external loads and may, therefore, significantly reduce or even eliminate deflection.

With service load behaviour improved, the use of high-strength steel reinforcement and high-strength concrete becomes both economical and structurally efficient. As we will see subsequently, only steel that can accommodate large initial elastic strains is suitable for prestressing concrete. The use of high-strength steel is, therefore, not only an advantage to prestressed concrete, it is a necessity. Prestressing results in lighter members, longer spans and an increase in the economical range of application of reinforced concrete.

Consider an unreinforced concrete beam of rectangular section, simply-supported over a span L and carrying a uniform load w, as shown in Figure 1.2a. When the tensile strength of concrete (f_t) is reached in the bottom fibre at mid-span, cracking and a sudden brittle failure will occur. If it is assumed that the concrete possesses zero tensile strength (i.e. $f_t = 0$), then no load can be carried and failure will occur at any load greater than zero. In this case, the collapse load is $w_u = 0$. An axial compressive force P applied to the beam, as shown in Figure 1.2b, induces a uniform compressive stress of intensity P/A on each cross-section. For failure to occur, the maximum moment caused by the external collapse load w_u must now induce an extreme fibre tensile stress equal in magnitude to P/A. In this case, the maximum moment is located at mid-span and, if linear-elastic material behaviour is assumed, simple beam theory gives (Figure 1.2b):

$$\frac{M}{Z} = \frac{w_u L^2}{8Z} = \frac{P}{A}$$

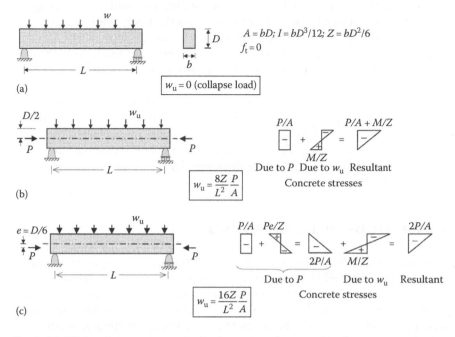

Figure 1.2 Effect of prestress on the load carrying capacity of a plain concrete beam. (a) Zero prestress. (b) Axial prestress (e = 0). (c) Eccentric prestress (e = D/6).

based on which the collapse load can be determined as:

$$w_u = \frac{8Z}{L^2}\frac{P}{A}$$

If the prestressing force P is applied at an eccentricity of $D/6$, as shown in Figure 1.2c, the compressive stress caused by P in the bottom fibre at mid-span is equal to:

$$\frac{P}{A} + \frac{Pe}{Z} = \frac{P}{A} + \frac{PD/6}{bD^2/6} = \frac{2P}{A}$$

and the external load at failure w_u must now produce a tensile stress of $2P/A$ in the bottom fibre. This can be evaluated as follows (Figure 1.2c):

$$\frac{M}{Z} = \frac{w_u L^2}{8Z} = \frac{2P}{A}$$

and re-arranging gives:

$$w_u = \frac{16Z}{L^2}\frac{P}{A}$$

By locating the prestress at an eccentricity of $D/6$, the load-carrying capacity of the unreinforced plain concrete beam is effectively doubled.

The eccentric prestress induces an internal bending moment Pe, which is opposite in sign to the moment caused by the external load. An improvement in behaviour is obtained by using a variable eccentricity of prestress along the member using a draped cable profile.

If the prestress *counter-moment Pe* is equal and opposite to the load induced moment along the full length of the beam, each cross-section is subjected only to axial compression, i.e. each section is subjected to a uniform compressive stress of P/A. No cracking can occur and, if the curvature on each section is zero, the beam does not deflect. This is known as the *balanced load stage*.

1.2 METHODS OF PRESTRESSING

As mentioned in the previous section, prestress is usually imparted to a concrete member by highly tensioned steel reinforcement (in the form of wire, strand or bar) reacting on the concrete. The high-strength prestressing steel is most often tensioned using hydraulic jacks. The tensioning operation may occur before or after the concrete is cast and, accordingly, prestressed members are classified as either *pretensioned* or *post-tensioned*. More information on prestressing systems and prestressing hardware is provided in Chapter 3.

1.2.1 Pretensioned concrete

Figure 1.3 illustrates the procedure for pretensioning a concrete member. The prestressing tendons are initially tensioned between fixed abutments and anchored. With the formwork in place, the concrete is cast around the highly stressed steel tendons and cured. When the concrete has reached its required strength, the wires are cut or otherwise released from the abutments. As the highly stressed steel attempts to contract, it is restrained by the concrete, and the concrete is compressed. Prestress is imparted to the concrete via bond between the steel and the concrete.

Pretensioned concrete members are often precast in pretensioning beds that are long enough to accommodate many identical units simultaneously. To decrease the construction cycle time, steam curing may be employed to facilitate rapid concrete strength gain, and the prestress is often transferred to the concrete within 24 hours of casting. Because the concrete is usually stressed at such an early age, elastic shortening of the concrete and subsequent creep strains tend to be high. This relatively high time-dependent shortening of the concrete causes a significant reduction in the tensile strain in the bonded prestressing steel and a relatively high loss of prestress occurs with time.

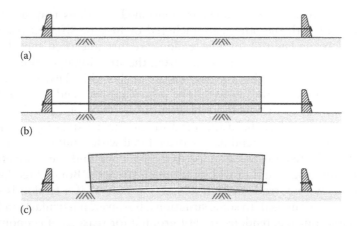

Figure 1.3 Pretensioning procedure. (a) Tendons stressed between abutments. (b) Concrete cast and cured. (c) Tendons released and prestress transferred.

1.2.2 Post-tensioned concrete

The procedure for post-tensioning a concrete member is shown in Figure 1.4. With the formwork in position, the concrete is cast around hollow ducts, which are fixed to any desired profile. The steel tendons are usually in place, unstressed in the ducts during the concrete pour, or alternatively may be threaded through the ducts at some later time. When the concrete has reached

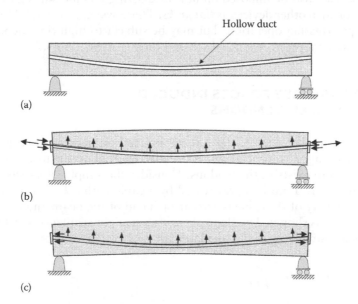

Figure 1.4 Post-tensioning procedure. (a) Concrete cast and cured. (b) Tendons stressed and prestress transferred. (c) Tendons anchored and subsequently grouted.

its required strength, the tendons are tensioned. Tendons may be stressed from one end with the other end anchored or may be stressed from both ends, as shown in Figure 1.4b. The tendons are then anchored at each stressing end. The concrete is compressed during the stressing operation, and the prestress is maintained after the tendons are anchored by bearing of the end anchorage plates onto the concrete. The post-tensioned tendons also impose a transverse force on the member wherever the direction of the cable changes.

After the tendons have been anchored and no further stressing is required, the ducts containing the tendons are often filled with grout under pressure. In this way, the tendons are bonded to the concrete and are more efficient in controlling cracks and providing ultimate strength. Bonded tendons are also less likely to corrode or lead to safety problems if a tendon is subsequently lost or damaged. In some situations, however, particularly in North America and Europe, tendons are not grouted for reasons of economy and remain permanently unbonded. In Australian practice, unbonded tendons are not permitted, except for slabs on ground [1].

Most in-situ prestressed concrete is post-tensioned. Relatively light and portable hydraulic jacks make on-site post-tensioning an attractive proposition. Post-tensioning is also used for segmental construction of large-span bridge girders.

1.2.3 Other methods of prestressing

Prestress may also be imposed on new or existing members using external tendons or such other devices as *flat jacks*. These systems are useful for temporary prestressing operations but may be subject to high time-dependent losses. External prestressing is discussed further in Section 3.7.

1.3 TRANSVERSE FORCES INDUCED BY DRAPED TENDONS

In addition to the longitudinal force P exerted on a prestressed member at the anchorages, transverse forces are also exerted on the member wherever curvature exists in the tendons. Consider the simply-supported beam shown in Figure 1.5a. It is prestressed by a cable with a *kink* at mid-span. The eccentricity of the cable is zero at each end of the beam and equal to e at mid-span, as shown. The slope of the two straight segments of cable is θ. Because θ is small, it can be calculated as:

$$\theta \approx \sin\theta \approx \tan\theta = \frac{e}{L/2} \tag{1.1}$$

In Figure 1.5b, the forces exerted by the tendon on the concrete are shown. At mid-span, the cable exerts an upward force R on the concrete

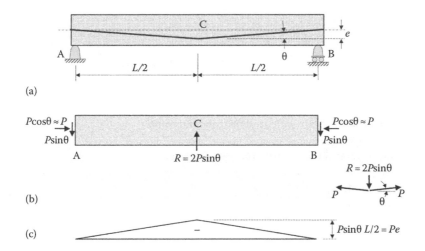

Figure 1.5 Forces and actions exerted by prestress on a beam with a centrally depressed tendon. (a) Elevation. (b) Forces imposed by prestress on concrete. (c) Bending moment diagram due to prestress.

equal to the sum of the vertical component of the prestressing force in the tendon on both sides of the *kink*. From statics:

$$R = 2P \sin\theta \approx \frac{4Pe}{L} \tag{1.2}$$

At each anchorage, the cable has a horizontal component of $P\cos\theta$ (which is approximately equal to P for small values of θ) and a vertical component equal to $P\sin\theta$ (approximated by $2Pe/L$).

Under this condition, the beam is said to be *self-stressed*. No external reactions are induced at the supports. However, the beam exhibits a non-zero curvature along its length and deflects upward owing to the internal bending moment caused by the prestress. As illustrated in Figure 1.5c, the internal bending moment at any section can be calculated from statics and is equal to the product of the prestressing force P and the eccentricity of the tendon at that cross-section.

If the prestressing cable has a curved profile, the cable exerts transverse forces on the concrete throughout its length. Consider the prestressed beam with the parabolic cable profile shown in Figure 1.6. With the x- and y-coordinate axes in the directions shown, the shape of the parabolic cable is:

$$y = -4e\left[\frac{x}{L} - \left(\frac{x}{L}\right)^2\right] \tag{1.3}$$

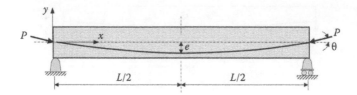

Figure 1.6 A simple beam with parabolic tendon profile.

and its slope and curvature are, respectively:

$$\frac{dy}{dx} = -\frac{4e}{L}\left(1 - \frac{2x}{L}\right) \tag{1.4}$$

and

$$\frac{d^2y}{dx^2} = +\frac{8e}{L^2} = \kappa_p \tag{1.5}$$

From Equation 1.4, the slope of the cable at each anchorage, i.e. when $x = 0$ and $x = L$, is:

$$\theta = \frac{dy}{dx} = \pm\frac{4e}{L} \tag{1.6}$$

and, provided the tendon slope is small, the horizontal and vertical components of the prestressing force at each anchorage may therefore be taken as P and $4Pe/L$, respectively.

Equation 1.5 indicates that the curvature of the parabolic cable is constant along its length. The curvature κ_p is the angular change in direction of the cable per unit length, as illustrated in Figure 1.7a. From the free-body diagram in Figure 1.7b, for small tendon curvatures, the cable exerts an upward transverse force $w_p = P\kappa_p$ per unit length over the full length of the cable. This upward force is an equivalent distributed

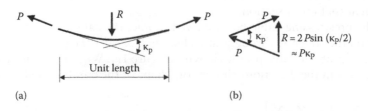

(a)   (b)

Figure 1.7 Forces on a curved cable of unit length. (a) Tendon segment of unit length. (b) Triangle of forces.

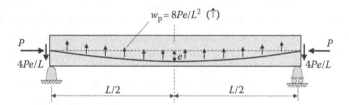

Figure 1.8 Forces exerted on a concrete beam by a tendon with a parabolic profile.

load along the member and, for a parabolic cable with the constant curvature of Equation 1.5, w_p is given by:

$$w_p = P\kappa_p = +\frac{8Pe}{L^2} \tag{1.7}$$

With the sign convention adopted in Figure 1.6, a positive value of w_p depicts an upward load. If the prestressing force is constant along the beam, which is never quite the case in practice, w_p is uniformly distributed and acts in an upward direction.

A free-body diagram of the concrete beam showing the forces exerted by the cable is illustrated in Figure 1.8. The zero reactions induced by the pre-stress imply that the beam is self-stressed. With the maximum eccentricity usually known, Equation 1.7 may be used to calculate the value of P required to cause an upward force w_p that exactly balances a selected portion of the external load. Under this *balanced load*, the beam exhibits no curvature and is subjected only to the longitudinal compressive force of magnitude P. This is the basis of a useful design approach, sensibly known as *load balancing*.

1.4 CALCULATION OF ELASTIC STRESSES

The components of stress on a prestressed cross-section caused by the pre-stress, the self-weight and the external loads are usually calculated using simple beam theory and assuming linear-elastic material behaviour. In addition, the properties of the gross concrete section are usually used in the calculations, provided the section is not cracked. Indeed, these assumptions have already been made in the calculations of the stresses illustrated in Figure 1.2 (Section 1.1).

Concrete, however, does not behave in a linear-elastic manner. At best, linear elastic calculations provide only an approximation of the state of stress on a concrete section immediately after the application of the load. Creep and shrinkage strains that gradually develop in the concrete usually cause a substantial redistribution of stresses with time, particularly on a section containing significant amounts of bonded reinforcement.

Elastic calculations are useful, however, in determining, for example if tensile stresses occur at service loads, and therefore if cracking is likely, or if compressive stresses are excessive and large time-dependent shortening may be expected. Elastic stress calculations may therefore be used to indicate potential serviceability problems.

If an elastic calculation indicates that cracking may occur at service loads, the cracked section analysis presented subsequently in Section 5.8.3 should be used to determine appropriate section properties for use in serviceability calculations. A more comprehensive picture of the variation of concrete stresses with time can be obtained using the time analyses described in Sections 5.7 and 5.9 to account for the time-dependent deformations caused by creep and shrinkage of the concrete.

In the following sub-sections, several different approaches for calculating *elastic* stresses on an uncracked concrete cross-section are described to provide insight into the effects of prestressing. Tensile (compressive) stresses are assumed to be positive (negative).

1.4.1 Combined load approach

The stress distributions on a cross-section caused by prestress, self-weight and the applied loads may be calculated separately and summed to obtain the combined stress distribution at any particular load stage. We will first consider the stresses caused by prestress and ignore all other loads. On a cross-section, such as that shown in Figure 1.9, equilibrium requires that

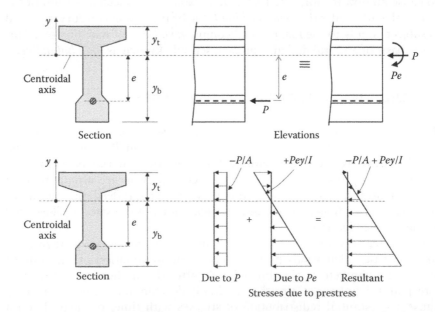

Figure 1.9 Concrete stress resultants and stresses caused by prestress.

the resultant of the concrete stresses is a compressive force that is equal and opposite to the tensile force in the steel tendon and located at the level of the steel, i.e. at an eccentricity e below the centroidal axis. This is statically equivalent to an axial compressive force P and a moment Pe located at the centroidal axis, as shown.

The stresses caused by the prestressing force of magnitude P and the hogging (−ve) moment Pe are also shown in Figure 1.9. The resultant stress induced by the prestress is given by:

$$\sigma = -\frac{P}{A} + \frac{Pey}{I}$$

(1.8)

where A and I are the area and second moment of area about the centroidal axis of the cross-section, respectively; and y is the distance from the centroidal axis (positive upwards).

It is common in elastic stress calculations to ignore the stiffening effect of the reinforcement and to use the properties of the gross cross-section. Although this simplification usually results in only small errors, it is not encouraged here. For cross-sections containing significant amounts of bonded steel reinforcement, the steel should be included in the determination of the properties of the transformed cross-section.

The elastic stresses caused by an applied positive moment M on the uncracked cross-section are:

$$\sigma = -\frac{My}{I}$$

(1.9)

and the combined stress distribution due to prestress and the applied moment is shown in Figure 1.10 and given by:

$$\sigma = -\frac{P}{A} + \frac{Pey}{I} - \frac{My}{I}$$

(1.10)

1.4.2 Internal couple concept

The resultant of the combined stress distribution shown in Figure 1.10 is a compressive force of magnitude P located at a distance ℓ above the level of the steel tendon, as shown in Figure 1.11. The compressive force in the concrete and the tensile force in the steel together form a couple, with magnitude equal to the applied bending moment and calculated as:

$$M = P\ell$$

(1.11)

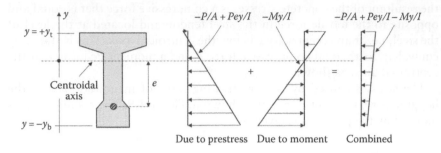

Figure 1.10 Combined concrete stresses.

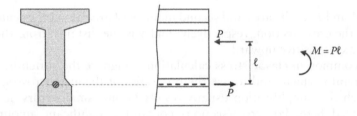

Figure 1.11 Internal couple.

When the applied moment $M = 0$, the lever arm ℓ is zero and the resultant concrete compressive force is located at the steel level. As M increases, the compressive stresses in the top fibres increase and those in the bottom fibres decrease, and the location of the resultant compressive force moves upward.

It is noted that provided the section is uncracked, the magnitude of P does not change appreciably as the applied moment increases and, as a consequence, the lever arm ℓ is almost directly proportional to the applied moment. If the magnitude and position of the resultant of the concrete stresses are known, the stress distribution can be readily calculated.

1.4.3 Load balancing approach

In Figure 1.8, the forces exerted on a prestressed beam by a parabolic tendon with equal end eccentricities are shown, and the uniformly distributed transverse load w_p is calculated from Equation 1.7. In Figure 1.12, all the loads acting on such a beam, including the external gravity loads w, are shown.

If $w = w_p$, the bending moment and shear force on each cross-section caused by the gravity load w are balanced by the equal and opposite values caused by w_p. With the transverse loads balanced, the beam is subjected only to the longitudinal prestress P applied at the anchorage. If the

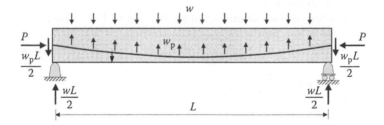

Figure 1.12 Forces on a concrete beam with a parabolic tendon profile.

anchorage is located at the centroid of the section, a uniform stress distribution of intensity P/A occurs on each section and the beam does not deflect.

If $w \neq w_p$, the bending moment M_{unbal} caused by the unbalanced load $(w - w_p)$ must be calculated and the resultant stress distribution (given by Equation 1.9) must be added to the stresses caused by the axial prestress $(-P/A)$.

1.4.4 Introductory example

The elastic stress distribution at mid-span of the simply-supported beam shown in Figure 1.13 is to be calculated. The beam spans 12 m and is post-tensioned by a single cable with zero eccentricity at each end and $e = 250$ mm at mid-span. The prestressing force in the tendon is here assumed to be constant along the length of the beam and equal to $P = 1760$ kN.

Each of the procedures discussed in the preceding sections are illustrated in the following calculations.

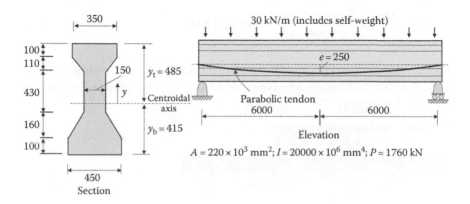

Figure 1.13 Beam details (Introductory example). Notes: P is assumed constant on every section; all dimensions are in mm.

Combined load approach:
The extreme fibre stresses at mid-span (σ_t, σ_b) due to P, Pe and M are calculated separately in the following and then summed.

At mid-span: $P = 1760$ kN; $Pe = 1760 \times 250 \times 10^{-3} = 440$ kNm and

$$M = \frac{wL^2}{8} = \frac{30 \times 12^2}{8} = 540 \text{ kNm}$$

Due to P:

$$\sigma_t = \sigma_b = -\frac{P}{A} = -\frac{1760 \times 10^3}{220 \times 10^3} = -8.0 \text{ MPa}$$

Due to Pe:

$$\sigma_t = \frac{Pey_t}{I} = \frac{440 \times 10^6 \times 485}{20,000 \times 10^6} = +10.67 \text{ MPa}$$

$$\sigma_b = -\frac{Pey_b}{I} = -\frac{440 \times 10^6 \times 415}{20,000 \times 10^6} = -9.13 \text{ MPa}$$

Due to M:

$$\sigma_t = -\frac{My_t}{I} = -\frac{540 \times 10^6 \times 485}{20,000 \times 10^6} = -13.10 \text{ MPa}$$

$$\sigma_b = \frac{My_b}{I} = \frac{540 \times 10^6 \times 415}{20,000 \times 10^6} = +11.21 \text{ MPa}$$

The corresponding concrete stress distributions and the combined elastic stress distribution on the concrete section at mid-span are shown in Figure 1.14.

Internal couple concept:
From Equation 1.11:

$$\ell = \frac{M}{P} = \frac{540 \times 10^6}{1760 \times 10^3} = 306.8 \text{ mm}$$

The resultant compressive force on the concrete section is 1760 kN and it is located $306.8 - 250 = 56.8$ mm above the centroidal axis. This is statically

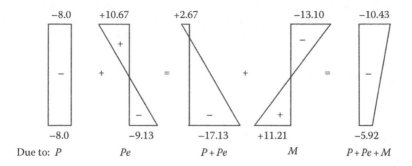

-8.0   +10.67   +2.67   -13.10   -10.43

-8.0   -9.13   -17.13   +11.21   -5.92

Due to: P Pe $P + Pe$ M $P + Pe + M$

Figure 1.14 Component stress distributions in introductory example.

equivalent to an axial compressive force of 1760 kN (applied at the centroid) plus a moment $M_{unbal} = 1760 \times 56.8 \times 10^{-3} = 100$ kNm. The extreme fibre concrete stresses are therefore:

$$\sigma_t = -\frac{P}{A} - \frac{M_{unbal}y_t}{I} = -\frac{1{,}760 \times 10^3}{220 \times 10^3} - \frac{100 \times 10^6 \times 485}{20{,}000 \times 10^6} = -10.43 \text{ MPa}$$

$$\sigma_b = -\frac{P}{A} + \frac{M_{unbal}y_b}{I} = -\frac{1{,}760 \times 10^3}{220 \times 10^3} + \frac{100 \times 10^6 \times 415}{20{,}000 \times 10^6} = -5.92 \text{ MPa}$$

and, of course, these are identical with the extreme fibre stresses calculated using the combined load approach and shown in Figure 1.14.

Load balancing approach:
The transverse force imposed on the concrete by the parabolic cable is obtained using Equation 1.7 as:

$$w_p = \frac{8Pe}{L^2} = \frac{8 \times 1{,}760 \times 10^3 \times 250}{12{,}000^2} = 24.44 \text{ kN/m (upward)}$$

The unbalanced load is therefore:

$$w_{unbal} = 30.0 - 24.44 = 5.56 \text{ kN/m (downward)}$$

and the resultant unbalanced moment at mid-span is:

$$M_{unbal} = \frac{w_{unbal}L^2}{8} = \frac{5.56 \times 12^2}{8} = 100 \text{ kN/m}$$

This is identical to the moment M_{unbal} calculated using the internal couple concept and, as determined previously, the elastic stresses at mid-span are obtained by adding the P/A stresses to those caused by M_{unbal}:

$$\sigma_t = -\frac{P}{A} - \frac{M_{unbal}y_t}{I} = -10.43 \text{ MPa}$$

$$\sigma_b = -\frac{P}{A} + \frac{M_{unbal}y_b}{I} = -5.92 \text{ MPa}$$

1.5 INTRODUCTION TO STRUCTURAL BEHAVIOUR – INITIAL TO ULTIMATE LOAD

The choice between reinforced and prestressed concrete for a particular structure is one of economics. Aesthetics may also influence the choice. For relatively short-span beams and slabs, reinforced concrete is usually the most economical alternative. As spans increase, however, reinforced concrete design is more and more controlled by the serviceability requirements. Strength and ductility can still be economically achieved but, in order to prevent excessive deflection, cross-sectional dimensions become uneconomically large. Excessive deflection is usually the governing *limit state*.

For medium- to long-span beams and slabs, the introduction of prestress improves both serviceability and economy. The optimum level of prestress depends on the span, the load history and the serviceability requirements. The level of prestress is often selected so that cracking at service loads does not occur. However, in many situations, there is no valid reason why controlled cracking should not be permitted. Insisting on enough prestress to eliminate cracking frequently results in unnecessarily high initial prestressing forces and, consequently, uneconomical designs. In addition, the high initial prestress often leads to excessively large camber and/or axial shortening. Members designed to remain uncracked at service loads are commonly termed *fully prestressed*.

In building structures, there are relatively few situations in which it is necessary to avoid cracking under the full service loads. In fact, the most economic design often results in significantly less prestress than is required for a *fully prestressed* member. Frequently, such members are designed to remain uncracked under the sustained or permanent load, with cracks opening and closing as the variable live load is applied and removed. Prestressed concrete members generally behave satisfactorily in the post-cracking load range, provided they contain sufficient bonded reinforcement to control the cracks. A cracked prestressed concrete section under service loads is significantly stiffer than a cracked reinforced concrete section of similar size and containing similar quantities of bonded reinforcement. Members that are designed to crack at the full service load are often called *partially prestressed*.

The elastic stress calculations presented in the previous section are applicable only if material behaviour is linear-elastic and the principle of superposition is valid. These conditions may be assumed to apply on a prestressed section prior to cracking, but only immediately after the loads are applied. As was mentioned in Section 1.4, the gradual development of creep and shrinkage strains in the concrete with time can cause a marked redistribution of stress between the bonded steel and the concrete on the cross-section. The greater the quantity of bonded reinforcement, the greater is the time-dependent redistribution of stress. This is demonstrated subsequently in Section 5.7.3 and discussed in Section 5.7.4. For the determination of the long-term stress and strain distributions, elastic stress calculations are not meaningful and may be misleading.

A typical moment versus instantaneous curvature relationship for a prestressed concrete cross-section is shown in Figure 1.15. Prior to the application of moment (i.e. when $M = 0$), if the prestressing force P acts at an eccentricity e from the centroidal axis of the uncracked cross-section, the curvature is $\kappa_0 = -Pe/(E_c I_{uncr})$, corresponding to point A in Figure 1.15. The curvature κ_0 is negative, because the internal moment caused by prestress is negative ($-Pe$). When the applied moment M is less than the cracking moment M_{cr}, the section is uncracked and the moment-curvature relationship is linear (from point A to point B in Figure 1.15)

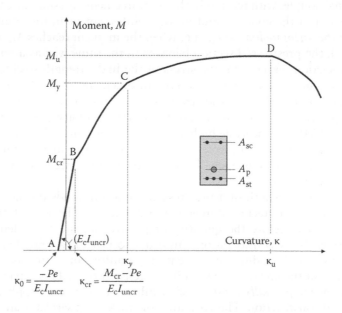

Figure 1.15 Typical moment versus instantaneous curvature relationship.

and $\kappa = (M - Pe)/(E_c I_{uncr}) \le \kappa_{cr}$. It is only in this region (i.e. when $M < M_{cr}$) that elastic stress calculations may be used and then only for short-term calculations.

If the external loads are sufficient to cause cracking (i.e. when the extreme fibre stress calculated from elastic analysis exceeds the tensile strength of concrete), the short-term behaviour becomes non-linear and the principle of superposition is no longer applicable.

As the applied moment on a cracked prestressed section increases (i.e. as the moment increases above M_{cr} from point B to point C in Figure 1.15), the crack height gradually increases from the tension surface towards the compression zone and the size of the uncracked part of the cross-section in compression above the crack decreases. This is different to the post-cracking behaviour of a non-prestressed reinforced concrete section, where, at first cracking, the crack suddenly propagates deep into the beam. The crack height and the depth of the concrete compression zone then remain approximately constant as the applied moment is subsequently varied.

As the moment on a prestressed concrete section increases further into the overload region (approaching point D in Figure 1.15), the material behaviour becomes increasingly non-linear. Permanent deformation occurs in the bonded prestressing tendons as the stress approaches its ultimate value, the non-prestressed conventional reinforcement yields (at or near point C where there is a change in direction of the moment curvature graph), and the compressive concrete in the ever decreasing region above the crack enters the non-linear range. The external moment is resisted by an internal couple, with tension in the reinforcement crossing the crack and compression in the concrete and in any reinforcement in the compressive zone. At the *ultimate load stage* (i.e. when the moment reaches M_u at a curvature κ_u), the prestressed section behaves in the same way as a reinforced concrete section, except that the stress in the high-strength steel tendon is very much higher than in conventional reinforcement. A significant portion of the very high steel stress and strain is due to the initial prestress P_i. For modern prestressing steels the initial stress in the tendon σ_{pi} ($=P_i/A_p$) is often about 1400 MPa. If the same higher strength steel were to be used without being initially prestressed, excessive deformation and unacceptably wide cracks may result at only a small fraction of the ultimate load (well below normal service loads).

The ultimate strength of a prestressed section depends on the quantity and strength of the steel reinforcement and tendons. The level of prestress, however, and therefore the quantity of prestressing steel are determined from serviceability considerations. In order to provide a suitable factor of safety for strength, additional conventional reinforcement may be required to supplement the prestressing steel in the tension zone. This is particularly so in the case of *partially prestressed* members and may even apply for fully prestressed construction. The avoidance of cracking at service loads and the satisfaction of selected elastic stress limits do not ensure adequate strength.

Strength must be determined from a rational analysis, which accounts for the non-linear material behaviour of both the steel and the concrete. Flexural strength analysis is described and illustrated in Chapter 6, and analyses for shear and torsional strength are presented in Chapter 7.

REFERENCE

1. AS3600-2009 (2009). *Australian Standard for Concrete Structures*. Standards Australia, Sydney, New South Wales, Australia.

Chapter 2

Design procedures and applied actions

2.1 LIMIT STATES DESIGN PHILOSOPHY

The broad design objective for a prestressed concrete structure is that it should satisfy the needs for which it was designed and built. In doing so, the structural designer must ensure that it is safe and serviceable, so that the chances of it failing during its design lifetime are sufficiently small. The structure must be strong enough and sufficiently ductile to resist, without collapsing, the overloads and environmental extremes that may be imposed on it. It must also be serviceable by performing satisfactorily under the day-to-day service loads without deforming, cracking or vibrating excessively. The two primary structural design objectives are therefore *strength* and *serviceability*.

Other structural design objectives are *stability* and *durability*. A structure must be stable and resist overturning or sliding, reinforcement must not corrode, concrete must resist abrasion and spalling, and the structure must not suffer a significant reduction of strength or serviceability with time. Further, it must have adequate fire protection, it must be robust, resist fatigue loading and satisfy any special requirements that are related to its intended use. A non-structural, but important, objective is *aesthetics* and, of course, an overarching design objective is *economy*. Ideally, the structure should be in harmony with, and enhance, the environment, and this often requires collaboration between the structural engineer, the environmental engineer, the architect and other members of the design team. The aim is to achieve, at minimum cost, an aesthetically pleasing and functional structure that satisfies the structural objectives of strength, serviceability, stability and durability.

For structural design calculations, the design objectives must be translated into quantitative terms called *design criteria*. For example, the maximum acceptable deflection for a particular beam or slab may be required, or the maximum crack width that can be tolerated in a concrete floor or wall. Also required are minimum numerical values for the strength of individual elements and connections. It is also necessary to identify and quantify appropriate design loads for the structure depending on the probability of their occurrence. Reasonable maximum values are required so that a suitable compromise is reached between the risk of overload and failure, and the requirement for economical construction.

Codes of practice specify design criteria that provide a suitable margin of safety (called the *safety index*) against a structure becoming unfit for service in any way. The specific form of the design criteria depends on the philosophy and method of design adopted by the code and the manner in which the inherent variability in both the load and the structural performance is considered. Modern design codes for structures have generally adopted the *limit states method* of design, whereby a structure must be designed to simultaneously satisfy a number of different *limit states* or design requirements, including adequate strength and serviceability. Minimum performance requirements are specified for each of these limit states and any one may become critical and govern the design of a particular member.

If a structure becomes unfit for service in any way, it is said to have entered a *limit state*. Limit states are the undesirable consequences associated with each possible mode of failure. In order to satisfy the design criteria set down in codes of practice, methods of design and analysis should be used, which are appropriate to the limit state being considered. For example, if the strength of a cross-section is to be calculated, *ultimate strength* analysis and design procedures are usually adopted. Collapse load methods of analysis and design (plastic methods) may be suitable for calculating the strength of ductile indeterminate structures. If the serviceability limit states of excessive deflection (or camber) or excessive cracking are considered, an analysis that accounts for the nonlinear and inelastic nature of concrete is usually required. The sources of these material nonlinearities include cracking, tension stiffening, creep and shrinkage. In addition, creep of the highly stressed, high strength prestressing steel (more commonly referred to as relaxation) may affect in-service structural behaviour.

Each limit state must be considered and designed for separately. Satisfaction of the requirements for one does not ensure satisfaction of the requirements for others. All undesirable consequences must be avoided. In this chapter, the design requirements for prestressed concrete in Australia are discussed, including the loads and load combinations for each limit state specified in the loading codes AS/NZS 1170 [1–5] and the relevant design criteria in the *Australian Standard for Concrete Structures* AS3600-2009 [6].

2.2 STRUCTURAL MODELLING AND ANALYSIS

2.2.1 Structural modelling

Structural modelling involves the development of simplified analytical models of the load distribution, the geometry of the structure and its supports, and the deformation of the structure. A concrete structure is extremely complex if it is considered from a microscopic point of view. Even from a more macroscopic viewpoint, the degree of complexity is very high. The cross-sectional dimensions vary along individual members, with relatively high tolerances on the specified dimensions being accepted. Concrete material properties depend on the degree of compaction of the concrete and the location in the structure, and vary significantly through the thickness of beams and slabs, and over the height of columns. The location of reinforcement and the concrete cover vary from point to point in a structure. Some regions are cracked and others are not. It is not possible to account for all these variations and uncertainties in the modelling and analysis of structures and, fortunately, it is not necessary.

An idealised analytical model of a concrete structure must be simple enough to allow the structural analysis to proceed and the mathematics to be tractable. It must also be accurate enough to provide a reasonable approximation of the behaviour of the real structure, including the flow of forces through the structure. This includes a reasonable quantitative estimate of the internal actions and deformations of the real structure.

As an example of a simplified structural model, consider the design of a two-way floor slab in a framed building under gravity loads. AS3600-2009 permits the floor to be divided into wide *design strips* centred on column lines in each orthogonal direction. Each one-way design strip, together with the columns immediately above and below the floor, is then modelled as a two-dimensional frame. The columns above and below the slab may be assumed to be fixed at their far ends. The horizontal design strip consists of the slab (of width L_t equal to the centre to centre distance of adjacent panels) plus any beam located in the column line in question. The span of the design strip L_o for a flat slab is as defined in Figure 2.1. The frame is then analysed, usually using a linear elastic frame analysis, and moments and shears in the design strip are determined. The design strip is next divided into the *column strip* (of width usually about $L_t/2$ and centred on the column line) and the *middle strip* (those portions of the design strip outside the column strip). The design strip moments are then distributed to the column and middle strip according to guidelines given in the Standard [6].

Depending on the degree of accuracy required, a range of possible analytical models of varying complexity is generally available. Different analytical models may need to be adopted depending on the limit state under consideration. For example, for the two-way floor slab, the simplified structural model consisting of one-way design strips spanning in orthogonal directions described above with actions determined from a linear elastic analysis may be appropriate for an assessment of flexural strength (provided the individual cross-sections are appropriately ductile). A more complex model, such as a finite element model that accounts for cracking, creep and shrinkage of concrete, may be more appropriate for an accurate prediction of deformation at service loads.

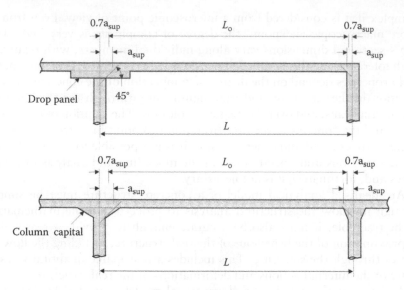

Figure 2.1 Support and span lengths for flat slabs [6].

Irrespective of the method chosen for the structural analysis or the type of structural model selected, when considering and interpreting the results of the analysis, AS3600-2009 requires that the simplifications, idealisations and assumptions implied in the analysis and inherent in the structural model be considered in relation to the real three-dimensional concrete structure.

2.2.2 Structural analysis

The structural analysis of an idealised structural model is the process by which the distribution of internal actions and the deformational response of the model are determined (see reference [7]). Classical methods of structural analysis were developed for hand calculation, but today the analysis of members and frames is commonly carried out in design offices using commercial software based on the stiffness method or the finite element approach. Most often, the analysis is based on the assumption that material behaviour is linear and elastic, even though the response of a concrete structure is not linear with respect to load, even under in-service conditions. Concrete cracks at relatively low load levels causing a nonlinear structural response. Under service loads, concrete undergoes creep and shrinkage deformations and, at the ultimate limit state, steel yields and concrete behaviour in compression is nonlinear and inelastic. Nevertheless, *linear elastic analysis* still forms the basis for the analysis of most concrete structures, with adjustments made in design to include both material and geometric nonlinearities when predicting either the ultimate strength or the in-service deformations.

Over the last few decades, significant advances have been made in the development of methods for including both material and geometric nonlinearities in the analysis of concrete structures. Today numerous commercial software packages are available for the *nonlinear analysis* of concrete structures under both service loads and ultimate loads. For many applications, these packages are unnecessarily complex and time consuming, but in some situations, nonlinear analysis is appropriate and its use for the analysis of complex structures is likely to increase.

If linear elastic analysis is used in the design for adequate strength, the internal actions caused by the design ultimate loads are determined at the critical cross-sections of each structural member. Each critical section is then designed to ensure that its ultimate strength exceeds the design ultimate actions. Although a linear elastic analysis of a particular structural model will give a single distribution of internal actions that is in equilibrium, it must be understood that this is not the distribution of actions in the real structure, where material behaviour under ultimate loads will be nonlinear and inelastic. The use of linear elastic analysis is only valid provided individual cross-sections and members are ductile, so that redistribution of internal actions can take place as the ultimate load is approached and the distribution of actions assumed in design can develop in the actual structure.

Alternative methods for the design for adequate strength include plastic methods of analysis (or collapse load analysis) and strut-and-tie modelling. These are simple forms of nonlinear analysis. In plastic analysis, plastic hinges with large rotational capacities are assumed to occur at the critical regions in a continuous beam or frame and either upper or lower bound estimates of the collapse load may be made. If plastic methods are used, ductility of the critical regions is an essential requirement. Strut-and-tie modelling is a lower bound method of plastic analysis where a designer selects an internal load path, and then designs the internal concrete struts and steel ties that have been selected to transfer the applied loads through the structure. A range of other methods of analysis, including methods based on stress analysis, are also available for use in strength design.

Linear elastic analysis and various forms of nonlinear analysis may also be used to predict the actions and deformations of a concrete structure at service loads, but account must be taken of the loss of stiffness caused by cracking (due to both the applied loads and the restraint to shrinkage and temperature changes), the tension stiffening effect and the time-dependent effects of creep and shrinkage. These sources of material nonlinearity complicate structural analysis at service loads. A comprehensive treatment of the serviceability of concrete structures is provided in reference [8], and

Chapter 5 outlines in some detail appropriate methods for the analysis of prestressed concrete structures at service loads.

Codes of practice, including AS3600-2009, provide guidance for design using linear elastic analysis, nonlinear analysis, plastic methods, strut-and-tie modelling, stress analysis and more approximate methods based on moment coefficients.

2.3 ACTIONS AND COMBINATIONS OF ACTIONS

2.3.1 General

In the design of concrete structures, the internal actions arising from appropriate combinations of the applied loads should be considered. The applied loads are called *actions* in AS/NZS 1170 [1–5] and these include:

Permanent actions (G)	Imposed actions (Q)	Wind actions (W)
Prestress (P)	Earthquake actions (E)	Earth pressure (F_e)
Liquid pressure (F_{lp})	Snow actions (F_{sn})	

In addition, possible accidental loads and loads arising during construction should be considered where they may adversely affect the various limit states' requirements. Other actions that may cause either stability, strength or serviceability failures include creep of concrete; shrinkage of concrete; other imposed deformations, such as may result from temperature changes and gradients, support settlements and foundation movements; and dynamic effects.

Permanent actions (often called *dead loads*) are generally defined as those loads imposed by both the structural and non-structural components of a structure that are permanent in nature. Permanent loads include the self-weight of the structure and the forces imposed by all walls, floors, roofs, ceilings, permanent partitions, service machinery and other permanent construction. They also include storage materials of a permanent nature. Permanent actions are usually fixed in position and can be estimated reasonably accurately from the mass of the relevant material or type of construction. For example, normal-weight concrete weighs about 24 kN/m^3 and lightweight concrete weighs between 15 and 20 kN/m^3. Unit weights of a range of materials are provided in Table 2.1.

Imposed actions (often called live loads) are loads that are attributed to the intended use or purpose of the structure. The *specified* imposed actions depend on the expected use or occupancy of the structure and usually includes allowances for impact and inertia loads (where applicable) and for possible overload. Both uniformly distributed and concentrated imposed actions for a variety of occupancies are specified in AS/NZS 1170.1 [2] and

Table 2.1 Weights of common construction materials
and building materials

Material	Weight (kN/m³)
Aluminium	26.7
Asphalt	21.2
Bitumen	10 to 14
Concrete – dense aggregate, unreinforced[a]	24.0
Concrete – light-weight aggregate	15 to 20
Fibre cement sheet (uncompressed/ compressed)	14.2/17.2
Glass – window	25.5
Granite	26.4
Iron, cast	70.7
Lead	111
Limestone (dense)	24.5
Masonry (solid brick)	19.0
Paper	6.9
Sand (dense and wet)	20.0
Sandstone	22.5
Steel	76.9
Timber – softwood	4.6 to 7
Timber – hardwood	8 to 11

[a] Add 0.6 for each 1% by volume of steel reinforcement.

values for some common usages are given in Table 2.2. The distributed and concentrated imposed actions should be considered separately and design carried out for the most adverse effect.

At the time the structure is being designed, the magnitude and distribution of the imposed actions are never known exactly, and it is by no means certain that the specified imposed actions will not be exceeded at some stage during the life of the structure. Imposed actions may or may not be present at any particular time. They are not constant and their position can vary. Although part of the imposed action is transient (short-term), some portion may be applied to the structure for extended periods (long-term) and have effects similar to permanent actions. Imposed actions also arise during construction due to stacking of building materials, the use of equipment or the construction procedure (such as the loads induced by floor-to-floor propping in multi-storey construction). These construction loads must always be anticipated and considered by the designer.

The specified wind, earthquake, snow and temperature loads depend on the geographical location and the relative importance of the structure (the mean return period). Wind loads also depend on the surrounding terrain and the height of the structure above the ground. Wind loads on structures are specified in AS/NZS1170.2 [3].

Table 2.2 Imposed actions specified in AS/NZS 1170.1 [2] for buildings

Type of activity/occupancy	Uniformly distributed load (kPa)	Concentrated load (kN)
Domestic/residential:		
Self-contained dwellings		
General areas	1.5	1.8
Balconies and habitable roofs[a]	2.0	1.8
Stairs and landings	2.0	2.7
Non-habitable roofs	0.5	1.4
Other		
General areas	2.0	1.8
Communal kitchens	3.0	2.7
Balconies and habitable roofs[a]	4.0	1.8
Offices:		
General use	3.0	2.7
Communal/commercial kitchens	3.0/5.0	2.7/4.5
Factories (general industrial)	5.0	4.5
Balconies and habitable roofs[a]	4.0	1.8
Areas where people may congregate:		
Dining rooms, lounges, restaurants	2.0	2.7
Public assembly areas (fixed seats)	4.0	2.7
Places of worship	4.0	2.7
Corridors, stairs, landings	5.0	4.5
Footpaths, terraces and plazas at ground level subject to wheeled vehicles	5.0	31
Museum floors and art galleries	4.0	4.5
Shopping areas (shop floors)	4.0	3.6
Warehouses and storage areas		
General storage	2.4 (per metre of storage height)	7.0
Compactus (up to 2 m height)	3.0 (per metre of storage height)	To be calculated
File room (office storage)	5.0	4.5
Paper storage	4.0 (per metre of storage height)	9.0
Cold storage	4.5 (per metre of storage height but ≥ 15.0)	9.0
Plant rooms	5.0	4.5
Parking garages, driveways, ramps for cars	2.5	13.0
Medium vehicle traffic areas (up to 10,000 kg)	5.0	31

[a] A habitable roof is a roof used for floor type activities.

2.3.2 Combinations of actions for the strength limit states

The actions used in the design for strength are the specified values, discussed in the previous section, multiplied by specified minimum load factors. With the built-in allowance for overloads, the specified actions are unlikely to be exceeded in the life of a structure. The load factors applied to each load type, together with factors of safety applied to the member's strength (discussed subsequently), ensure that the probability of strength failure is extremely low. The load factors depend on the type of action (load) and the load combination under consideration. For example, the load factors associated with permanent actions are generally less than those for imposed or wind actions, because the permanent action is known more reliably and therefore is less likely to be exceeded. Load factors for the strength limit states are generally greater than 1.0. The exception is where one load type opposes another load type. For example, when considering uplift from wind on a roof, the load factor on the weight of the roof is less than 1.0.

The load factors specified in AS/NZS1170.0 [1] for basic combinations of actions for the ultimate limit states are summarised below:

Permanent actions only:	$1.35G$	(2.1)
Permanent plus imposed actions:	$1.2G + 1.5Q$	(2.2)
Permanent plus long-term imposed actions:	$1.2G + 1.5\psi_l Q$	(2.3)
Permanent plus wind plus imposed actions:	$1.2G + W_u + \psi_c Q$	(2.4)
Permanent action plus wind action reversal:	$0.9G + W_u$	(2.5)
Permanent plus earthquake plus imposed actions:	$G + E_u + \psi_E Q$	(2.6)
Permanent plus earth pressure plus imposed actions:	$1.2G + S_u + \psi_c Q$	(2.7)

where S_u represents the ultimate values of various actions, including snow actions (with $S_u = F_{sn}$ for snow determined in accordance with AS/NZS1170.3 [4]) and earth pressure (with $S_u = 1.5\, \Gamma_e$).

Each factored combination of actions must be considered and the most severe should be used in the design for strength.

Where applicable, a prestressing force acting as an applied load should be included with a load factor of unity in all the above load combinations, except for the case of permanent actions plus prestressing force at transfer, where the more severe of $1.15G + 1.15P$ and $0.9G + 1.15P$ shall be used.

In the combinations of actions of Equations 2.1 through 2.7:

ψ_ℓ is that fraction of the imposed action deemed to be applied for long periods of time and for uniformly distributed imposed actions: ψ_ℓ = 0.4 for residential and domestic areas, office areas, parking areas, retail areas and roofs used for floor type applications; ψ_ℓ = 0.6 for storage and other areas; ψ_ℓ = 0.0 for all other roofs; and ψ_ℓ = 1.0 for long-term installed machinery;

ψ_c is a combination factor for imposed actions with the same numerical values as ψ_ℓ, except that ψ_c = 1.20 for long-term installed machinery;

ψ_E is the earthquake combination factor for imposed actions and for uniformly distributed imposed actions: ψ_E = 0.3 for residential and domestic areas, office areas, parking areas, retail areas and roofs used for floor type applications; ψ_E = 0.6 for storage and other areas; ψ_E = 0.0 for all other roofs; and ψ_E = 1.0 for long-term installed machinery.

2.3.3 Combinations of actions for the stability limit states

All structures should be designed such that the factor of safety against instability due to overturning, uplift or sliding is suitably high. AS3600-2009 [6] requires that the structure remains stable under the most severe of the action combinations for the strength limit states (see Section 2.3.2). The actions causing instability should be separated from those tending to resist it. The *design action effect* is then calculated from the actions tending to cause a destabilising effect using the most severe of the action combinations for the strength limit state given in Equations 2.1 to 2.7. The *design resistance effect* is calculated as 0.9 times the permanent actions tending to cause a stabilising effect (i.e. resisting instability). The structure should be so proportioned that its design resistance effect is not less than the design action effect.

Consider, for example the case of a standard cantilever retaining wall. The overturning moment caused by both the lateral earth pressure and the lateral thrust of any permanent and imposed action surcharge is calculated using Equation 2.7. To provide a suitable margin of safety against stability failure, the overturning moment should not be greater than 0.9 times the sum of the restoring moments caused by the self-weight of the wall, the weight of the backfill and any other permanent surcharge above the wall.

2.3.4 Combinations of actions for the serviceability limit states

The design loads to be used in serviceability calculations are the day-to-day *service loads* and these may be less than the *specified actions*. For example, the specified imposed actions Q have a built-in allowance for overload and impact. There is a low probability that they will be exceeded. It is usually not necessary, therefore, to ensure acceptable deflections and crack widths under the full specified loads. Use of the actual load combinations under *normal* conditions of service (i.e. the *expected* loads) is more appropriate.

Often, codes of practice differentiate between the *specified* loads and the *expected* loads. Depending on the type of structure, the expected loads may be significantly less than the specified loads. If the aim of the serviceability calculation is to produce a *best estimate* of the likely behaviour, then expected loads should be considered. Often the magnitudes of the expected loads are not defined and are deemed to be a matter for engineering judgment. In AS/NZS1170.0 [1], the expected imposed actions to be used in short-term or instantaneous serviceability calculations are $\psi_s Q$ and for long-term calculations the sustained part of the imposed actions should be taken as $\psi_\ell Q$. The short-term factor for uniformly distributed imposed actions is $\psi_s = 0.7$ for residential and domestic areas, office areas, parking areas, retail areas and roofs; and $\psi_s = 1.0$ for storage and other areas and for installed machinery. The long-term factor ψ_ℓ is as defined in Section 2.3.2. If the aim is to satisfy a particular serviceability limit state and the consequences of failure to do so are high, then the specified loads may be more appropriate than the expected loads. Once again, the decision should be based on engineering judgment and experience.

With regard to the permanent actions, the expected and the specified values are the same, and the specified permanent actions should be used in all serviceability calculations. Accordingly, AS/NZS1170.1 [2] specifies that combinations of actions for the serviceability limit states shall be those appropriate to the serviceability condition being considered and may include one or a number of the following actions: (1) permanent action G; (2) short-term imposed action $\psi_s Q$; (3) long-term imposed action $\psi_\ell Q$; (4) wind action for serviceability W_s; (5) earthquake action for serviceability E_{serv}; (6) prestressing force P; and (7) other serviceability actions.

2.4 DESIGN FOR THE STRENGTH LIMIT STATES

2.4.1 General

The *design strength* of a member or connection must always be greater than the *design action effect* produced by the most severe factored load combination (as outlined in Section 2.3.2). On a particular cross-section, the design action effect may be an axial load N^*, a shear force V^*, a bending moment M^*, a twisting moment T^* or a combinations of these. The design strength of a cross-section is a conservative estimate of the actual strength. In modern concrete codes, one of two alternative design philosophies is used to determine the design strength. The first involves the use of *strength* (or *capacity*) *reduction factors* and the second approach involves the use of *partial safety factors for material strengths*. AS3600-2009 adopts the former and the procedures set out in the Standard are summarised in the following sections.

2.4.2 Strength checks for use with linear elastic analysis, simplified methods of analysis and for statically determinate structures

When linear elastic analysis or other simplified methods are used to determine the internal actions in a member, or for statically determinate members, the *design strength* or *design capacity* R_d of a critical cross-section is taken as the product of the *ultimate strength* R_u and a strength reduction factor ϕ. The ultimate strength R_u is determined from ultimate strength theory, and procedures for the determination of R_u are outlined in Section 6.3 for bending, in Section 7.2.4 for shear, in Section 7.3.3 for torsion and in Section 13.3 for combined axial force and bending. The strength reduction factor ϕ (called the capacity reduction factor in AS3600-2009) is a factor of safety introduced to account for the variability of the material properties controlling strength and the likelihood of underperformance. It also accounts for possible variations in steel positions and concrete dimensions, inaccuracies in design procedures and workmanship, the degree of ductility and the required reliability of the member.

The *design action effects* E_d are the internal actions calculated from the critical combination of factored design loads for the strength limit state (see Section 2.3.2).

For all critical cross-sections, the design requirement is:

$$R_d = \phi\, R_u \geq E_d \tag{2.8}$$

where the capacity reduction factors specified in AS3600-2009 are summarised in Table 2.3.

In Table 2.3, at Item (b), k_{uo} is the ratio d_n/d_o, where d_n is the depth to the neutral axis from the extreme compressive fibre at the ultimate moment and d_o is the distance from the extreme compressive fibre of the concrete to the centroid of the outermost layer of tensile reinforcement or tendons (but not less than 0.8 times the overall depth of the cross-section). The determination of k_{uo} is illustrated in Chapter 6.

For cross-sections of flexural members containing normal ductility reinforcement and/or tendons, $\phi = 0.8$ when $k_{uo} \leq 0.36$ and the section is deemed to be ductile. When $k_{uo} > 0.36$, the section is deemed to be non-ductile and the capacity reduction factor decreases linearly with k_{uo} to a minimum value of 0.6. In this way, AS3600-2009 discourages the use of non-ductile cross-section but does not prevent it.

2.4.3 Strength checks for use with other methods of structural analysis

Linear elastic stress analysis: The structure is analysed for the critical combination of factored actions (see Section 2.3.2) using linear stress analysis,

Table 2.3 Capacity reduction factors (ϕ) specified in AS3600-2009 [6]

Type of action effect	Value of ϕ
(a) Axial force without bending:	
Axial tension:	
(i) with Class N reinforcement and/or tendons	0.8
(ii) with Class L reinforcement	0.64
Axial compression	0.6
(b) Bending without axial tension or compression:	
(i) with Class N reinforcement and/or tendons	$0.6 \leq (1.19 - 13k_{uo}/12) \leq 0.8$
(ii) with Class L reinforcement	$0.6 \leq (1.19 - 13k_{uo}/12) \leq 0.64$
(c) Bending with axial tension:	
(i) with Class N reinforcement and/or tendons	$\phi + [(0.8 - \phi)(N_u/N_{uot})]$ and ϕ is obtained from Item (b)(i)
(ii) with Class L reinforcement	$\phi + [(0.64 - \phi)(N_u/N_{uot})]$ and ϕ is obtained from Item (b)(ii)
(d) Bending with axial compression:	
(i) $N_u \geq N_{ub}$	0.6
(ii) $N_u < N_{ub}$	$0.6 + (\phi - 0.6)(1 - N_u/N_{ub})$ and ϕ is obtained from Item (b)
(e) Shear and/or torsion	0.7
(f) Bearing	0.6
(g) Bending, shear and compression in plain concrete	0.6
(h) Bending, shear and tension in fixings	0.6

Note: For any member containing Class L reinforcement, with or without other grades of reinforcement or tendons, the maximum value of ϕ is 0.64.

typically using the finite element method, and compressive and tensile stresses are limited to maximum allowable values. The calculated principal compressive stresses in the concrete at any point $\sigma_{2,cal}$ should satisfy the following condition:

$$\sigma_{2,cal} \leq \phi_s \beta(0.9f_c') \qquad (2.9)$$

while steel stresses in the reinforcement or tendons carrying internal tensile forces must not exceed $\phi_s f_{sy}$ or $\phi_s f_{py}$, as appropriate.

The stress reduction factor ϕ_s in Equation 2.9 is introduced to provide an appropriate margin of safety and $\phi_s = 0.6$ for concrete in compression, $\phi_s = 0.8$ for ductile reinforcement (Class N) and tendons, and $\phi_s = 0.64$ for non-ductile reinforcement (Class L). The term β is an effective compressive strength factor and, for unconfined concrete, $\beta = 1.0$, when the principal tensile stress does not exceed the uniaxial tensile strength of concrete (f'_{ct}), otherwise $\beta = 0.6$. For confined concrete, a higher value of β can be used, up to a maximum value of $\beta = 2.0$. In Equation 2.9, the design strength of unconfined concrete in compression is taken as $0.9f'_c$.

Strut-and-tie analysis: When the forces in the struts and ties of an appropriate model are determined for the critical factored load combination for the strength limit states, separate checks are made on the adequacy of each strut, each tie and on the nodes that connect them using:

$$\phi_{st}\, R_u \geq E_d \tag{2.10}$$

where ϕ_{st} is the strength reduction factor associated with strut-and-tie modeling and $\phi_{st} = 0.6$ for concrete in compression and $\phi_{st} = 0.8$ for steel in tension. Note that non-ductile reinforcement is not permitted when using strut-and-tie modelling, or any other form of plastic analysis. Examples of design of the anchorage zone of a post-tensioned members using strut-and-tie modeling are illustrated in Chapter 8.

Nonlinear analysis of framed structures: When this method of analysis is used, account is taken of the nonlinear behaviour of the steel and concrete in the various regions of the structure as the collapse load is approached. Account may also be taken of the change of geometry of the structure and the consequent secondary actions that develop. The rotation of plastic hinges, the post peak softening of regions as internal actions redistribute and the development of alternative load paths as the geometry of the structure changes are all accounted for in the analysis.

A trial structure is analysed and the capacity of the entire structure $R_{u.sys}$ is determined using mean values of all material properties rather than the characteristic values that are used in the design for strength using other methods of analysis. The strength check involves the satisfaction of:

$$\phi_{sys}\, R_{u.sys} \geq E_d \tag{2.11}$$

where the design action effect E_d is now the critical combination of factored actions as outlined in Section 2.3.2.

For ductile structures, where the deflection and deformations at peak loads are an order of magnitude greater than at service loads and where yielding of the reinforcement and tendons occurs well before the peak load, the system strength reduction factor $\phi_{sys} = 0.7$. In all other structures, $\phi_{sys} = 0.5$.

Nonlinear stress analysis at collapse: The design check required here is also made using Equation 2.11, except that nonlinear stress analysis is used to determine the load capacity of the entire structure or component.

2.5 DESIGN FOR THE SERVICEABILITY LIMIT STATES

2.5.1 General

When designing for serviceability, the designer must ensure that the structure behaves satisfactorily and can perform its intended function at service loads. Deflection (or camber) must not be excessive, cracks must be adequately controlled and no portion of the structure should suffer excessive vibration. Design for serviceability usually first involves a linear elastic analysis of the structure to determine the internal actions under the most critical load combination for the serviceability limit states (see Section 2.3.4) and then the determination of the deformation of the structure accounting for the nonlinear and inelastic behaviour of concrete under in-service conditions.

The design for serviceability is possibly the most difficult and least well understood aspect of the design of concrete structures. Service load behaviour depends primarily on the properties of the concrete, which are often not known reliably. Moreover, concrete behaves in a nonlinear manner at service loads. The nonlinear behaviour of concrete that complicates serviceability calculations is caused by cracking, tension stiffening, creep and shrinkage.

In modern concrete structures, serviceability failures are relatively common. The tendency towards higher strength materials and the use of ultimate strength design procedures for the proportioning of structures has led to shallower, more slender elements and, consequently, an increase in deformations at service loads. As far back as 1967 [9], the most common cause of damage in concrete structures was due to excessive slab deflections. If the incidence of serviceability failure is to decrease, the design for serviceability must play a more significant part in routine structural design and the structural designer must resort more often to analytical tools that are more accurate than those found in most building codes. The analytical models for the estimation of in-service deformations outlined in Chapter 5 provide designers with reliable and rational means for predicting both the short-term and time-dependent deformations in prestressed concrete structures. A comprehensive treatment of the in-service and time-dependent analysis of concrete structures is provided in Reference [8].

The level of prestress in beams and slabs is generally selected to satisfy the serviceability requirements. The control of cracking in a prestressed concrete structure is usually achieved by limiting the stress increment in the bonded reinforcement (caused by the application of the full service load) to some appropriately low value and ensuring that the bonded reinforcement is suitably distributed.

2.5.2 Deflection limits

The design for serviceability, particularly the control of deflections, is frequently the primary consideration when determining the cross-sectional dimensions of beams and floor slabs in concrete structures. This is particularly so in the case of slabs, as they are typically thin in relation to their spans and are therefore deflection sensitive. It is stiffness rather than strength that usually controls the design of slabs.

AS3600-2009 specifies two basic approaches for deflection control. The first and simplest approach is deflection control by the satisfaction of a minimum depth requirement or a maximum span-to-depth ratio. The second approach is deflection control by the calculation of deflection (or camber) using appropriate models of material and structural behaviour. This calculated deflection should not exceed the *deflection limits* that are appropriate to the structure and its intended use. The deflection limits should be selected by the designer and are often a matter of engineering judgment.

AS3600-2009 gives general guidance for both the selection of the maximum deflection limits and the calculation of deflection. However, the simplified procedures for calculating the deflections of beams and slabs are necessarily design-oriented and simple to use, involving crude approximations of the complex effects of cracking, tension stiffening, concrete creep, concrete shrinkage and load history. They have been developed and calibrated for simply-supported reinforced concrete beams [10] and often produce inaccurate and unconservative predictions when applied to lightly reinforced concrete slabs [11]. In addition, AS3600-2009 does not provide real guidance on how to adequately model the time-dependent effects of creep and shrinkage in deflection calculations.

There are three main types of deflection problem that may affect the serviceability of a concrete structure:

1. Where excessive deflection causes either aesthetic or functional problems;
2. Where excessive deflection results in unintended load paths or damage to either structural or non-structural elements attached to the member; and
3. Where dynamic effects due to insufficient stiffness cause discomfort to occupants.

Examples of deflection problems of Type 1 include visually unacceptable sagging (or hogging) of slabs and beams and ponding of water on roofs. Type 1 problems are generally overcome by limiting the magnitude of the final long-term deflection (here called *total deflection*) to some appropriately low value. The total deflection of a beam or slab in a building is usually the sum of the short-term and time-dependent deflections caused by the permanent actions (including self-weight), the prestress (if any), the expected imposed actions and the load independent effects of shrinkage and temperature change.

When the total deflection exceeds about span/250 below the horizontal, it may become visually unacceptable. Total deflection limits that are appropriate for the particular member and its intended function must be selected by the designer. For example, a total deflection limit of span/250 may be appropriate for the floor of a car park, but would be entirely inadequate for a gymnasium floor that is required to remain essentially plane under service conditions and where functional problems arise at very small total deflections. AS3600-2009 requires that a limit on the total deflection be selected that is appropriate to the structure and its intended use, but that limit should not be greater than span/250 for a span supported at both ends and span/125 for a cantilever (see Table 2.4).

Examples of Type 2 problems include, among others, deflection-induced damage to ceiling or floor finishes, cracking of masonry walls and other brittle partitions, improper functioning of sliding windows and doors, tilting of storage racking and so on. To avoid these problems, a limit must be placed on that part of the total deflection that occurs after the attachment of the non-structural elements in question, that is the *incremental deflection*. This incremental deflection is the sum of the long-term deflection due to all the sustained loads and shrinkage, the short-term deflection due to the transitory imposed actions and the short-term deflection due to any permanent actions applied to the structure after the attachment of the non-structural elements under consideration, together with any temperature-induced deflection.

AS3600-2009 limits the incremental deflection for members supporting masonry partitions or other brittle finishes to between span/500 and span/1000 depending on the provisions made to minimise the effect of movement (see Table 2.4). Incremental deflections of span/500 can, in fact, cause cracking in supported masonry walls, particularly when doorways or corners prevent arching and when no provisions are made to minimise the effect of movement.

Type 3 deflection problems include the perceptible *springy* vertical motion of floor systems and other vibration related problems. Very little quantitative information for controlling this type of deflection problem is available in codes of practice. For a member subjected to vehicular or pedestrian traffic, AS3600-2009 [6] requires that the deflection caused by imposed actions be less than span/800. For a floor that is not supporting or attached to non-structural elements likely to be damaged by large deflection, ACI 318M-11 [12] places a limit of span/360 on the short-term deflection due to imposed actions. These limits provide a minimum requirement on the stiffness of members that may, in some cases, be sufficient to avoid Type 3 problems. Such problems are potentially the

Table 2.4 Limits for calculated vertical deflections of beams and slabs with effective span L_{ef} AS3600-2009 [6]

Type of member	Deflection to be considered	Deflection limitation for spans (Notes 1, 2, 3, 4 and 6)	Deflection limitation for cantilevers (Notes 4, 5 and 6)
All members	The total deflection	$L_{ef}/250$	$L_{ef}/125$
Members supporting masonry partitions	The deflection that occurs after the addition or attachment of the partitions	$L_{ef}/500$ where provision is made to minimise the effect of movement, otherwise $L_{ef}/1000$	$L_{ef}/250$ where provision is made to minimise the effect of movement, otherwise $L_{ef}/500$
Members supporting other brittle finishes	The deflection that occurs after the addition or attachment of the finish	Manufacturer's specification but not more than $L_{ef}/500$	Manufacturer's specification but not more than $L_{ef}/250$
Members subjected to vehicular or pedestrian traffic	The imposed action (including dynamic impact) deflection	$L_{ef}/800$	$L_{ef}/400$
Transfer members (Note 7)	Total deflection	$L_{ef}/500$ where provision is made to minimise the effect of deflection of the transfer member on the supported structure, otherwise $L_{ef}/1000$	$L_{ef}/250$

Notes:

1. The effective span L_{ef} is the lesser of the centre-to-centre distance between the supports and the clear span plus the member depth for a beam or slab; or the clear span plus half the member depth for a cantilever.
2. In general, deflection limits should be applied to all spanning directions. This includes, but is not limited to, each individual member and the diagonal spans across each design panel. For flat slabs with uniform loadings, only the column strip deflections in each direction need be checked.
3. If the location of masonry partitions or other brittle finishes is known and fixed, these deflection limits need only be applied to the length of the member supporting them. Otherwise, the more general requirements of Note 2 should be followed.
4. Deflection limits given may not safeguard against ponding.
5. For cantilevers, the deflection limitations given in this table apply only if the rotation at the support is included in the calculation of deflection.
6. Consideration should be given by the designer to the cumulative effect of deflections, and this should be taken into account when selecting a deflection limit.
7. When checking the deflections of transfer members and structures, allowance should be made in the design of the supported members and structure for the deflection of the supporting members. This will normally involve allowance for settling supports and may require continuous bottom reinforcement at settling columns.

most common for prestressed concrete floors, where load balancing is often employed to produce a nearly horizontal floor under the permanent actions and the bulk of the final deflection is due to the transient imposed actions. Such structures are generally uncracked at service loads, the total deflection is small and Types 1 and 2 deflection problems are easily avoided.

2.5.3 Vibration control

Where a structure supports vibrating machinery or is subjected to any other significant dynamic load (such as pedestrian traffic), or where a structure may be subjected to ground motion caused by earthquake, blast or adjacent road or rail traffic, vibration control becomes an important design requirement. This is particularly so for slender structures, such as tall buildings or long-span beams or slabs. Vibration is best controlled by isolating the structure from the source of vibration. Where this is not possible, vibration may be controlled by limiting the frequency of the fundamental mode of vibration of the structure to a value that is significantly different from the frequency of the source of vibration. When a structure is subjected only to pedestrian traffic, 5 Hz is often taken as the minimum frequency of the fundamental mode of vibration of a beam or slab [13,14]. For detailed design guidance on dealing with floor vibrations reference should be made to specialist literature, such as References [15,16].

2.5.4 Crack width limits

In the design of a prestressed concrete structure, the calculation of crack widths is rarely required. Crack control is deemed to be provided by appropriate detailing of the reinforcement and by limiting the stress in the reinforcement crossing the crack to some appropriately low value (see Section 5.12). The limiting steel stress depends on the maximum acceptable crack width for the structure and that in turn depends on the structural requirements and the local environment. For example, for the control of flexural cracking in partially-prestressed concrete beams, AS3600-2009 requires, rather conservatively, that the increment of stress in the steel tendon near the tension face is limited to 200 MPa as the load increases from its value when the extreme concrete tensile fibre is at zero stress to the full short-term service load. In addition, the centre to centre spacing of reinforcement (including bonded tendons) must not exceed 300 mm.

Recommended maximum crack widths are given in Table 2.5 (taken from Reference [8]).

Table 2.5 Recommended maximum final design crack width, w_{cr}* [8]

Environment	Design requirement	Maximum final crack width, w_{cr}* (mm)
Sheltered environment (where crack widths will not adversely affect durability)	Aesthetic requirement:	
	Where cracking could adversely affect the appearance of the structure	
	Close in buildings	0.3
	Distant in buildings	0.5
	Where cracking will not be visible and aesthetics is not important	0.7
Exposed environment	Durability requirement:	
	Where wide cracks could lead to corrosion of reinforcement	0.3
Aggressive environment	Durability requirement:	
	Where wide cracks could lead to corrosion of reinforcement	0.30 (when $c^\# \geq 50$ mm) 0.25 (otherwise)

Note: $c^\#$ is the concrete cover to the nearest steel reinforcement.

2.6 DESIGN FOR DURABILITY

A durable structure is one that meets the design requirements for service-ability, strength and stability throughout its design working life, without significant loss of function and without excessive unforeseen maintenance. Design for durability is an important part of the design process. The design life for most concrete structures is typically 50–100 years, but prestressed concrete structures are often required to have a design life in excess of 100 years, particularly for large bridges and monumental structures. It is important to ensure that the concrete and the steel do not deteriorate significantly during the design life of the structure, so that maintenance and repair costs are kept to a minimum. Of course, for a structure in service for 100 years, some maintenance costs are inevitable, but excessive repair and maintenance can result in uneconomical life-cycle costs.

Accurate and reliable models for the deterioration of concrete with time and for the initiation and propagation of corrosion in the steel reinforcement and tendons are difficult to codify. According to AS3600-2009, the design for durability begins with the selection of an *exposure classification* for the various regions and surfaces of the structure (see Table 2.6). This is then followed by the satisfaction of a range of deemed-to-comply design requirements, including requirements relating to concrete quality (both in terms of minimum compressive strength and restrictions on the chemical content), concrete cover to the steel reinforcement (and tendons) and curing of the concrete.

Table 2.6 Exposure classifications specified in AS3600-2009 [6] for reinforced and prestressed concrete

Surface and exposure environment	Exposure classification (Notes 1 and 7)
1. Surface of members in contact with the ground:	
(a) Members protected by a damp proof membrane	A1
(b) Residential footings in non-aggressive soils	A1
(c) Other members in non-aggressive soils	A2
(d) Members in aggressive soils	U
2. Surface of members in interior environments:	
(a) Fully enclosed within a building except for a brief period of weather during construction	
(i) Residential	A1
(ii) Non-residential	A2
(b) In industrial buildings, where the member is subject to repeated wetting and drying	B1
3. Surface of members in above-ground exterior environments in areas that are:	
(a) Inland (>50 km from coastline) environment being:	
(i) Non-industrial and arid climatic zone	A1
(ii) Non-industrial and temperate climatic zone	A2
(iii) Non-industrial and tropical climatic zone	B1
(iv) Industrial (see Note 2) and any climatic zone	B1
(b) Near coastal (1 km to 50 km from coastline), any climatic zone	B1
(c) Coastal (see Note 3) and any climatic zone	B2
4. Surface of members in water:	
(a) In freshwater	B1
(b) In soft or running water	U
5. Surface of marine structures in seawater:	
(a) permanently submerged	B2
(b) In spray zone (see Note 4)	C1
(c) In tidal/splash zone (see Note 5)	C2
6. Surface of members in other environments (i.e. any exposure environment not specified in Items 1 to 5 above (see Note 6)	U

Notes:

1. The classifications apply to reinforced and prestressed concrete members. In this context, reinforced concrete includes any concrete containing metals that rely on the concrete for protection against environmental degradation. Plain concrete members containing metallic embedments should be treated as reinforced members when considering durability.
2. Industrial refers to areas that are within 3 km of industries that discharge atmospheric pollutants.
3. For the purpose of this table, the coastal zone includes locations within 1 km of the shoreline of large expanses of saltwater. Where there are strong prevailing winds or vigorous surf, the distance should be increased beyond 1 km and higher levels of protection should be considered.
4. The spray zone is the zone from 1 m above wave crest level.
5. The tidal/splash zone is the zone 1 m below lowest astronomical tide (LAT) and up to 1 m above highest astronomical tide (HAT) on vertical structures, and all exposed soffits of horizontal structures over the sea.
6. Further guidance on measures appropriate in exposure classification U may be obtained from AS 3735.
7. In this table, classifications A1, A2, B1, B2, C1 and C2 represent increasing degrees of severity of exposure, while classification U represents an exposure environment not specified in this Table but for which a degree of severity of exposure should be appropriately assessed. Protective surface coatings may be taken into account in the assessment of the exposure classification.

Good quality, well-compacted concrete, with low permeability and adequate cover to the reinforcement and tendons are the essential prerequisites for durable structures. In Tables 2.7, the minimum strength grade for concrete (see Section 4.2.3) specified in AS3600-2009 [6] is provided for each of the exposure classes in Table 2.6 for a design life of 50 years. In Table 2.8, the minimum concrete cover necessary to protect the steel reinforcement and tendons against corrosion is specified for situations where standard formwork and standard compaction of concrete are used.

These durability requirements may result in higher concrete strength and higher cover to the reinforcement and tendons than are required for other aspects of the structural design. In addition, design requirements are also specified in the code to ensure structures resist unacceptable deterioration due to abrasion from traffic, cycles of freezing and thawing and contact with aggressive soils.

Table 2.7 Minimum strength and curing requirements specified in AS3600-2009 [6] for concrete

Exposure classification	Minimum compressive strength, f'_c (MPa)	Minimum period of continuous wet curing (immediately after finishing)	Minimum average compressive strength at the time of stripping of forms (MPa)
A1	20	3 days	15
A2	25		
B1	32	7 days	20
B2	40		25
C1	50		32
C2	50		

Table 2.8 Required cover specified in AS3600-2009 [6] where standard formwork and compaction are used

Exposure classification	Required cover c (mm)				
	Characteristic compressive strength f'_c (MPa)				
	20	25	32	40	≥50
A1	20	20	20	20	20
A2	(50)	30	25	20	20
B1		(60)	40	30	25
B2			(65)	45	35
C1				(70)	50
C2					65

2.7 DESIGN FOR FIRE RESISTANCE

In addition to satisfying the other design requirements, prestressed concrete structures must be able to fulfil their required functions when exposed to fire for at least a specified period. This means that a structure should maintain its *structural adequacy* (load carrying capacity) for a specified period called the *fire resistance period* (FRP). In the case of walls and slabs, the structure must also maintain its *integrity* for a specified period so as to prevent the passage of flames or hot gases through the structure. In addition, a wall or slab must maintain its ability to prevent ignition of combustible material in the compartment beyond the surface exposed to the fire (*structural insulation*).

The Building Code of Australia and AS3600-2009 specify minimum fire resistance periods (FRP) for concrete structures and structural components, and a range of deemed-to-comply provisions that ensure their satisfaction. Minimum distances from the exposed surface of a structural member to the nearest reinforcing bar or tendon are specified to ensure structural adequacy of beams, slabs, walls and columns for various fire resistance periods. Minimum thicknesses of slabs and walls are also specified for various fire resistance periods to ensure integrity and adequate insulation.

For example, Table 2.9 provides possible combinations of minimum values of the beam width b at the level of the bottom tensile reinforcement and the average axis distance a_m for the bottom tensile reinforcing bars in beams exposed to fire for various fire resistance periods. Similarly, Table 2.10 provides possible combinations of minimum values for the smaller cross-sectional dimension b of a column cross-section (or the diameter of a circular column) and the axis distance a_s for the reinforcing bars in columns exposed to fire. The axis distance is the minimum distance from the axis of a reinforcing bar to the nearest concrete surface exposed to fire.

Table 2.11 shows the minimum effective thickness $(D_s)_{min}$ of a slab required for insulation. Also shown in Table 2.11 is the minimum axis distance a_s to the lowest layer of bottom tensile reinforcement in the slab. The minimum axis distances specified in Tables 2.9 through 2.11 should be increased by 10 mm for prestressing tendons.

In design, the cover to the reinforcing bars and tendons must satisfy both the requirements for fire and the requirements for durability.

2.8 DESIGN FOR ROBUSTNESS

Robustness is the requirement that a structure should be able to withstand damage to an element without total collapse of the structure or a significant part of the structure in the vicinity of the damaged element.

Table 2.9 Minimum dimensions a_m and b (in mm) specified in AS3600-2009 [6] for beams exposed to fire

FRP for structural adequacy (min)	Possible combinations of a_m and b for simply-supported beams							
	Combination 1		Combination 2		Combination 3		Combination 4	
	a_m	b	a_m	b	a_m	b	a_m	b
30	25	80	20	120	15	160	15	200
60	40	120	35	160	30	200	25	300
90	55	150	45	200	40	300	35	400
120	65	200	60	240	55	300	50	500
180	80	240	70	300	65	400	60	600
240	90	280	80	350	75	500	70	700

FRP for structural adequacy (minutes)	Possible combinations of a_m and b for continuous beams							
	Combination 1		Combination 2		Combination 3		Combination 4	
	a_m	b	a_m	b	a_m	b	a_m	b
30	15	80	12	160	—	—	—	—
60	25	120	12	200	—	—	—	—
90	35	150	25	250	—	—	—	—
120	45	200	35	300	35	450	30	500
180	60	240	50	400	50	550	40	600
240	75	280	60	500	60	650	50	700

Note: For beams with only one layer of reinforcement, the minimum axis distance of the corner bar (or tendon) to the side of the beam should be increased by 10 mm, except, where the value of b is greater than that given in Combination 4, no increase is required. The minimum axis distances should be increased by 10 mm for pre-stressing tendons.

A structure should be so designed that should a local accident occur, then the damage should be contained within an area local to the accident. Should one member be removed, for example the remainder of the structure should hang together and not precipitate a progressive collapse. For this requirement to be satisfied, the members and materials of construction must have adequate ductility. In particular, the reinforcement and tendons assumed in design to constitute the ties in the structure, must be highly ductile. Robustness reduces the consequences of gross errors or of local structural failures.

AS/NZS1170.0 [1] requires that structures be designed and detailed such that adjacent parts of the structure are tied together in both the horizontal and vertical planes so that the structure can withstand an event without being damaged to an extent disproportionate to that event. In addition, structures should be able to resist lateral loads applied

Design procedures and applied actions 45

Table 2.10 Minimum dimensions b and a_s (in mm) specified in AS3600-2009 [6] for columns in a fire

Standard fire resistance (min)	Columns exposed on more than one side						Columns exposed on one side	
	$\mu_{fi} = 0.2$ (Note 1)		$\mu_{fi} = 0.5$ (Note 1)		$\mu_{fi} = 0.7$ (Note 1)		$\mu_{fi} = 0.7$ (Note 1)	
	a_s (Note 2)	b	a_s (Note 2)	b	a_s (Note 2)	b	a_s (Note 2)	b
30	25	200	25	200	32	200	25	155
					27	300		
60	25	200	36	200	46	250	25	155
			31	300	40	350		
90	31	200	45	300	53	350	25	155
	25	300	38	400	40 (Note 3)	450 (Note 3)		
120	40	250	45 (Note 3)	350 (Note 3)	57 (Note 3)	350 (Note 3)	35	175
	35	350	40 (Note 3)	450 (Note 3)	51 (Note 3)	450 (Note 3)		
180	45 (Note 3)	350 (Note 3)	63 (Note 3)	350 (Note 3)	70 (Note 3)	450 (Note 3)	55	230
240	61 (Note 3)	350 (Note 3)	73 (Note 3)	450 (Note 3)			70	295

Notes:

1. μ_{fi} is the ratio of design axial load in a fire N_f^* to design resistance of the column at normal temperatures ϕN_u.
2. The minimum axis distances should be increased by 10 mm for prestressing tendons.
3. These combinations are for columns with a minimum of 8 bars.

Table 2.11 Minimum dimensions D_s and a_s (in mm) specified in AS3600-2009 [6] for a slab exposed to fire

Fire resistance Period (min)	Minimum effective slab thickness, (D_s)min	Minimum axis distance a_s to the lowest layer of bottom tensile reinforcement			Continuous slabs (one- and two-way)
		Simply-supported slabs			
		One-way slab	Edge-supported two-way slab $(L_y \geq L_x)$		
			$L_y/L_x \leq 1.5$	$1.5 < L_y/L_x \leq 2.0$	
30	60	10	10	10	10
60	80	20	10	15	10
90	100	30	15	20	15
120	120	40	20	25	20
180	150	55	30	40	30
240	175	65	40	50	40

Note: The minimum axis distances should be increased by 10 mm for prestressing tendons.

simultaneously at each floor level not less than the following percentages of $(G + \psi_c Q)$ for each level (storey):

1. 1.0% for structures taller than 15 m above the ground; and
2. 1.5% for all other structures.

The direction of the lateral load should be the direction that produces the most adverse effect in the member under consideration.

In addition, AS/NZS1170.0 [1] requires that all parts of the structures be interconnected, with each connection able to transmit at least $0.05 \times (G + \psi_c Q)$ for that connection.

REFERENCES

1. AS/NZS1170.0:2002. (2002). Structural design actions – General Principles. Amendments 1–5 (2003–2011), Standards Australia, Sydney, New South Wales, Australia.
2. AS/NZS1170.1:2002. (2002). Structural design actions – Permanent, imposed and other actions. Amendments 1,2 (2005,2009), Standards Australia, Sydney, New South Wales, Australia.
3. AS/NZS1170.2:2011. (2011). Structural design actions – Wind actions. Standards Australia, Sydney, New South Wales, Australia.
4. AS/NZS1170.3:2003. (2003). Structural design actions – Snow and ice actions. Standards Australia, Sydney, New South Wales, Australia.
5. AS1170.4:2007. (2007). Structural design actions – Earthquake actions in Australia. Standards Australia, Sydney, New South Wales, Australia.

6. AS3600–2009. (2009). *Australian Standard for Concrete Structures.* Standards Australia, Sydney, New South Wales, Australia.
7. Ranzi, G. and Gilbert, R.I. (2015). *Structural Analysis – Principles, Methods and Modelling.* Boca Raton, FL: CRC Press.
8. Gilbert, R.I. and Ranzi, G. (2011). *Time-Dependent Behaviour of Concrete Structures.* London, U.K.: Spon Press.
9. Mayer, H. and Rüsch, H. (1967). Building damage caused by deflection of reinforced concrete building components. *Technical Translation 1412,* National Research Council Ottawa, Canada (Deutscher Auschuss für Stahlbeton, Heft 193, Berlin, West Germany, 1967).
10. Branson, D.E. (1965). Instantaneous and time-dependent deflections of simple and continuous reinforced concrete beams. HPR Report No. 7, Part 1, Alabama Highway Department, Bureau of Public Roads, Montgomery, AL.
11. Gilbert, R.I. (2007). Tension stiffening in lightly reinforced concrete slabs. *Journal of Structural Engineering, ASCE,* 133(6), 899–903.
12. ACI318M-11. (2011). Building code requirements for structural concrete and commentary. American Concrete Institute, Detroit, MI.
13. Irwin, A.W. (1978). Human response to dynamic motion of structures. *The Structural Engineer,* 56A(9), 237–244.
14. Mickleborough, N.C. and Gilbert, R.I. (1986). Control of concrete floor slab vibration by *L/D* limits. *Proceedings of the 10th Australasian Conference on the Mechanics of Structures and Materials,* University of Adelaide, Adelaide, South Australia, Australia.
15. Smith, A.L., Hicks, S.J. and Devine, P.J. (2009). *Design of Floors for Vibration: A New Approach,* Revised edition (SCI P354). Steel Construction Institute, Ascot, U.K.
16. Willford, M.R. and Young, P. (2006) A design guide for footfall induced vibration of structures (CCIP-016). The Concrete Centre, Surrey, U.K.

Chapter 3

Prestressing systems

3.1 INTRODUCTION

Prestressing systems used in the manufacture of prestressed concrete have developed over the years, mainly through research and development by specialist companies associated with the design and execution of prestressed concrete structures. These companies are often involved in other associated works, including soil and rock anchors, lifting of heavy structures, cable-stayed bridges and suspension bridges that require specialist expertise and patented materials, equipment and design. Information on the products of each company is generally available directly from the respective company or from its websites.

This chapter describes and illustrates the basic forms of prestressing and the components used for prestressing. These include the steel tendons used to prestress the concrete, namely the wire, strand and bar, together with the required anchorages, ducts and couplers. The specialised equipment required for stressing and for grouting of the ducts in post-tensioning applications is illustrated, and the basic principles and concepts of the various systems are provided, including illustrations of the individual prestressing operations. The material properties of both the concrete and steel used in the design of prestressed concrete are detailed and discussed in Chapter 4.

3.2 TYPES OF PRESTRESSING STEEL

There are three basic types of high-strength steel commonly used as tendons in prestressed concrete construction:

1. Cold-drawn stress-relieved round wire;
2. Stress-relieved strand; and
3. High-strength alloy steel bars.

The term *tendon* is generally defined as the wire, strand or bar (or any discrete group of wires, strands or bars) that is intended to be either pretensioned or post-tensioned [1].

Wires are cold-drawn solid steel elements, circular in cross-section, with diameter usually in the range of 2.5 to 12.5 mm. Cold-drawn wires are produced by drawing hot-rolled medium to high carbon steel rods through dies to produce wires of the required diameter. The drawing process cold works the steel, thereby altering its mechanical properties and increasing its strength. The wires are then stress-relieved by a process of continuous heat treatment and straightening to improve ductility and produce the required material properties (such as low relaxation). The typical characteristic minimum breaking stress f_{pb} for wires is in the range 1650 to 1700 MPa. Wires are sometimes indented or crimped to improve their bond characteristics. Wire diameters vary from country to country, but in Australia the most common wire diameters are 5 mm and 7 mm. In recent years, the use of wires in prestressed concrete construction has declined, with the seven-wire strand being preferred in most applications.

Stress-relieved strand is the most commonly used prestressing steel. Both 7-wire and 19-wire strands are available. Seven-wire strand consists of six wires tightly wound around a seventh, slightly larger diameter, central core wire, as shown in Figure 3.1a. The pitch of the six spirally wound wires is between 12 and 18 times the nominal strand diameter. The nominal diameters of the 7-wire strands in general use are in the range 8 mm to 18 mm, with typical characteristic minimum breaking

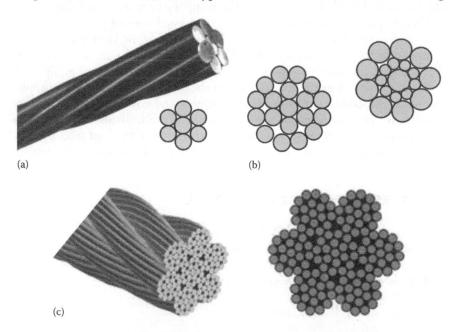

(a)   (b)

(c)

Figure 3.1 Types of strand. (a) 7-wire strand. (b) 19-wire strand – alternative cross-sections. (c) Cable consisting of seven 19-wire strands.

stresses in the range 1700 to 1900 MPa. 7-wire strand is widely used in both pretensioned and post-tensioned applications. 19-wire strand consists of two layers of 9 wires or alternatively two layers of 6 and 12 wires tightly wound spirally around a central wire. The pitch of the spirally wound wires is 12 to 22 times the nominal strand diameter. The nominal diameters of 19-wire strands in general use are in the range 17 mm to 22 mm and typical cross-sections are shown in Figure 3.1b. 19-wire strand is used in post-tensioned applications, but because of its relatively low surface area to volume ratio, it is not recommended for pretensioned applications, where the transfer of prestress relies on the surface area of the strand available for bond to the concrete.

A strand may be compacted by drawing it through a compacting die, thereby reducing its diameter, while maintaining the same cross-sectional area of steel. Compacting strand also facilitates the gripping of the strand at its anchorage.

The mechanical properties of the strand are slightly different from those of the wire from which it is made. This is because the stranded wires tend to straighten slightly when subjected to tension, thus reducing the apparent elastic modulus. For design purposes, the yield stress of stress-relieved strand is often taken as $0.85\, f_{pb}$ and the elastic modulus is often taken to be $E_p = 195 \times 10^3$ MPa.

Cables consist of a group of tendons often formed by multi-wire strands woven together as shown in Figure 3.1c. Stay cables used extensively in cable-stayed and suspension bridges are generally made from strands.

High strength alloy steel bars are hot-rolled with alloying elements introduced into the steel making process. Some bars are ribbed to improve bond. Bars are single straight lengths of solid steel of greater diameter than wire, with diameters typically in the range 26 mm to 75 mm, and with typical characteristic minimum breaking stresses in the range 1000 to 1050 MPa.

3.3 PRETENSIONING

As the name implies, *pretensioning* involves the tensioning of steel strands prior to casting of the concrete and was introduced in Section 1.2.1. The prestressing operation requires an appropriate tensioning bed for the precast elements, bulk heads at both ends to anchor the individual strands and formwork for the precast concrete elements. A typical pretensioning bed is shown in Figure 3.2a. Pretensioning is often carried out in a factory environment where the advantages of quality control and mass production can be achieved. Pretensioning the strands can be achieved by stressing multiple strands or wires simultaneously or by stressing each strand or wire individually. Set-ups for multi- and single-strand stressing are shown in Figure 3.2a and b, respectively.

(a)   (b)

Figure 3.2 Pretensioning beds. (a) Multi-strand pretensioning. (b) Single-strand pretensioning. (Courtesy of VSL International Limited, Hong Kong.)

The application of prestress to the structure or structural element, for most practical cases, involves the use of a hydraulic jack to stress either single strand tendons or groups of strands in a tendon (multi-strand tendons). Prestressing jacks are hydraulically operated by pumping oil under pressure into a piston device, thereby elongating the tendon and increasing the tension in the tendon. When the required tendon elongation is achieved, each end of the tendon is anchored to the bulkhead using wedges that grip the strand at the anchorage. The following sequence of operations is typical for multi-strand pretensioning:

1. The strands are laid out in a pretensioning bed, as illustrated in Figure 3.2, where ram jacks are positioned next to the bulkhead and are extended to ensure that sufficient distance is available for the deformations to take place at transfer. The stroke of the ram jacks must be longer than the desired elongation of the strands.
2. The formwork moulds are closed and the strands are then stressed.
3. Concrete is poured.
4. When the concrete reaches its transfer strength, the load from the strands is transferred to the concrete by allowing the ram jacks to gradually retract and the load is transferred to the concrete members. The strands are cut after the ram jacks are fully retracted. The prestressing force is transferred by a combination of friction and bond between the concrete and steel.

Whilst not in common use now, deflecting or harping of the strands, if required in design, can be achieved by either anchoring the strand or wire in the bed of the unit or by the use of a hydraulic ram or harping device to hold the pretensioned strand in the desired position whilst the concrete is cast and cured, prior to the transfer of the prestressing force to the precast elements. The harping device deflects the strands and provides a varying eccentricity of the prestressing force within the concrete member.

3.4 POST-TENSIONING

Post-tensioning of concrete was introduced in Section 1.2.2 and is used in a wide range of structures to apply prestress on-site. Post-tensioning offers significant flexibility in the way the prestress is applied to a structure, with the tendon profiles readily adjusted to suit the applied loading and the support conditions. Post-tensioning lends itself to *stage stressing*, whereby increments of prestress are applied as required at different stages of construction as the external loads progressively increase.

Post-tensioned systems consist of corrugated galvanised steel or plastic ducts (with grout vents for bonded tendons), prestressing strands, anchorages and grout for bonded tendons. The post-tensioned tendon profile is achieved by fixing the ducts to temporary supports (often attached to the non-prestressed reinforcement in a beam) at appropriate intervals within the formwork. For slabs on ground, the strands are generally supported on bar chairs, as can be seen in Figure 3.3. The ducts that house the prestressing tendons may be fabricated from corrugated steel sheathing or, in more recent developments, plastic ducting, as shown in Figure 3.4a and b, respectively.

Figure 3.5 illustrates a schematic of the layout of a post-tensioning strand in a typical continuous floor slab. The details would also apply for a continuous beam. The prestressing tendon follows a profile determined from

Figure 3.3 Examples of post-tensioned slabs before concrete casting. (Courtesy of VSL Australia, Sydney.)

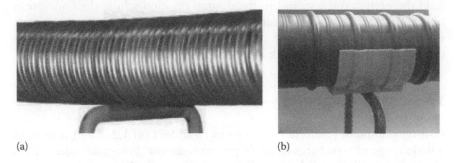

(a)   (b)

Figure 3.4 Post-tensioning ducts. (a) Corrugated metal duct. (b) Plastic ducting.
(Courtesy of VSL International Limited, Hong Kong.)

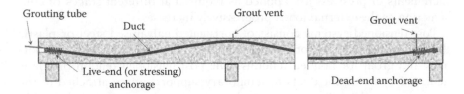

Figure 3.5 Tendon layout and details in a continuous post-tensioned slab.

the design loading and the location and type of supports. After casting, the concrete is allowed to cure until it reaches the required transfer strength. Depending on the system being used, or the requirement of the structural design, an initial prestressing force is sometimes applied when the concrete compressive strength reaches about 10 MPa (to facilitate removal of forms), with the strands re-stressed up to the initial jacking force at a later stage when the concrete has gained its required strength at transfer.

In many parts of the world, including Australia, it is a usual practice for the ducts to be grouted after the post-tensioning operation has been completed. Grout is pumped into the duct at one end under pressure. Grout vents are located at various locations along the duct (as shown in Figure 3.5) to ensure, during the grouting operation, that the wet grout completely fills the duct.

After the grout has set, the post-tensioned tendon is effectively bonded to the surrounding concrete. The grout serves several purposes including higher utilisation of the prestressing steel in bending under ultimate limit state conditions, better corrosion protection of the tendon and, importantly, the prevention of failure of the entire tendon due to localised damage at the anchorage or an accidental cutting of the strand. Further discussion of the advantages and disadvantages of both bonded and unbonded prestressed concrete is presented in Section 3.5.

The prestress is applied by a hydraulic jack (Figure 3.6a and b) reacting against the concrete at the stressing anchorage located at one end of

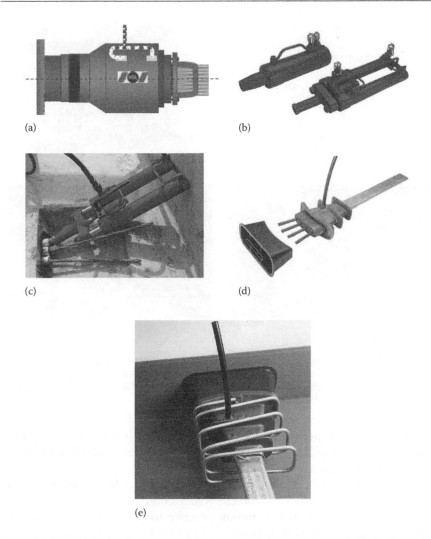

(a)

(b)

(c)

(d)

(e)

Figure 3.6 Typical hydraulic jacks and live-end anchorage details. (a) Hydraulic jacks for multi-strand stressing. (b) Hydraulic jacks for individual strand stressing. (c) Post-tensioning individual strands in a slab duct. (d) Live end anchorage components. (e) Live end anchorage with confining steel and grout vent. (Courtesy of VSL International Limited, Hong Kong; VSL Australia, Sydney.)

the member, usually referred to as the live-end. A small hand-held jack, stressing the individual strands in a slab duct, is shown in Figure 3.6c. The live end of a post-tensioning anchorage system has several basic components comprising an anchor head, associated wedges required to anchor the strands and an anchorage casting or bearing plate. While these anchorages come in many shapes and sizes, the load transfer mechanism of these anchorages remains essentially the same. The stressing operation involves

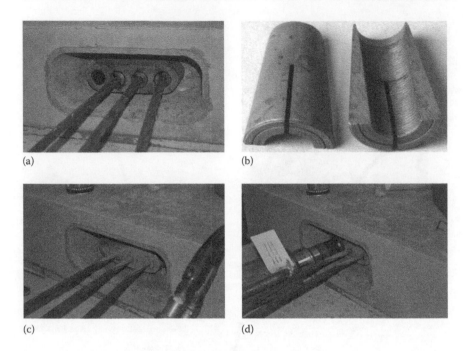

Figure 3.7 Post-tensioning of strands. (a) Strands before post-tensioning. (b) Anchorage
wedge components. (c) Painting of strands before post-tensioning. (d) Stressing
the first of the painted strands. (Courtesy of VSL International Limited,
Hong Kong; VSL Australia, Sydney.)

the hydraulic jack pulling the strands protruding behind the anchorage
until the required jacking force is reached. Typical live end anchorages for
a flat ducted tendon are shown in Figure 3.6d and e.

Prestressing strands at the live end of a slab tendon before post-tensioning are shown in Figure 3.7a, and the wedges used to clamp the strand are
shown in Figure 3.7b. It is common practice to paint the strands before
post-tensioning (as shown in Figure 3.7c) to enable the elongation of each
strand to be readily measured after the stressing operation (Figure 3.7d).
After jacking, the post-tensioned strands are anchored by the wedges in the
anchor head, and the load is transferred from the jack to the structure via
the anchor casting or bearing plate.

Although the live anchorage can also be used at an external non-
stressing end, when stressing is only required from one end of the member, the non-stressing end often takes the form of an internal dead end
anchorage where the ends of the strands are cast in the concrete. Whilst
many forms of this anchorage exist, the principle is to create a pas-
sive anchorage block by either spreading out the exposed strand bundle to form local anchor nodules/bulbs at the extremities beyond the
duct (Figure 3.8a) or anchoring the strands by means of swaged barrels

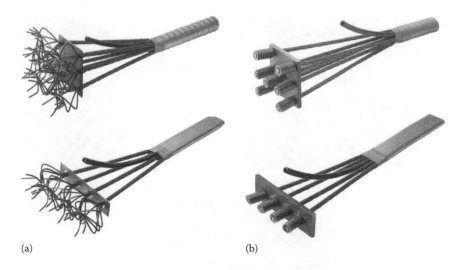

(a)   (b)

Figure 3.8 Dead-end anchorage arrangements. (a) Strands with crimped wires (onion end). (b) Swaged barrel and plate. (Courtesy of VSL Australia, Sydney.)

clamped on the strands and bearing against a steel plate (Figure 3.8b). The duct is sealed to prevent the ingress of concrete during construction. The tendon is stressed only after the surrounding concrete has reached its required transfer strength.

Typical anchorage systems for use with multi-strand arrangements are shown in Figure 3.9 and a typical post-tensioning installation is shown in Figure 3.10a, with the end anchorage after completion of the prestressing operation as shown in Figure 3.10b. Figure 3.10c shows the grinding of the strands after the stressing is completed.

Tendon couplers and intermediate anchorages can be used to connect tendons within a member. Typical examples of coupling and intermediate anchorages are shown in Figure 3.11.

A well-designed grout mix and properly grouted tendons are important for the durability of the structure. The success of a grouting operation

Figure 3.9 Typical multi-strand tendon anchorages. (Courtesy of VSL Australia, Sydney.)

(a)

(b)

(c)

Figure 3.10 Multi-strand jack and anchorage operations. (a) Jacking the strands. (b) Anchorage after prestressing. (c) Cutting the strands after completion of post-tensioning. (Courtesy of VSL Australia, Sydney.)

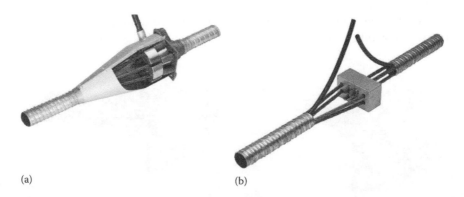

(a)

(b)

Figure 3.11 Coupling and intermediate anchorages for multi-strand systems. (a) Coupling anchorage. (b) Intermediate anchorage. (Courtesy of VSL Australia, Sydney.)

Figure 3.12 Grout vents and caps. (Courtesy of VSL International Limited, Hong Kong.)

depends on many factors, including the correct placement of the grout vents for the injection of grout and for expelling the air in the duct at the grout outlets. Vents are required at the high points of the tendon profile to expel the air in the ducts as the grout is injected into the duct at the tendon end or at the anchorage point of the tendon. Vents at or near the high points allow air and water to be removed from the crest of the duct profile. Grout emitted from the vent at the far end of the duct signals that the duct is completely filled with grout. The use of temporary or permanent grout caps ensures complete filling of the anchorages and permits the verification of the grouting at a later stage. Grout vents are typically located at the high points of the duct over the interior supports, as well as at the end anchorages (as illustrated in Figure 3.5). Figure 3.12 shows typical grout vents and details at a live-end anchorage and at a point along the duct. The ducts into which the grout is injected must be sufficiently large to allow easy installation of the strands and unimpeded flow of grout during the grouting operation.

An air pressure test is usually undertaken before grouting the duct to ensure the possibility of grout leakage is minimised. Grouting follows a standard procedure and, to be effective, requires experienced personnel. The grout is pumped into duct inlet in a continuous uninterrupted fashion. As the grout emerges from the vent, the vent is not closed until the emerging grout has the same consistency and viscosity as the grout being pumped into the inlet. Intermediate vents along the tendon are then closed in sequence after ensuring that the grout has the required consistency and viscosity.

3.5 BONDED AND UNBONDED POST-TENSIONED CONSTRUCTION

In unbonded post-tensioned construction, the strands are not grouted inside the ducts, and remain unbonded from the surrounding concrete throughout the life of the structure. This permits the strands to move locally relative to the structural concrete member. There is no strain compatibility between the prestressing steel and the surrounding concrete. To ensure the strands are able to move relatively freely within the duct, each strand is usually coated with lithium grease, or equivalent, and is located within an external plastic sheathing to provide corrosion protection. The force from the tensioned strands is transferred to the structural member at the end anchorages.

There are advantages and disadvantages of bonded or unbonded construction and the use of either is dependent on design and construction requirements. However, in Australian practice, the disadvantages of unbonded construction are considered to outweigh the advantages and AS3600-2009 [1] does not permit the use of unbonded post-tensioning, except for slabs on ground.

Durability is an important consideration for all forms of construction. Therefore, the provision of active corrosion protection is of significant importance. By grouting the tendons, an alkaline environment is provided around the steel, thus providing active corrosion protection (passivation).

Bonded prestressing steel ensures that any change in strain at the tendon level is the same in both the tendon and the surrounding concrete. At overloads, as the concrete member deforms and the strain at the tendon level increases, the full capacity of the bonded tendon can be realised and the ultimate capacity of the cross-section is increased substantially by grouting. The steel is capable of developing additional force, due to bond, in a relatively short distance. The effect of tendon or anchorage failure is localised after grouting, and the remainder of the tendon is largely unaffected and remains functional. Bonded tendons are also better than unbonded tendons for controlling cracking and for resisting progressive collapse if local failure occurs.

With appropriate design consideration, the prestressing forces in the unbonded tendons can theoretically be adjusted throughout the life of the structure. Tendons may be able to be inspected, re-stressed or even replaced. For example, some tendons for nuclear works are unbonded, as they need to be monitored and, as necessary, re-stressed. In unbonded construction, since the prestressing force in the tendon is transmitted to the beam only at the end anchorages, there is an almost uniform distribution of strain in the tendon under load. Changes in the force in the tendon are only possible due to friction and due to deformation of the member, thereby increasing the overall length of the tendon between anchorages. At overloads, the full strength of an unbonded tendon may not be achieved and the ultimate strength of unbonded construction is therefore, generally less than that of bonded construction. The anchorage of the unbonded tendons is therefore, a critical component, since the entire prestressing force is transmitted at this point throughout the life of the structure.

3.6 CIRCULAR PRESTRESSING

The term *circular prestressing* is applied to structures with a circular form such as cylindrical water tanks, liquid and natural gas tanks, storage silos, tunnels, digesters and nuclear containment vessels. In general, the term is applied when the direction of the prestress at any point is circumferential, i.e. in the direction of the tangent to the circumference of the circular prestressed concrete surface structure. The circular prestressing compresses the structure to counteract the tensile bursting forces or loads from within the structure. Circular prestressing is also appropriate in cylindrical shells. The ring beams around the edges of long-span shell structures develop significant tension forces and these can be balanced by prestressing.

Circular prestressing can take several forms, depending on the process of prestressing and the type of structure. Circumferential prestressing may be applied using individual tendons and multiple anchorages, or by using continuous wrapping, whereby a single tendon is wrapped around the circular structure. For tanks and silos it is common practice to have buttresses in the walls, permitting easier detailing, installation and stressing. Figure 3.13 shows the prestressing buttress of a circularly prestressed tank.

Figure 3.13 Prestressing anchorages at a buttress of a prestressed tank. (Courtesy of VSL International Limited, Hong Kong.)

3.7 EXTERNAL PRESTRESSING

Whilst the standard internal prestressing discussed in previous sections remains the basic procedure for the majority of structures, external prestressing of concrete structures has become much more popular with certain forms of structural members. As the name suggests, external prestressing is the application of prestress from a prestressing tendon or cable placed externally to the concrete elements. The cable or tendon may be placed on the outside the structure or, in the case of box-type girders, inside the structure. Since the prestressing steel is located external to the concrete, there is no bond between the structural concrete and the prestressing components (unlike pretensioned concrete and bonded post-tensioned concrete). Early prestressed bridges were often steel structures with external steel bars used to impart the prestress (to stiffen and to strengthen the structure), but these structures are not considered here.

Since the tendons are external and unbonded to the concrete structure, they can be removed and, if required, replaced at any time during the life of the structure.

A significant use of external prestressing is in the construction of concrete box girder bridge decks. The external tendons are typically anchored in the concrete diaphragms within the box and are deviated at carefully designed saddles located at the bottom of the structure at the mid-spans and at the top of the structure at the supports. These deviators can be made of steel pipes or void formers that are integrated with the concrete box section. Figure 3.14 shows the external tendons inside a box girder bridge, including a saddle to locate the tendons near the bottom of the section at mid-span.

As the tendons are placed outside the concrete section, pouring of concrete in the web is made easier and, as the web compression area is not reduced by the voids created by internal tendon ducts, the web thickness can be kept to a minimum.

Figure 3.14 External tendons in bridge box girder section. (Courtesy of VSL International Limited, Hong Kong.)

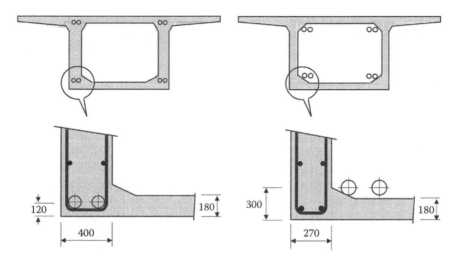

Figure 3.15 Illustration of web thickness reduction possible with external prestressing compared to internal prestressing (dimensions in mm).

Figure 3.15 illustrates a typical reduction in web thickness of a concrete box girder due to external prestressing. External prestressing cables also generally have lower prestress losses. Another major advantage of external prestressing is that it can be used in new structures as well as strengthening or retrofitting of existing structures.

The disadvantages of external prestressing include the reduction in tendon eccentricity when the tendons are to be kept inside the concrete box structure, i.e. above the bottom flange slab of the box section, and the slight additional costs of providing replaceable anchorages, ducts and deviation saddles for external tendons.

REFERENCE

1. AS3600-2009. (2009). *Australian Standard for Concrete Structures*, Standards Australia, Sydney, New South Wales, Australia.

Chapter 4

Material properties

4.1 INTRODUCTION

The deformation of a prestressed concrete member throughout the full range of loading depends on the properties and behaviour of the constituent materials. In order to satisfy the design objective of adequate structural strength, the ultimate strengths of both concrete and steel need to be known. In addition, factors affecting material strength and the nonlinear behaviour of each material in the overload range must be considered. In order to check for serviceability, the instantaneous and time-dependent properties of concrete and steel at typical in-service stress levels are required.

As was mentioned in Chapter 1, the prestressing force in a prestressed concrete member gradually decreases with time. This *loss* of prestress, which is usually 10% to 25% of the initial value, is mainly caused by creep and shrinkage strains that develop with time in the concrete at the level of the bonded steel, as well as relaxation of the tendons. Reasonable estimates of the creep and shrinkage characteristics of concrete and procedures for the *time analysis* of prestressed structures are essential for an accurate prediction and a clear understanding of in-service behaviour. The loss of prestress caused by relaxation of the prestressing steel is caused by creep in the tendon. With the relatively low relaxation of modern prestressing steels, however, this component of prestress loss is usually relatively small (less than 5%).

The intention in this chapter is to present a broad outline of material behaviour and to provide sufficient quantitative information on material properties to complete most design tasks, with specific reference to Australian guidelines [1–8] and remarks based on ACI recommendations [9–12].

4.2 CONCRETE

More comprehensive treatments of the properties of concrete and the factors affecting them are given by others, including Neville [13] and Metha and Monteiro [14].

4.2.1 Composition of concrete

Concrete is a mixture of cement, water and aggregates. It may also contain one or more chemical admixtures. Within hours of mixing and placing, concrete sets and begins to develop strength and stiffness as a result of chemical reactions between the cement and water. These reactions are known as hydration. Calcium silicates in the cement react with water to produce calcium silicate hydrate and calcium hydroxide. The resultant alkalinity of the concrete helps to provide corrosion protection for the reinforcement.

The relative proportions of cement, water and aggregates may vary considerably depending on the chemical properties of each component and the desired properties of the concrete. A typical mix used for prestressed concrete by weight might be coarse aggregate 45%, fine aggregate 30%, cement 18% and water 7%.

In most countries, several different types of Portland cement are available, including general purpose cements, high early strength cements, low heat of hydration cements and various cements that provide enhanced sulphate resistance. In order to alter and improve the properties of concrete, other cementitious materials may be used to replace part of the Portland cement and the use of blended cements is now commonplace. These cement replacement materials include silica fume, fly ash, blast furnace slag and natural pozzolans.

The ratio of water to cement by weight required to hydrate the cement completely is about 0.25, although larger quantities of water are often required in practice in order to produce a workable mix. For the concrete typically used in prestressed structures, the water-to-cement ratio is about 0.4. It is desirable to use as little water as possible, since water not used in the hydration reaction causes voids in the cement paste that reduce the strength and increase the permeability of the concrete.

Chemical admixtures are widely used to improve one or more properties of the concrete. High-strength concretes with low water-to-cement ratios are made more workable by the inclusion of superplasticisers in the mix. These polymers greatly improve the flow of the wet concrete, and allow very high-strength and low-permeability concrete to be used with conventional construction techniques.

The rock and sand aggregates used in concrete should be inert and properly graded. Expansive and porous aggregates should not be used, and aggregates containing organic matter or other deleterious substances, such as salts or sulphates, should also be avoided.

4.2.2 Strength of concrete

In structural design, the quality of concrete is usually controlled by the specification of a minimum *characteristic compressive strength* at 28 days, denoted as f'_c. The characteristic strength is the strength that is exceeded by 95% of the uniaxial compressive strength measurements taken from standard compression tests. In Australia and the United States, such tests are performed on concrete cylinders. Cylinders are generally either 150 mm diameter by 300 mm long or 100 mm diameter by 200 mm long. In Europe and the United Kingdom, 150 mm concrete cubes are used in standard compression tests. Because the restraining effect at the loading surfaces is greater for the cube than for the longer cylinder, strength measurements taken from cubes are higher than those taken from cylinders. The ratio between cylinder and cube strengths $R_{cyl/cu}$ is about 0.8 for low-strength concrete (i.e. cylinder strengths of 20 to 30 MPa) and increases as the strength increases. The following expression for $R_{cyl/cu}$ is often used [13]:

$$R_{cyl/cu} = 0.76 + 0.2\log_{10}\left(\frac{\sigma_{cu}}{19.6}\right) \tag{4.1}$$

where σ_{cu} is the cube strength in MPa. Throughout this book, f_c' refers to the specified characteristic compressive strength relating to standard cylinder tests.

In practice, the concrete used in prestressed construction is usually of better quality and higher strength than that required for ordinary reinforced concrete structures. Values of f_c' in the range of 40 to 65 MPa are most often used, but higher strengths are not uncommon. Indeed, prestressed members fabricated from reactive powder concretes with compressive strengths in excess of 150 MPa have been used in a wide variety of structures.

The forces imposed on a prestressed concrete section are relatively large and the use of high-strength concrete keeps section dimensions to a minimum. High-strength concrete also has obvious advantages in the anchorage zone of post-tensioned members where bearing stresses are large, and in pretensioned members where a higher bond strength better facilitates the transfer of prestress.

As the compressive strength of concrete increases, so too does its tensile strength. The use of higher strength concrete may therefore, delay (or even prevent) the onset of cracking in a member. High-strength concrete is considerably stiffer than low-strength concrete. The elastic modulus is higher and elastic deformations due to both the prestress and the external loads are smaller. In addition, high-strength concrete generally creeps less than low-strength concrete. This results in smaller losses of prestress and smaller long-term deformations.

The effect of concrete strength on the shape of the stress-strain curve for concrete in uniaxial compression is shown in Figure 4.1. The modulus of elasticity (the initial slope of the tangent to the ascending portion of each curve) increases with increasing strength and each curve reaches its maximum stress at a strain in the range of 0.0015 to 0.00285.

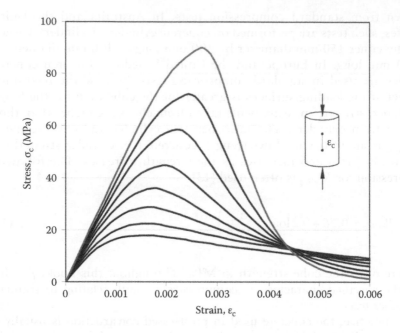

Figure 4.1 Effect of strength on the shape of the uniaxial compressive stress-strain curve for concrete.

The shape of the unloading portion of each curve (after the peak stress has been reached) depends, among other things, on the characteristics of the testing machine. By applying deformation to a specimen, instead of load, in a testing machine that is stiff enough to absorb the energy of a failing specimen, an extensive unloading branch of the stress-strain curve can be obtained. Concrete can undergo very large compressive strains and still carry load. This deformability of concrete tends to decrease with increasing strength.

The strength of properly placed and well compacted concrete depends primarily on the water-to-cement ratio, the size of the specimen, the size, strength and stiffness of the aggregate, the cement type, the curing conditions and the age of the concrete. The strength of concrete increases as the water-to-cement ratio decreases.

The compressive strength of concrete increases with time. A rapid initial strength gain (in the first day or so after casting) is followed by a decreasing rate of strength gain thereafter. The rate of development of strength with time depends on the type of curing and the type of cement. In prestressed concrete construction, a rapid initial gain in strength is usually desirable so that the prestress may be applied to the structure as early as possible. This is particularly so for precast pretensioned production. Steam curing and high early strength cement are often used for this purpose.

A prediction of the characteristic strength of concrete at any time t based on the measured or specified 28 day characteristic strength may be made using the following equation [10]:

$$f_c'(t) = \frac{t}{\alpha + \beta t} f_c'(28)$$ (4.2)

where α and β are constants that depend on the cement type and the method of curing. For general purpose Portland cement, $\alpha = 4.0$ and $\beta = 0.85$ for moist cured concrete and $\alpha = 1.0$ and $\beta = 0.95$ for steam cured concrete. For high early strength cement, $\alpha = 2.3$ and $\beta = 0.92$ for moist cured concrete and $\alpha = 0.7$ and $\beta = 0.98$ for steam cured concrete.

The strength of concrete in tension is an order of magnitude less than the compressive strength and is far less reliably known. A reasonable estimate is required, however, in order to anticipate the onset of cracking and predict service-load behaviour in the post-cracking range. The characteristic flexural tensile strength of concrete $f_{ct.f}'$ (or modulus of rupture) is determined from the maximum extreme fibre tensile stresses calculated from the results of standard flexural strength tests on plain concrete prisms and usually lies within the range of $0.6\sqrt{f_c'}$ to $1.0\sqrt{f_c'}$ (in MPa). Because of the relatively large scatter of measured tensile strengths, the lower end of this range is usually specified in building codes, including AS3600-2009 [1]. In direct tension, where the tensile stress is uniform (or nearly so) over the section, the characteristic direct tensile strength of concrete f_{ct}' is usually about 60% of the characteristic flexural tensile strength. For lightweight aggregate concrete, these tensile strengths should be reduced by a factor of about 0.67.

In practice, concrete is often subjected to multi-axial states of stress. For example, a state of biaxial stress exists in the web of a beam, or in a shear wall or a deep beam. Triaxial stress states exist within connections, in confined columns, in two-way slabs and in other parts of a structure. A number of pioneering studies of the behaviour of concrete under multi-axial states of stress (including References [15] to [17]) led to the formulation of material modelling laws now used routinely in finite element software for the nonlinear stress analysis of complex concrete structures. A typical biaxial strength envelope is shown in Figure 4.2, where σ_1 and σ_2 are the orthogonal stresses and σ_{cu} is the uniaxial compressive strength of concrete.

The strength of concrete under biaxial compression is greater than for uniaxial compression. Transverse compression improves the longitudinal compressive strength by confining the concrete, thereby delaying (or preventing) the propagation of internal microcracks. Figure 4.2 also shows that transverse compression reduces the tensile strength of concrete, due mainly to the Poisson's ratio effect. Similarly, transverse tension reduces the compressive strength. In triaxial compression, both the strength of concrete and the strain at which the peak stress is reached are greatly increased and

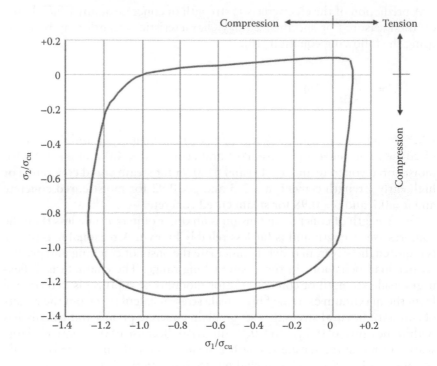

Figure 4.2 Typical biaxial strength envelope for concrete.

even small confining pressures can increase strength significantly. Correctly detailed transverse reinforcement provides confinement to produce a tri-axial stress state in the compressive zone of columns and beams, thereby improving both strength and ductility.

4.2.3 Strength specifications in AS3600-2009

4.2.3.1 Characteristic compressive strength

The strength of concrete is specified in AS3600-2009 [1] in terms of the lower characteristic compressive cylinder strength at 28 days, f'_c. This is the value of compressive strength exceeded by 95% of all standard cylinders tested in accordance with AS1012.9 [2] at age 28 days after curing under standard laboratory conditions.

The *standard strength grades* (expressed in terms of the characteristic compressive strength) are 20, 25, 32, 40, 50, 65, 80 and 100 MPa. While the normal strength grades (20 to 50 MPa) may be considered as *Normal Class Concrete* specified only in terms of characteristic compressive strength, the high strength grades (65 to 100 MPa) are *Special Class Concretes* requiring additional specifications (such as workability, water/binder ratios, binder content, admixtures, water reducing agents, aggregate type, shrinkage requirements and more).

4.2.3.2 Mean in situ compressive strength

The mean compressive strength of sample cylinders at 28 days for a strength grade of 20 MPa is about 25% higher than the lower characteristic strength (f_c') and about 10% higher than f_c' for a strength grade of 100 MPa. The *in situ compressive strength* of concrete (i.e. the compressive strength of the concrete in the structure on site) is taken to be 90% of the cylinder strength. The mean value of the cylinder strength f_{cm} and the mean value of the in situ strength f_{cmi} corresponding to the standard strength grades are shown in Table 4.1.

4.2.3.3 Tensile strength

The uniaxial tensile strength f_{ct} is defined in AS3600-2009 [1] as the maximum stress that concrete can withstand when subjected to uniaxial tension. Direct uniaxial tensile tests are difficult to perform and tensile strength is usually measured via either flexural tests on prisms in accordance with AS1012.11 [4] or indirect splitting tests on cylinders as specified in AS1012.10 [3]. In flexure, the apparent tensile stress at the extreme tensile fibre of the critical cross-section under the peak load is calculated assuming linear elastic behaviour and is taken to be the flexural tensile strength $f_{ct.f}$. The flexural tensile strength $f_{ct.f}$ is significantly higher than f_{ct} due to the strain gradient and the post-peak unloading portion of the stress-strain curve for concrete in tension. Typically, f_{ct} is about 50% to 60% of $f_{ct.f}$. The indirect tensile strength measured from a split cylinder test $f_{ct.sp}$ is also higher than f_{ct} (usually by about 10%) due to the confining effect of the bearing plate in the standard test.

AS3600-2009 permits f_{ct} to be determined from either the measured values of $f_{ct.f}$ or $f_{ct.sp}$ using the relationships $f_{ct} = 0.6\,f_{ct.f}$ or $f_{ct} = 0.9\,f_{ct.sp}$. For design purposes, where standard curing is specified, and in the absence of more accurate data from testing, the lower characteristic 28 day flexural tensile strength $f_{ct.f}'$ and the lower characteristic 28 day uniaxial tensile stress f_{ct}' may be taken as:

$$f_{ct.f}' = 0.6\sqrt{f_c'} \tag{4.3}$$

Table 4.1 Mean cylinder and in situ concrete compressive strengths at 28 days specified in AS3600-2009 [1]

Characteristic cylinder strength, f_c' (MPa)	20	25	32	40	50	65	80	100
Mean cylinder strength, f_{cm} (MPa)	25	31	39	48	59	76	91	110
Mean in situ strength, f_{cmi} (MPa)	22	28	35	43	53	68	82	99

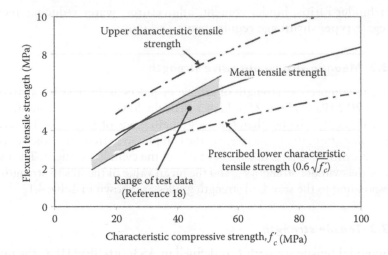

Figure 4.3 Comparison on prescribed values for flexural tensile strength with test data [18].

and

$$f'_{ct} = 0.36\sqrt{f'_c} \tag{4.4}$$

According to AS3600-2009 [1], the mean and upper characteristic values of the flexural and the uniaxial tensile strength may be estimated by multiplying the lower characteristic values by 1.4 and 1.8, respectively. A comparison between the prescribed values for flexural tensile strength and test data is shown in Figure 4.3 (taken from Reference [18]).

4.2.3.4 Stress–strain curves for concrete in compression

AS3600-2009 [1] specifies that, if required, the stress-strain curve for concrete may be determined from test data or may be assumed to be 'of curvilinear form defined by recognised simplified equations'. The standard further specifies that, for design purposes, the shape of the uniaxial stress-strain curve used to model in situ concrete should be adjusted so that the maximum stress is $0.9 f'_c$. This is in recognition that on site conditions of compaction, curing and exposure may not be as benign as those for a cylinder prepared, cured and tested in a laboratory environment. Where mean in-situ values rather than characteristic values are required, the shape of the stress-strain curve used to model in situ concrete should be adjusted so that the maximum stress is f_{cmi}.

Numerous equations describing the curvilinear stress-strain relationship for concrete in compression are available in the literature. Thorenfeldt et al. [19]

showed that the stress-strain curve for conventional and high strength concretes can be represented by:

$$\sigma_c = \sigma_{cp} \frac{n\eta}{n-1+\eta^{nk}} \tag{4.5}$$

where η is the ratio of the concrete strain ε_c corresponding to the concrete stress σ_c and the strain ε_{cp} corresponding to the peak in-situ stress σ_{cp}, i.e. $\eta = \varepsilon_c/\varepsilon_{cp}$. The term n is a curve fitting factor given by $n = E_c/(E_c - E_{cp})$, where E_c is the modulus of elasticity of the concrete and E_{cp} is the secant modulus corresponding to peak stress (i.e. $E_{cp} = \sigma_{cp}/\varepsilon_{cp}$). The factor k is a decay factor for the post-peak response and increases with concrete strength. When $\varepsilon \leq \varepsilon_{cp}$, $k = 1$ and, when $\varepsilon > \varepsilon_{cp}$, Collins and Porasz [20] proposed that $k = 0.67 + \sigma_{cp}/62 \geq 1.0$, where σ_{cp} is expressed in MPa.

Based on data by Setunge [21] for Australian concretes, Attard and Stewart [22] recommended that the strain at peak stress may be taken as $\varepsilon_{cp} = 4.11\sigma_{cp}^{0.75}/E_c$, where σ_{cp} and E_c are in MPa. The family of stress-strain curves (one curve for each standard strength grade) obtained from Equation 4.5 (with $\sigma_{cp} = 0.9f_c'$) is plotted in Figure 4.1. When modelling the mean stress-strain relationship of in-situ concrete, σ_{cp} should be taken as f_{cmi}.

4.2.4 Deformation of concrete

4.2.4.1 Discussion

The deformation of a loaded concrete specimen is both instantaneous and time-dependent. If the load is sustained, the deformation of the specimen gradually increases with time and may eventually be several times larger than the instantaneous value.

The gradual development of strain with time is caused by creep and shrinkage. Creep strain is produced by sustained stress. Shrinkage is independent of stress and results primarily from the loss of water as the concrete dries and from chemical reactions in the hardened concrete. Creep and shrinkage cause increases in axial deformation and curvature on reinforced and prestressed concrete cross-sections, losses of prestress, local redistribution of stress between the concrete and the steel reinforcement, and redistribution of internal actions in statically indeterminate members. Creep and shrinkage are often the cause of excessive deflection (or camber) and excessive shortening of prestressed members. In addition, shrinkage may cause unsightly cracking that could lead to serviceability or durability problems. On a more positive note, creep relieves concrete of stress concentrations and imparts a measure of deformability to concrete.

Researchers have been investigating the time-dependent deformation of concrete ever since it was first observed and reported over a century ago,

and a great deal of literature has been written on the topic. Detailed sum-
maries of the time-dependent properties of concrete and the factors that
affect them are contained in texts by Neville [13,23], Neville et al. [24],
Gilbert [25], Gilbert and Ranzi [26] and Ghali et al. [27,28] and in technical documents such as those from ACI Committee 209 [10–12].

The time-varying deformation of concrete may be illustrated by considering a concrete specimen subjected to a constant sustained stress. At any
time t, the total concrete strain $\varepsilon_c(t)$ in an uncracked, uniaxially loaded
specimen consists of a number of components that include the instanta-
neous strain $\varepsilon_{ce}(t)$, the creep strain $\varepsilon_{cc}(t)$, the shrinkage strain $\varepsilon_{cs}(t)$ and the
temperature strain $\varepsilon_T(t)$. Although not strictly correct, it is usually accept-
able to assume that all four components are independent and may be calculated separately and combined to obtain the total strain.

When calculating the in-service behaviour of a concrete structure at constant temperature, it is usual to express the concrete strain at a point as the
sum of the instantaneous, creep and shrinkage components:

$$\varepsilon_c(t) = \varepsilon_{ce}(t) + \varepsilon_{cc}(t) + \varepsilon_{cs}(t) \tag{4.6}$$

The strain components in a drying specimen held at constant temperature
and subjected to a constant sustained compressive stress σ_{c0} first applied at
time τ_0 are illustrated in Figure 4.4. Immediately after the concrete sets or
at the end of moist curing ($t = \tau_d$ in Figure 4.4), shrinkage strain begins to
develop and continues to increase at a decreasing rate. On application of

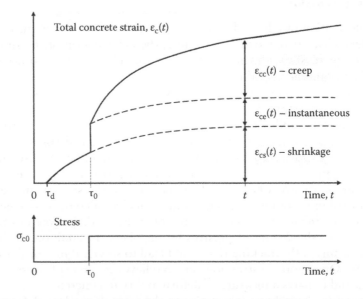

Figure 4.4 Concrete strain versus time for a specimen subjected to constant sustained stress.

the stress, a sudden jump in the strain diagram (instantaneous strain) is followed by an additional gradual increase in strain due to creep.

The prediction of the time-dependent behaviour of a concrete member requires an accurate estimate of each of these strain components at critical locations. This requires knowledge of the stress history, in addition to accurate data for the material properties. The stress history depends both on the applied load and on the boundary conditions of the member. Calculations are complicated by the restraint to creep and shrinkage provided by both the bonded reinforcement and the external supports, and the continuously varying concrete stress history that inevitably results.

The material properties that influence each of the strain components depicted in Figure 4.4 are described in the following sections. Methods for predicting the time-dependent behaviour of prestressed concrete cross-sections and members are discussed in Chapter 5.

4.2.4.2 Instantaneous strain

The magnitude of the instantaneous strain $\varepsilon_{ce}(t)$ caused by either compressive or tensile stress depends on the magnitude of the applied stress, the rate at which the stress is applied, the age of the concrete when the stress was applied and the stress-instantaneous strain relationship for the concrete. Consider the uniaxial instantaneous strain versus compressive stress curve shown in Figure 4.5. When the applied stress is less than about half the compressive strength, the curve is essentially linear and the instantaneous strain is usually considered to be elastic (fully recoverable). In this low-stress range, the secant modulus E_c does not vary significantly with stress and is only

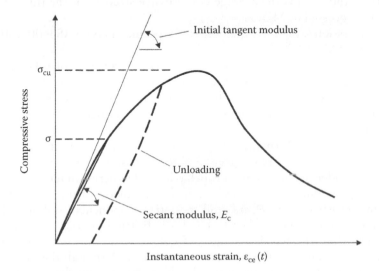

Figure 4.5 Typical compressive stress-instantaneous strain curve.

slightly smaller than the initial tangent modulus. At higher stress levels, the stress-strain curve becomes significantly nonlinear and a significant proportion of the instantaneous strain is irrecoverable upon unloading.

In concrete structures, compressive concrete stresses caused by the day-to-day service loads rarely exceed half of the compressive strength. It is therefore, reasonable to assume that the instantaneous behaviour of concrete at service loads is linear-elastic and that instantaneous strain is given by:

$$\varepsilon_{ce}(t) = \frac{\sigma(t)}{E_c} \tag{4.7}$$

The value of the elastic modulus E_c increases with time as the concrete gains strength and stiffness. It also depends on the rate of application of the stress and increases as the loading rate increases. For most practical purposes, these variations are usually ignored and it is common practice to assume that E_c is constant with time and equal to its initial value calculated at the time of first loading τ_0. For stress levels less than about $0.4f_c'$, and for stresses applied over a relatively short period (say up to 5 min), a numerical estimate of the elastic modulus for normal strength concretes (20 to 50 MPa) may be obtained from the well-known expression proposed by Pauw [29]:

$$E_c = \rho^{1.5}0.043\sqrt{f_c(\tau_0)} \quad \text{(in MPa)} \tag{4.8}$$

where ρ is the density of concrete (about 2400 kg/m^3 for normal weight concrete) and $f_c(\tau_0)$ is the average compressive strength at the time of first loading expressed in MPa.

An expression similar to Equation 4.8 is specified in both AS3600-2009 [1] and ACI 318M-11 [9].

When the stress is applied more slowly, say over a period of 1 day, significant additional deformation occurs owing to the rapid early development of creep. For the estimation of short-term deformation in such a case, it is recommended that the elastic modulus given by Equation 4.8 be reduced by about 20% [25].

The in-service performance of a concrete structure is very much affected by the concrete's inability to carry significant tension. It is therefore, necessary to consider the instantaneous behaviour of concrete in tension, as well as in compression. Prior to cracking, the instantaneous strain of concrete in tension consists of both elastic and inelastic components. In design, however, concrete is usually taken to be elastic-brittle in tension, i.e. at stress levels less than the tensile strength of concrete, the instantaneous strain versus stress relationship is assumed to be linear. Although the magnitude of the elastic modulus in tension is likely to differ from that in compression, it is usual to assume that both values are equal. Prior to cracking, the instantaneous strain in tension may be calculated using Equation 4.7 When the tensile strength is reached, cracking occurs and the concrete stress perpendicular to the crack is usually assumed to be zero. In reality, if the rate of tensile deformation is controlled, and crack widths are small, concrete can carry some tension across a crack due to friction that exists on the rough mating surfaces of the crack.

Poisson's ratio for concrete ν generally lies within the range of 0.15 to 0.22 and for most practical purposes may be taken equal to 0.2.

4.2.4.3 Creep strain

For concrete subjected to a constant sustained stress, the gradual development of creep strain is illustrated in Figure 4.4. In the period immediately after first loading, creep develops rapidly, but the rate of increase slows appreciably with time. Creep has traditionally been thought to approach a limiting value as the time after first loading approaches infinity, but more recent research suggests that creep continues to increase indefinitely, albeit at a slower rate. After several years under load, the rate of change of creep with time is small. Creep of concrete has its origins in the hardened cement paste and is caused by a number of different mechanisms. A comprehensive treatment of creep in plain concrete is given by Neville et al. [24].

Many factors influence the magnitude and rate of development of creep. Some are properties of the concrete mix, while others depend on the environmental and loading conditions. In general, the capacity of concrete to creep decreases as the concrete quality increases. At a particular stress level, creep in higher-strength concrete is less than that in lower-strength concrete. Creep decreases as the water-to-cement ratio is reduced. An increase in either the aggregate content or the maximum aggregate size reduces creep, as does the use of a stiffer aggregate type.

Creep also depends on the environment. Creep increases as the relative humidity decreases. Creep is therefore, greater when accompanied by drying. Creep is also greater in thin members with large surface area-to-volume ratios, such as slabs and walls. However, the dependence of creep on both the relative humidity and the size and shape of the specimen decreases as the concrete strength increases. Near the surface of a member, creep takes place in a drying environment and is therefore, greater than in regions remote from a drying surface. In addition to the relative humidity, creep is dependent on the ambient temperature. A temperature rise increases the deformability of the cement paste and accelerates drying, and thus increases creep. The dependence of creep on temperature is more pronounced at elevated temperatures and is far less significant for temperature variations between 0°C and 20°C. However, creep in concrete at a mean temperature of 40°C is perhaps 25% higher than that at 20°C [30].

In addition to the environment and the characteristics of the concrete mix, creep depends on the loading history, in particular the magnitude and duration of the stress and the age of the concrete when the stress is first applied. The age at first loading τ_0 has a marked influence on the final magnitude of creep. Concrete loaded at an early age creeps more than concrete loaded at a later age. Concrete is therefore, a time-hardening material, although even in very old concrete the tendency to creep never entirely disappears [31].

When the sustained concrete stress is less than about $0.5f'_c$, creep is approximately proportional to stress and is known as *linear creep*. At higher stress levels creep increases at a faster rate and becomes nonlinear with respect to stress. This nonlinear behaviour of creep at high stress levels is thought to be related to an increase in micro-cracking. Compressive stresses rarely exceed $0.5f'_c$ in concrete structures at service loads, and creep may be taken as proportional to stress in most situations in the design for serviceability. From a structural design point of view, therefore, nonlinear creep is of little relevance.

Creep strain is made up of a recoverable component, called the delayed elastic strain $\varepsilon_{cc.d}(t)$ and an irrecoverable component called flow $\varepsilon_{cc.f}(t)$. These components are illustrated by the creep strain versus time curve in Figure 4.6a caused by the stress history shown in Figure 4.6b. The recoverable creep is thought to be caused by the elastic aggregate acting on the viscous cement paste after the applied stress is removed. If a concrete specimen is unloaded after a long period under load, the magnitude of the recoverable creep is of the order of 40% to 50% of the elastic strain (between 10% and 20% of the total creep strain). Although the delayed elastic strain is observed only as recovery when the load is removed, it is generally believed to be of the same magnitude under load and to develop rapidly in the period immediately after loading. Rüsch et al. [30] suggested that the shape of the delayed elastic strain curve is independent of the age or dimensions of the specimen and is unaffected by the composition of the concrete.

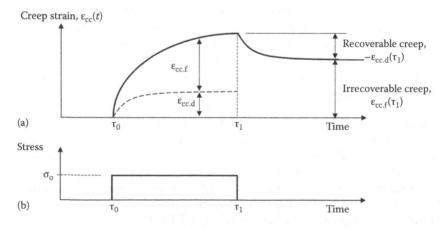

Figure 4.6 Recoverable and irrecoverable creep components. (a) Creep strain history. (b) Stress history.

The capacity of concrete to creep is usually measured in terms of the creep coefficient $\varphi_{cc}(t,\tau)$. In a concrete specimen subjected to a constant sustained compressive stress $\sigma_c(\tau)$, first applied at age τ, the creep coefficient at time t is the ratio of creep strain to instantaneous strain and is given by:

$$\varphi_{cc}(t,\tau) = \frac{\varepsilon_{cc}(t,\tau)}{\varepsilon_{ce}(\tau)} \tag{4.9}$$

and the creep strain at time t caused by a constant sustained stress $\sigma_c(\tau)$ first applied at age τ is:

$$\varepsilon_{cc}(t,\tau) = \varphi_{cc}(t,\tau)\varepsilon_{ce}(\tau) = \varphi_{cc}(t,\tau)\frac{\sigma_c(\tau)}{E_c(\tau)} \tag{4.10}$$

where $E_c(\tau)$ is the elastic modulus at time τ. For concrete subjected to a constant sustained stress, knowledge of the creep coefficient allows the rapid determination of the creep strain at any time using Equation 4.10.

Since both the creep and the instantaneous strain components are proportional to stress for compressive stress levels less than about $0.5f_c'$, the creep coefficient $\varphi_{cc}(t,\tau)$ is a pure time function and is independent of the applied stress. The creep coefficient increases with time at a decreasing rate. Although there is some evidence that the creep coefficient increases indefinitely, the final creep coefficient $\varphi_{cc}^*(\tau) = \varepsilon_{cc}^*(\tau)/\varepsilon_{ce}(\tau)$ is often taken as the 30 year value and its magnitude usually falls within the range of 1.5–4.0. A number of the well-known methods for predicting the creep coefficient were described and compared in References [25,26]. The approach specified in AS3600-2009 [1] for making numerical estimates of $\varphi_{cc}(t,\tau)$ is presented in Section 4.2.5.3.

The final creep coefficient is a useful measure of the capacity of concrete to creep. Since creep strain depends on the age of the concrete at the time of first loading, so too does the creep coefficient. This effect of ageing is illustrated in Figure 4.7. The magnitude of the final creep coefficient $\varphi_{cc}^*(\tau)$ decreases as the age at first loading τ increases:

$$\varphi_{cc}^*(\tau_i) > \varphi_{cc}^*(\tau_j) \quad \text{for} \quad \tau_i < \tau_j$$

This time-hardening or ageing of concrete complicates the calculation of creep strain caused by a time-varying stress history.

Another frequently used time function is known as *specific creep* $C(t,\tau)$, defined as the proportionality factor relating stress to linear creep:

$$\varepsilon_{cc}(t,\tau) = C(t,\tau)\sigma_c(\tau) \tag{4.11}$$

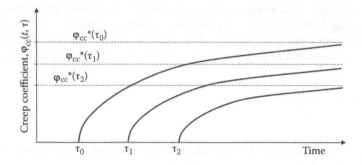

Figure 4.7 Effect of age at first loading on the creep coefficient.

or

$$C(t,\tau) = \frac{\varepsilon_{cc}(t,\tau)}{\sigma_c(\tau)} \tag{4.12}$$

$C(t, \tau)$ is the creep strain at time t produced by a sustained *unit* stress first applied at age τ.

From Equations 4.9 through 4.11, the relationship between the creep coefficient and specific creep is:

$$\varphi_{cc}(t,\tau) = C(t,\tau)E_c(\tau) \tag{4.13}$$

The sum of the instantaneous and creep strains at time t produced by a sustained unit stress applied at τ is defined as the *creep function $J(t,\tau)$* and is given by:

$$J(t,\tau) = \frac{1}{E_c(\tau)} + C(t,\tau) = \frac{1}{E_c(\tau)}\left[1 + \varphi_{cc}(t,\tau)\right] \tag{4.14}$$

The stress-produced strains (i.e. the instantaneous plus creep strains) caused by a constant sustained stress $\sigma_c(\tau)$ first applied at age τ (also called the stress-dependent strains) may therefore, be determined from:

$$\varepsilon_{ce}(t) + \varepsilon_{cc}(t,\tau) = J(t,\tau)\sigma_c(\tau) = \frac{\sigma_c(\tau)}{E_c(\tau)}\left[1 + \varphi_{cc}(t,\tau)\right] = \frac{\sigma_c(\tau)}{E_e(t,\tau)} \tag{4.15}$$

where $E_e(t, \tau))$ is known as the *effective modulus* and is given by:

$$E_e(t,\tau) = \frac{E_c(\tau)}{1 + \varphi_{cc}(t,\tau)} \tag{4.16}$$

If the stress is gradually applied to the concrete, rather than abruptly applied, the subsequent creep strain is reduced, because the concrete ages during the period of application of the stress. This can be accommodated analytically by the use of a reduced or adjusted creep coefficient. For an increment of stress $\Delta\sigma_c(\tau)$ applied to the concrete gradually, beginning at time τ, the load-dependent strain may be obtained by modifying Equation 4.15 as follows:

$$\varepsilon_{ce}(t) + \varepsilon_{cc}(t, \tau) = \frac{\Delta\sigma_c(\tau)}{E_c(\tau)}\left[1 + \chi(t, \tau)\varphi_{cc}(t, \tau)\right] = \frac{\Delta\sigma_c(\tau)}{\bar{E}_e(t, \tau)} \tag{4.17}$$

where:

$$\bar{E}_e(t, \tau) = \frac{E_c(\tau)}{1 + \chi(t, \tau)\varphi_{cc}(t, \tau)} \tag{4.18}$$

$\bar{E}_e(t, \tau)$ is called the age-adjusted effective modulus, and $\chi(t,\tau)$ is an ageing coefficient first introduced by Trost [32] and later developed by Dilger and Neville [33] and Bazant [34].

Like the creep coefficient, the ageing coefficient depends on the rate of application of the gradually applied stress and the age at first loading and varies between about 0.4 and 1.0. Methods for the determination of the ageing coefficient are available, for example, in *fib* Model Code 2010 [35]. Gilbert and Ranzi [26] showed that for concrete first loaded at early ages ($\tau_0 < 20$ days) and where the applied load is sustained, the final long-term ageing coefficient may be taken as $\chi^* = 0.65$. In situations where the deformation is held constant and the concrete stress relaxes, the final long-term ageing coefficient may be taken as $\chi^* = 0.8$.

The previous discussions have been concerned with the creep of concrete in compression. However, the creep of concrete in tension is also of interest in a number of practical situations, for example when studying the effects of restrained or differential shrinkage. Tensile creep also plays a significant role in the analysis of suspended reinforced concrete slabs at service loads where stress levels are generally low and, typically, much of the slab is initially uncracked.

Comparatively, little attention has been devoted to the study of tensile creep [12] and only limited experimental results are available in the literature [36]. Some researchers have multiplied the creep coefficients measured for compressive stresses by factors in the range of 1 to 3 to produce equivalent coefficients describing tensile creep, e.g. Chu and Carreira [37] and Bazant and Oh [38].

It appears that the mechanisms of creep in tension are different to those in compression. The magnitudes of both tensile and compressive creep increase when loaded at earlier ages. However, the rate of change of tensile creep with time does not decrease in the same manner as for compressive creep, with the development of tensile creep being more linear [36]. Drying tends to increase tensile creep in a similar manner to compressive creep, and tensile creep is in part recoverable upon removal of the load. Further research is needed to provide clear design guidance. In this book, it is assumed that the magnitude and rate of development of tensile creep are similar to that of compressive creep at the same low stress levels. Although not strictly correct, this assumption simplifies calculations and does not usually introduce serious inaccuracies.

4.2.4.4 Shrinkage strain

Shrinkage of concrete is the time-dependent strain in an unloaded and unrestrained specimen at constant temperature. Shrinkage is often divided into several components, including *plastic* shrinkage, *chemical* shrinkage, *thermal* shrinkage and *drying* shrinkage. Plastic shrinkage occurs in the wet concrete before setting, whereas chemical, thermal and drying shrinkage all occur in the hardened concrete after setting. Some high strength concretes are prone to *plastic shrinkage*, which may result in significant cracking before and during the setting process. This cracking occurs due to capillary tension in the pore water and is best prevented by taking measures during construction to avoid the rapid evaporation of bleed water. Before the concrete has set, the bond between the plastic concrete and the reinforcement has not yet developed, and the steel is ineffective in controlling plastic shrinkage cracking.

Drying shrinkage is the reduction in volume caused principally by the loss of water during the drying process. It increases with time at a gradually decreasing rate and takes place in the months and years after setting. The magnitude and rate of development of drying shrinkage depend on all the factors that affect the drying of concrete, including the relative humidity, the size and shape of the member and the mix characteristics, in particular, the type and quantity of the binder, the water content and water-to-cement ratio, the ratio of fine-to-coarse aggregate and the type of aggregate.

Chemical shrinkage results from various chemical reactions within the cement paste and includes hydration shrinkage, which is related to the degree of hydration of the binder in a sealed specimen with no moisture exchange. Chemical shrinkage (often called *autogenous shrinkage*) occurs rapidly in the days and weeks after casting and is less dependent on the environment and the size of the specimen than drying shrinkage.

Thermal shrinkage is the contraction that results in the first few hours (or days) after setting as the heat of hydration gradually dissipates. The term *endogenous shrinkage* is sometimes used to refer to that part of the shrinkage of the hardened concrete that is not associated with drying (i.e. the sum of autogenous and thermal shrinkage).

The shrinkage strain ε_{cs} is usually considered to be the sum of the drying shrinkage component (which is the reduction in volume caused principally by the loss of water during the drying process) and the endogenous shrinkage component. Drying shrinkage in high strength concrete is smaller than in normal strength concrete due to the smaller quantities of free water after hydration. However, thermal and chemical shrinkage may be significantly higher. Although drying and endogenous shrinkage are quite different in nature, there is often no need to distinguish between them from a structural engineering point of view.

Shrinkage increases with time at a decreasing rate, as illustrated in Figure 4.4. Shrinkage is assumed to approach a final value ε_{cs}^* as time approaches infinity.

Drying shrinkage is affected by all the factors that affect the drying of concrete, in particular the water content and the water-cement ratio of the mix, the size and shape of the member and the ambient relative humidity. All else being equal, drying shrinkage increases when the water-cement ratio increases, the relative humidity decreases and the ratio of the exposed surface area to volume increases. Temperature rises accelerate drying and therefore, increase shrinkage. By contrast, endogenous shrinkage increases as the cement content increases and the water-cement ratio decreases. In addition, endogenous shrinkage is not significantly affected by the ambient relative humidity.

The effect of a member's size on drying shrinkage should be emphasised. For a thin member, such as a slab, the drying process may be essentially complete after several years, but for the interior of a larger member, the drying process may continue throughout its lifetime. For uncracked mass concrete structures, there is no significant drying (shrinkage) except for about 300 mm from each exposed surface. By contrast, the chemical shrinkage is less affected by the size and shape of the specimen.

Shrinkage is also affected by the volume and type of aggregate. Aggregate provides restraint to shrinkage of the cement paste, so that an increase in the aggregate content reduces shrinkage. Shrinkage is also smaller when stiffer aggregates are used, i.e. aggregates with higher elastic moduli. Thus shrinkage is considerably higher in lightweight concrete than in normal weight concrete (by up to 50%).

The approach specified in AS3600-2009 [1] for making numerical estimates of shrinkage strain is presented in Section 4.2.5.4.

4.2.5 Deformational characteristics specified in AS3600-2009

4.2.5.1 Introduction

Great accuracy in the prediction of the creep coefficient and the shrinkage strain is not possible. The variability of these material characteristics is high. Design predictions are most often made using one of many numerical methods that are available for predicting the creep coefficient and shrinkage strain. These methods vary in complexity, ranging from relatively complicated methods, involving the determination of numerous coefficients that account for the many factors affecting creep and shrinkage, to much simpler procedures. A description of and comparison between some of the more well known methods is provided in ACI 209.2R-08 [12]. Although the properties of concrete vary from country to country as the mix characteristics and environmental conditions vary, the agreement between the procedures for estimating both creep and shrinkage is still remarkably poor, particularly for shrinkage. In addition, the comparisons between predictive models show that the accuracy of a particular model is not directly proportional to its complexity, and predictions made using several of the best known methods differ widely.

In the following sections, the relatively simple models contained in AS3600-2009 [1] for predicting the elastic modulus, the creep coefficient and the shrinkage strain for concretes with a compressive strength in the range of 20 MPa $\leq f_c' \leq$ 100 MPa are presented. These models were originally developed by Gilbert [39].

4.2.5.2 Elastic modulus

For stress levels less than about 0.4 f_{cm} for normal strength concrete ($f_c' \leq$ 50 MPa – see Table 4.1) and about 0.6 f_{cm} for high strength concrete ($50 < f_c' \leq$ 100 MPa), and for stresses applied over a relatively short period (say up to 5 min), a numerical estimate of the in situ elastic modulus may be made from:

$$E_c = \rho^{1.5} 0.043 \sqrt{f_{cmi}} \, (\text{in MPa}) \quad \text{when } f_{cmi} \leq 40 \text{ MPa} \tag{4.19}$$

$$E_c = \rho^{1.5} [0.024 \sqrt{f_{cmi}} + 0.12] \, (\text{in MPa}) \quad \text{when } 40 < f_{cmi} \leq 100 \text{ MPa} \tag{4.20}$$

where ρ is the density of the concrete in kg/m^3 (not less than 2400 kg/m^3 for normal weight concrete) and f_{cmi} is the mean in situ compressive strength in MPa at the time of first loading.

Table 4.2 The elastic modulus for in situ concrete E_c [1]

f'_c (MPa)	20	25	32	40	50	65	80	100
f_{cmi} (MPa)	22.5	27.9	35.4	43.7	53.7	68.2	81.9	99.0
E_c (MPa)	24,000	26,700	30,100	32,750	34,800	37,400	39,650	42,200

Table 4.3 Increase in elastic modulus with age of concrete $t - E_c(t)/E_c$ [26]

Cement type	Age of concrete in days (t)					
	3	7	28	90	360	30,000
Ordinary Portland cement	0.68	0.83	1.0	1.09	1.15	1.20
High early strength cement	0.77	0.88	1.0	1.06	1.09	1.13

Equation 4.19 was originally proposed by Pauw [29], as previously introduced in Equation 4.8. Values for E_c obtained using Equations 4.19 and 4.20 for in situ normal weight concrete ($\rho = 2400$ kg/m³) at age 28 days for different values of f'_c are given in Table 4.2. The mean in situ strength compressive strength f_{cmi} in Table 4.2 is taken to be 90% of the standard *mean* cylinder strength and for 100 MPa concrete the mean in situ strength is actually smaller than the characteristic cylinder strength f'_c.

The magnitude of E_c prescribed by AS3600-2009 [1] has an accuracy of ±20% depending, among other things, on the aggregate type and quantity, the aggregate to binder ratio, the curing regime and the rate of application of the load. The values of E_c for concretes made with limestone or sandstone aggregates are likely to be lower than that specified in Equations 4.19 and 4.20, and for concretes with basalt aggregates, E_c may be somewhat higher.

Equations 4.19 and 4.20 can be used to determine the elastic modulus at any age, provided the mean in situ compressive strength at that age is used. Typical variations in E_c with time t are shown in Table 4.3.

4.2.5.3 Creep coefficient

In Section 4.2.4.3, the creep coefficient at time t associated with a constant stress first applied at age τ was defined as the ratio of the creep strain at time t to the (initial) elastic strain. AS3600-2009 [1] recognises that the most accurate way of determining the final creep coefficient is by testing or by using results obtained from measurements on similar local concretes. However, testing is often not a practical option for the structural designer and a relatively simple approach is provided in AS3600-2009. The approach does not account for such factors as aggregate type, cement type, cement replacement materials and more, but it does provide a ball-park estimate of the creep coefficient for concrete that is suitable for routine use in structural design.

The creep coefficient at any time φ_{cc} may be calculated from:

$$\varphi_{cc} = k_2 \, k_3 \, k_4 \, k_5 \, \varphi_{cc.b} \qquad (4.21)$$

The basic creep coefficient $\varphi_{cc.b}$ is the mean value of the final creep coefficient for a specimen loaded at age 28 days under a constant stress of $0.4f'_c$ and may be determined by tests in accordance with AS1012.16 [6] or, in the absence of test data, may be taken as the value given in Table 4.4.

The factor k_2 in Equation 4.21 describes the development of creep with time. It depends on the hypothetical thickness t_h, the environment and the time after loading and is given in Figure 4.8. The hypothetical thickness is defined as $t_h = 2A_g/u_e$, where A_g is the cross-sectional area of the member and u_e is that portion of the section perimeter exposed to the atmosphere plus half the total perimeter of any voids contained within the section.

The factor k_3 depends on the age at first loading τ (in days) and is given by:

$$k_3 = \frac{2.7}{1 + \log(\tau)} \quad \text{(for } \tau > 1 \text{ day)} \qquad (4.22)$$

The factor k_4 accounts for the environment, with $k_4 = 0.7$ for an arid environment, $k_4 = 0.65$ for an interior environment, $k_4 = 0.60$ for a

Table 4.4 The basic creep coefficient $\varphi_{cc.b}$

f'_c (MPa)	20	25	32	40	50	65	80	100
$\varphi_{cc.b}$	5.2	4.2	3.4	2.8	2.4	2.0	1.7	1.5

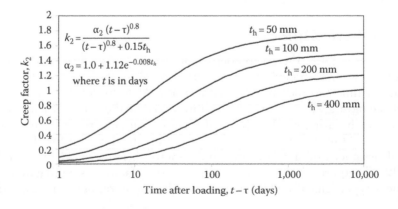

Figure 4.8 The factor k_2 versus time [39].

temperate inland environment and $k_4 = 0.5$ for a tropical or near-coastal environment. The factor k_5 accounts for the reduced influence of both relative humidity and specimen size on the creep of concrete as the concrete strength increases (or more precisely, as the water-binder ratio decreases) and is calculated as follows:

$$k_5 = 1.0 \quad \text{when } f_c' \leq 50 \text{ MPa} \tag{4.23}$$

$$k_5 = (2.0 - \alpha_3) - 0.02 \, (1.0 - \alpha_3) f_c' \quad \text{when } 50 \text{ MPa} < f_c' \leq 100 \text{ MPa} \tag{4.24}$$

where $\alpha_3 = 0.7/(k_4\alpha_2)$ and α_2 is specified in Figure 4.8.

A family of creep coefficient versus duration of loading curves obtained using Equation 4.21 is shown in Figure 4.9 for a concrete specimen located in a temperate inland environment, with a hypothetical thickness $t_h = 150$ mm, concrete strength $f_c' = 40$ MPa and loaded at different ages τ.

The final creep coefficients φ_{cc}^* (after 30 years) predicted by the above method are given in Table 4.5 for concrete first loaded at 28 days, for characteristic strengths of 25 to 100 MPa, for three hypothetical thicknesses ($t_h = 100$, 200 and 400 mm) and for concrete located in different environments.

It must be emphasised that creep of concrete is highly variable with significant differences in the measured creep strains in seemingly identical specimens tested under identical conditions (both in terms of load and environment). The creep coefficient predicted by Equation 4.21 should be taken as an average value with a range of $\pm 30\%$. The upper end of this range is likely for concrete subjected to prolonged periods of temperatures in excess of 25°C or sustained stress levels in excess of $0.4 f_c'$.

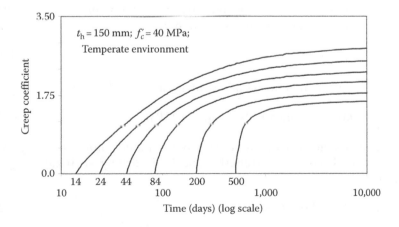

Figure 4.9 Typical creep coefficient versus time curves (from Equation 4.21).

Table 4.5 Final creep coefficients (after 30 years) φ_{cc}^* for concrete first loaded at 28 days

	Final creep coefficient, φ_{cc}^*											
	Arid environment			Interior environment			Temperate inland environment			Tropical and near-coastal environment		
f_c'	t_h (mm)			t_h (mm)			t_h (mm)			t_h (mm)		
(MPa)	100	200	400	100	200	400	100	200	400	100	200	400
25	4.83	3.91	3.28	4.49	3.63	3.04	4.14	3.35	2.81	3.45	2.79	2.34
32	3.91	3.16	2.65	3.63	2.94	2.46	3.35	2.71	2.27	2.79	2.26	1.89
40	3.22	2.61	2.18	2.99	2.42	2.03	2.76	2.23	1.87	2.30	1.86	1.56
50	2.76	2.23	1.87	2.56	2.07	1.74	2.37	1.91	1.60	1.97	1.60	1.34
65	2.07	1.76	1.54	1.96	1.67	1.46	1.84	1.57	1.38	1.61	1.39	1.23
80	1.56	1.41	1.29	1.51	1.36	1.25	1.45	1.32	1.22	1.34	1.23	1.14
100	1.15	1.14	1.12	1.15	1.14	1.12	1.15	1.14	1.12	1.15	1.14	1.12

4.2.5.4 Shrinkage strain

The model for estimating the magnitude of shrinkage strain in normal and high strength concrete specified in AS3600-2009 [1] divides the total shrinkage strain ε_{cs} into two components, the autogenous shrinkage ε_{cse} and the drying shrinkage ε_{csd}, as given by:

$$\varepsilon_{cs} = \varepsilon_{cse} + \varepsilon_{csd} \tag{4.25}$$

The autogenous shrinkage specified in AS3600-2009 [1] includes an allowance for early thermal shrinkage and is assumed to develop relatively rapidly and to increase with concrete strength. At any time t (in days) after casting, the autogenous shrinkage is given by:

$$\varepsilon_{cse} = \varepsilon_{cse}^*(1.0 - e^{-0.1t}) \tag{4.26}$$

where ε_{cse}^* is the final autogenous shrinkage and may be taken as:

$$\varepsilon_{cse}^* = (0.06f_c' - 1.0) \times 50 \times 10^{-6} \quad (f_c' \text{ in MPa}) \tag{4.27}$$

Drying shrinkage develops more slowly than autogenous shrinkage and decreases with concrete strength. The basic drying shrinkage $\varepsilon_{csd.b}$ is given by:

$$\varepsilon_{csd.b} = (1.0 - 0.008f_c') \times \varepsilon_{csd.b}^* \tag{4.28}$$

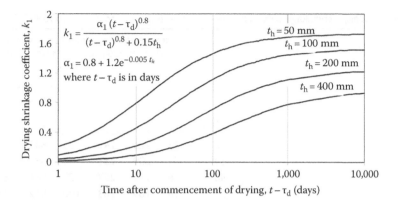

Figure 4.10 Drying shrinkage strain coefficient k_1 for various values of t_h [39].

where $\varepsilon_{csd.b}^*$ depends on the quality of the local aggregates and may be taken as 800×10^{-6} for concrete supplied in Sydney and Brisbane (where the aggregate quality is known to be good), 900×10^{-6} in Melbourne and 1000×10^{-6} elsewhere (where the aggregate quality is less certain).

At any time after the commencement of drying $(t - \tau_d)$, the drying shrinkage may be taken as:

$$\varepsilon_{csd} = k_1 k_4 \varepsilon_{csd.b} \tag{4.29}$$

where k_1 describes the development of drying shrinkage with time and is given in Figure 4.10.

The factor k_4 depends on the environment and is equal to 0.7 for an arid environment, 0.65 for an interior environment, 0.6 for a temperate inland environment and 0.5 for a tropical or near-coastal environment.

As expressed in Equation 4.25, the design shrinkage at any time is therefore, the sum of the autogenous shrinkage (Equation 4.26) and the drying shrinkage (Equation 4.29). The proposed model provides good agreement with available shrinkage measurements on Australian concretes. For specimens located in arid, temperate and tropical environments with average quality aggregate (i.e. with $\varepsilon_{csd.b}^* = 1000 \times 10^{-6}$) and with a hypothetical thickness $t_h = 200$ mm, the shrinkage strain components predicted by the above model at 28 days after the commencement of drying and after 30 years (i.e. at $t - \tau_d = 28$ and $t - \tau_d = 10,950$ days) are given in Table 4.6.

The method outlined in AS3600-2009 [1] requires only a short calculation, but designers should be aware that the estimate of ε_{cs} is within a range of ±30%. Comparisons of the predictions made using the method

Table 4.6 Design shrinkage strain components (t_h = 200 mm and $\varepsilon_{csd.b}^*$ = 1000 × 10⁻⁶)

| | | Shrinkage strain ε_{cs} (×10⁻⁶) and shrinkage strain components ε_{cse} (×10⁻⁶) and ε_{csd} (×10⁻⁶) | | | | | |
| | | $t - \tau_d$ = 28 days | | | $t - \tau_d$ = 10,950 days (30 years) | | |
Environment	f_c' (MPa)	ε_{cse}	ε_{csd}	ε_{cs}	ε_{cse}	ε_{csd}	ε_{cs}
Arid	25	23	225	249	25	683	708
	32	43	209	253	46	635	681
	40	66	191	257	70	581	651
	50	94	169	263	100	512	612
	65	136	135	271	145	410	555
	80	178	101	280	190	307	497
	100	235	56	291	250	171	421
Temperate inland	25	23	193	217	25	586	611
	32	43	180	223	46	545	591
	40	66	164	230	70	498	568
	50	94	145	239	100	439	539
	65	136	116	252	145	351	496
	80	178	87	265	190	264	454
	100	235	48	283	250	146	396
Tropical, near-coastal and coastal	25	23	161	184	25	488	513
	32	43	150	193	46	454	500
	40	66	137	203	70	415	485
	50	94	121	215	100	366	466
	65	136	97	233	145	293	438
	80	178	72	251	190	220	410
	100	235	40	275	250	122	372

with shrinkage measured over 30 years and reported by Brooks [40] are shown in Figure 4.11.

Because details of the mix proportions and aggregate type are not included in the prediction model for shrinkage, a better estimate of the final design shrinkage strain in a particular member can be obtained from the results of a standard 56 day shrinkage test, carried out in accordance with AS1012.13 [5]. The standard test is conducted on a 75 mm × 75 mm × 280 mm prism, which is moist cured for 7 days and then allowed to dry at a relative humidity of 50%. For the standard test specimen, the hypothetical thickness is t_h = 37.5 mm, k_4 = 0.7 and, from Figure 4.10, at 56 days k_1 = 1.466.

To illustrate the procedure, we will consider an example in which the final design shrinkage strain ε_{cs} after 30 years is required for a 200 mm thick slab with f_c' = 40 MPa and located in a temperate inland environment. The slab

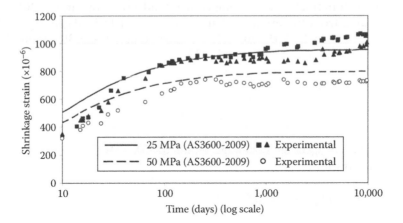

Figure 4.11 Predicted versus experimental shrinkage-time curves [40].

will be drying from both the top and bottom surfaces. Let us assume that standard 56 day shrinkage tests have been carried out on the concrete and the shrinkage strain measured after 56 days is $\varepsilon_{cs}(56) = 650 \times 10^{-6}$. All the autogenous shrinkage will have occurred within the 56 days period of the test and for 40 MPa concrete, Equation 4.27 gives:

$$\varepsilon_{cse}(56) = \varepsilon_{cse}^* = (0.06f_c' - 1.0) \times 50 \times 10^{-6} = 70 \times 10^{-6}$$

According to AS3600-2009 [1], the autogenous shrinkage that would have occurred within the first 7 days of wet curing is $\varepsilon_{cse}(7) = \varepsilon_{cse}^*(1.0 - e^{-0.1 \times 7}) = 0.5\varepsilon_{cse}^* = 35 \times 10^{-6}$. The remaining autogenous shrinkage that occurred after 7 days, and therefore, included in the 56 day shrinkage measurement, is $\varepsilon_{cse}(56) - \varepsilon_{cse}(7) = 35 \times 10^{-6}$. The drying shrinkage strain at 56 days is therefore:

$$\varepsilon_{csd}(56) = \varepsilon_{cs}(56) - \varepsilon_{cse}(56) = 650 \times 10^{-6} - 35 \times 10^{-6} = 615 \times 10^{-6}$$

and, with $k_1 = 1.466$ and $k_4 = 0.7$, the final basic drying shrinkage $\varepsilon_{csd.b}^*$ can now be determined from Equations 4.28 and 4.29:

$$\varepsilon_{csd}(56) = k_1 k_4 (1.0 - 0.008\, f_c') \times \varepsilon_{csd.b}^*$$

$$= 1.466 \times 0.7 \times (1 - 0.008 \times 40) \times \varepsilon_{csd.b}^* = 615 \times 10^{-6}$$

Therefore: $\varepsilon_{csd.b}^* = 881 \times 10^{-6}$.

For a 200 mm thick slab in a temperate inland environment: $t_h = 200$ mm, $k_4 = 0.6$ and, from Equation 4.29 at 30 years ($t = 10,950$ days), $k_1 = 1.22$. With $\varepsilon^*_{csd.b} = 881 \times 10^{-6}$, the final drying shrinkage (at age 30 years) ε^*_{csd} is obtained from Equations 4.28 and 4.29:

$$\varepsilon^*_{csd} = k_1 k_4 (1.0 - 0.008 f'_c) \times \varepsilon^*_{csd.b}$$

$$= 1.22 \times 0.6 \times (1 - 0.008 \times 40) \times 881 \times 10^{-6} = 439 \times 10^{-6}$$

With the autogenous shrinkage equal to 70×10^{-6}, the final design shrinkage strain is:

$$\varepsilon^*_{cs} = \varepsilon^*_{cse} + \varepsilon^*_{csd} = 509 \times 10^{-6}$$

4.2.5.5 Thermal expansion

The coefficient of thermal expansion for concrete depends on the coefficient of thermal expansion of the coarse aggregate and on the mix proportions. For most types of coarse aggregate, the coefficient lies within the range of $5 \times 10^{-6}/°C$ to $13 \times 10^{-6}/°C$ [13]. For design purposes and in the absence of more detailed information (test data), AS3600-2009 [1] specifies a coefficient of thermal expansion for concrete of $10 \times 10^{-6}/°C \pm 20\%$.

4.3 STEEL REINFORCEMENT

The strength of a reinforced or prestressed concrete element in bending, shear, torsion or direct tension depends on the properties of the steel reinforcement and tendons, and it is necessary to adequately model the various types of steel reinforcement and their material properties.

Steel reinforcement is used in concrete structures to provide strength, ductility and serviceability. Steel reinforcement can also be strategically placed to reduce both immediate and time-dependent deformations. Adequate quantities of bonded reinforcement will also provide crack control, wherever cracks occur in the concrete.

4.3.1 General

Conventional, non-prestressed reinforcement in the form of bars, cold-drawn wires or welded wire mesh is used in prestressed concrete structures for the same reasons as it is used in conventional reinforced concrete construction, including:

- To provide additional tensile strength and ductility in regions of the structure where sufficient tensile strength and ductility are not provided by the prestressing steel. Non-prestressed, longitudinal bars, for example, are often included in the tension zone of beams to supplement the prestressing steel and increase the flexural strength. Non-prestressed reinforcement in the form of stirrups is most frequently used to carry the diagonal tension caused by shear and torsion in the webs of prestressed concrete beams.

- To control flexural cracks at service loads in prestressed concrete beams and slabs where some degree of cracking under full service loads is expected.
- To control cracking induced by restraint to shrinkage and temperature changes in regions and directions of low (or no) prestress.
- To carry compressive forces in regions where the concrete alone may not be adequate, such as in columns or in the compressive zone of heavily reinforced beams.
- Lateral ties or helices are used to provide restraint to bars in compression (i.e. to prevent lateral buckling of compressive reinforcement prior to the attainment of full strength) and to provide confinement for the compressive concrete in columns, beams and connections, thereby increasing both the strength and deformability of the confined concrete.
- To reduce long-term deflection and shortening due to creep and shrinkage by the inclusion of longitudinal bars in the compression region of the member.
- To provide resistance to the transverse tension that develops in the anchorage zone of post-tensioned members and to assist the concrete to carry the high bearing stresses immediately behind the anchorage plates.
- To reinforce the overhanging flanges in T-, I- or L-shaped cross-sections in both the longitudinal and transverse directions.

Types and sizes of non-prestressed reinforcement vary from country to country. In Australia, for example, the two most commonly used types of reinforcing bar are the following:

1. Grade R250N reinforcing bars with a characteristic yield stress of $f_{sy} = 250$ MPa are hot-rolled plain round bars of 6 or 10 mm diameter (designated R6 and R10 bars) and are commonly used for fitments, such as ties and stirrups; and
2. Grade D500N bars with a characteristic yield stress $f_{sy} = 500$ MPa are hot-rolled deformed bars with sizes ranging from 12 to 40 mm diameter (in 4 mm increments).

Regularly spaced rib-shaped deformations on the surface of a deformed bar improve the bond between the concrete and the steel and, greatly improve the anchorage potential of the bar. It is for this reason that

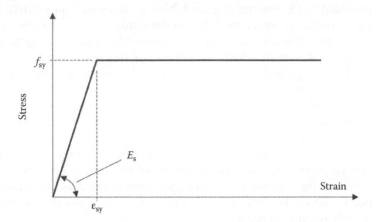

Figure 4.12 Idealised stress-strain relationship for non-prestressed steel.

deformed bars are used as longitudinal reinforcement in most reinforced and prestressed concrete members.

In design calculations, non-prestressed steel is usually assumed to be elastic-plastic, as shown in Figure 4.12. Before yielding, the reinforcement is elastic, with steel stress σ_s proportional to the steel strain ε_s, i.e. $\sigma_s = E_s \varepsilon_s$, where E_s is the elastic modulus of the steel. After yielding, the stress-strain curve is usually assumed to be horizontal (perfectly plastic) and the steel stress $\sigma_s = f_{sy}$ at all values of strain exceeding the strain at first yield $\varepsilon_{sy} = f_{sy}/E_s$. The yield stress f_{sy} is taken to be the *strength* of the material and strain hardening is most often ignored. The stress-strain curve in compression is assumed to be similar to that in tension.

4.3.2 Specification in AS3600-2009

4.3.2.1 Strength and ductility

Conventional steel reinforcement in Australia must comply with the requirements of AS/NZS 4671 [7] and is specified in terms of its strength grade and ductility class. In design, the characteristic yield stress f_{sy} of the reinforcement should not be taken greater than the value specified in Table 4.7. Two ductility classes are recognised by AS3600-2009, Class N (normal ductility) and Class L (low ductility).

The ductility Class is specified in terms of the minimum characteristic value of uniform elongation ε_{su} and the minimum tensile strength to yield stress ratio f_{su}/f_{sy}. The uniform elongation ε_{su} is the strain at peak stress f_{su}. For steel to be classified as Class N, its uniform elongation must exceed 0.05 and its tensile strength to yield stress ratio must exceed 1.08. For steel

Table 4.7 Yield strength and ductility class of Australian reinforcement

Reinforcement		f_{sy} (MPa)	Minimum ε_{su}	Minimum f_{su}/f_{sy}	Ductility class
Type (Note 1)	Grade				
Plain bar	R250N	250	0.05	1.08	N
Deformed bar	D500L (Note 2)	500	0.015	1.03	L
	D500N	500	0.05	1.08	N
Welded wire mesh (plain,	D500L	500	0.015	1.03	L
deformed of indented)	D500N	500	0.05	1.08	N

Notes:
1. All reinforcement types must comply with the requirement of AS/NZS4671 [7].
2. Class L deformed bars may only be used as fitments.

to be classified as Class L, its uniform elongation must exceed 0.015 and its tensile strength to yield stress ratio must exceed 1.03.

The minimum values of $\varepsilon_{su} = 0.015$ for Class L reinforcement is significantly lower than the lowest value permitted in Europe and, accordingly, AS3600-2009 [1] specifies that Class L reinforcement should not be used in situations where the reinforcement is required to undergo large plastic deformation under strength limit state conditions (i.e. strains in excess of 0.015). Other limitations on the use of Class L reinforcement are placed elsewhere in AS3600-2009, including the following:

1. a reduced strength reduction factor (compared to Class N reinforcement) to compensate for its relatively low ductility;
2. it is not to be used if plastic methods of design are used; and
3. it is not to be used if the analysis has relied on some measure of moment redistribution.

4.3.2.2 Elastic modulus

The modulus of elasticity of reinforcing steel E_s is the slope of the initial elastic part of the stress-strain curve, when the stress is less than f_{sy}, and, in the absence of test data, may be taken as equal to 200×10^3 MPa, irrespective of the type and Ductility Class of the steel. Alternatively, E_s may be determined from standard tests. The elastic modulus in compression is taken to be identical to that in tension.

4.3.2.3 Stress–strain curves

The shape of the stress-strain curve for reinforcement may be determined from tests. A typical curve for a 500N quenched and self-tempered bar is

shown in Figure 4.13a. We have already seen that conventional steel reinforcement is most often assumed to be elastic-perfectly plastic. Alternatively,
any recognised simplified equation suitably calibrated to approximate the
shape of the actual curve may be used. The idealised bilinear relationship
shown in Figure 4.13b may be a suitable simplification.

For nonlinear and other refined methods of analysis, actual stress-strain
curves, using mean rather than characteristic values, should be used. The
actual curves for several steels are shown in Figure 4.13c. In reality, the
actual yield stress of a reinforcing bar is usually significantly higher than
the guaranteed minimum indicated in Table 4.7.

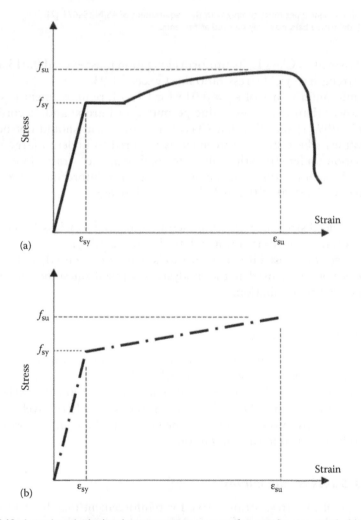

Figure 4.13 Actual and idealised stress strain curves for reinforcing steel. (a) Actual
stress-strain curve of a quenched and self-tempered bar. (b) Idealised stress-
strain curve of a quenched and self-tempered bar.   (Continued)

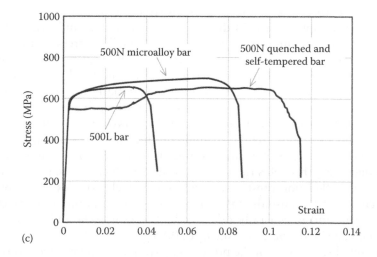

(c)

Figure 4.13 (Continued) Actual and idealised stress strain curves for reinforcing steel. (c) Actual stress-strain curves.

4.3.2.4 Coefficient of thermal expansion

In the absence of test data, the coefficient of thermal expansion of reinforcement is specified in AS3600-2009 (1) as $12 \times 10^{-6}/°C$.

4.4 STEEL USED FOR PRESTRESSING

4.4.1 General

The shortening of the concrete caused by creep and shrinkage in a prestressed member causes a corresponding shortening of the prestressing steel that is physically attached to the concrete either by bond or by anchorages at the ends of the tendon. This shortening can be significant and usually results in a loss of stress in the steel of between 150 and 300 MPa. In addition, creep in the highly stressed prestressing steel also causes a loss of stress through relaxation. Significant additional losses of prestress can result from other sources, such as friction along a post-tensioned tendon or draw-in at an anchorage at the time of prestressing.

For an efficient and practical design, the total loss of prestress should be a relatively small portion of the initial prestressing force. The steel used to prestress concrete must therefore, be capable of carrying a very high initial stress. A tensile strength of between 1000 and 1900 MPa is typical for modern prestressing steels. The early attempts to prestress concrete with low-strength steels failed because almost the entire prestressing force was rapidly lost owing to the time-dependent deformations of the poor-quality concrete in use at that time.

There are three basic forms of high-strength prestressing steels (as detailed in Chapter 3): cold-drawn, stress-relieved round wire; stress-relieved strand; and high-strength alloy steel bars. The stress-strain curves for the various types of prestressing steel exhibit similar characteristics, as illustrated in Figure 4.14. There is no well-defined yield point (as exists for some lower strength steels). Each curve is initially linear elastic (with an elastic modulus E_p similar to that for lower strength steels) and with a relatively high proportional limit. When the curves become nonlinear as deformation increases, the stress gradually increases monotonically until the steel fractures. The elongation at fracture is usually between 3.5% and 7%. High strength steel is therefore, considerably less ductile than conventional, hot-rolled non-prestressed reinforcing steel. For design purposes, the *yield stress* f_{py} is usually taken as the stress corresponding to the 0.1% offset strain and is generally taken to be between 80% and 85% of the minimum tensile strength (i.e. $0.8 f_{pb}$ to $0.85 f_{pb}$).

The initial stress level in the prestressing steel after the prestress is transferred to the concrete is usually high, often in the range of 70% to 80% of the tensile strength of the material. At such high stress levels, high strength steel creeps. At lower stress levels, such as is typical for non-prestressed steel, the creep of steel is negligible. If a tendon is stretched and held at a constant length (constant strain), the development of creep strain in the steel is exhibited as a loss of elastic strain, and hence a loss of stress. This loss of stress in a specimen subjected to constant strain is known as *relaxation*. Creep, and hence relaxation, in steel is highly dependent on the stress level and increases at an increasing rate as the stress level increases. Relaxation

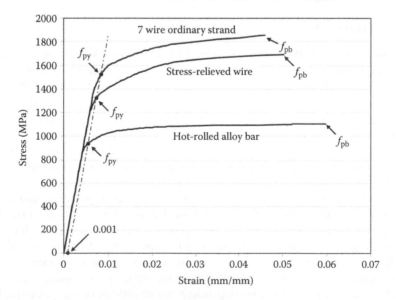

Figure 4.14 Typical stress-strain curves for tendons.

in steel also increases rapidly as temperature increases. In recent years, low relaxation steel has normally been used in order to minimise the losses of prestress resulting from relaxation.

4.4.2 Specification in AS3600-2009

4.4.2.1 Strength and ductility

Prestressing steels in Australia must comply with the requirements of AS/NZS4672.1 [8]. The terms used to define the strength and ductility of prestressing steel in AS3600-2009 are illustrated on the typical stress-strain curve shown in Figure 4.15. The characteristic breaking strength is f_{pb} and typical values are given in Table 4.8, together with the corresponding breaking force in the tendon P_{pb}. In design calculations, f_{pb} is taken as the strength of the tendon. In practice, the breaking stress of 95% of all test samples will exceed f_{pb}. The strain corresponding to f_{pb} is the uniform elongation ε_{pu} and, when expressed as a percentage, it is given the symbol A_{gt} in AS4672.1 [8]. The tensile strength and ductility of some commonly used types and sizes of Australian prestressing steel are given in Table 4.8.

The yield stress f_{py} is taken as the 0.1% proof stress and may be determined by testing. In the absence of test data, the prescribed values given in Table 4.8 may be used. The prescribed values of f_{py} are specified in AS3600-2009 [1] as a fraction of f_{pb}. For example, f_{py} may be taken as $0.80\,f_{pb}$ for as-drawn wire,

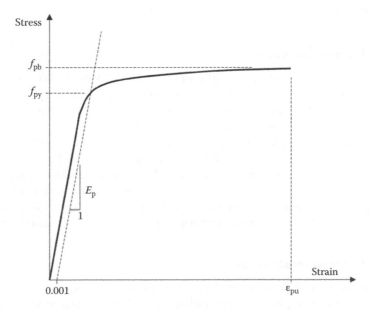

Figure 4.15 Stress-strain curve for prestressing steel.

Table 4.8 Typical sizes, strength and ductility characteristics of prestressing steels [8]

Type	Diameter (mm)	Area (mm²)	f_{pb} (MPa)	f_{py} (MPa)	P_{pb} (kN)	ε_{pu}
As-drawn wire	5.0	19.6	1700	1360	33.3	0.035
	7.0	38.5	1670	1336	64.3	0.035
Stress-relieved wire	5.0	19.9	1700	1411	33.8	0.035
	7.0	38.5	1670	1386	64.3	0.035
7-wire ordinary strand	9.5	55.0	1850	1517	102	0.035
	12.7	98.6	1870	1533	184	0.035
	15.2	140	1790	1468	250	0.035
7-wire compacted strand	15.2	165	1820	1492	300	0.035
	18.0	223	1700	1394	379	0.035
Hot-rolled alloy bars (supper grade)	26	562	1030	834	579	0.06
	29	693	1030	834	714	0.06
	32	840	1030	834	865	0.06
	36	995	1030	834	1025	0.06
	40	1232	1030	834	1269	0.06
	56	2428	1030	834	2501	0.06
	75	4371	1030	834	4502	0.06

Source: Data from AS/NZS4672(1)-2007, Steel prestressing materials Part 1: General requirements. Standards Australia, Sydney, New South Wales, Australia, 2007.

0.83 f_{pb} for stress-relieved wire, 0.82 f_{pb} for all grades of strand, 0.81 f_{pb} for hot-rolled super grade bars and 0.89 f_{pb} for hot-rolled ribbed bars.

4.4.2.2 Elastic modulus

The prescribed values of the modulus of elasticity in AS3600-2009 [1] for different tendon types are taken from AS/NZS4672.1 [8]. For all types of wire, $E_p = 205 \pm 10$ GPa, while for strand $E_p = 195 \pm 5$ GPa and for bar $E_p = 200 \pm 10$ GPa. Alternatively, the elastic modulus can be obtained by measuring the elongation of sample pieces of tendon in direct tension tests. AS3600-2009 [1] states that the prescribed or measured values of elastic modulus may vary by up to ±10% and possibly more when a multi-strand or multi-wire tendon is stressed as a single cable. Variations in elastic modulus of the tendon will affect the calculated extension of the tendon during the stressing operation, and this should be considered appropriately both in design and during construction.

4.4.2.3 Stress–strain curve

For nonlinear and other refined methods of analysis, actual stress-strain curves for the steel, using mean rather than characteristic values, should be used. The actual shape of the stress-strain curve for tendons may be determined from tests. Typical curves for various types of tendons were shown in Figure 4.14. Alternatively, a simplified equation, such as that described by Loov [41], suitably calibrated to approximate the shape of the actual curve may be used in design.

For design and construction purposes, the jacking and other forces are generally obtained from the manufacturer's literature, and the actual stress-strain curve of the material supplied should be used to calculate the elongation during jacking.

4.4.2.4 Steel relaxation

Although AS/NZS4672.1 [8] recognises both normal-relaxation and low-relaxation wires and strand, AS3600-2009 [1] only provides information on low-relaxation steels. AS3600-2009 [1] specifies that the *design relaxation R* for low-relaxation wire, low-relaxation strand and alloy-steel bars (as a percentage of the initial prestress) is to be determined from:

$$R = k_4 \, k_5 \, k_6 \, R_b \tag{4.30}$$

where k_4 is the coefficient depending on the duration of the prestressing force and is given by $k_4 = \log [5.4(t)^{1/6}]$; t is the time after prestressing (in days); k_5 is the coefficient that is dependent on the stress in the tendon as a proportion of the characteristic minimum breaking strength of the tendon f_{pb} and is given in Figure 4.16; k_6 depends on the average temperature $T(°C)$ over the time period t and may be taken as $k_6 = T/20$ but not less than 1.0; and R_b is the basic relaxation of the tendon and is defined as the loss of

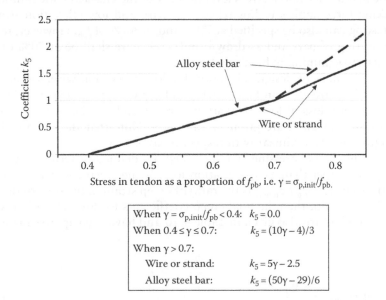

Figure 4.16 Coefficient k_5.

Table 4.9 Relaxation (%) at 1000 hrs for wire, strand and bar (*T* = 20°C)

Type of tendon	Tendon stress as a proportion of f_{pb}		
	0.6	0.7	0.8
Stress-relieved wire – low-relaxation	1.0	2.0	3.0[a]
– normal-relaxation	3.5	6.5	10.0
Quenched and tempered wire – low-relaxation	1.0	2.0	4.5[a]
– normal-relaxation	2.0	4.0	9.0
Strand – low-relaxation		2.5	3.5[a]
– normal-relaxation		8.0	12.0
Hot-rolled bar	1.5	4.0[a]	—

[a]   These are the values to be used as the basic relaxation R_b in Equation 4.30.

stress in a tendon initially stressed to 0.8 f_{pb} and held at constant strain for a period of 1000 h at 20°C.

R_b is expressed as a percentage of the initial stress and is determined in accordance with AS/NZS 4672.1 [8]. Maximum values of the relaxation at 1000 hours at various initial stress levels and at 20°C are specified in AS/NZS 4672.1 [8] and are given in Table 4.9. Values for both normal- and low-relaxation wires and strands are provided in Table 4.9, but Equation 4.30 does not apply to normal-relaxation steels. Test values may be significantly lower than these maximum values.

The mandatory initial stress level of 80% of the characteristic minimum breaking force applies to low-relaxation wire and low-relaxation strand, but testing can also be specified at 70% and/or 60% of f_{pb}. However, relaxation testing of bars and as-drawn (mill coil) wire shall be at 70% of the characteristic minimum breaking force.

Typical values of design relaxation for low-relaxation wire, strand and bars at 20°C are given in Figure 4.17, and typical final (30 year) values of the relaxation loss of low- relaxation wire, strand and bars at an average temperature of 20°C are given in Table 4.9. Note that the creep of high strength steels is nonlinear with respect to stress.

Creep in the prestressing steel may also be defined in terms of a creep coefficient rather than as a relaxation loss. If the creep coefficient for the prestressing steel $\varphi_p(t, \sigma_{p,init})$ is the ratio of creep strain in the steel to the initial elastic strain, then the final creep coefficients for low relaxation wire, strand and bar are also given in Table 4.10 and have been approximated as follows:

$$\varphi_p(t, \sigma_{p,init}) = \frac{R}{1-R} \tag{4.31}$$

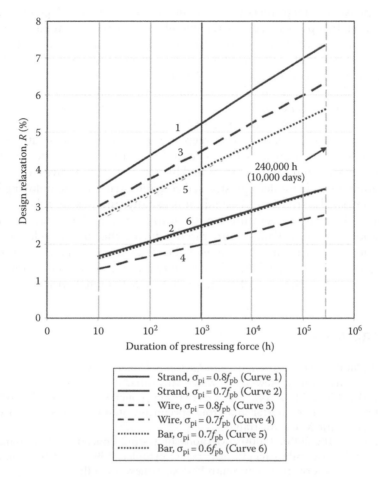

Figure 4.17 Design relaxation for low relaxation stress-relieved wire, strand and alloy bars.

Table 4.10 Long-term relaxation losses and corresponding final creep coefficients for low relaxation wire, strand and bar ($T = 20°C$)

		Tendon stress as a proportion of f_p		
	Type of tendon	0.6	0.7	0.8
Wire	Relaxation loss, R (%)	1.9	2.8	4.2
	Creep coefficient, $\varphi_p(t,\sigma_{p,init})$	0.019	0.029	0.044
Strand	Relaxation loss, R (%)	2.4	3.5	5.3
	Creep coefficient, $\varphi_p(t,\sigma_{p,init})$	0.025	0.036	0.056
Bar	Relaxation loss, R (%)	3.7	5.6	10.3
	Creep coefficient, $\varphi_p(t,\sigma_{p,init})$	0.038	0.059	0.115

As emphasised, creep (relaxation) of the prestressing steel depends on the stress level. In a prestressed concrete member, the stress in a tendon is gradually reduced with time due to creep and shrinkage in the concrete. This gradual decrease of stress results in a reduction of creep in the steel and hence smaller relaxation losses. To determine relaxation losses in a concrete structure therefore, the final relaxation loss obtained from Equation 4.30 (or Table 4.10) should be multiplied by a reduction factor λ_r that accounts for the time-dependent shortening of the concrete due to creep and shrinkage. The factor λ_r depends on the creep and shrinkage characteristics of the concrete, the initial prestressing force and the stress in the concrete at the level of the steel and can be determined by iteration [28]. However, because relaxation losses in modern prestressed concrete structures (employing low-relaxation steels) are relatively small, it is usually sufficient to take $\lambda_r \approx 0.8$.

When elevated temperatures exist during curing (i.e. steam curing), relaxation is increased and occurs rapidly during the curing cycle. For low relaxation steel in a concrete member subjected to an initial period of steam curing, it is recommended that the design relaxation should be significantly greater than the value given by Equation 4.30 (calculated with $T = 20°C$).

REFERENCES

1. AS3600-2009. *Australian Standard for Concrete Structures*. Standards Australia, Sydney, New South Wales, Australia.
2. AS1012.9: 2014. (2014). Methods of testing concrete – Compressive strength tests – concrete, mortar and grout specimens. Standards Australia, Sydney, New South Wales, Australia.
3. AS1012.10: 2000. (2000). Methods of testing concrete – Determination of indirect tensile strength of concrete cylinders ('Brazil' or splitting test). Standards Australia, New South Wales, Sydney, Australia.
4. AS1012.11: 2000. (2000). Methods of testing concrete – Determination of the modulus of rupture. Standards Australia, Sydney, New South Wales, Australia.
5. AS1012.13: 1992. (1992). Methods of testing concrete – Determination of the drying shrinkage of concrete for samples prepared in the field or in the laboratory. Standards Australia, Sydney, New South Wales, Australia.
6. AS1012.16: 1996. (1996). Methods of testing concrete – Determination of creep of concrete cylinders in compression. Standards Australia, Sydney, New South Wales, Australia.
7. AS/NZS4671-2001. (April 2001). Steel reinforcing materials. Incorporating Amendment 1:2003, Standards Australia, Sydney, New South Wales, Australia.
8. AS/NZS4672(1)-2007. (2007). Steel prestressing materials Part 1: General requirements. Standards Australia, Sydney, New South Wales, Australia.
9. ACI318M-11. (2011). Building code requirements for reinforced concrete and Commentary. American Concrete Institute, Detroit, MI.

10. ACI209R-92. (1992). Prediction of creep, shrinkage and temperature effects in concrete structures. ACI Committee 209, American Concrete Institute, reapproved 2008, Detroit, MI.
11. ACI209.1R-05. (2005). Report on factors affecting shrinkage and creep of hardened concrete. ACI Committee 209, American Concrete Institute, Detroit, MI.
12. ACI Committee 209. (2008). Guide for modeling and calculating shrinkage and creep in hardened concrete (ACI 209.2R-08). American Concrete Institute, Farmington Hills, MI, 44pp.
13. Neville, A.M. (1996). *Properties of Concrete*, 4th edn. London, U.K.: Wiley.
14. Metha, P.K. and Monteiro, P.J. (2014). *Concrete: Microstructure, Properties and Materials*, 4th edn. New York: McGraw-Hill Education.
15. Kupfer, H.B., Hilsdorf, H.K. and Rüsch, H. (1975). Behaviour of concrete under biaxial stresses. *ACI Journal*, 66, 656–666.
16. Tasuji, M.E., Slate, F.O. and Nilson, A.H. (1978). Stress-strain response and fracture of concrete in biaxial loading. *ACI Journal*, 75, 306–312.
17. Darwin, D. and Pecknold, D.A. (1977). Nonlinear biaxial stress-strain law for concrete. *Journal of the Engineering Mechanics Division, ASCE*, 103, 229–241.
18. Raphael, J.M. (1984). Tensile strength of concrete. *ACI Journal*, 81(2), 158–165.
19. Thorenfeldt, E., Tomaszewicz, A. and Jensen, J.J. (June 1987). Mechanical properties of high strength concrete and application in design. *International Symposium on Utilization of High Strength Concrete*, Stavanger, Norway, pp. 149–159.
20. Collins, M.P. and Porasz, A. (1989). Shear strength for high strength concrete, Bulletin No. 193. *Design Aspects of High Strength Concrete, Comité Euro-International du Béton (CEB)*, 1989, pp. 75–83.
21. Setunge, S. (1992). Structural properties of very high strength concrete. PhD Thesis, Monash University, Melbourne, Victoria, Australia.
22. Attard, M.M. and Stewart, M.G. (1998). An improved stress block model for high strength concrete. Research Report No. 154.10.1997, Department of Civil, Surveying and Environmental Engineering, University of Newcastle, Callaghan, New South Wales, Australia, 42pp.
23. Neville, A.M. (1970). *Creep of Concrete: Plain, Reinforced and Prestressed*. Amsterdam, the Netherlands: North-Holland.
24. Neville, A.M., Dilger, W.H. and Brooks, J.J. (1983). *Creep of Plain and Structural Concrete*. London, U.K.: Construction Press.
25. Gilbert, R.I. (1988). *Time Effects in Concrete Structures*. Amsterdam, the Netherlands: Elsevier.
26. Gilbert, R.I. and Ranzi, G. (2011). *Time-Dependent Behaviour of Concrete Structures*. London, U.K.. Spon Press, 426pp.
27. Ghali, A. and Favre, R. (1986). *Concrete Structures: Stresses and Deformations*. London, U.K.: Chapman & Hall.
28. Ghali, A., Favre, R. and Eldbadry, M. (2002). *Concrete Structures: Stresses and Deformations*. 3rd edn. London, U.K.: Spon Press, 584pp.
29. Pauw, A. (1960). Static modulus of elasticity of concrete as affected by density. *ACI Journal*, 57, 679–687.

30. Rüsch, H., Jungwirth, D. and Hilsdorf, H.K. (1983). *Creep and Shrinkage – Their Effect on the Behaviour of Concrete Structures*. New York: Springer-Verlag, 284pp.
31. Trost, H. (1978). Creep and creep recovery of very old concrete. *RILEM Colloquium on Creep of Concrete*, Leeds, U.K.
32. Trost, H. (1967). Auswirkungen des Superpositionsprinzips auf Kriech- und Relaxations Probleme bei Beton und Spannbeton. *Beton- und Stahlbetonbau*, 62, I 230-8, 261-9.
33. Dilger, W. and Neville, A.M. (1971). *Method of Creep Analysis of Structural Members*, ACI SP 27–17. Detroit, MI: ACI, pp. 349–379.
34. Bazant, Z.P. (April 1972). Prediction of concrete creep effects using age-adjusted effective modulus method. *ACI Journal*, 69, 212–217.
35. FIB (2013). *Fib Model Code for Concrete Structures 2010*. Fib – International Federation for Structural Concrete, Ernst & Sohn, Lausanne, Switzerland, 434pp.
36. Ostergaard, L., Lange, D.A., Altouabat, S.A. and Stang, H. (2001). Tensile basic creep of early-age concrete under constant load. *Cement and Concrete Research*, 31, 1895–1899.
37. Chu, K.-H. and Carreira, D.J. (1986). Time-dependent cyclic deflections in R/C beams. *Journal of Structural Engineering, ASCE*, 112(5), 943–959.
38. Bazant, Z.P. and Oh, B.H. (1984). Deformation of progressively cracking reinforced concrete beams. *ACI Journal*, 81(3), 268–278.
39. Gilbert, R.I. (2002). Creep and shrinkage models for high strength concrete – Proposals for inclusion in AS3600. *Australian Journal of Structural Engineering*, 4(2), 95–106.
40. Brooks, J.J. (2005). 30-year creep and shrinkage of concrete. *Magazine of Concrete Research*, 57(9), 545–556.
41. Loov, R.E. (1988). A general equation for the steel stress for bonded prestressed concrete members. *Journal of the Prestressed Concrete Institute*, 33, 108–137.

Chapter 5

Design for serviceability

5.1 INTRODUCTION

The level of prestress and the layout of the tendons in a member are usually determined from the serviceability requirements for that member. For example, if a water-tight and crack-free slab is required, tension in the slab must be eliminated or limited to some appropriately low value. If, on the other hand, the deflection under a particular service load is to be minimised, a load-balancing approach may be used to determine the prestressing force and cable drape (see Section 1.4.3).

For the serviceability requirements to be satisfied in each region of a member at all times after first loading, a reasonably accurate estimate of the magnitude of prestress is needed in design. This requires reliable procedures for the determination of both the instantaneous and time-dependent losses of prestress. Instantaneous losses of prestress occur during the stressing (and anchoring) operation and include elastic shortening of the concrete, friction along a post-tensioned cable and slip at the anchorages. As mentioned in the previous chapters, the time-dependent losses of prestress are caused by creep and shrinkage of the concrete and stress relaxation in the steel. Procedures for calculating both the instantaneous and time-dependent losses of prestress are presented in Section 5.10.

There are two critical stages in the design of prestressed concrete for serviceability. The first stage is immediately after the prestress is transferred to the concrete, when the prestress is at a maximum and the external load is usually at a minimum. The instantaneous losses have taken place, but no time-dependent losses have yet occurred. At this stage, the concrete is usually young and the concrete strength may be relatively low. The prestressing force immediately after transfer at a particular section is designated here as P_i. The second critical stage is after the time-dependent losses have taken place and the full-service load is applied, i.e. when the prestressing force is at a minimum and the external service load is at a maximum. The prestressing force at this stage is referred to as the *effective* prestress and designated P_e.

At each of these stages (and at all intermediate stages), it is necessary to ensure that both the strength and the serviceability requirements of the member are satisfied. Strength depends on the cross-sectional area and position of both the steel tendons and the non-prestressed reinforcement. However, it is not strength that determines the level of prestress, but serviceability. When the prestressing force and the amount and the distribution of the prestressing steel have been determined, the flexural strength may be readily increased, if necessary, by the addition of non-prestressed conventional reinforcement. This is discussed in more detail in Chapter 6. Shear strength may be improved by the addition of transverse stirrups (as discussed in Chapter 7). As will be seen throughout this chapter, the presence of bonded conventional reinforcement also greatly influences both the short- and long-term behaviour at service loads, both for cracked and uncracked prestressed members. The design for strength and serviceability therefore, cannot be performed independently, as the implications of one affect the other.

General design requirements for the serviceability limit states, including load combinations for serviceability, were discussed in Chapter 2. It is necessary to ensure that the instantaneous and time-dependent deflection and the axial shortening under service loads are acceptably small and that cracking, if it occurs, is well controlled by suitably detailed bonded reinforcement. To determine the in-service behaviour of a member, it is therefore necessary to establish the extent of cracking, if any, by checking the magnitude of elastic tensile stresses. If a member remains uncracked (i.e. the maximum tensile stress at all stages is less than the tensile strength of concrete), the properties of the uncracked section may be used in all deflection and camber calculations (see Sections 5.6 and 5.7). If cracking occurs, a cracked section analysis may be performed to determine the properties of the cracked section and the post-cracking behaviour of the member (see Sections 5.8 and 5.9).

5.2 CONCRETE STRESSES BOTH AT TRANSFER AND UNDER FULL SERVICE LOADS

In the past, codes of practice have set mandatory maximum limits on the magnitude of the concrete stresses, both tensile and compressive. In reality, concrete stresses, calculated by a linear elastic analysis, are often not even close to those that exist after a short period of creep and shrinkage, particularly in members containing significant quantities of bonded reinforcement. It makes little sense to limit concrete stresses in compression and tension, unless they are determined based on nonlinear analysis, in which the time-varying constitutive relationship for concrete is accurately modelled. Even if nonlinear analysis is undertaken, limiting the concrete stresses in compression to a maximum prescribed value, or making sure the concrete tensile stresses are less than the tensile strength of concrete, does not ensure either adequate strength of a structural member or satisfactory behaviour at service loads.

Notwithstanding the above, some codes still classify prestressed members in terms of the calculated maximum tensile stress f_{ct} in the concrete in the precompressed tensile zone of a member. For example, ACI318M-11 [1] classifies prestressed members as:

(a) *Uncracked* (Class U) if $f_{ct} \leq 0.62\sqrt{f_c'}$
(b) *Transitional* (Class T) if $0.62\sqrt{f_c'} < f_{ct} \leq 1.0\sqrt{f_c'}$
(c) *Cracked* (Class C) if $f_{ct} > 1.0\sqrt{f_c'}$

where f_{ct} and f_c' are expressed in MPa. As we shall see subsequently in Section 5.7.4, limiting the maximum concrete tensile stress calculated in an elastic analysis to $0.62\sqrt{f_c'}$ certainly does not mean that the member will remain uncracked.

ACI318M-11 [1] imposes a limit of $0.6f_{cp}'$ on the calculated extreme fibre compressive stress at transfer, except that this limit can be increased to $0.7f_{cp}'$ at the ends of simply-supported members (where f_{cp}' is the specified characteristic strength of concrete at transfer). ACI318M-11 [1] also requires that where the concrete tensile stress exceeds $0.5\sqrt{f_{cp}'}$ at the ends of a simply-supported member, or $0.25\sqrt{f_{cp}'}$ elsewhere, additional bonded reinforcement should be provided in the tensile zone to resist the total tensile force computed with the assumption of an uncracked cross-section.

In addition, for Class U and Class T flexural members, ACI318M-11 [1] specifies that the extreme fibre compressive stress, calculated assuming uncracked cross-sectional properties and after all losses of prestress, should not exceed the following:

Due to prestress plus sustained load: $0.45f_c'$
Due to prestress plus total load: $0.6f_c'$

These limits are imposed to decrease the probability of fatigue failure in beams subjected to repeated loads and to avoid the development of non-linear creep that develops under high compressive stresses.

AS3600-2009 [2] places no mandatory limits on concrete stresses at transfer (or at any other time) although it does imply that the sustained compressive stress at the level of the tendon at any time should not exceed $0.5f_c'$. The choice of stress limit is left entirely to the designer and should be based on the appropriate serviceability requirements. There is much to recommend this approach. As we have already stated, satisfaction of any set of stress limits does not guarantee that the serviceability requirements of the structure will be satisfied. Camber and deflection calculations are still required and, if cracking occurs, crack widths must remain acceptably small. It is therefore, appropriate to discuss the reasons for and the implications of selecting particular stress limits.

Firstly, consider whether or not stress limits are required at transfer, i.e. when the prestressing force is at its maximum value and the time-dependent losses have not yet occurred. There can be little doubt that the magnitudes of both compressive and tensile concrete stresses at transfer need to be carefully considered. It is important that the concrete compressive stress at the steel level at transfer does not exceed about $0.5f'_{cp}$. At higher stress levels, large nonlinear creep strains develop with time, resulting in large creep deformation and high losses of prestress. Designers must also check strength at transfer and the satisfaction of the above compressive stress limit will usually, although not necessarily, lead to an adequate factor of safety against compressive failure at transfer.

It is also important to ensure that cracking does not occur immediately after transfer in locations where there is no (or insufficient) bonded reinforcement. The regions of a member that are subjected to tension at transfer are often those that are later subjected to compression when the full service load is applied. If these regions are unreinforced and uncontrolled cracking is permitted at transfer, an immediate serviceability problem exists. When the region is later compressed, cracks may not close completely, local spalling may occur and even a loss of shear strength could result. If cracking is permitted at transfer, bonded reinforcement should be provided to carry all the tension and to ensure that the cracks are fine and well controlled. For the calculation of concrete stresses immediately after transfer, an elastic analysis using gross cross-sectional properties is usually satisfactory.

In some cases, concrete stresses may need to be checked under full service loads when all prestress losses have taken place. If cracking is to be avoided, concrete tensile stresses must not exceed the tensile strength of concrete. Care should be taken when calculating the maximum tensile stress to accurately account for the load-independent tension induced by restraint to shrinkage or temperature effects. However, even if the tensile stress does reach the tensile strength of concrete and some minor cracking occurs, the cracks will be well controlled and the resulting loss of stiffness will not be significant, provided sufficient bonded reinforcement or tendons are provided near the tensile face.

For many prestressed concrete situations, there are no valid reasons why cracking should be avoided at service loads and, therefore, no reason why a limit should be placed on the maximum tensile stress in the concrete. Indeed, in modern prestressed concrete building structures, many members crack under normal service loads. If cracking does occur, the resulting loss of stiffness must be accounted for in deflection calculations and a nonlinear cracked section analysis is required to determine behaviour in the post-cracking range. Crack widths must also be controlled. Crack control may be achieved by limiting both the spacing of the bonded reinforcement and the change of stress in the reinforcement after cracking (see Section 5.12).

Under full service loads, which occur infrequently, there is often no practical reason why compressive stress limits should be imposed. Separate checks for flexural strength, ductility and shear strength are obviously necessary. Some members, such as trough girders or inverted T-beams, are prone to high concrete compressive stresses under full service loads and, in the design of these members, care should be taken to limit the extreme fibre compressive stress at service loads. If a large portion of the total service load is permanent, compressive stress levels in excess of about $0.5f_c'$ should be avoided. To reduce the probability of fatigue failure in uncracked or lightly cracked members subjected to repeated loads, and to avoid nonlinear creep deformations, it is recommended that appropriate limits are placed on the maximum compressive stress under service loads, after all losses have taken place. The limits specified in ACI318M-11 [1], and discussed above, are appropriate.

The primary objective in selecting concrete stress limits is to obtain a serviceable structure. As was discussed in Section 1.4, elastic stress calculations are not strictly applicable to prestressed concrete. Creep and shrinkage cause a gradual transfer of compression from the concrete to the bonded steel. Nevertheless, elastic stress calculations may indicate potential serviceability problems and the satisfaction of concrete tensile stress limits is a useful procedure to control the extent of cracking. It should be understood, however, that the satisfaction of a set of elastic concrete stress limits does not, in itself, ensure serviceability and it certainly does not ensure adequate strength. The designer must check both strength and serviceability separately, irrespective of the stress limits selected.

5.3 MAXIMUM JACKING FORCES

To prevent overstressing during the jacking operation, AS3600-2009 [2] specifies that the maximum tensile force that can be applied to a tendon should not exceed:

1. For pretensioned tendons $\qquad$ $0.80 f_{pb} A_p$
2. For stress-relieved post-tensioned tendons $\qquad$ $0.85 f_{pb} A_p$
3. For post-tensioned tendons and bars not stress-relieved $\quad 0.75 f_{pb} A_p$

When tensioning a tendon, the stressing procedure should ensure that the force in the tendon increases at a uniform rate. The prestressing force should be measured at the jack to an accuracy of ±3% [2] and the tendon extension during tensioning should also be measured. After tensioning the tendon, the prestressing force should be transferred gradually to the concrete. A check should also be made to ensure that the measured extension of each tendon agrees with the calculated extension based on the measured prestressing force and a knowledge of the cable profile and the load-extension curve for the tendon [2]. Any difference between the two figures greater than 10% must be investigated. Differences could be due to problems arising for a variety of reasons, including blockage of the duct due to the ingress of cement paste, the wrong size tendon being used, slip at the dead-end anchorage, variations of tendon profile and hence different friction losses in the duct or anchorage from the assumed values and variations in strand properties, including differences in strand due to worn dies used in drawing the strand wires during manufacture.

5.4 DETERMINATION OF PRESTRESS AND ECCENTRICITY IN FLEXURAL MEMBERS

There are a number of possible starting points for the determination of the prestressing force P and eccentricity e required at a particular cross-section. The starting point depends on the particular serviceability requirements for the member. The prestressing force and the cable layout for a member may be selected to minimise deflection under some portion of the applied load, i.e. a load-balancing approach to design. With such an approach, cracking may occur when the applied load is substantially different from the selected balanced load, such as at transfer or under the full service loads after all losses, and this possibility needs to be checked and accounted for in serviceability calculations.

The quantities P and e are often determined to satisfy preselected stress limits. Cracking may or may not be permitted under service loads. As was mentioned in the previous section, satisfaction of concrete stress limits does not necessarily ensure that deflection, camber and axial shortening are within acceptable limits. Separate checks are required for each of these serviceability limit states.

5.4.1 Satisfaction of stress limits

Numerous design approaches have been proposed for the satisfaction of concrete stress limits, including analytical and graphical techniques, e.g. Magnel [3], Lin [4] and Warner and Faulkes [5]. A simple and convenient approach is described here.

If the member is required to remain uncracked throughout, suitable stress limits should be selected for the tensile stress at transfer F_{tp} and the tensile stress under full load F_t. In addition, limits should also be placed on the concrete compressive stress at transfer F_{cp} and under full loads F_c. If cracking under the full loads is permitted, the stress limit F_t is relaxed and the remaining three limits are enforced.

In Figure 5.1, the uncracked cross-section of a beam at the critical moment location is shown, together with the concrete stresses at transfer

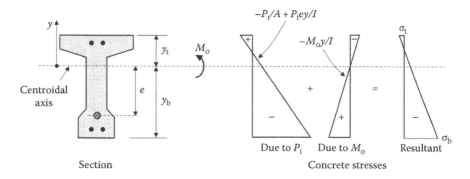

Figure 5.1 Concrete stresses at transfer.

caused by the initial prestress of magnitude P_i (located at an eccentricity e below the centroidal axis of the concrete section) and by the external moment M_o resulting from the loads acting at transfer. Often self-weight is the only load (other than prestress) acting at transfer. In Figure 5.1, we have assumed the cross-section is uncracked and that concrete stresses are calculated assuming linear elastic material behaviour.

At transfer, the concrete stress in the top fibre must not exceed the tensile stress limit F_{tp}. If tensile (compressive) stress is assumed to be positive (negative), we have:

$$\sigma_t = -\frac{P_i}{A} + \frac{P_i e y_t}{I} - \frac{M_o y_t}{I} \leq F_{tp}$$

Rearranging and introducing the term $\alpha_t = A y_t / I = A / Z_t$, we get:

$$P_i \leq \frac{A F_{tp} + \alpha_t M_o}{\alpha_t e - 1} \tag{5.1}$$

where A is the area of the transformed cross-section; I is the second moment of area of the transformed section about the centroidal axis; and Z_t is the elastic modulus of the cross-section with respect to the top fibre (equal to I/y_t).

Similarly, the concrete stress in the bottom fibre must be greater than the negative compressive stress limit at transfer:

$$\sigma_b = -\frac{P_i}{A} - \frac{P_i e y_b}{I} + \frac{M_o y_b}{I} \geq F_{cp}$$

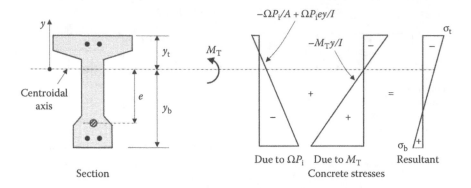

Figure 5.2 Concrete stresses under full loads (after all prestress losses).

Rearranging and introducing the term $\alpha_b = Ay_b/I = A/Z_b$, we get:

$$P_i \le \frac{-AF_{cp} + \alpha_b M_o}{\alpha_b e + 1} \tag{5.2}$$

where Z_b is the elastic modulus of the cross-section with respect to the bottom fibre ($=I/y_b$) and the compressive stress limit F_{cp} is a negative quantity.

Figure 5.2 shows the concrete stresses on an uncracked cross-section caused by both the effective prestressing force acting on the concrete after all losses have taken place and the applied moment M_T resulting from the full service load. The effective prestressing force acting on the concrete part of the cross-section is taken as ΩP_i, where Ω depends on the time-dependent loss of prestress in the tendon and the amount of force transferred from the concrete into the bonded non-prestressed reinforcement as it restrains the development of creep and shrinkage in the concrete with time. For cross-sections containing no conventional reinforcement Ω is typically about 0.80, but may be significantly smaller for sections containing conventional reinforcement (see Section 5.7.4, where in Tables 5.1 and 5.2, Ω varies between 0.410 and 0.839 depending on the amount and position of bonded reinforcement).

For an uncracked member under full service loads, the concrete stress in the bottom fibre must be less than the selected tensile stress limit F_t:

$$\sigma_b = -\frac{\Omega P_i}{A} - \frac{\Omega P_i e y_b}{I} + \frac{M_T y_b}{I} \le F_t$$

and rearranging gives:

$$P_i \ge \frac{-AF_t + \alpha_b M_T}{\Omega(\alpha_b e + 1)} \tag{5.3}$$

The compressive stress in the top fibre must also satisfy the appropriate stress limit F_c:

$$\sigma_t = -\frac{\Omega P_i}{A} + \frac{\Omega P_i e y_t}{I} - \frac{M_T y_t}{I} \geq F_c$$

and rearranging gives:

$$P_i \geq \frac{A F_c + \alpha_t M_T}{\Omega(\alpha_t e - 1)} \tag{5.4}$$

If the eccentricity e at the cross-section is known, satisfaction of Equations 5.1 and 5.2 will ensure that the desired stress limits at transfer are not exceeded. Equations 5.1 and 5.2 provide an upper limit on the magnitude of P_i. Equations 5.3 and 5.4 provide a lower limit on the magnitude of P_i. Satisfaction of all four equations will ensure that the selected stress limits at transfer and under full loads are all satisfied.

If a particular cross-section is too small, it may not be possible to satisfy all four stress limits and either a larger cross-section can be selected or the offending stress limit(s) can be relaxed and the effect of this variation assessed separately in design. Although separate checks are required to ensure satisfaction of the strength and serviceability requirements, Equations 5.1 through 5.4 provide a useful starting point in design for sizing both the cross-sectional dimensions and the prestressing details.

If the maximum value of P_i that satisfies Equation 5.2 is the same as the minimum value required to satisfy Equation 5.3, information is obtained about the properties of the smallest cross-section that can be selected to ensure satisfaction of both the stress limits F_{cp} at transfer and F_t under full loads (i.e. the smallest sized cross-section that will ensure that cracking does not occur under full service loads). Equating the right hand sides of Equations 5.2 and 5.3, we get the following expression for the section modulus (Z_b) of the minimum sized cross-section:

$$(Z_b)_{min} = \frac{M_T - \Omega M_o}{F_t - \Omega F_{cp}} \tag{5.5}$$

Similarly, if we equate the right hand sides of Equations 5.1 and 5.4, we get the following expression for the section modulus (Z_t) of the smallest sized cross-section required to satisfy both the stress limits F_{tp} at transfer and F_c under full loads (i.e. the smallest sized cross-section to ensure that cracking does not occur under full service loads):

$$(Z_t)_{min} = \frac{M_T - \Omega M_o}{\Omega F_{tp} - F_c} \tag{5.6}$$

It must be remembered that F_{cp} and F_c represent compressive stress limits and are negative quantities. Equations 5.5 and 5.6 are useful starting points in the selection of an initial cross-section.

In order to make use of Equations 5.5 and 5.6 in preliminary design, an estimate of the time-dependent loss of prestress in the concrete must be made. Usually, a first estimate of Ω of about 0.8 is reasonable if low-relaxation prestressing steel is used and the cross-section does not contain significant quantities of non-prestressed steel. As already mentioned, if the cross-section contains significant quantities of bonded reinforcement, Ω may be significantly smaller. However, if this is the case, it may not be necessary to enforce a no cracking requirement, and Equations 5.5 and 5.6 would no longer be relevant. Any initial estimate of Ω must be checked after the prestress, the eccentricity and the quantity of bonded reinforcement have been determined. A suitable procedure for determining the time-dependent loss of stress in the concrete is described in Sections 5.7 and 5.9.

AS3600-2009 [2] assumes that flexural cracking is controlled if the maximum tensile concrete stress in a member does not exceed $0.25\sqrt{f_c'}$, when subjected to short-term service loads. Alternatively, for a limited amount of well-controlled cracking, AS3600-2009 [2] sets the tensile stress limit to $0.6\sqrt{f_c'}$, provided bonded reinforcement is placed near the tensile face at a bar spacing not exceeding 300 mm. In general, however, if cracking is permitted under full service loads, a tensile stress limit F_t is not specified, and Equations 5.3 and 5.5 do not apply. Tensile and compressive stress limits at transfer are usually enforced and, therefore, Equations 5.1 and 5.2 are still applicable and continue to provide an upper limit on the level of prestress. The only minimum limit on the level of prestress is that imposed by Equation 5.4 and, for most practical cases, this does not influence the design.

When there is no need to satisfy a tensile stress limit under full loads, any level of prestress that satisfies Equations 5.1, 5.2 and 5.4 may be used, including $P_i = 0$ (which corresponds to a reinforced concrete member). Often members that are designed to crack under the full service loads are proportioned so that no tension exists in the concrete under the sustained load. It is the variable live load that causes cracks to open and close as the variable load is applied and removed. The selection of prestress in such a case can still be made conveniently using Equation 5.3, if the maximum total service moment M_T is replaced in Equation 5.3 by the sustained or permanent moment M_{sus}.

If cracking occurs, the cross-section required for the cracked prestressed member may need to be larger than that required for a fully-prestressed uncracked member for a particular deflection limit. In addition, the quantity of non-prestressed reinforcement is usually significantly greater. Often, however, the reduction in prestressing costs more than compensates for the additional concrete and non-prestressed reinforcement costs and cracked *partially-prestressed* members are the most economical structural solution in a wide range of applications.

EXAMPLE 5.1

A one-way slab is simply supported over a span $L = 12$ m and is to be designed to carry a maximum superimposed service load of $w_s = 5$ kPa (kN/m^2) in addition to its own self-weight. The slab is post-tensioned by regularly spaced tendons with parabolic profiles (with zero eccentricity at each support and a maximum eccentricity at mid-span). Each tendon contains four 12.7 mm diameter strands in a flat duct. The material properties are as follows:

$f'_{cp} = 25$ MPa; $f'_c = 40$ MPa; $E_{cp} = 26{,}700$ MPa; $E_c = 32{,}750$ MPa;
$f_{pb} = 1{,}870$ MPa

Assume that the losses of prestress at mid-span immediately after transfer are 8% and the time-dependent losses due to creep, shrinkage and relaxation are 15%. Determine the prestressing force and eccentricity required to satisfy the following concrete stress limits:

At transfer: $F_{tp} = 0.25\sqrt{f'_{cp}} = 1.25$ MPa and $F_{cp} = -0.5f'_{cp} = -12.5$ MPa

After all losses: $F_t = 0.25\sqrt{f'_c} = 1.58$ MPa and $F_c = -0.5f'_c = -20.0$ MPa

Also determine the required number and spacing of tendons and the initial deflection of the slab at mid-span immediately after transfer.

In order to obtain an estimate of the slab self-weight (which is the only load other than the prestress at transfer), a trial slab thickness of 300 mm (span/40) is assumed initially. Assuming the concrete weighs 24 kN/m^3, the self-weight is:

$w_{sw} = 24 \times 0.3 = 7.2$ kN/m^2

and the moments at mid-span of the slab both at transfer and under the full service load (evaluated for a 1 m wide strip of slab) are as follows:

$$M_o = \frac{w_{sw}L^2}{8} = \frac{7.2 \times 12^2}{8} = 129.6 \text{ kNm/m}$$

$$M_T = \frac{(w_{sw} + w_s)L^2}{8} = \frac{(7.2 + 5.0) \times 12^2}{8} = 219.6 \text{ kNm/m}$$

From Equation 5.5:

$$(Z_b)_{min} = \frac{M_T - \Omega M_o}{F_t - \Omega F_{cp}} = \frac{(219.6 - 0.85 \times 129.6) \times 10^6}{1.58 - [0.85 \times (-12.5)]} = 8.97 \times 10^6 \ mm^3/m$$

and from Equation 5.6:

$$(Z_t)_{min} = \frac{M_T - \Omega M_o}{\Omega F_{tp} - F_c} = \frac{(219.6 - 0.85 \times 129.6) \times 10^6}{(0.85 \times 1.25) - (-20.0)} = 5.20 \times 10^6 \ mm^3/m$$

For the rectangular slab cross-section, the minimum section modulus must exceed $(Z_b)_{min} = 8.97 \times 10^6 \ mm^3/m$ and the corresponding minimum slab depth is therefore:

$$D_{min} = \sqrt{6(Z_b)_{min} / 1000} = 232 \ mm$$

If we select a slab thickness $D = 230$ mm: the revised self-weight is $w_{sw} = 5.52$ kN/m²; the revised moments are $M_o = 99.4$ kNm/m and $M_T = 189.4$ kNm/m; and the revised section properties are $(Z_b)_{min} = 8.60 \times 10^6 \ mm^3/m$ and $D_{min} = 227$ mm.

Taking $D = 230$ mm, the relevant section properties are:

$A = 230 \times 10^3 \ mm^2/m$, $I = 1014 \times 10^6 \ mm^4/m$, $Z_b = Z_t = 8.817 \times 10^6 \ mm^3/m$

and $\alpha_b = \alpha_t = 26.1 \times 10^{-3} \ mm^{-1}$

If we take the minimum concrete cover to the strand as 30 mm, the eccentricity at mid-span of the 12.7 mm diameter strands is:

$e = e_{max} = D/2 - 30 - 0.5 \times 12.7 = 78.7$ mm

From Equation 5.1:

$$P_i \leq \frac{AF_{tp} + \alpha_t M_o}{\alpha_t e - 1} = \frac{230 \times 10^3 \times 1.25 + 26.1 \times 10^{-3} \times 99.4 \times 10^6}{(26.1 \times 10^{-3} \times 78.7) - 1}$$

$$= 2734 \ kN/m$$

From Equation 5.2:

$$P_i \leq \frac{-AF_{cp} + \alpha_b M_o}{\alpha_b e + 1} = \frac{-230 \times 10^3 \times (-12.5) + 26.1 \times 10^{-3} \times 99.4 \times 10^6}{(26.1 \times 10^{-3} \times 78.7) - 1}$$

$$= 1791 \ kN/m$$

The prestressing force immediately after transfer P_i must not exceed 1791 kN/m.

From Equation 5.3:

$$P_i \geq \frac{-AF_t + \alpha_b M_T}{\Omega(\alpha_b e + 1)} = \frac{-230 \times 10^3 \times 1.58 + 26.1 \times 10^{-3} \times 189.4 \times 10^6}{0.85 \times (26.1 \times 10^{-3} \times 78.7 + 1)}$$

$$= 1764 \text{ kN/m}$$

From Equation 5.4:

$$P_i \geq \frac{AF_c + \alpha_t M_T}{\Omega(\alpha_t e - 1)} = \frac{230 \times 10^3 \times (-20) + 26.1 \times 10^{-3} \times 189.4 \times 10^6}{0.85 \times (26.1 \times 10^{-3} \times 78.7 - 1)}$$

$$= 383 \text{ kN/m}$$

The minimum prestressing force P_i is therefore, 1764 kN/m, and this value is used in the following calculations. With 8% immediate losses between mid-span and the jacking point at one end of the span, the required jacking force is:

$$P_j = \frac{P_i}{(1 - 0.08)} = \frac{1764}{0.92} = 1917 \text{ kN/m}$$

From Table 4.8, a 12.7 mm diameter 7-wire low-relaxation strand has a cross-sectional area of $A_p = 98.6$ mm^2 and a minimum breaking load of $f_{pb} A_p = 184.4$ kN. According to AS3600-2009 [2], the maximum jacking force in a strand is $0.85 f_{pb} A_p = 156.7$ kN (see Section 5.3). A flat duct containing four 12.7 mm strands can therefore, be stressed with a maximum jacking force of 4 × 156.7 = 626.9 kN.

The minimum number of ducts required in each metre width of slab is therefore:

$$\frac{P_j}{626.9} = \frac{1917}{626.9} = 3.06$$

and the maximum spacing between cables is therefore, 1000/3.06 = 327 mm.

A 4 strand tendon every 320 mm is specified with a jacking force per duct of 1917 × 0.32 = 613 kN.

For this slab, provided the initial estimates of losses are correct, the properties of the uncracked cross-section can be used in all deflection calculations, since stress limits have been selected to ensure that cracking does not occur either at transfer or under the full service loads. At transfer, $E_{cp} = 26,700$ MPa, $I = 1,014 \times 10^6$ mm^4/m and the uniformly distributed upward load caused by the parabolic tendons with drape equal to 0.0787 m is obtained from Equation 1.7 as:

$$w_p = \frac{8P_e e}{L^2} = \frac{8 \times 1764 \times 0.0787}{12^2} = 7.71\,\text{kN/m}$$

The resultant upward load is $w_p - w_{sw} = 7.71 - 5.52 = 2.19$ kN/m and the initial deflection (camber) at mid-span at transfer (before any creep and shrinkage deformations have taken place) is:

$$v_i = \frac{5}{384}\frac{(w_p - w_{sw})L^4}{E_{cp}I} = \frac{5 \times 2.19 \times 12,000^4}{384 \times 26,700 \times 1,014 \times 10^6}$$

$$= 21.8\,\text{mm (upwards)}$$

This may or may not be acceptable depending on the serviceability (deflection) requirements for the slab.

To complete this design, the effects of creep and shrinkage will need to be considered under the permanent loads. The final long-term deflections will need to be calculated and checked against the deflection criteria and the actual long-term losses must be checked. A reliable procedure for undertaking these calculations is outlined in Section 5.11.4. Of course, the ultimate strength in bending and in shear will also need to be checked and the anchorage zones must be designed.

5.4.2 Load balancing

Using the load-balancing approach, the effective prestress after losses P_e and the eccentricity e are selected such that the transverse load imposed by the prestress w_p balances a selected portion of the external load. The effective prestress P_e in a parabolic cable of drape e required to balance a uniformly distributed external load w_b is obtained using Equation 1.7 as follows:

$$P_e = \frac{w_b L^2}{8e} \tag{5.7}$$

Concrete stresses are checked under the remaining unbalanced service loads to identify regions of possible cracking and regions of high compression. Deflection under the unbalanced loads may need to be calculated and controlled. Losses are calculated and stresses immediately after transfer are also checked. Having determined the amount and layout of the prestressing steel (and the prestressing force) to satisfy serviceability requirements, the design for adequate strength can then proceed. Load balancing is widely used for the design of indeterminate members and also for simple determinate beams and slabs. It is only strictly applicable, however, prior to cracking when the member behaves linearly and the principle of superposition, on which load balancing relies, is valid.

EXAMPLE 5.2

Reconsider the 12 m span, 230 mm thick one-way slab of Example 5.1 and determine the prestress required to balance the slab self-weight (5.52 kN/m²). The parabolic tendons have zero eccentricity at each support and e = 78.7 mm at mid-span. As in Example 5.1, the time-dependent losses at mid-span are 15% and the instantaneous losses between mid-span and the jacking end are 8%.

With w_b = 5.52 kN/m² and e = 0.0787 m, Equation 5.7 gives:

$$P_e = \frac{w_b L^2}{8e} = \frac{5.52 \times 12^2}{8 \times 0.0787} = 1263 \text{ kN/m}$$

The prestressing force at mid-span immediately after transfer and the jacking force are:

$$P_i = \frac{P_e}{(1 - 0.15)} = \frac{1263}{0.85} = 1485 \text{ kN/m}$$

and

$$P_j = \frac{P_i}{(1 - 0.08)} = \frac{1485}{0.92} = 1614 \text{ kN/m}$$

As in Example 5.1, the maximum jacking force in a duct containing four strands is $0.85 f_{pb} A_p$ = 626.9 kN and so the required maximum duct spacing is 388 mm.

Using a 4 strand tendon every 380 mm, the jacking force per tendon is:

$$1614 \times 0.38 = 613 \text{ kN}$$

In Example 5.1, the tensile stress limit under full loads was F_t = 1.58 MPa. In this example, the prestress is significantly lower and, therefore, F_t will be exceeded. The bottom fibre stress at mid-span after all losses and under the full service loads (i.e. when M_T = 189.4 kNm/m as determined in Example 5.1) is:

$$\sigma_b = -\frac{P_e}{A} - \frac{P_e e}{Z_b} + \frac{M_T}{Z_b} = -\frac{1263 \times 10^3}{230 \times 10^3} - \frac{1263 \times 10^3 \times 78.7}{8.817 \times 10^6} + \frac{189.4 \times 10^6}{8.817 \times 10^6}$$

$$= 4.72 \text{ MPa}$$

and this will almost certainly cause cracking. The resulting loss of stiffness must be included in subsequent deflection calculations using the procedures outlined in Sections 5.11.3 and 5.11.4. In addition, the smaller quantity of prestressing steel required in this example, in comparison with the slab in Example 5.1, will result in reduced flexural strength. A layer of non-prestressed bottom reinforcement may be required to satisfy strength requirements.

5.5 CABLE PROFILES

When the prestressing force and eccentricity are determined for the critical sections, the location of the cable at every section along the member must be specified. For a member that has been designed using concrete stress limits, the tendons may be located so that the stress limits are observed on every cross-section. Equations 5.1 through 5.4 may be used to establish a range of values for eccentricity at any particular cross-section that satisfies the selected stress limits.

At any cross-section, if M_o and M_T are the moments caused by the external loads at transfer and under full service loads, respectively, and P_i and P_e are the prestressing forces before and after the time-dependent losses at the same section, the extreme fibre stresses must satisfy the following:

$$-\frac{P_i}{A} + \frac{(P_i e - M_o)}{Z_t} \le F_{tp} \tag{5.8}$$

$$-\frac{P_i}{A} - \frac{(P_i e - M_o)}{Z_b} \ge F_{cp} \tag{5.9}$$

$$-\frac{P_e}{A} - \frac{(P_e e - M_T)}{Z_b} \le F_t \tag{5.10}$$

$$-\frac{P_e}{A} + \frac{(P_e e - M_T)}{Z_t} \ge F_c \tag{5.11}$$

Equations 5.8 through 5.11 are equivalent to Equations 5.1 through 5.4 and can be rearranged to provide limits on the tendon eccentricity, as follows:

$$e \le \frac{M_o}{P_i} + \frac{Z_t F_{tp}}{P_i} + \frac{1}{\alpha_t} \tag{5.12}$$

$$e \le \frac{M_o}{P_i} - \frac{Z_b F_{cp}}{P_i} - \frac{1}{\alpha_t} \tag{5.13}$$

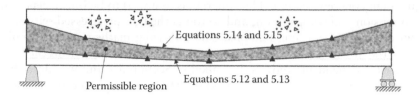

Figure 5.3 Typical permissible region for determination of cable profile.

$$e \geq \frac{M_T}{P_e} - \frac{Z_b F_t}{P_e} - \frac{1}{\alpha_b} \tag{5.14}$$

$$e \geq \frac{M_T}{P_e} + \frac{Z_t F_c}{P_e} + \frac{1}{\alpha_t} \tag{5.15}$$

It should be remembered that F_{cp} and F_c are negative numbers and that $\alpha_t = A/Z_t$ and $\alpha_b = A/Z_b$.

After P_i and P_e have been determined at the critical sections, the friction, draw-in and time-dependent losses along the member are estimated (see Sections 5.10.2 and 5.10.3), and the corresponding prestressing forces at intermediate sections are calculated. At each intermediate section, the maximum eccentricity that will satisfy both stress limits at transfer is obtained from either Equation 5.12 or 5.13. The minimum eccentricity required to satisfy the tensile and compressive stress limits under full loads is obtained from either Equation 5.14 or 5.15. A region of the member is thus established in which the line of action of the resulting prestressing force should be located. Such a permissible region is shown in Figure 5.3. Relatively few intermediate sections need to be considered to determine an acceptable cable profile.

When the prestress and eccentricity at the critical sections are selected using the load-balancing approach, the cable profile should match, as closely as practicable, the bending moment diagram caused by the balanced load. For cracked, partially-prestressed members, Equations 5.12 and 5.13 are usually applicable and allow the maximum eccentricity to be defined. The cable profile should then be selected according to the loading type and the bending moment diagram.

5.6 SHORT-TERM ANALYSIS OF UNCRACKED CROSS-SECTIONS

5.6.1 General

The short-term behaviour of an uncracked prestressed concrete cross-section can be determined by transforming the bonded reinforcement into equivalent areas of concrete and performing an elastic analysis on the equivalent concrete section. The concrete is assumed to be linear-elastic in both tension and compression, and so too is the non-prestressed reinforcement and the prestressing tendons. The following mathematical formulation of the short-term analysis of an uncracked cross-section forms the basis of the time-dependent analysis described in Section 5.7 and was described by Gilbert and Ranzi [6]. The procedure can be applied to cross-sections with a vertical axis of symmetry, such as those shown in Figure 5.4.

The contribution of each reinforcing bar or bonded prestressing tendon is included in the calculations according to its location within the cross-section, as shown in Figure 5.5.

The numbers of *layers* of non-prestressed and prestressed reinforcement are m_s and m_p, respectively. In Figure 5.5b, $m_s = 3$ and $m_p = 2$. The properties of each layer of non-prestressed reinforcement are defined by its area, elastic modulus and location with respect to the arbitrarily chosen x-axis and are labelled as $A_{s(i)}$, $E_{s(i)}$ and $y_{s(i)}$, respectively, where $i = 1, ..., m_s$. Similarly, $A_{p(i)}$, $E_{p(i)}$ and $y_{p(i)}$ represent, respectively, the area, elastic modulus and location of the prestressing steel with respect to the x-axis and $i = 1, ..., m_p$.

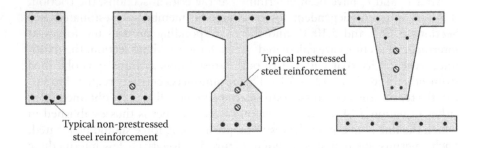

Figure 5.4 Typical reinforced and prestressed concrete sections.

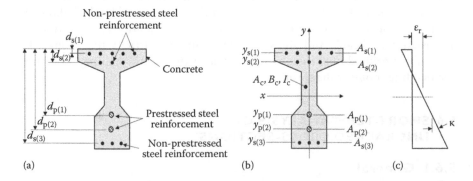

Figure 5.5 Generic cross-section, arrangement of reinforcement and strain. (a) Cross-section. (b) Reinforcement bar and tendon areas ($A_{s(i)}$ and $A_{p(i)}$ located $y_{s(i)}$ and $y_{p(i)}$ from the reference axis (x-axis). (c) Strain.

The geometric properties of the concrete part of the cross-section are A_c, B_c and I_c, where A_c is the concrete area, B_c is the first moment of area of the concrete about the x-axis and I_c is the second moment of area of the concrete about the x-axis. In this analysis, the orientation of the x and y axes are as shown in Figure 5.5.

If the cross-section is subjected to an axial force N_{ext} applied at the origin of the x and y axes and a bending moment M_{ext} applied about the x-axis, the strain diagram is as shown in Figure 5.5c and the strain at any distance y above the reference axis is given by:

$$\varepsilon = \varepsilon_r - y\kappa \tag{5.16}$$

The two unknowns of the problem, i.e. ε_r and κ, are then determined by enforcing horizontal and rotational equilibrium at the cross-section:

$$N_{int} = N_{ext} \tag{5.17}$$

and

$$M_{int} = M_{ext} \tag{5.18}$$

where N_{int} and M_{int} are the internal axial force and moment, respectively, given by:

$$N_{int} = \int_A \sigma \, dA \tag{5.19}$$

and

$$M_{int} = \int_A -y\sigma \, dA \tag{5.20}$$

When the two unknowns (ε_r and κ) are calculated from the two equilibrium equations (Equations 5.17 and 5.18) and the strain is determined using Equation 5.16, the stresses in the concrete and steel may be obtained from the appropriate constitutive relationships. The internal actions are then readily determined from the stresses using Equations 5.19 and 5.20.

This procedure forms the basis of both the short-term analyses presented in the remainder of this section and the long-term analysis presented in Section 5.7.

5.6.2 Short-term cross-sectional analysis

In order to determine the stresses and deformations immediately after first loading or immediately after transfer at τ_0, linear-elastic stress-strain relationships for the concrete and the steel are usually adopted and these can be expressed by:

$$\sigma_{c,0} = E_{c,0}\varepsilon_0 \tag{5.21}$$

$$\sigma_{s(i),0} = E_{s(i)}\varepsilon_0 \tag{5.22}$$

$$\sigma_{p(i),0} = E_{p(i)}\left(\varepsilon_0 + \varepsilon_{p(i),\text{init}}\right) \quad \text{if } A_{p(i)} \text{ is bonded} \tag{5.23}$$

$$\sigma_{p(i),0} = E_{p(i)}\varepsilon_{p(i),\text{init}} \quad \text{if } A_{p(i)} \text{ is unbonded} \tag{5.24}$$

in which $\sigma_{c,0}$, $\sigma_{s(i),0}$ and $\sigma_{p(i),0}$ represent the stresses in the concrete, in the i-th layer of non-prestressed reinforcement (with $i = 1, ..., m_s$) and in the i-th layer of prestressing steel (with $i = 1, ..., m_p$), respectively, immediately after first loading at time τ_0, while $\varepsilon_{p(i),\text{init}}$ is the initial strain in the i-th layer of prestressing steel produced by the initial tensile prestressing force $P_{\text{init}(i)}$ and is given by:

$$\varepsilon_{p(i),\text{init}} = \frac{P_{\text{init}(i)}}{A_{p(i)}E_{p(i)}} \tag{5.25}$$

For a post-tensioned cross-section, $P_{\text{init}(i)}$ is the prestressing force immediately after stressing the tendon and, for a pretensioned member, $P_{\text{init}(i)}$ is the prestressing force immediately before transfer. With this method, the prestressing force is included in the analysis by means of an induced strain $\varepsilon_{p(i),\text{init}}$, rather than an external action [7,8].

For unbonded tendons, Equation 5.24 is strictly only applicable immediately before transfer. After transfer, as the beam deforms under load, the strain in the tendon will increase. However, at service loads, the change in geometry of the member is relatively small and the resulting change in the tendon strain is usually less than 0.5% of $\varepsilon_{p(i),\text{init}}$ and can be ignored.

At time τ_0, the internal axial force resisted by the cross-section and the internal moment about the reference axis are denoted as $N_{\text{int},0}$ and $M_{\text{int},0}$, respectively. The internal axial force $N_{\text{int},0}$ is the sum of the axial forces resisted by the component materials forming the cross-section and is given by:

$$N_{\text{int},0} = N_{c,0} + N_{s,0} + N_{p,0} \tag{5.26}$$

where $N_{c,0}$, $N_{s,0}$ and $N_{p,0}$ represent the axial forces resisted by the concrete, the non-prestressed reinforcement and the prestressing steel, respectively. The axial force resisted by the concrete is calculated from:

$$N_{c,0} = \int_{A_c} \sigma_{c,0}\, dA = \int_{A_c} E_{c,0}\varepsilon_0\, dA = \int_{A_c} E_{c,0}(\varepsilon_{r,0} - y\kappa_0)\, dA = A_c E_{c,0}\varepsilon_{r,0} - B_c E_{c,0}\kappa_0$$

$$(5.27)$$

The axial force resisted by the non-prestressed steel is:

$$N_{s,0} = \sum_{i=1}^{m_s}\left(A_{s(i)}E_{s(i)}\right)\left(\varepsilon_{r,0} - y_{s(i)}\kappa_0\right) = \sum_{i=1}^{m_s}\left(A_{s(i)}E_{s(i)}\right)\varepsilon_{r,0} - \sum_{i=1}^{m_s}\left(y_{s(i)}A_{s(i)}E_{s(i)}\right)\kappa_0$$

$$(5.28)$$

and the axial force resisted by the prestressing steel if bonded is given in Equation 5.29, and if unbonded, may be approximated by Equation 5.30:

$$N_{p,0} = \sum_{i=1}^{m_p}\left(A_{p(i)}E_{p(i)}\right)\varepsilon_{r,0} - \sum_{i=1}^{m_p}\left(y_{p(i)}A_{p(i)}E_{p(i)}\right)\kappa_0 + \sum_{i=1}^{m_p}\left(A_{p(i)}E_{p(i)}\varepsilon_{p(i),\text{init}}\right) \quad (5.29)$$

$$N_{p,0} = \sum_{i=1}^{m_p}\left(A_{p(i)}E_{p(i)}\varepsilon_{p(i),\text{init}}\right) \qquad (5.30)$$

The additional subscripts '0' used for the strain at the level of the reference axis ($\varepsilon_{r,0}$) and the curvature (κ_0) highlight that these are calculated at time τ_0 after the application of $N_{\text{ext},0}$ and $M_{\text{ext},0}$ and after the transfer of prestress.

By substituting Equations 5.27 through 5.30 into Equation 5.26, the equation for $N_{\text{int},0}$ is expressed in terms of the actual geometry and elastic moduli of the materials forming the cross-section.

When the prestressing steel is bonded to the concrete:

$$N_{\text{int},0} = \left(A_c E_{c,0} + \sum_{i=1}^{m_s} A_{s(i)}E_{s(i)} + \sum_{i=1}^{m_p} A_{p(i)}E_{p(i)} \right)\varepsilon_{r,0}$$

$$- \left(B_c E_{c,0} + \sum_{i=1}^{m_s} y_{s(i)}A_{s(i)}E_{s(i)} + \sum_{i=1}^{m_p} y_{p(i)}A_{p(i)}E_{p(i)} \right)\kappa_0 + \sum_{i=1}^{m_p}\left(A_{p(i)}E_{p(i)}\varepsilon_{p(i),\text{init}}\right)$$

$$= R_{A,0}\varepsilon_{r,0} - R_{B,0}\kappa_0 + \sum_{i=1}^{m_p}\left(A_{p(i)}E_{p(i)}\varepsilon_{p(i),\text{init}}\right) \qquad (5.31)$$

When the prestressing steel is unbonded:

$$N_{int,0} = \left(A_c E_{c,0} + \sum_{i=1}^{m_s} A_{s(i)} E_{s(i)} \right) \varepsilon_{r,0} - \left(B_c E_{c,0} + \sum_{i=1}^{m_s} y_{s(i)} A_{s(i)} E_{s(i)} \right) \kappa_0$$

$$+ \sum_{i=1}^{m_p} \left(A_{p(i)} E_{p(i)} \varepsilon_{p(i),init} \right)$$

$$= R_{A,0} \varepsilon_{r,0} - R_{B,0} \kappa_0 + \sum_{i=1}^{m_p} \left(A_{p(i)} E_{p(i)} \varepsilon_{p(i),init} \right) \tag{5.32}$$

In Equations 5.31 and 5.32, $R_{A,0}$ and $R_{B,0}$ represent, respectively, the axial rigidity and the stiffness related to the first moment of area about the reference axis calculated at time τ_0 and, for a cross-section containing bonded tendons, are given by:

$$R_{A,0} = A_c E_{c,0} + \sum_{i=1}^{m_s} A_{s(i)} E_{s(i)} + \sum_{i=1}^{m_p} A_{p(i)} E_{p(i)} \tag{5.33}$$

$$R_{B,0} = B_c E_{c,0} + \sum_{i=1}^{m_s} y_{s(i)} A_{s(i)} E_{s(i)} + \sum_{i=1}^{m_p} y_{p(i)} A_{p(i)} E_{p(i)} \tag{5.34}$$

For unbonded construction, the contribution of the prestressing steel $A_{p(i)}$ to the rigidities $R_{A,0}$ and $R_{B,0}$ is ignored.

Similarly, the equation for $M_{int,0}$ may be expressed by Equations 5.35 and 5.36.

When the prestressing steel is bonded to the concrete:

$$M_{int,0} = -\left(B_c E_{c,0} + \sum_{i=1}^{m_s} y_{s(i)} A_{s(i)} E_{s(i)} + \sum_{i=1}^{m_p} y_{p(i)} A_{p(i)} E_{p(i)} \right) \varepsilon_{r,0}$$

$$+ \left(I_c E_{c,0} + \sum_{i=1}^{m_s} y_{s(i)}^2 A_{s(i)} E_{s(i)} + \sum_{i=1}^{m_p} y_{p(i)}^2 A_{p(i)} E_{p(i)} \right) \kappa_0 - \sum_{i=1}^{m_p} \left(y_{p(i)} A_{p(i)} E_{p(i)} \varepsilon_{p(i),init} \right)$$

$$= -R_{B,0} \varepsilon_{r,0} + R_{I,0} \kappa_0 - \sum_{i=1}^{m_p} \left(y_{p(i)} A_{p(i)} E_{p(i)} \varepsilon_{p(i),init} \right) \tag{5.35}$$

When the prestressing steel is not bonded to the concrete:

$$M_{int,0} = -\left(B_c E_{c,0} + \sum_{i=1}^{m_s} y_{s(i)} A_{s(i)} E_{s(i)}\right)\varepsilon_{r,0}$$

$$+\left(I_c E_{c,0} + \sum_{i=1}^{m_s} y_{s(i)}^2 A_{s(i)} E_{s(i)}\right)\kappa_0 - \sum_{i=1}^{m_p}\left(y_{p(i)} A_{p(i)} E_{p(i)} \varepsilon_{p(i),init}\right)$$

$$= -R_{B,0}\varepsilon_{r,0} + R_{I,0}\kappa_0 - \sum_{i=1}^{m_p}\left(y_{p(i)} A_{p(i)} E_{p(i)} \varepsilon_{p(i),init}\right) \tag{5.36}$$

where $R_{I,0}$ is the flexural rigidity at time τ_0 and, for a cross-section containing bonded tendons, is given by:

$$R_{I,0} = I_c E_{c,0} + \sum_{i=1}^{m_s} y_{s(i)}^2 A_{s(i)} E_{s(i)} + \sum_{i=1}^{m_p} y_{p(i)}^2 A_{p(i)} E_{p(i)} \tag{5.37}$$

For unbonded construction, the contribution of the prestressing steel $A_{p(i)}$ to the flexural rigidity $R_{I,0}$ is ignored.

Substituting the expressions for $N_{int,0}$ and $M_{int,0}$ (Equations 5.31, 5.32 and 5.35 and 5.36) into Equations 5.17 and 5.18 produces the system of equilibrium equations that may be written in compact form as:

$$\mathbf{r}_{ext,0} = \mathbf{D}_0 \boldsymbol{\varepsilon}_0 + \mathbf{f}_{p,init} \tag{5.38}$$

where:

$$\mathbf{r}_{ext,0} = \begin{bmatrix} N_{ext,0} \\ M_{ext,0} \end{bmatrix} \tag{5.39}$$

$$\mathbf{D}_0 = \begin{bmatrix} R_{A,0} & -R_{B,0} \\ -R_{B,0} & R_{I,0} \end{bmatrix} \tag{5.40}$$

$$\boldsymbol{\varepsilon}_0 = \begin{bmatrix} \varepsilon_{r,0} \\ \kappa_0 \end{bmatrix} \tag{5.41}$$

$$\mathbf{f}_{p,init} = \sum_{i=1}^{m_p}\begin{bmatrix} A_{p(i)} E_{p(i)} \varepsilon_{p(i),init} \\ -y_{p(i)} A_{p(i)} E_{p(i)} \varepsilon_{p(i),init} \end{bmatrix} = \sum_{i=1}^{m_p}\begin{bmatrix} P_{init(i)} \\ -y_{p(i)} P_{init(i)} \end{bmatrix} = \begin{bmatrix} P_{init} \\ M_{init} \end{bmatrix} \tag{5.42}$$

The vector $\mathbf{r}_{\text{ext},0}$ is the vector of the external actions at first loading (at time τ_0), i.e. axial force $N_{\text{ext},0}$ and moment $M_{\text{ext},0}$; the matrix $\mathbf{D}_0$ contains the cross-sectional material and geometric properties calculated at τ_0; the strain vector $\boldsymbol{\varepsilon}_0$ contains the unknown independent variables describing the strain diagram at time τ_0 ($\varepsilon_{r,0}$ and κ_0); and the vector $\mathbf{f}_{\text{p,init}}$ contains the actions caused by the initial prestressing.

The vector $\boldsymbol{\varepsilon}_0$ is readily obtained by solving the equilibrium equations (Equation 5.38) giving:

$$\boldsymbol{\varepsilon}_0 = \mathbf{D}_0^{-1}\left(\mathbf{r}_{\text{ext},0} - \mathbf{f}_{\text{p,init}}\right) = \mathbf{F}_0\left(\mathbf{r}_{\text{ext},0} - \mathbf{f}_{\text{p,init}}\right) \tag{5.43}$$

where:

$$\mathbf{F}_0 = \frac{1}{R_{A,0}R_{I,0} - R_{B,0}^2}\begin{bmatrix} R_{I,0} & R_{B,0} \\ R_{B,0} & R_{A,0} \end{bmatrix} \tag{5.44}$$

The stress distribution related to the concrete and reinforcement can then be calculated from the constitutive equations (Equations 5.21 through 5.24) re-expressed here as:

$$\sigma_{c,0} = E_{c,0}\varepsilon_0 = E_{c,0}[1 - y]\boldsymbol{\varepsilon}_0 \tag{5.45}$$

$$\sigma_{s(i),0} = E_{s(i)}\varepsilon_{s(i),0} = E_{s(i)}[1 - y_{s(i)}]\boldsymbol{\varepsilon}_0 \tag{5.46}$$

If $A_{p(i)}$ is bonded:

$$\sigma_{p(i),0} = E_{p(i)}\left(\varepsilon_{p(i),0} + \varepsilon_{p(i),\text{init}}\right) = E_{p(i)}[1 - y_{p(i)}]\boldsymbol{\varepsilon}_0 + E_{p(i)}\varepsilon_{p(i),\text{init}} \tag{5.47}$$

If $A_{p(i)}$ is unbonded:

$$\sigma_{p(i),0} = E_{p(i)}\varepsilon_{p(i),\text{init}} \tag{5.48}$$

where:

$$\varepsilon_0 = \varepsilon_{r,0} - y\kappa_0 = [1 - y]\boldsymbol{\varepsilon}_0$$

Although this procedure is presented here assuming linear-elastic material properties, it is quite general and is also applicable to nonlinear material representations, in which case the integrals of Equations 5.19 and 5.20 might have to be evaluated numerically. However, when calculating the

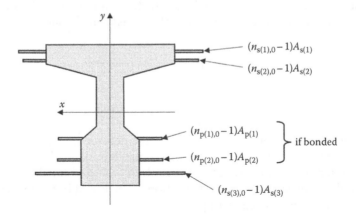

$(n_{s(1),0} - 1)A_{s(1)}$

$(n_{s(2),0} - 1)A_{s(2)}$

$(n_{p(1),0} - 1)A_{p(1)}$ $\left.\begin{array}{c} \\ \\ \end{array}\right\}$ if bonded

$(n_{p(2),0} - 1)A_{p(2)}$

$(n_{s(3),0} - 1)A_{s(3)}$

Figure 5.6 Transformed section with bonded reinforcement transformed into equivalent areas of concrete.

short-term response of uncracked reinforced and prestressed concrete cross-sections under typical in-service loads, material behaviour is essentially linear-elastic.

In reinforced and prestressed concrete design, it is common to calculate the cross-sectional properties by transforming the section into equivalent areas of one of the constituent materials. For example, for the cross-section of Figure 5.5a, the transformed concrete cross-section for the short-term analysis is shown in Figure 5.6, with the area of each layer of bonded steel reinforcement and tendons ($A_{s(i)}$ and $A_{p(i)}$, respectively) transformed into equivalent areas of concrete ($n_{s(i),0} A_{s(i)}$ and $n_{p(i),0} A_{p(i)}$, respectively), where $n_{s(i),0} = E_{s(i)}/E_{c,0}$ is the modular ratio of the i-th layer of non-prestressed steel and $n_{p(i),0} = E_{p(i)}/E_{c,0}$ is the modular ratio of the i-th layer of prestressing steel.

For the transformed section of Figure 5.6, the cross-sectional rigidities defined in Equations 5.33, 5.34 and 5.37 can be re-calculated as:

$$R_{A,0} = A_0 E_{c,0} \tag{5.49}$$

$$R_{B,0} = B_0 E_{c,0} \tag{5.50}$$

$$R_{I,0} = I_0 E_{c,0} \tag{5.51}$$

where A_0 is the area of the transformed concrete section, and B_0 and I_0 are the first and second moments of the transformed area about the reference x-axis at first loading.

Substituting Equations 5.49 to 5.51 into Equation 5.44 enables $\mathbf{F}_0$ to be expressed in terms of the properties of the transformed concrete section as follows:

$$\mathbf{F}_0 = \frac{1}{E_{c,0}(A_0 I_0 - B_0^2)} \begin{bmatrix} I_0 & B_0 \\ B_0 & A_0 \end{bmatrix} \tag{5.52}$$

The two approaches proposed for the calculation of the cross-sectional rigidities, i.e. the one based on Equations 5.33, 5.34 and 5.37 and the one relying on the properties of the transformed section (Equations 5.49 through 5.51), are equivalent. The procedure based on the transformed section (Equations 5.49 through 5.51) is often preferred for the analysis of reinforced and prestressed concrete sections. The use of both approaches is illustrated in the following example.

EXAMPLE 5.3

The short-term behaviour of the post-tensioned beam cross-section shown in Figure 5.7a is to be determined immediately after transfer. The section contains a single unbonded cable containing ten 12.7 mm diameter strands (from Table 4.8, $A_p = 10 \times 98.6 = 986$ mm^2 and $f_{pb} = 1870$ MPa) located within a 60 mm diameter duct, and two layers of non-prestressed reinforcement, as shown. The force in the prestressing steel is $P_i = P_{init} = 1350$ kN. The external moment acting on the cross-section at transfer is $M_o = 100$ kNm (=M_{ext}). The elastic moduli for concrete and steel are $E_c = 30 \times 10^3$ MPa and $E_s = E_p = 200 \times 10^3$ MPa, from which $n_s = E_s/E_c = 6.67$ and $n_p = E_p/E_c = 6.67$.

In this example, the reference x-axis is taken at mid-depth. The transformed section is shown in Figure 5.7b. Because the prestressing steel is not bonded to the concrete, it does not form part of the transformed section.

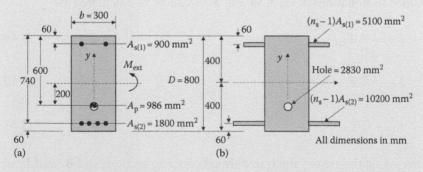

Figure 5.7 Post-tensioned cross-section (Example 5.3). (a) Section. (b) Transformed section.

In addition, the hole created in the concrete section by the hollow duct must also be taken into account. The properties of the transformed section with respect to the reference x-axis are:

$$A_0 = bD + (n_s - 1)A_{s(1)} + (n_s - 1)A_{s(2)} - A_{hole}$$

$$= 300 \times 800 + (6.67 - 1) \times 900 + (6.67 - 1) \times 1800 - 2830$$

$$= 252,500 \text{ mm}^2$$

$$B_0 = bDy_c + (n_s - 1)A_{s(1)}y_{s(1)} + (n_s - 1)A_{s(2)}y_{s(2)} - A_{hole}y_{hole}$$

$$= 300 \times 800 \times 0 + (6.67 - 1) \times 900 \times (+340)$$

$$+ (6.67 - 1) \times 1800 \times (-340) - 2830 \times (-200)$$

$$= -1.168 \times 10^6 \text{mm}^3$$

$$I_0 = \frac{bD^3}{12} + (n_s - 1)A_{s(1)}y_{s(1)}^2 + (n_s - 1)A_{s(2)}y_{s(2)}^2 - A_{hole}y_{hole}^2$$

$$= \frac{300 \times 800^3}{12} + (6.67 - 1) \times 900 \times 340^2$$

$$+ (6.67 - 1) \times 1,800 \times (-340)^2 - 2830 \times (-200)^2$$

$$= 14,455 \times 10^6 \text{mm}^4$$

From Equation 5.52:

$$\mathbf{F}_0 = \frac{1}{30,000 \times (252,500 \times 14,455 \times 10^6 - (-1.168 \times 10^6)^2)}$$

$$\times \begin{bmatrix} 14,455 \times 10^6 & -1.168 \times 10^6 \\ -1.168 \times 10^6 & 252,470 \end{bmatrix}$$

$$= \begin{bmatrix} 132.1 \times 10^{-12} & -10.67 \times 10^{-15} \\ -10.67 \times 10^{-15} & 2.307 \times 10^{-15} \end{bmatrix}$$

From Equation 5.39, the vector of internal actions is:

$$\mathbf{r}_{ext,0} = \begin{bmatrix} N_{ext,0} \\ M_{ext,0} \end{bmatrix} = \begin{bmatrix} 0 \\ 100 \times 10^6 \end{bmatrix}$$

and from Equation 5.25, the initial strain in the prestressing steel due to the initial prestressing force is:

$$\varepsilon_{p,init} = \frac{P_{init}}{A_p E_p} = \frac{1,350 \times 10^3}{986 \times 200,000} = 0.006846$$

The vector of internal actions caused by the initial prestress $\mathbf{f}_{p,init}$, containing the initial prestressing force $(A_p E_p \varepsilon_{p,init})$ and its moment about the reference x-axis $(-y_p A_p E_p \varepsilon_{p,init})$, is given by Equation 5.42:

$$\mathbf{f}_{p,init} = \begin{bmatrix} P_{init} \\ -y_p P_{init} \end{bmatrix} = \begin{bmatrix} 1350 \times 10^3 \\ -(-200) \times 1350 \times 10^3 \end{bmatrix} = \begin{bmatrix} 1350 \times 10^3 \\ 270 \times 10^6 \end{bmatrix}$$

where the dimension y_p is the distance from the x-axis to the centroid of the prestressing steel. The strain vector ε_0 containing the unknown strain variables is determined from Equation 5.43:

$$\varepsilon_0 = \mathbf{F}_0 \left(\mathbf{r}_{ext,0} - \mathbf{f}_{p,init} \right)$$

$$= \begin{bmatrix} 132.1 \times 10^{-12} & -10.67 \times 10^{-15} \\ -10.67 \times 10^{-15} & 2.307 \times 10^{-15} \end{bmatrix} \left(\begin{bmatrix} 0 \\ 100 \times 10^6 \end{bmatrix} - \begin{bmatrix} 1350 \times 10^3 \\ 270 \times 10^6 \end{bmatrix} \right)$$

$$= \begin{bmatrix} -176.5 \times 10^{-6} \\ -0.3778 \times 10^6 \end{bmatrix}$$

The strain at the reference axis and the curvature are therefore:

$$\varepsilon_{r,0} = -176.5 \times 10^{-6} \quad \text{and} \quad \kappa_0 = -0.3778 \times 10^{-6} \text{ mm}^{-1}$$

and, from Equation 5.16, the top fibre strain (at $y = +400$ mm) and the bottom fibre strain (at $y = -400$ mm) are:

$$\varepsilon_{0(top)} = \varepsilon_{r,0} - 400 \times \kappa_0 = [-176.5 - 400 \times (-0.3778)] \times 10^{-6}$$
$$= -25.4 \times 10^{-6}$$

$$\varepsilon_{0(btm)} = \varepsilon_{r,0} - (-400) \times \kappa_0 = [-176.5 + 400 \times (-0.3778)] \times 10^{-6}$$
$$= -327.6 \times 10^{-6}$$

The top and bottom fibre stresses in the concrete and the stresses in the two layers of reinforcement and in the prestressing steel are obtained from Equations 5.45 to 5.48:

$$\sigma_{c,0(top)} = E_{c,0}[1 - y_{top}]\varepsilon_0 = E_{c,0}\varepsilon_{0(top)} = 30,000 \times (-25.4 \times 10^{-6})$$
$$= -0.762 \text{ MPa}$$

$$\sigma_{c,0(btm)} = E_{c,0}[1 - y_{btm}]\varepsilon_0 = E_{c,0}\varepsilon_{0(btm)} = 30,000 \times (-327.6 \times 10^{-6})$$
$$= -9.828 \text{ MPa}$$

$$\sigma_{s(1),0} = E_s[1 - y_{s(1)}]\varepsilon_0 = 200,000 \times [1 - 340]\begin{bmatrix} -176.5 \times 10^{-6} \\ -0.3778 \times 10^{-6} \end{bmatrix} = -9.61\,\text{MPa}$$

$$\sigma_{s(2),0} = E_s[1 - y_{s(2)}]\varepsilon_0 = 200,000 \times [1 + 340]\begin{bmatrix} -176.5 \times 10^{-6} \\ -0.3778 \times 10^{-6} \end{bmatrix} = -61.0\,\text{MPa}$$

$$\sigma_{p,0} = E_p\varepsilon_{p,init} = 200,000 \times 0.006846 = +1369\,\text{MPa}$$

The distributions of strain and stress on the cross-section immediately after transfer are shown in Figure 5.8.

Alternatively, instead of analysing the transformed cross-section, the cross-sectional rigidities could have been calculated using Equations 5.33, 5.34 and 5.37. The properties of the concrete part of the cross-section (with respect to the x-axis) are:

$$A_c = bD - A_{s(1)} - A_{s(2)} - A_{hole}$$

$$= 300 \times 800 - 900 - 1,800 - 2,830$$

$$= 234,470\,\text{mm}^2$$

$$B_c = bDy_c - A_{s(1)}y_{s(1)} - A_{s(2)}y_{s(2)} - A_{hole}y_{hole}$$

$$= 300 \times 800 \times 0 - 900 \times 340 - 1800 \times (-340) - 2830 \times (-200)$$

$$= 872,000\,\text{mm}^3$$

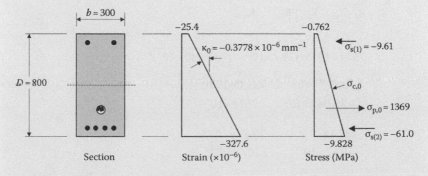

Figure 5.8 Strains and stresses immediately after transfer (Example 5.3).

$$I_c = bD^3/12 - A_{s(1)}y_{s(1)}^2 - A_{s(2)}y_{s(2)}^2 - A_{hole}y_{hole}^2$$

$$= 300 \times 800^3/12 - 900 \times 340^2 - 1800 \times (-340)^2 - 2830 \times (-200)^2$$

$$= 12,375 \times 10^6 \text{ mm}^4$$

For this member with unbonded tendons, Equations 5.33, 5.34 and 5.37 give:

$$R_{A,0} = A_c E_{c,0} + \sum_{i=1}^{m_s} A_{s(i)} E_{s(i)}$$

$$= 234,470 \times 30,000 + (900 + 1,800) \times 200,000$$

$$= 7,574 \times 10^6 \text{ N}$$

$$R_{B,0} = B_c E_{c,0} + \sum_{i=1}^{m_s} y_{s(i)} A_{s(i)} E_{s(i)}$$

$$= 872,000 \times 30,000 + [340 \times 900 + (-340) \times 1,800] \times 200,000$$

$$= -35,040 \times 10^6 \text{ Nmm}$$

$$R_{I,0} = I_c E_{c,0} + \sum_{i=1}^{m_s} y_{s(i)}^2 A_{s(i)} E_{s(i)}$$

$$= 12,375 \times 10^6 \times 30,000 + [340^2 \times 900 + (-340)^2 \times 1,800] \times 200,000)$$

$$= 433.7 \times 10^{12} \text{ Nmm}^2$$

and from Equation 5.44:

$$\mathbf{F}_0 = \frac{1}{R_{A,0}R_{I,0} - R_{B,0}^2} \begin{bmatrix} R_{I,0} & R_{B,0} \\ R_{B,0} & R_{A,0} \end{bmatrix}$$

$$= \frac{1}{7574 \times 10^6 \times 433.7 \times 10^{12} - (-35,040 \times 10^6)^2}$$

$$\times \begin{bmatrix} 433.7 \times 10^{12} & -35,040 \times 10^6 \\ -35,040 \times 10^6 & 7574 \times 10^6 \end{bmatrix}$$

$$= \begin{bmatrix} 132.1 \times 10^{-12} & -10.67 \times 10^{-15} \\ -10.67 \times 10^{-15} & 2.307 \times 10^{-15} \end{bmatrix}$$

This is identical to the matrix $\mathbf{F}_0$ obtained earlier from Equation 5.52.

EXAMPLE 5.4

Determine the instantaneous stress and strain distributions on the precast pretensioned concrete section shown in Figure 5.9. The cross-section is that of a Girder Type 3 from the AS5100.5-2004 [9] guidelines, with the area of the gross section $A_{gross} = 317 \times 10^3$ mm² and the second moment of area of the gross section about the centroidal axis $I_{gross} = 49,900 \times 10^6$ mm⁴. The centroid of the gross cross-sectional area is located 602 mm below its top fibre, i.e. $d_c = 602$ mm.

The section is subjected to a compressive axial force $N_{ext,0} = -100$ kN and a sagging moment of $M_{ext,0} = +1000$ kNm applied with respect to the reference x-axis, that is taken in this example to be 300 mm below the top fibre of the cross-section.

Assuming all materials are linear-elastic with $E_{c,0} = 32$ GPa and $E_s = E_p = 200$ GPa. The modular ratios of the reinforcing steel and the prestressing steel are therefore, $n_{s(i),0} = n_{p(i),0} = 6.25$. The initial prestressing forces applied to the three layers of tendons ($A_{p(1)} = 300$ mm², $A_{p(2)} = 500$ mm² and $A_{p(3)} = 800$ mm², respectively) prior to the transfer of prestress are $P_{init(1)} = 375$ kN, $P_{init(2)} = 625$ kN and $P_{init(3)} = 1000$ kN.

The distances of the steel layers from the reference axis are $y_{s(1)} = +240$ mm, $y_{s(2)} = -790$ mm, $y_{p(1)} = -580$ mm, $y_{p(2)} = -645$ mm and $y_{p(3)} = -710$ mm. From Equation 5.25, the initial strains in the prestressing steel layers prior to the transfer of prestress to the concrete are:

$$\varepsilon_{p(1),init} = \frac{P_{init(1)}}{A_{p(1)}E_p} = \frac{375 \times 10^3}{300 \times 200,000} = 0.00625$$

$$\varepsilon_{p(2),init} = \frac{P_{init(2)}}{A_{p(2)}E_p} = \frac{625 \times 10^3}{500 \times 200,000} = 0.00625$$

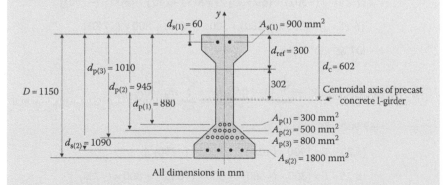

All dimensions in mm

Figure 5.9 Precast prestressed concrete section (Example 5.4).

$$\varepsilon_{p(3),init} = \frac{P_{init(3)}}{A_{p(3)}E_p} = \frac{1,000 \times 10^3}{800 \times 200,000} = 0.00625$$

From Equation 5.39, the vector of internal actions at first loading is:

$$\mathbf{r}_{ext,0} = \begin{bmatrix} N_{ext,0} \\ M_{ext,0} \end{bmatrix} = \begin{bmatrix} -100 \times 10^3 \text{ N} \\ 1000 \times 10^6 \text{ Nmm} \end{bmatrix}$$

and, from Equation 5.42, the vector of initial prestressing forces is:

$$\mathbf{f}_{p,init} = \sum_{i=1}^{m_p} \begin{bmatrix} P_{init(i)} \\ -y_{p(i)}P_{init(i)} \end{bmatrix} = \begin{bmatrix} 375 \times 10^3 \\ -(-580) \times 375 \times 10^3 \end{bmatrix} + \begin{bmatrix} 625 \times 10^3 \\ -(-645) \times 625 \times 10^3 \end{bmatrix}$$

$$+ \begin{bmatrix} 1000 \times 10^3 \\ -(-710) \times 1000 \times 10^3 \end{bmatrix} = \begin{bmatrix} 2000 \times 10^3 \text{N} \\ 1330.6 \times 10^6 \text{Nmm} \end{bmatrix}$$

With the centroid of the concrete cross-section 302 mm below the reference axis (i.e. $y_c = -302$ mm), the properties of the transformed section (with the steel transformed into equivalent areas of concrete) with respect to the reference x-axis are:

$$A_0 = A_{gross} + \sum_{i=1}^{2}(n_s - 1)A_{s(i)} + \sum_{i=1}^{3}(n_p - 1)A_{p(i)}$$

$$= 317 \times 10^3 + (6.25 - 1) \times (900 + 1800)$$

$$+ (6.25 - 1) \times (300 + 500 + 800)$$

$$= 339,575 \text{ mm}^2$$

$$B_0 = A_{gross}y_c + \sum_{i=1}^{2}(n_s - 1)A_{s(i)}y_{s(i)} + \sum_{i=1}^{3}(n_p - 1)A_{p(i)}y_{p(i)}$$

$$= 317 \times 10^3 \times (-302) + (6.25 - 1) \times [900 \times (+240) + 1800 \times (-790)]$$

$$+ (6.25 - 1) \times [300 \times (-580) + 500 \times (-645) + 800 \times (-710)]$$

$$= -107.65 \times 10^6 \text{ mm}^3$$

$$I_0 = I_{gross} + A_{gross}y_c^2 + \sum_{i=1}^{2}(n_s - 1)A_{s(i)}y_{s(i)}^2 + \sum_{i=1}^{3}(n_p - 1)A_{p(i)}y_{p(i)}^2$$

$$= 49,900 \times 10^6 + 317 \times 10^3 \times (-302)^2$$

$$+ (6.25 - 1) \times [900 \times (+240)^2 + 1800 \times (-790)^2]$$

$$+ (6.25 - 1) \times [300 \times (-580)^2 + 500 \times (-645)^2 + 800 \times (-710)^2]$$

$$= 88,721 \times 10^6 \text{mm}^4$$

From Equation 5.52:

$$\mathbf{F}_0 = \frac{1}{32000 \times (339575 \times 88721 \times 10^6 - (-107.65 \times 10^6)^2)}$$

$$\times \begin{bmatrix} 88721 \times 10^6 & -107.65 \times 10^6 \\ -107.65 \times 10^6 & 339575 \end{bmatrix}$$

$$= \begin{bmatrix} 149.6 \times 10^{-12} & -181.5 \times 10^{-15} \\ -181.5 \times 10^{-15} & 572.4 \times 10^{-18} \end{bmatrix}$$

and the strain vector ε_0 containing the unknown strain variables is determined from Equation 5.43:

$$\varepsilon_0 = \mathbf{F}_0 \left(\mathbf{r}_{ext,0} - \mathbf{f}_{p,init} \right)$$

$$= \begin{bmatrix} 149.6 \times 10^{-12} & -181.5 \times 10^{-15} \\ -181.5 \times 10^{-15} & 572.4 \times 10^{-18} \end{bmatrix} \left(\begin{bmatrix} -100 \times 10^3 \\ 1000 \times 10^6 \end{bmatrix} - \begin{bmatrix} 2000 \times 10^3 \\ 1330.6 \times 10^6 \end{bmatrix} \right)$$

$$= \begin{bmatrix} -254.1 \times 10^{-6} \\ +0.1918 \times 10^6 \end{bmatrix}$$

The strain at the reference axis and the curvature are therefore:

$$\varepsilon_{r,0} = -254.1 \times 10^{-6} \quad \text{and} \quad \kappa_0 = +0.1918 \times 10^{-6} \text{ mm}^{-1}$$

and, from Equation 5.16, the top fibre strain (at $y = +300$ mm) and the bottom fibre strain (at $y = -850$ mm) are:

$$\varepsilon_{0(top)} = \varepsilon_{r,0} - 300 \times \kappa_0 = (-254.1 - 300 \times 0.1918) \times 10^{-6}$$

$$= -311.6 \times 10^{-6}$$

$$\varepsilon_{0(btm)} = \varepsilon_{r,0} - (-850) \times \kappa_0 = (-254.1 + 850 \times 0.1918) \times 10^{-6}$$

$$= -91.0 \times 10^{-6}$$

The top and bottom fibre stresses in the concrete and the stresses in the two layers of reinforcement and in the prestressing steel are obtained from Equations 5.45 through 5.48:

$$\sigma_{c,0(top)} = E_{c,0}\varepsilon_{0(top)} = 32,000 \times (-311.6 \times 10^{-6}) = -9.97 \text{ MPa}$$

$$\sigma_{c,0(btm)} = E_{c,0}\varepsilon_{0(btm)} = 32,000 \times (-91.0 \times 10^{-6}) = -2.91 \text{ MPa}$$

$$\sigma_{s(1),0} = E_s[1 - y_{s(1)}]\varepsilon_0 = 200,000 \times [1 - 240]\begin{bmatrix} -254.1 \times 10^{-6} \\ +0.1918 \times 10^{-6} \end{bmatrix} = -60.0 \text{ MPa}$$

$$\sigma_{s(2),0} = E_s[1 - y_{s(2)}]\varepsilon_0 = 200,000 \times [1 + 790]\begin{bmatrix} -254.1 \times 10^{-6} \\ +0.1918 \times 10^{-6} \end{bmatrix} = -20.5 \text{ MPa}$$

$$\sigma_{p(1),0} = E_p[1 - y_{p(1)}]\varepsilon_0 + E_p\varepsilon_{p(1),init}$$

$$= 200,000 \times [1 + 580]\begin{bmatrix} -254.1 \times 10^{-6} \\ +0.1918 \times 10^{-6} \end{bmatrix} + 200,000 \times 0.00625$$

$$= +1,221.4 \text{ MPa}$$

$$\sigma_{p(2),0} = 200,000 \times [1 + 645]\begin{bmatrix} -254.1 \times 10^{-6} \\ +0.1918 \times 10^{-6} \end{bmatrix} + 200,000 \times 0.00625$$

$$= +1,223.9 \text{ MPa}$$

$$\sigma_{p(3),0} = 200,000 \times [1 + 710]\begin{bmatrix} -254.1 \times 10^{-6} \\ +0.1918 \times 10^{-6} \end{bmatrix} + 200,000 \times 0.00625$$

$$= +1,226.4 \text{ MPa}$$

The distributions of strain and stress on the cross-section immediately after transfer are shown in Figure 5.10.

As already outlined in Example 5.3, the cross-sectional rigidities included in $\mathbf{F}_0$ can also be calculated using Equations 5.33, 5.34 and 5.37. The properties of the concrete part of the cross-section (with respect to the x-axis) are:

$$A_c = A_{gross} - A_{s(1)} - A_{s(2)} - A_{p(1)} - A_{p(2)} - A_{p(3)}$$

$$= 317 \times 10^3 - 900 - 1,800 - 300 - 500 - 800 = 312,700 \text{ mm}^2$$

$$B_c = A_{gross}y_c - A_{s(1)}y_{s(1)} - A_{s(2)}y_{s(2)} - A_{p(1)}y_{p(1)} - A_{p(2)}y_{p(2)} - A_{p(3)}y_{p(3)}$$

$$= -93.46 \times 10^6 \text{ mm}^3$$

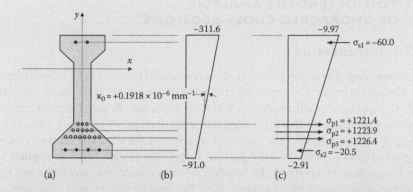

Figure 5.10 Strain and stress diagrams (Example 5.4). (a) Cross-section. (b) Strain ($\times 10^{-6}$). (c) Stress (MPa).

$$I_c = I_{gross} + A_{gross}y_c^2 - A_{s(1)}y_{s(1)}^2 - A_{s(2)}y_{s(2)}^2 - A_{p(1)}y_{p(1)}^2 - A_{p(2)}y_{p(2)}^2 - A_{p(3)}y_{p(3)}^2$$

$$= 76,924 \times 10^6 \, mm^4$$

For this member with bonded tendons, Equations 5.33, 5.34 and 5.37 give:

$$R_{A,0} = A_c E_{c,0} + \sum_{i=1}^{2} A_{s(i)} E_{s(i)} + \sum_{i=1}^{3} A_{p(i)} E_{p(i)}$$

$$= 312,700 \times 32,000 + (900 + 1,800) \times 200,000$$

$$+ (300 + 500 + 800) \times 200,000 = 10,866 \times 10^6 \, N$$

$$R_{B,0} = B_c E_{c,0} + \sum_{i=1}^{2} y_{s(i)} A_{s(i)} E_{s(i)} + \sum_{i=1}^{3} y_{p(i)} A_{p(i)} E_{p(i)} = -3445 \times 10^9 \, Nmm$$

$$R_{I,0} = I_c E_{c,0} + \sum_{i=1}^{2} y_{s(i)}^2 A_{s(i)} E_{s(i)} + \sum_{i=1}^{3} y_{p(i)}^2 A_{p(i)} E_{p(i)} = 2839 \times 10^{12} \, Nmm^2$$

and from Equation 5.44:

$$F_0 = \frac{1}{R_{A,0}R_{I,0} - R_{B,0}^2} \begin{bmatrix} R_{I,0} & R_{B,0} \\ R_{B,0} & R_{A,0} \end{bmatrix} = \begin{bmatrix} 149.6 \times 10^{-12} & -181.5 \times 10^{-15} \\ -181.5 \times 10^{-15} & 572.4 \times 10^{-18} \end{bmatrix}$$

This is identical to the matrix F_0 obtained earlier from Equation 5.52.

5.7 TIME-DEPENDENT ANALYSIS
OF UNCRACKED CROSS-SECTIONS

5.7.1 Introduction

The time-dependent deformation of a prestressed member is greatly affected by the quantity and location of the bonded reinforcement (both conventional non-prestressed reinforcement and tendons). Bonded reinforcement provides restraint to the time-dependent shortening of concrete caused by creep and shrinkage. As the concrete creeps and shrinks, the reinforcement is gradually compressed. An equal and opposite tensile force is applied to the concrete at the level of the bonded reinforcement, thereby reducing the compression caused by prestress. It is the tensile forces that are applied gradually at each level of bonded reinforcement that result in significant time-dependent changes in curvature and deflection. A reliable estimate of these forces is essential if meaningful predictions of long-term behaviour are required.

Procedures specified in codes of practice for predicting losses of prestress due to creep and shrinkage are usually too simplified to be reliable and often lead to significant error, particularly for members containing non-prestressed reinforcement. In the following section, a simple analytical technique is presented for estimating the time-dependent behaviour of a general prestressed cross-section of any shape and containing any number of levels of prestressed and non-prestressed reinforcement. The procedure has been described in more detail in Gilbert and Ranzi [6], and makes use of the age-adjusted effective modulus method to model the effects of creep in concrete.

5.7.2 The age-adjusted effective modulus method

The age-adjusted effective modulus for concrete is often used to account for the creep strain that develops in concrete due to a gradually applied stress and was introduced in Section 4.2.4.3 (see Equations 4.17 and 4.18). The stress history shown in Figure 5.11 is typical of the change in stress that

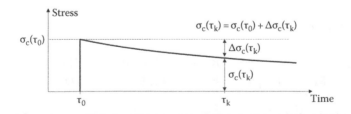

Figure 5.11 A gradually reducing stress history.

occurs with time at many points in a reinforced or prestressed concrete member containing bonded reinforcement (or other forms of restraint) when subjected to a sustained load.

For concrete subjected to the stress history of Figure 5.11, the total strain at time τ_k may be expressed as the sum of the instantaneous and creep strains produced by $\sigma_c(\tau_0)$ (see Equation 4.15), the instantaneous and creep strains produced by the gradually applied stress increment $\Delta\sigma_c(\tau_k)$ (see Equation 4.17) and the shrinkage strain as follows:

$$\varepsilon(\tau_k) = \frac{\sigma_c(\tau_0)}{E_{c,0}}\left[1 + \varphi_{cc}(\tau_k,\tau_0)\right] + \frac{\Delta\sigma_c(\tau_k)}{E_{c,0}}\left[1 + \chi(\tau_k,\tau_0)\varphi_{cc}(\tau_k,\tau_0)\right] + \varepsilon_{cs}(\tau_k)$$

$$= \frac{\sigma_c(\tau_0)}{E_e(\tau_k,\tau_0)} + \frac{\Delta\sigma_c(\tau_k)}{\overline{E}_e(t,\tau_0)} + \varepsilon_{cs}(\tau_k) \tag{5.53}$$

where $E_e(\tau_k,\tau_0)$ is the *effective modulus* (Equation 4.16), and $\overline{E}_e(\tau_k,\tau_0)$ is the *age-adjusted effective modulus* (Equation 4.18), both reproduced here for convenience:

$$E_e(\tau_k,\tau_0) = \frac{E_{c,0}}{1 + \varphi_{cc}(\tau_k,\tau_0)} \tag{5.54}$$

$$\overline{E}_e(\tau_k,\tau_0) = \frac{E_{c,0}}{1 + \chi(\tau_k,\tau_0)\varphi_{cc}(\tau_k,\tau_0)} \tag{5.55}$$

In the remainder of this chapter, the simplified notation $E_{e,k}$ and $\overline{E}_{e,k}$ will be used instead of $E_e(\tau_k,\tau_0)$ and $\overline{E}_e(\tau_k,\tau_0)$, respectively.

For members subjected to constant sustained loads, where the change in concrete stress is caused by the restraint to creep and shrinkage provided by bonded reinforcement, the aging coefficient may be approximated by $\chi(\tau_k,\tau_0) = 0.65$, when τ_k exceeds about 100 days [6].

Equation 5.53 can be written in terms of the concrete stress at first loading, i.e. $\sigma_c(\tau_0)$ $(=\sigma_{c,0})$, and the concrete stress at time τ_k, i.e. $\sigma_c(\tau_k)$ $(=\sigma_{c,k})$, as follows:

$$\varepsilon(\tau_k) = \varepsilon_k = \frac{\sigma_{c,0}}{E_{e,k}} + \frac{\sigma_{c,k} - \sigma_{c,0}}{\overline{E}_{e,k}} + \varepsilon_{cs,k}$$

$$= \frac{\sigma_{c,0}\varphi_{cc}(\tau_k,\tau_0)[1 - \chi(\tau_k,\tau_0)]}{E_{c,0}} + \frac{\sigma_{c,k}[1 + \chi(\tau_k,\tau_0)\varphi_{cc}(\tau_k,\tau_0)]}{E_{c,0}} + \varepsilon_{cs,k} \tag{5.56}$$

and rearranging Equation 5.56 gives:

$$\sigma_{c,k} = \bar{E}_{e,k}(\varepsilon_k - \varepsilon_{cs,k}) + \sigma_{c,0}\bar{F}_{e,0} \tag{5.57}$$

where:

$$\bar{F}_{e,0} = \varphi_{cc}(\tau_k, \tau_0)\frac{[\chi(\tau_k, \tau_0) - 1]}{[1 + \chi(\tau_k, \tau_0)\varphi_{cc}(\tau_k, \tau_0)]} \tag{5.58}$$

Equation 5.57 is a stress-strain time relationship for concrete and can be conveniently used to determine the long-term deformations of a wide variety of concrete structures. The method of analysis is known as the *Age-adjusted Effective Modulus Method*, AEMM [6].

5.7.3 Long-term analysis of an uncracked cross-section subjected to combined axial force and bending using AEMM

Cross-sectional analysis using Equation 5.57 as the constitutive relationship for concrete provides an effective tool for determining how stresses and strains vary with time due to creep and shrinkage of the concrete and relaxation of the prestressing steel. For this purpose, two instants in time are identified, as shown in Figure 5.12. One time instant is the time at first loading, i.e. $t = \tau_0$, and one represents the instant in time at which stresses and strains need to be evaluated, i.e. $t = \tau_k$. It is usually convenient to measure time in days starting from the time when the concrete is poured.

During the time interval Δt_k ($=\tau_k - \tau_0$), creep and shrinkage strains develop in the concrete and relaxation occurs in the tendons. The gradual change of concrete strain with time causes changes of stress in the bonded reinforcement. In general, as the concrete shortens due to compressive creep and shrinkage, the reinforcement is compressed, and there is a gradual increase in the compressive stress in the non-prestressed reinforcement and a gradual loss of prestress in any bonded tendons. To maintain equilibrium,

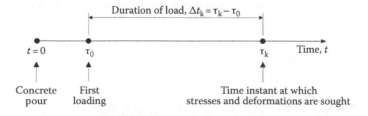

Figure 5.12 Relevant instants in time (AEMM).

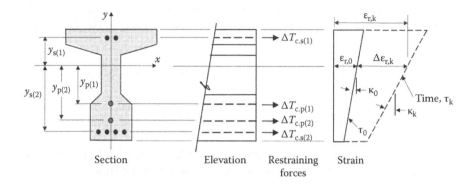

Figure 5.13 Time-dependent actions and deformations.

the gradual change of force in the steel at each bonded reinforcement level is opposed by an equal and opposite restraining force on the concrete, as shown in Figure 5.13.

These gradually applied restraining forces ($\Delta T_{c.s(i)}$ and $\Delta T_{c.p(i)}$) are usually tensile and, for a prestressed or partially-prestressed cross-section, tend to relieve the concrete of its initial compression. The loss of prestress in the concrete is therefore, often significantly more than the loss of prestress in the tendons.

The resultants of the creep and shrinkage induced internal restraining forces on the concrete are an increment of axial force $\Delta N(\tau_k)$ and an increment of moment about the reference axis $\Delta M(\tau_k)$ given by:

$$\Delta N(\tau_k) = \sum_{i=1}^{m_s} \Delta T_{c.s(i)} + \sum_{i=1}^{m_p} \Delta T_{c.p(i)} \tag{5.59}$$

and

$$\Delta M(\tau_k) = -\sum_{i=1}^{m_s} \Delta T_{c.s(i)} y_{s(i)} - \sum_{i=1}^{m_p} \Delta T_{c.p(i)} y_{p(i)} \tag{5.60}$$

Equal and opposite actions, i.e. $-\Delta N(\tau_k)$ and $-\Delta M(\tau_k)$, are applied to the bonded steel parts of the cross-section.

The strain at time τ_k at any distance y from the reference axis (i.e. the x-axis in Figure 5.13) may be expressed in terms of the strain at the reference axis $\varepsilon_{r,k}$ and the curvature κ_k:

$$\varepsilon_k = \varepsilon_{r,k} - y\, \kappa_k \tag{5.61}$$

The magnitude of the change of strain $\Delta\varepsilon_k (=\varepsilon_k - \varepsilon_0)$ that occurs with time at any point on the cross-section is the sum of each of the following components:

1. the free shrinkage strain $\varepsilon_{cs}(\tau_k) = \varepsilon_{cs,k}$ (which is usually considered to be uniform over the section);
2. the unrestrained creep strain caused by the initial concrete stress $\sigma_{c,0}$ existing at the beginning of the time period, that is, $\varepsilon_{cr,k} = \phi_{cc}(\tau_k,\tau_0)\,\sigma_{c,0}/E_{c,0}$; and
3. the creep and elastic strain caused by $\Delta N(\tau_k)$ and $\Delta M(\tau_k)$ gradually applied to the concrete cross-section throughout the time period.

In the time analysis to determine stresses and deformations at τ_k, the steel reinforcement and prestressing tendons are assumed to be linear-elastic (as for the short-term analysis) and the constitutive relationship for the concrete is that given by Equation 5.57. The stress-strain relationships for each material at τ_0 and at τ_k are therefore as specified below.

At τ_0:

$$\sigma_{c,0} = E_{c,0}\varepsilon_0 \tag{5.62}$$

$$\sigma_{s(i),0} = E_{s(i)}\varepsilon_0 \tag{5.63}$$

$$\sigma_{p(i),0} = E_{p(i)}\left(\varepsilon_0 + \varepsilon_{p(i),\text{init}}\right) \tag{5.64}$$

At τ_k:

$$\sigma_{c,k} = \bar{E}_{e,k}\left(\varepsilon_k - \varepsilon_{cs,k}\right) + \bar{F}_{e,0}\sigma_{c,0} \tag{5.65}$$

$$\sigma_{s(i),k} = E_{s(i)}\varepsilon_k \tag{5.66}$$

$$\sigma_{p(i),k} = E_{p(i)}\left(\varepsilon_k + \varepsilon_{p(i),\text{init}} - \varepsilon_{p.\text{rel}(i),k}\right) \tag{5.67}$$

where $\bar{F}_{e,0}$ is given by Equation 5.58, $\bar{E}_{e,k}$ is the age-adjusted effective modulus at $t = \tau_k$ (Equation 5.55) and $\varepsilon_{p.\text{rel}(i),k}$ is the tensile creep strain that has developed in the i-th prestressing tendon at time τ_k (often referred to as the *relaxation strain*) and may be calculated from:

$$\varepsilon_{p.\text{rel}(i),k} = \frac{\sigma_{p(i),0}}{E_{p(i)}}\varphi_{p(i)} = \varepsilon_{p(i),0}\varphi_{p(i)} \tag{5.68}$$

where $\varphi_{p(i)}$ is the creep coefficient for the prestressing steel at time τ_k due to an initial stress $\sigma_{p(i),0}$ in the i-th prestressing tendon just after transfer (as given in Table 4.10).

The governing equations describing the long-term behaviour of a cross-section are obtained by enforcing equilibrium at the cross-section at time τ_k following the approach already presented in the previous section for the instantaneous analysis at time τ_0 (Equations 5.21 through 5.52). Restating the equilibrium equations (Equations 5.17 through 5.20) at time τ_k gives:

$$\mathbf{r}_{ext,k} = \mathbf{r}_{int,k} \tag{5.69}$$

where:

$$\mathbf{r}_{ext,k} = \begin{bmatrix} N_{ext,k} \\ M_{ext,k} \end{bmatrix} \tag{5.70}$$

and

$$\mathbf{r}_{int,k} = \begin{bmatrix} N_{int,k} \\ M_{int,k} \end{bmatrix} \tag{5.71}$$

and $N_{int,k}$ and $M_{int,k}$ are the internal axial force and moment resisted by the cross-section at time τ_k, while $N_{ext,k}$ and $M_{ext,k}$ are the external applied actions at this time. As in Equation 5.26, the axial force $N_{int,k}$ is the sum of the axial forces carried by the concrete, reinforcement and tendons:

$$N_{int,k} = N_{c,k} + N_{s,k} + N_{p,k} \tag{5.72}$$

Considering the time-dependent constitutive relationship for the concrete (Equation 5.65), the axial force resisted by the concrete at time τ_k can be expressed as:

$$N_{c,k} = \int_{A_c} \sigma_{c,k}\, \mathrm{d}A = \int_{A_c} \left[\overline{E}_{e,k}\left(\varepsilon_{r,k} - y\kappa_k - \varepsilon_{cs,k}\right) + \overline{F}_{e,0}\sigma_{c,0} \right] \mathrm{d}A$$

$$= A_c \overline{E}_{e,k}\varepsilon_{r,k} - B_c \overline{E}_{e,k}\kappa_k - A_c \overline{E}_{e,k}\varepsilon_{cs,k} + \overline{F}_{e,0}N_{c,0} \tag{5.73}$$

where $\varepsilon_{r,k}$ and κ_k are the strain at the level of the reference axis and the curvature at time τ_k, respectively, while $N_{c,0}$ is the axial force resisted by the concrete at time τ_0. For the time analysis, $N_{c,0}$ is assumed to be known having been determined from the instantaneous analysis and may be calculated from Equation 5.27.

Using the constitutive equations for the steel (Equations 5.66 and 5.67), the axial forces carried by the reinforcing bars and the prestressing steel at time τ_k are calculated as:

$$N_{s,k} = \sum_{i=1}^{m_s}\left(A_{s(i)}E_{s(i)}\right)\varepsilon_{r,k} - \sum_{i=1}^{m_s}\left(y_{s(i)}A_{s(i)}E_{s(i)}\right)\kappa_k \qquad (5.74)$$

$$N_{p,k} = \sum_{i=1}^{m_p}\left(A_{p(i)}E_{p(i)}\right)\varepsilon_{r,k} - \sum_{i=1}^{m_p}\left(y_{p(i)}A_{p(i)}E_{p(i)}\right)\kappa_k$$

$$+ \sum_{i=1}^{m_p}\left[A_{p(i)}E_{p(i)}\left(\varepsilon_{p(i),init} - \varepsilon_{p.rel(i),k}\right)\right] \qquad (5.75)$$

By substituting Equations 5.73, 5.74 and 5.75 into Equation 5.72, the internal axial force is given by:

$$N_{int,k} = \left(A_c\bar{E}_{e,k} + \sum_{i=1}^{m_s}A_{s(i)}E_{s(i)} + \sum_{i=1}^{m_p}A_{p(i)}E_{p(i)}\right)\varepsilon_{r,k} - \left(B_c\bar{E}_{e,k} + \sum_{i=1}^{m_s}y_{s(i)}A_{s(i)}E_{s(i)}\right.$$

$$\left. + \sum_{i=1}^{m_p}y_{p(i)}A_{p(i)}E_{p(i)}\right)\kappa_k - A_c\bar{E}_{e,k}\varepsilon_{cs,k} + \bar{F}_{e,0}N_{c,0}$$

$$+ \sum_{i=1}^{m_p}\left[A_{p(i)}E_{p(i)}\left(\varepsilon_{p(i),init} - \varepsilon_{p.rel(i),k}\right)\right]$$

$$= R_{A,k}\varepsilon_{r,k} - R_{B,k}\kappa_k - A_c\bar{E}_{e,k}\varepsilon_{cs,k} + \bar{F}_{e,0}N_{c,0} + \sum_{i=1}^{m_p}\left[A_{p(i)}E_{p(i)}\left(\varepsilon_{p(i),init} - \varepsilon_{p.rel(i),k}\right)\right]$$

$$(5.76)$$

where the axial rigidity and the stiffness related to the first moment of area calculated at time τ_k have been referred to as $R_{A,k}$ and $R_{B,k}$, respectively, and are given by:

$$R_{A,k} = A_c\bar{E}_{e,k} + \sum_{i=1}^{m_s}A_{s(i)}E_{s(i)} + \sum_{i=1}^{m_p}A_{p(i)}E_{p(i)} \qquad (5.77)$$

$$R_{B,k} = B_c\bar{E}_{e,k} + \sum_{i=1}^{m_s}y_{s(i)}A_{s(i)}E_{s(i)} + \sum_{i=1}^{m_p}y_{p(i)}A_{p(i)}E_{p(i)} \qquad (5.78)$$

In a similar manner, the internal moment $M_{\text{int},k}$ resisted by the cross-section at time τ_k can be expressed as:

$$
\begin{aligned}
M_{\text{int},k} = &-\left(B_c \overline{E}_{e,k} + \sum_{i=1}^{m_s} y_{s(i)} A_{s(i)} E_{s(i)} + \sum_{i=1}^{m_p} y_{p(i)} A_{p(i)} E_{p(i)} \right) \varepsilon_{r,k} \\
&+ \left(I_c \overline{E}_{e,k} + \sum_{i=1}^{m_s} y_{s(i)}^2 A_{s(i)} E_{s(i)} + \sum_{i=1}^{m_p} y_{p(i)}^2 A_{p(i)} E_{p(i)} \right) \kappa_k + B_c \overline{E}_{e,k} \varepsilon_{cs,k} + \overline{F}_{e,0} M_{c,0} \\
&- \sum_{i=1}^{m_p} \left[y_{p(i)} A_{p(i)} E_{p(i)} \left(\varepsilon_{p(i),\text{init}} - \varepsilon_{p.\text{rel}(i),k} \right) \right] \\
= &-R_{B,k} \varepsilon_{r,k} + R_{I,k} \kappa_k + B_c \overline{E}_{e,k} \varepsilon_{cs,k} + \overline{F}_{e,0} M_{c,0} \\
&- \sum_{i=1}^{m_p} \left[y_{p(i)} A_{p(i)} E_{p(i)} \left(\varepsilon_{p(i),\text{init}} - \varepsilon_{p.\text{rel}(i),k} \right) \right]
\end{aligned}
\tag{5.79}
$$

where the flexural rigidity $R_{I,k}$ of the cross-section calculated at time τ_k is given by:

$$
R_{I,k} = I_c \overline{E}_{e,k} + \sum_{i=1}^{m_s} y_{s(i)}^2 A_{s(i)} E_{s(i)} + \sum_{i=1}^{m_p} y_{p(i)}^2 A_{p(i)} E_{p(i)}
\tag{5.80}
$$

and $M_{c,0}$ is the moment resisted by the concrete component at time τ_0. From the instantaneous analysis:

$$
M_{c,0} = -\int_{A_c} y \sigma_{c,0}\, dA = -\int_{A_c} y E_{c,0} \left(\varepsilon_{r,0} - y \kappa_0 \right) dA = -B_c E_{c,0} \varepsilon_{r,0} + I_c E_{c,0} \kappa_0
\tag{5.81}
$$

After substituting Equations 5.76 and 5.79 into Equation 5.69, the equilibrium equations at time τ_k may be written in compact form as:

$$
r_{\text{ext},k} = D_k \varepsilon_k + f_{cr,k} - f_{cs,k} + f_{p,\text{init}} - f_{p.\text{rel},k}
\tag{5.82}
$$

where:

$$
\varepsilon_k = \begin{bmatrix} \varepsilon_{r,k} \\ \kappa_k \end{bmatrix}
\tag{5.83}
$$

and

$$\mathbf{D}_k = \begin{bmatrix} R_{A,k} & -R_{B,k} \\ -R_{B,k} & R_{I,k} \end{bmatrix}$$

(5.84)

The vector $\mathbf{f}_{cr,k}$ represents the unrestrained creep produced by the stress $\sigma_{c,0}$ resisted by the concrete at time τ_0 and is given by:

$$\mathbf{f}_{cr,k} = \bar{F}_{e,0} \begin{bmatrix} N_{c,0} \\ M_{c,0} \end{bmatrix} = \bar{F}_{e,0} E_{c,0} \begin{bmatrix} A_c \varepsilon_{r,0} - B_c \kappa_0 \\ -B_c \varepsilon_{r,0} + I_c \kappa_0 \end{bmatrix}$$

(5.85)

and $\bar{F}_{e,0}$ is given in Equation 5.58. The vector $\mathbf{f}_{cs,k}$ accounts for the uniform (unrestrained) shrinkage strain that develops in the concrete over the time period and is given by:

$$\mathbf{f}_{cs,k} = \begin{bmatrix} A_c \\ -B_c \end{bmatrix} \bar{E}_{e,k} \varepsilon_{cs,k}$$

(5.86)

The vector $\mathbf{f}_{p,init}$ in Equation 5.82 accounts for the initial prestress and the vector $\mathbf{f}_{p.rel,k}$ accounts for the resultant actions caused by the loss of prestress in the tendon due to relaxation (calculated in terms of P_i). These are given by:

$$\mathbf{f}_{p,init} = \sum_{i=1}^{m_p} \begin{bmatrix} P_{init(i)} \\ -y_{p(i)} P_{init(i)} \end{bmatrix}$$

(5.87)

and

$$\mathbf{f}_{p.rel,k} = \sum_{i=1}^{m_p} \begin{bmatrix} P_{i(i)} \varphi_{p(i)} \\ -y_{p(i)} P_{i(i)} \varphi_{p(i)} \end{bmatrix}$$

(5.88)

Equation 5.82 can be solved for ε_k as:

$$\varepsilon_k = \mathbf{D}_k^{-1} \left(\mathbf{r}_{ext,k} - \mathbf{f}_{cr,k} + \mathbf{f}_{cs,k} - \mathbf{f}_{p,init} + \mathbf{f}_{p.rel,k} \right) = \mathbf{F}_k \left(\mathbf{r}_{ext,k} - \mathbf{f}_{cr,k} + \mathbf{f}_{cs,k} - \mathbf{f}_{p,init} + \mathbf{f}_{p.rel,k} \right)$$

(5.89)

where:

$$\mathbf{F}_k = \frac{1}{R_{A,k} R_{I,k} - R_{B,k}^2} \begin{bmatrix} R_{I,k} & R_{B,k} \\ R_{B,k} & R_{A,k} \end{bmatrix}$$

(5.90)

The stress distribution at time τ_k can then be calculated as follows:

$$\sigma_{c,k} = \bar{E}_{e,k}\left(\varepsilon_k - \varepsilon_{cs,k}\right) + \bar{F}_{e,0}\sigma_{c,0} = \bar{E}_{e,k}\left\{\left[1-y\right]\varepsilon_k - \varepsilon_{cs,k}\right\} + \bar{F}_{e,0}\sigma_{c,0} \qquad (5.91)$$

$$\sigma_{s(i),k} = E_{s(i)}\varepsilon_k = E_{s(i)}\left[1-y_{s(i)}\right]\varepsilon_k \qquad (5.92)$$

$$\sigma_{p(i),k} = E_{p(i)}\left(\varepsilon_k + \varepsilon_{p(i),\text{init}} - \varepsilon_{p.\text{rel}(i),k}\right)$$

$$= E_{p(i)}\left[1-y_{p(i)}\right]\varepsilon_k + E_{p(i)}\varepsilon_{p(i),\text{init}} - E_{p(i)}\varepsilon_{p.\text{rel}(i),k} \qquad (5.93)$$

where at any point y from the reference axis $\varepsilon_k = \varepsilon_{r,k} - y\kappa_k = [1-y]\varepsilon_k$.

The cross-sectional rigidities, i.e. $R_{A,k}$, $R_{B,k}$ and $R_{I,k}$, required for the solution at time τ_k can also be calculated from the properties of the age-adjusted transformed section, obtained by transforming the bonded steel areas (reinforcement and tendons) into equivalent areas of the aged concrete at time τ_k, as follows:

$$R_{A,k} = \bar{A}_k\bar{E}_{e,k} \qquad (5.94)$$

$$R_{B,k} = \bar{B}_k\bar{E}_{e,k} \qquad (5.95)$$

$$R_{I,k} = \bar{I}_k\bar{E}_{e,k} \qquad (5.96)$$

where $\bar{E}_{e,k}$ is the age-adjusted effective modulus, $\bar{A}_k$ is the area of the age-adjusted transformed section, and $\bar{B}_k$ and $\bar{I}_k$ are the first and second moments of the area of the age-adjusted transformed section about the reference axis. For the determination of $\bar{A}_k$, $\bar{B}_k$ and $\bar{I}_k$, the areas of the bonded steel are transformed into equivalent areas of concrete by multiplying by the age-adjusted modular ratio $\bar{n}_{es(i),k} = E_{s(i)}/\bar{E}_{e,k}$ or $\bar{n}_{ep(i),k} = E_{p(i)}/\bar{E}_{e,k}$, as appropriate.

Based on Equations 5.94 through 5.96, the expression for F_k (in Equation 5.90) can be re-written as:

$$F_k = \frac{1}{\bar{E}_{e,k}(\bar{A}_k\bar{I}_k - \bar{B}_k^2)}\begin{bmatrix} \bar{I}_k & \bar{B}_k \\ \bar{B}_k & \bar{A}_k \end{bmatrix} \qquad (5.97)$$

The calculation of the time-dependent stresses and deformations using the above procedure is illustrated in Examples 5.5 and 5.6.

EXAMPLE 5.5

For the post-tensioned concrete cross-section shown in Figure 5.7, the strain and stress distributions at τ_0 were calculated in Example 5.3, immediately after the transfer of prestress and the application of an external bending moment $M_{ext,0} = 100$ kNm (see Figure 5.8). Soon after transfer, the post-tensioned duct was filled with grout, thereby bonding the tendon to the concrete and ensuring compatibility of concrete and steel strains at all times after τ_0. If the applied moment remains constant during the time interval τ_0 to τ_k (i.e. $M_{ext,k} = M_{ext,0}$), calculate the strain and stress distributions at time τ_k using the age-adjusted effective modulus method.

As in Example 5.3, $E_{c,0} = 30$ GPa; $E_s = E_p = 200$ GPa; $n_{s,0} = E_s/E_{c,0} = 6.67$; $n_{p,0} = E_p/E_{c,0} = 6.67$; $f_{pb} = 1870$ MPa and, with $P_i = 1350$ kN, the initial stress and strain in the tendon are $\sigma_{pi} = P_i/A_p = 1369$ MPa $= 0.73 f_{pb}$ and $\varepsilon_{pi} = \sigma_{pi}/E_p = 0.006846$, respectively. Take $\varphi_{cc}(\tau_k,\tau_0) = 2.5$, $\chi(\tau_k,\tau_0) = 0.65$, $\varepsilon_{cs}(\tau_k) = -600 \times 10^{-6}$, the relaxation loss $R = 4\%$ and, from Equation 4.31, $\varphi_p = 0.042$. (Note that with $\sigma_{pi} = 0.73 f_{pb}$, Table 4.10 gives the final relaxation loss for low-relaxation strand at about 4.0%).

From Example 5.3, the strain at the reference axis and the curvature at τ_0 are:

$$\varepsilon_{r,0} = -176.5 \times 10^{-6} \quad \text{and} \quad \kappa_0 = -0.3778 \times 10^{-6} \text{ mm}^{-1}$$

and the vector of external actions on the cross-section at τ_k (expressed in N and Nmm) is:

$$\mathbf{r}_{ext,k} = \begin{bmatrix} N_{ext,k} \\ M_{ext,k} \end{bmatrix} = \begin{bmatrix} 0 \\ 100 \times 10^6 \end{bmatrix}$$

From Equations 5.55:

$$\bar{E}_{e,k} = \frac{E_{c,0}}{1 + \chi(\tau_k,\tau_0)\varphi_{cc}(\tau_k,\tau_0)} = \frac{30,000}{1 + 0.65 \times 2.5} = 11,429 \text{ MPa and therefore}$$

$$\bar{n}_{es,k} = \bar{n}_{ep,k} = 17.5$$

From Equation 5.58:

$$\bar{F}_{e,0} = \frac{\varphi_{cc}(\tau_k,\tau_0)\left[\chi(\tau_k,\tau_0)-1\right]}{1 + \chi(\tau_k,\tau_0)\varphi_{cc}(\tau_k,\tau_0)} = \frac{2.5 \times (0.65 - 1.0)}{1.0 + 0.65 \times 2.5} = -0.333$$

With the duct fully-grouted, the properties of the concrete part of the section are:

$$A_c = bD - A_{s(1)} - A_{s(2)} - A_p = 300 \times 800 - 900 - 1800 - 986 = 236,314 \text{ mm}^2$$

$$B_c = bDy_c - A_{s(1)}y_{s(1)} - A_{s(2)}y_{s(2)} - A_p y_p$$

$$= 300 \times 800 \times 0 - 900 \times 340 - 1,800 \times (-340) - 986 \times (-200) = 503,200 \text{ mm}^3$$

$$I_c = bD^3/12 - A_{s(1)}y_{s(1)}^2 - A_{s(2)}y_{s(2)}^2 - A_p y_p^2$$

$$= 300 \times 800^3/12 - 900 \times 340^2 - 1800 \times (-340)^2 - 986 \times (-200)^2$$

$$= 12,448 \times 10^6 \text{ mm}^4$$

and the properties of the age-adjusted transformed cross-section are:

$$\bar{A}_k = bD + (\bar{n}_{es,k} - 1)A_{s(1)} + (\bar{n}_{es,k} - 1)A_{s(2)} + (\bar{n}_{ep,k} - 1)A_p$$

$$= 300 \times 800 + (17.5 - 1) \times 900 + (17.5 - 1) \times 1800 + (17.5 - 1) \times 986$$

$$= 300,819 \text{ mm}^2$$

$$\bar{B}_k = bDy_c + (\bar{n}_{es,k} - 1)A_{s(1)}y_{s(1)} + (\bar{n}_{es,k} - 1)A_{s(2)}y_{s(2)} + (\bar{n}_{ep,k} - 1)A_p y_p$$

$$= 300 \times 800 \times 0 + (17.5 - 1) \times [900 \times (+340) + 1800 \times (-340)]$$

$$+ (17.5 - 1) \times 986 \times (-200)$$

$$= -8.303 \times 10^6 \text{ mm}^3$$

$$\bar{I}_k = \frac{bD^3}{12} + (\bar{n}_{es,k} - 1)[A_{s(1)}y_{s(1)}^2 + A_{s(2)}y_{s(2)}^2] + (\bar{n}_{ep,k} - 1)A_p y_p^2$$

$$= \frac{300 \times 800^3}{12} + (17.5 - 1)[900 \times 340^2 + 1800 \times (-340)^2]$$

$$+ (17.5 - 1) \times 986 \times (-200)^2$$

$$= 18,600 \times 10^6 \text{ mm}^4$$

From Equation 5.85:

$$\mathbf{f}_{cr,k} = \bar{F}_{e,0} E_{c,0} \begin{bmatrix} A_c \varepsilon_{r,0} - B_c \kappa_0 \\ -B_c \varepsilon_{r,0} + I_c \kappa_0 \end{bmatrix}$$

$$= -0.333 \times 30,000 \begin{bmatrix} 236314 \times (-176.5 \times 10^{-6}) - 503200 \times (-0.3778 \times 10^{-6}) \\ -503200 \times (-176.5 \times 10^{-6}) + 12448 \times 10^6 \times (-0.3778 \times 10^{-6}) \end{bmatrix}$$

$$= \begin{bmatrix} +415.2 \times 10^3 \text{ N} \\ +46.14 \times 10^6 \text{ Nmm} \end{bmatrix}$$

and from Equation 5.86:

$$\mathbf{f}_{cs,k} = \begin{bmatrix} A_c \\ -B_c \end{bmatrix} \bar{E}_{e,k} \varepsilon_{cs,k} = \begin{bmatrix} 236329 \times 11429 \times \left(-600 \times 10^{-6}\right) \\ -503200 \times 11429 \times \left(-600 \times 10^{-6}\right) \end{bmatrix}$$

$$= \begin{bmatrix} -1621 \times 10^3 \quad N \\ 3.45 \times 10^6 \quad Nmm \end{bmatrix}$$

The relaxation strain in the prestressing tendon is:

$$\varepsilon_{p.rel,k} = \varepsilon_{pi} \; \varphi_p = 0.006846 \times 0.042 = 0.0002875$$

and the vectors of internal actions caused by the initial prestress and by relaxation are given by Equations 5.87 and 5.88, respectively:

$$\mathbf{f}_{p,init} = \begin{bmatrix} P_{init} \\ -P_{init} y_p \end{bmatrix} = \begin{bmatrix} 1350 \times 10^3 \; N \\ 270 \times 10^6 \; Nmm \end{bmatrix}$$

$$\mathbf{f}_{p.rel,k} = \begin{bmatrix} P_i \varphi_p \\ -y_p P_i \varphi_p \end{bmatrix} = \begin{bmatrix} 1350 \times 10^3 \times 0.042 \\ 270 \times 10^6 \times 0.042 \end{bmatrix} = \begin{bmatrix} 56.7 \times 10^3 \; N \\ 11.34 \times 10^6 \; Nmm \end{bmatrix}$$

Note that for this post-tensioned cross-section $P_i = P_{init}$.
Equation 5.97 gives:

$$\mathbf{F}_k = \frac{1}{\bar{E}_{e,k} (\bar{A}_k \bar{I}_k - \bar{B}_k^2)} \begin{bmatrix} \bar{I}_k & \bar{B}_k \\ \bar{B}_k & \bar{A}_k \end{bmatrix}$$

$$= \frac{1}{11429 \times (300,819 \times 18,600 \times 10^6 - (-8.303 \times 10^6)^2)}$$

$$\begin{bmatrix} 18,600 \times 10^6 & -8.303 \times 10^6 \\ -8.303 \times 10^6 & 300,819 \end{bmatrix}$$

$$= \begin{bmatrix} 294.5 \times 10^{-12} \, N^{-1} & -131.5 \times 10^{-15} \, N^{-1} mm^{-1} \\ -131.5 \times 10^{-15} \, N^{-1} mm^{-1} & 4.763 \times 10^{-15} \, N^{-1} mm^{-2} \end{bmatrix}$$

and the strain vector ε_k at time τ_k is determined using Equation 5.89:

$$\varepsilon_k = F_k \left(r_{ext,k} - f_{cr,k} + f_{cs,k} - f_{p,init} + f_{p,rel,k} \right)$$

$$= \begin{bmatrix} 294.5 \times 10^{-12} & -131.5 \times 10^{-15} \\ -131.5 \times 10^{-15} & 4.763 \times 10^{-15} \end{bmatrix} \begin{bmatrix} (0 - 415.2 - 1621 - 1350 + 56.7) \times 10^3 \\ (100 - 46.14 + 3.45 - 270 + 11.34) \times 10^6 \end{bmatrix}$$

$$= \begin{bmatrix} -953.9 \times 10^{-6} \\ -0.5214 \times 10^{-6} \, \text{mm}^{-1} \end{bmatrix}$$

The strain at the reference axis and the curvature at time τ_k are, respectively:

$$\varepsilon_{r,k} = -953.9 \times 10^{-6} \quad \text{and} \quad \kappa_k = -0.5214 \times 10^{-6} \, \text{mm}^{-1}$$

From Equation 5.61, the top ($y = +400$ mm) and bottom ($y = -400$ mm) fibre strains are:

$$\varepsilon_{k(top)} = \varepsilon_{r,k} - 400 \times \kappa_k = [-953.9 - 400 \times (-0.5214)] \times 10^{-6}$$

$$= -745.4 \times 10^{-6}$$

$$\varepsilon_{k(btm)} = \varepsilon_{r,k} - (-400) \times \kappa_k = [-953.9 + 400 \times (-0.5214)] \times 10^{-6}$$

$$= -1162 \times 10^{-6}$$

The concrete stress distribution at time τ_k is calculated using Equation 5.91:

$$\sigma_{c,k(top)} = \overline{E}_{e,k} \left(\varepsilon_{k(top)} - \varepsilon_{cs,k} \right) + \overline{F}_{e,0} \sigma_{c,0(top)}$$

$$= 11,429 \times [-745.4 - (-600)] \times 10^{-6} + (-0.333) \times (-0.762)$$

$$= -1.41 \, \text{MPa}$$

$$\sigma_{c,k(btm)} = \overline{E}_{e,k} \left(\varepsilon_{k(btm)} - \varepsilon_{cs,k} \right) + \overline{F}_{e,0} \sigma_{c,0(top)}$$

$$= 11,429 \times [-1162 - (-600)] \times 10^{-6} + (-0.333) \times (-9.828)$$

$$= -3.15 \, \text{MPa}$$

and, from Equation 5.92, the stresses in the non-prestressed reinforcement are:

$$\sigma_{s(1),k} = E_{s(1)} \left[1 - y_{s(1)} \right] \varepsilon_k = 200 \times 10^3 \left[1 - 340 \right] \begin{bmatrix} -953.9 \times 10^{-6} \\ -0.5214 \times 10^{-6} \end{bmatrix}$$

$$= -155 \, \text{MPa}$$

$$\sigma_{s(2),k} = E_s \Big[1 - y_{s(2)} \Big] \varepsilon_k = 200 \times 10^3 \Big[1 - (-340) \Big] \begin{bmatrix} -953.9 \times 10^{-6} \\ -0.5214 \times 10^{-6} \end{bmatrix}$$

$$= -226 \, \text{MPa}$$

The final stress in the prestressing steel at time τ_k is given by Equation 5.93:

$$\sigma_{p,k} = E_p \Big[1 - y_p \Big] \varepsilon_k + E_p \big(\varepsilon_{p,init} - \varepsilon_{p,rel,k} \big)$$

$$= 200 \times 10^3 \Big[1 - (-200) \Big] \begin{bmatrix} -953.9 \times 10^{-6} \\ -0.5214 \times 10^{-6} \end{bmatrix}$$

$$+ 200 \times 10^3 \times (0.006846 - 0.0002875)$$

$$= +1100 \, \text{MPa}$$

The stress and strain distributions at τ_0 (from Example 5.3) and at τ_k are shown in Figure 5.14.

Note that the time-dependent loss of prestress in the tendon is 19.7%, but the time-dependent loss in the compressive stress in the concrete at the bottom fibre is 67.9%. With time, much of the compressive force exerted by the tendon on the cross-section has been transferred from the concrete to the bonded steel reinforcement as a direct result of creep and shrinkage in the concrete. The very significant change in stress in the non-prestressed reinforcement with time is typical of the behaviour of uncracked regions of reinforced and prestressed concrete members in compression.

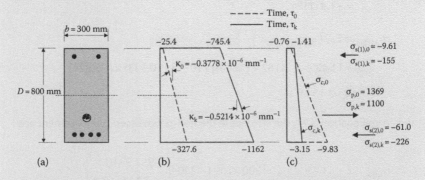

Figure 5.14 Strains and stresses at times at τ_0 and τ_k (Example 5.5). (a) Section. (b) Strain ($\times 10^{-6}$). (c) Stress (MPa).

EXAMPLE 5.6

A time-dependent analysis of the pretensioned concrete cross-section of Example 5.4 (see Figure 5.9) is to be undertaken using the age-adjusted effective modulus method. The strain and stress distributions at τ_0 immediately after the application of an axial force $N_{ext,0} = -100$ kN and a bending moment of $M_{ext,0} = 1000$ kNm were calculated in Example 5.4 and shown in Figure 5.10. If the applied actions remain constant during the time interval τ_0 to τ_k, determine the strain and stress distributions at time τ_k.

As in Example 5.4, $E_{c,0} = 32$ GPa, $E_s = E_p = 200$ GPa and, therefore, $n_{s(i),0} = n_{p(i),0} = 6.25$. For the time interval τ_0 to τ_k, the creep and shrinkage input for the concrete is $\varphi_{cc}(\tau_k,\tau_0) = 2.0$, $\chi(\tau_k,\tau_0) = 0.65$, $\varepsilon_{cs}(\tau_k) = -400 \times 10^{-6}$ and, for the prestressing steel, $\varphi_{p(1)} = \varphi_{p(2)} = \varphi_{p(3)} = 0.03$.

From Example 5.4, the strain at the reference axis and the curvature at τ_0 are:

$$\varepsilon_{r,0} = -254.1 \times 10^{-6} \quad \text{and} \quad \kappa_0 = +0.1918 \times 10^{-6} \text{ mm}^{-1}.$$

and the vector of external actions on the cross-section at τ_k is:

$$r_{ext,k} = \begin{bmatrix} N_{ext,k} \\ M_{ext,k} \end{bmatrix} = \begin{bmatrix} -100 \times 10^3 \\ 1000 \times 10^6 \end{bmatrix}$$

From Equations 5.55:

$$\bar{E}_{e,k} = \frac{E_{c,0}}{1 + \chi(\tau_k,\tau_0)\varphi_{cc}(\tau_k,\tau_0)} = \frac{32000}{1 + 0.65 \times 2.0}$$

$$= 13{,}910 \text{ MPa from which}: \bar{n}_{es,k} = \bar{n}_{ep,k} = 14.37$$

From Equation 5.58:

$$\bar{F}_{e,0} = \frac{\varphi_{cc}(\tau_k,\tau_0)\left[\chi(\tau_k,\tau_0) - 1\right]}{1 + \chi(\tau_k,\tau_0)\varphi_{cc}(\tau_k,\tau_0)} = \frac{2.0 \times (0.65 - 1.0)}{1.0 + 0.65 \times 2.0} = -0.304$$

The properties of the concrete part of the cross-section are:

$$A_c = A_{gross} - A_{s(1)} - A_{s(2)} - A_{p(1)} - A_{p(2)} - A_{p(3)} = 312{,}700 \text{ mm}^2$$

$$B_c = A_{gross}(d_c - d_{ref}) - \left(A_{s(1)}y_{s(1)} + A_{s(2)}y_{s(2)}\right) - \left(A_{p(1)}y_{p(1)} + A_{p(2)}y_{p(2)} + A_{p(3)}y_{p(3)}\right)$$

$$= -93.46 \times 10^6 \text{ mm}^3$$

$$I_c = I_{gross} + A_{gross}(d_c - d_{ref})^2 - (A_{s(1)}y_{s(1)}^2 + A_{s(2)}y_{s(2)}^2)$$

$$- (A_{p(1)}y_{p(1)}^2 + A_{p(2)}y_{p(2)}^2 + A_{p(3)}y_{p(3)}^2)$$

$$= 76,920 \times 10^6 \, mm^4$$

and the properties of the age-adjusted transformed section in equivalent concrete areas are:

$$\bar{A}_k = A_{gross} + \bar{n}_{es,k} - 1)(A_{s(1)} + A_{s(2)}) + (\bar{n}_{ep,k} - 1)(A_{p(1)} + A_{p(2)} + A_{p(3)})$$

$$= 374.5 \times 10^3 \, mm^2$$

$$\bar{B}_k = A_{gross}(d_c - d_{ref}) + (\bar{n}_{es,k} - 1)(A_{s(1)}y_{s(1)} + A_{s(2)}y_{s(2)})$$

$$+ (\bar{n}_{ep,k} - 1)(A_{p(1)}y_{p(1)} + A_{p(2)}y_{p(2)} + A_{p(3)}y_{p(3)})$$

$$= -126.1 \times 10^6 \, mm^3$$

$$\bar{I}_k = I_{gross} + A_{gross}(d_c - d_{ref})^2 + (\bar{n}_{es,k} - 1)(A_{s(1)}y_{s(1)}^2 + A_{s(2)}y_{s(2)}^2)$$

$$+ (\bar{n}_{ep,k} - 1)(A_{p(1)}y_{p(1)}^2 + A_{p(2)}y_{p(2)}^2 + A_{p(3)}y_{p(3)}^2)$$

$$= 104,060 \times 10^6 \, mm^4$$

From Equation 5.85:

$$\mathbf{f}_{cr,k} = \begin{bmatrix} +599.1 \times 10^3 \, N \\ +87.55 \times 10^6 \, Nmm \end{bmatrix}$$

and from Equation 5.86:

$$\mathbf{f}_{cs,k} = \begin{bmatrix} -1740 \times 10^3 \, N \\ -520.1 \times 10^6 \, Nmm \end{bmatrix}$$

The relaxation strains in the prestressing tendons are:

$$\varepsilon_{p(1).rel,k} = \varepsilon_{p(1),0} \, \varphi_{p(1)} = 0.0001832$$

$$\varepsilon_{p(2).rel,k} = \varepsilon_{p(2),0} \, \varphi_{p(2)} = 0.0001836$$

$$\varepsilon_{p(3).rel,k} = \varepsilon_{p(3),0} \, \varphi_{p(3)} = 0.0001840$$

and the vectors of internal actions caused by the initial prestress and by relaxation are given by Equations 5.87 and 5.88, respectively:

$$\mathbf{f}_{p,init} = \sum_{i=1}^{3} \begin{bmatrix} P_{init(i)} \\ -y_{p(i)}P_{init(i)} \end{bmatrix} = \begin{bmatrix} 2000 \times 10^3 \text{ N} \\ 1330.6 \times 10^6 \text{ Nmm} \end{bmatrix}$$

$$\mathbf{f}_{p.rel,k} = \sum_{i=1}^{3} \begin{bmatrix} P_{i(i)}\varphi_{p(i)} \\ -y_{p(i)}P_{i(i)}\varphi_{p(i)} \end{bmatrix} = \begin{bmatrix} 58.79 \times 10^3 \text{ N} \\ 39.12 \times 10^6 \text{ Nmm} \end{bmatrix}$$

Note that in this pretensioned member $P_{i(i)}$ is less than $P_{init(i)}$ due to elastic shortening losses.

Equation 5.97 gives:

$$\mathbf{F}_k = \frac{1}{\overline{E}_{e,k}(\overline{A}_k \overline{I}_k - \overline{B}_k^2)} \begin{bmatrix} \overline{I}_k & \overline{B}_k \\ \overline{B}_k & \overline{A}_k \end{bmatrix}$$

$$= \begin{bmatrix} 324.2 \times 10^{-12} \text{ N}^{-1} & -392.9 \times 10^{-15} \text{ N}^{-1}\text{mm}^{-1} \\ -392.9 \times 10^{-15} \text{ N}^{-1}\text{mm}^{-1} & 1.167 \times 10^{-15} \text{ N}^{-1}\text{mm}^{-2} \end{bmatrix}$$

The strain ε_k at time τ_k is determined using Equation 5.89:

$$\varepsilon_k = \mathbf{F}_k \left(\mathbf{r}_{ext,k} - \mathbf{f}_{cr,k} + \mathbf{f}_{cs,k} - \mathbf{f}_{p,init} + \mathbf{f}_{p.rel,k} \right)$$

$$= \begin{bmatrix} 324.2 \times 10^{-12} & -392.9 \times 10^{-15} \\ -392.9 \times 10^{-15} & 1.167 \times 10^{-15} \end{bmatrix} \begin{bmatrix} (-100 - 599.1 - 1740 - 2000 + 58.79) \times 10^3 \\ (1000 - 87.55 - 520.1 - 1330.6 + 39.12) \times 10^6 \end{bmatrix}$$

$$= \begin{bmatrix} -1066.9 \times 10^{-6} \\ +0.6719 \times 10^{-6} \text{ mm}^{-1} \end{bmatrix}$$

The strain at the reference axis and the curvature at time τ_k are respectively:

$$\varepsilon_{r,k} = -1066.9 \times 10^{-6} \quad \text{and} \quad \kappa_k = +0.6719 \times 10^{-6} \text{ mm}^{-1}$$

From Equation 5.61, the top ($y = +300$ mm) and bottom ($y = -850$ mm) fibre strains are:

$$\varepsilon_{k(top)} = \varepsilon_{r,k} - 300 \times \kappa_k = -1269 \times 10^{-6}$$

$$\varepsilon_{k(btm)} = \varepsilon_{r,k} + 850 \times \kappa_k = -495.8 \times 10^{-6}$$

The concrete stress distribution at time τ_k is calculated using Equation 5.91:

$$\sigma_{c,k(top)} = \overline{E}_{e,k}\left(\varepsilon_{k(top)} - \varepsilon_{cs,k}\right) + \overline{F}_{e,0}\sigma_{c,0(top)} = -9.05\,\text{MPa}$$

$$\sigma_{c,k(btm)} = \overline{E}_{e,k}\left(\varepsilon_{k(btm)} - \varepsilon_{cs,k}\right) + \overline{F}_{e,0}\sigma_{c,0(top)} = -0.45\,\text{MPa}$$

and, from Equation 5.92, the stresses in the non-prestressed reinforcement are:

$$\sigma_{s(1),k} = E_s\left[1 \quad -y_{s(1)}\right]\varepsilon_k = -245.6\,\text{MPa}$$

$$\sigma_{s(2),k} = E_s\left[1 \quad -y_{s(2)}\right]\varepsilon_k = -107.2\,\text{MPa}$$

The final stress in the prestressing steel at time τ_k is given by Equation 5.93:

$$\sigma_{p(1),k} = E_p\left[1 \quad -y_{p(1)}\right]\varepsilon_k + E_p(\varepsilon_{p(1),init} - \varepsilon_{p(1).rel,k}) = 1{,}077.9\,\text{MPa}$$

$$\sigma_{p(2),k} = E_p\left[1 \quad -y_{p(2)}\right]\varepsilon_k + E_p(\varepsilon_{p(2),init} - \varepsilon_{p(2).rel,k}) = 1{,}086.6\,\text{MPa}$$

$$\sigma_{p(3),k} = E_p\left[1 \quad -y_{p(3)}\right]\varepsilon_k + E_p(\varepsilon_{p(3),init} - \varepsilon_{p(3).rel,k}) = 1{,}095.2\,\text{MPa}$$

The stress and strain distributions at τ_0 (obtained in Example 5.4) and at τ_k are shown in Figure 5.15.

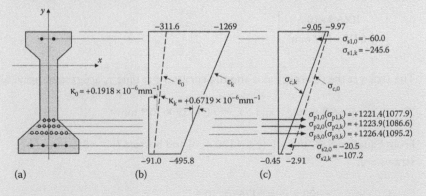

Figure 5.15 Strains and stresses at times at τ_0 and τ_k (Example 5.6). (a) Cross-section. (b) Strain ($\times 10^{-6}$). (c) Stress (MPa).

5.7.4 Discussion

The results of several time analyses on the cross-section shown in Figure 5.16 are presented in Tables 5.1 and 5.2. The geometry of the cross-section and the material properties are similar to the cross-section of Examples 5.3 and 5.5. The effects of varying the quantities of the compressive and tensile non-prestressed reinforcement ($A_{s(1)}$ and $A_{s(2)}$, respectively) on the time-dependent deformation can be seen for three different values of sustained bending moment. At M_{ext} = 100 kNm, the initial concrete stress distribution is approximately triangular with higher compressive stresses in the bottom fibres (as determined in Example 5.3). At M_{ext} = 270 kNm, the initial concrete stress distribution is approximately uniform over the depth of the section and the curvature is small. At M_{ext} = 440 kNm, the initial stress distribution is again triangular with high compressive stresses in the top fibres. In each case, the prestressing force in the tendon immediately after transfer is P_i = 1350 kN and therefore, σ_{pi} = 1369 MPa.

In Tables 5.1 and 5.2, $\Delta F_{s(1)}$, $\Delta F_{s(2)}$ and ΔF_p are the compressive changes of force that gradually occur in the non-prestressed steel and the tendons with time. To maintain equilibrium, equal and opposite tensile forces are gradually imposed on the concrete as the bonded reinforcement restrains the time-dependent creep and shrinkage strains in the concrete. In Section 5.4.1, we defined the final compressive force acting on the concrete as ΩP_i and here:

$$\Omega = 1 + \frac{\Delta F_{s(1)} + \Delta F_{s(2)} + \Delta F_p}{P_i} \qquad (5.98)$$

remembering that $\Delta F_{s(1)}$, $\Delta F_{s(2)}$ and ΔF_p are all negative. Values of Ω for each analysis are also given in Tables 5.1 and 5.2.

From the results in Table 5.1, the effect of increasing the quantity of non-prestressed tensile reinforcement $A_{s(2)}$ is to increase the change in positive or sagging curvature with time. The increase is most pronounced when the

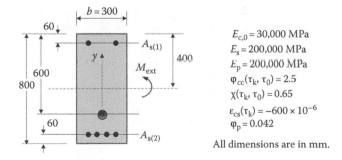

Figure 5.16 Post-tensioned cross-section with tendon bonded after transfer.

Table 5.1 Effect of varying the bottom steel $A_{s(2)}$ (with $A_{s(1)} = 0$)

M_{ext} (kNm)	$A_{s(2)}$ (mm²)	$\varepsilon_{r,0}$ (×10⁻⁶)	κ_0 (×10⁻⁶) mm⁻¹	$\Delta F_{s(1)}$ (kN)	$\Delta F_{s(2)}$ (kN)	ΔF_p (kN)	Ω (Equation 5.98)	$\varepsilon_{r,k}$ (×10⁻⁶)	κ_k (×10⁻⁶) mm⁻¹
100	0	−191	−0.455	0	0	−326	0.759	−1142	−1.122
	1800	−178	−0.372	0	−282	−265	0.595	−1011	−0.228
	3600	−167	−0.306	0	−408	−230	0.527	−935	+0.285
270	0	−190	−0.008	0	0	−271	0.799	−1161	+0.359
	1800	−182	+0.038	0	−206	−230	0.677	−1071	+0.967
	3600	−177	+0.075	0	−302	−206	0.624	−1019	+1.318
440	0	−189	+0.438	0	0	−217	0.839	−1179	+1.840
	1800	−187	+0.448	0	−130	−195	0.759	−1131	+2.162
	3600	−186	+0.455	0	−197	−182	0.719	−1104	+2.350

Table 5.2 Effect of varying the top steel $A_{s(1)}$ (with $A_{s(2)}$ = 1800 mm²)

M_{ext} (kNm)	$A_{s(1)}$ (mm²)	$\varepsilon_{r,0}$ (×10⁻⁶)	κ_0 (×10⁻⁶) mm⁻¹	$\Delta F_{s(1)}$ (kN)	$\Delta F_{s(2)}$ (kN)	ΔF_p (kN)	Ω (Equation 5.98)	$\varepsilon_{r,k}$ (×10⁻⁶)	κ_k (×10⁻⁶) mm⁻¹
100	0	−178	−0.372	0	−282	−265	0.595	−1011	−0.228
	900	−176	−0.378	−131	−297	−265	0.487	−954	−0.521
	1800	−176	−0.383	−223	−308	−266	0.410	−912	−0.734
270	0	−182	+0.038	0	−206	−230	0.677	−1071	+0.967
	900	−178	+0.014	−176	−228	−230	0.530	−984	+0.522
	1800	−175	−0.007	−294	−244	−231	0.430	−922	+0.200
440	0	−187	+0.448	0	−130	−195	0.759	−1131	+2.162
	900	−180	+0.406	−221	−159	−195	0.574	−1015	+1.565
	1800	−174	+0.370	−366	−179	−196	0.451	−931	+1.134

initial concrete compressive stress at the level of the steel is high, i.e. when the sustained moment is relatively small and the section is initially subjected to a negative or hogging curvature. When $A_{s(2)}= 3600$ mm^2 and $M_{ext} = 100$ kNm (row 3 in Table 5.1), κ_k is positive despite the significant initial negative curvature κ_0. Table 5.1 also indicates that the addition of non-prestressed steel in the tensile zone will reduce the time-dependent camber, which often causes problems in precast members subjected to low sustained loads. For sections on which M_{ext} is sufficient to cause an initial positive curvature, such as when $M_{ext} = 440$ kNm in Table 5.1, an increase in $A_{s(2)}$ causes an increase in time-dependent curvature and hence an increase in final deflection.

The inclusion of non-prestressed steel in the top of the section $A_{s(1)}$ increases the change in negative curvature with time, as indicated in Table 5.2. For sections where the initial curvature is positive, such as when $M_{ext} = 440$ kNm, the inclusion of $A_{s(1)}$ reduces the time-dependent change in positive curvature (and hence the time-dependent deflection of the member). However, when κ_0 is negative, such as when $M_{ext} = 100$ kNm, the inclusion $A_{s(1)}$ can cause an increase in negative curvature with time and hence an increase in upward camber of the member with time.

The significant unloading of the concrete with time on the sections containing non-prestressed reinforcement should be noted. For example, when $M_{ext} = 270$ kNm and $A_{s(1)} = A_{s(2)} = 1800$ mm^2 (i.e. equal quantities of top and bottom non-prestressed reinforcement), the concrete is subjected to a total gradually applied tensile force of $-(\Delta F_{s(1)} + \Delta F_{s(2)} + \Delta F_p) = 769$ kN as shown in Table 5.2. This means that 57% of the initial compression in the concrete (the initial prestressing force) is transferred into the bonded reinforcement with time. The bottom fibre concrete compressive stress reduces from -5.32 to -0.99 MPa. The loss of prestress in the tendon, however, is only 231 kN (17.1%).

It is evident that an accurate picture of the time-dependent behaviour of a prestressed concrete cross-section cannot be obtained unless the restraint provided to creep and shrinkage by the non-prestressed steel is adequately accounted for. It is also evident that the presence of non-prestressed reinforcement significantly reduces the cracking moment with time and may in fact relieve the concrete of much of its initial compression.

5.8 SHORT-TERM ANALYSIS OF CRACKED CROSS-SECTIONS

5.8.1 General

In the cross-sectional analyses in Sections 5.6 and 5.7, it was assumed that concrete can carry the imposed stresses, both compressive and tensile. However, concrete is not able to carry large tensile stresses. If the tensile stress at a point reaches the tensile strength of concrete (Equations 4.3 and 4.4), cracking occurs. On a cracked cross-section, it is reasonable to assume that tensile stress of any magnitude cannot be carried normal to the crack surface at any time after cracking and tensile forces can only be carried across a crack by steel reinforcement. Therefore, on a cracked cross-section, internal actions can be carried only by the steel reinforcement (and tendons) and the uncracked parts of the concrete section.

In members subjected only to axial tension, caused either by external loads or by restraint to shrinkage or temperature change, *full-depth* cracks occur when the tensile stress reaches the tensile strength of the concrete at a particular location (i.e. at each crack location, the entire cross-section is cracked). When the axial tension is caused by restraint to shrinkage, cracking causes a loss of stiffness and a consequent decrease in the internal tension. The crack width and the magnitude of the restraining force, as well as the spacing between cracks, depend on the amount of bonded reinforcement. The steel carries the entire tensile force across each crack, but between the cracks in a member subjected to axial tension, the concrete continues to carry tensile stress due to the bond between the steel and the concrete, and hence the tensile concrete between the cracks continues to contribute to the member stiffness. This is known as the *tension stiffening effect*.

In a flexural member, cracking occurs when the tensile stress produced by the external moment at a particular section overcomes the compression caused by prestress, and the extreme fibre stress reaches the tensile strength of concrete. *Primary cracks* develop at a reasonably regular spacing on the tensile side of the member. The bending moment at which cracking first occurs is the *cracking moment* M_{cr}. If the applied moment at any time is greater than the cracking moment, cracking will occur and, at each crack, the concrete below the neutral axis on the cracked section is ineffective. In the previous section, we saw that the initial compressive stress in the concrete due to prestress is gradually relieved by creep and shrinkage and so the cracking moment decreases with time.

A loss of stiffness occurs at first cracking and the short-term moment-curvature relationship becomes nonlinear. The height of primary cracks h_o depends on the quantity of tensile reinforcement and the magnitude of any axial force or prestress. For reinforced concrete members in pure bending with no axial force, the height of the primary cracks h_o immediately after cracking is usually relatively high (0.6 to 0.9 times the depth of the member depending on the quantity and position of tensile steel) and remains approximately constant under increasing bending moments until either the steel reinforcement yields or the concrete stress-strain relationship in the compressive region becomes nonlinear. For prestressed members and members subjected to bending plus axial compression, h_o may be relatively small initially and gradually increases as the applied moment increases.

Immediately after first cracking, the intact concrete between adjacent primary cracks carries considerable tensile force, mainly in the direction of the reinforcement, due to the bond between the steel and the concrete. The average tensile stress in the concrete may be a significant percentage of the tensile strength of concrete. The steel stress is a maximum at a crack, where the steel carries the entire tensile force, and drops to a minimum between the cracks. The bending stiffness of the member is considerably greater than that based on a fully-cracked section, where concrete in tension is assumed to carry zero stress. This *tension stiffening* effect is particularly significant in lightly reinforced concrete slabs under service loads.

For prestressed concrete members, or reinforced members in combined bending and compression, the effect of tension stiffening is less pronounced because the loss of stiffness caused by cracking is less significant. As the applied moment increases, the depth of the primary cracks increases gradually (in contrast to the sudden crack propagation in a reinforced member in pure bending) and the depth of the concrete compressive zone is significantly greater than would be the case if no axial prestress was present.

5.8.2 Assumptions

The Euler-Bernoulli assumption that plane sections remain plane is not strictly true for a cross-section in the cracked region of a beam. However, if strains are measured over a gauge length containing several primary cracks, the *average* strain diagram may be assumed to be linear over the depth of a cracked cross-section.

The analysis presented here is based on the following assumptions:

1. Plane sections remain plane and, as a consequence, the strain distribution is linear over the depth of the section.
2. Perfect bond exists between the non-prestressed steel reinforcement and the concrete, and between the bonded tendons and the concrete, i.e. the bonded steel and concrete strains are assumed to be compatible. This is usually a reasonable assumption at service loads in members containing deformed steel reinforcing bars and strands.
3. Strain in unbonded tendons is assumed to be unaffected by deformation of the concrete cross-section.
4. Tensile stress in the concrete is ignored, and therefore, the tensile concrete does not contribute to the cross-sectional properties.
5. Material behaviour is linear-elastic. This includes concrete in compression, and both the non-prestressed and prestressed reinforcement.

5.8.3 Analysis

In the short-term analysis of fully-cracked prestressed concrete cross-sections at first loading (at time τ_0), it is assumed that the axial force and bending moment about the x-axis ($N_{ext,0}$ and $M_{ext,0}$, respectively) produce tension of sufficient magnitude to cause cracking in the *bottom* fibres of the cross-section and compression at the top of the section.

Consider the cracked prestressed concrete cross-section shown in Figure 5.17. The section is symmetric about the y-axis and the orthogonal x-axis

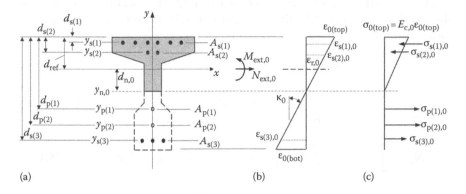

Figure 5.17 Fully-cracked prestressed concrete cross-section. (a) Cross-section. (b) Strain diagram. (c) Stress.

is selected as the reference axis. Also shown in Figure 5.17 are the initial stress and strain distributions when the section is subjected to combined external bending and axial force ($M_{ext.0}$ and $N_{ext.0}$) sufficient to cause cracking in the bottom fibres.

As for the analysis of an uncracked cross-section, the properties of each layer of non-prestressed reinforcement are defined by its area, elastic modulus and location with respect to the arbitrarily chosen x-axis, i.e. $A_{s(i)}$, $E_{s(i)}$ and $y_{s(i)}$ ($=d_{ref} - d_{st(i)}$), respectively. Similarly, $A_{p(i)}$, $E_{p(i)}$ and $y_{p(i)}$ ($=d_{ref} - d_{p(i)}$) represent the area, elastic modulus and location of the prestressing steel with respect to the x-axis, respectively.

The strain at any distance y from the reference x-axis at time τ_0 is given by:

$$\varepsilon_0 = \varepsilon_{r,0} - y\kappa_0 \tag{5.99}$$

and the stresses in the concrete and in the bonded reinforcement are:

$$\sigma_{c,0} = E_{c,0}\varepsilon_0 = E_{c,0}(\varepsilon_{r,0} - y\kappa_0) \quad \text{for } y \geq y_{n,0}(=-d_{n,0}) \tag{5.100}$$

$$\sigma_{c,0} = 0 \quad \text{for } y < y_{n,0} \tag{5.101}$$

$$\sigma_{s(i),0} = E_{s(i)}\varepsilon_0 = E_{s(i)}(\varepsilon_{r,0} - y_{s(i)}\kappa_0) \tag{5.102}$$

$$\sigma_{p(i),0} = E_{p(i)}\left(\varepsilon_0 + \varepsilon_{p(i),init}\right) = E_{p(i)}\left(\varepsilon_{r,0} - y_{p(i)}\kappa_0 + \varepsilon_{p(i),init}\right) \tag{5.103}$$

where $y_{n,0}$ is the y-coordinate of the neutral axis, as shown in Figure 5.17a, and $\varepsilon_{p(i),init}$ is the strain in the i-th layer of prestressing steel immediately before the transfer of prestress to the concrete as expressed in Equation 5.25.

The internal axial force $N_{\text{int},0}$ on the cracked cross-section is the sum of the axial forces resisted by the various materials forming the cross-section and is given by:

$$N_{\text{int},0} = N_{c,0} + N_{s,0} + N_{p,0} \tag{5.104}$$

where $N_{c,0}$, $N_{s,0}$ and $N_{p,0}$ are as previously given for an uncracked section in Equations 5.27 through 5.30 and re-expressed here as:

$$N_{c,0} = \int_{A_c} \sigma_{c,0}\,\mathrm{d}A = \int_{A_c} E_{c,0}(\varepsilon_{r,0} - y\kappa_0)\,\mathrm{d}A = A_c E_{c,0}\varepsilon_{r,0} - B_c E_{c,0}\kappa_0 \tag{5.105}$$

$$N_{s,0} = R_{A,s}\varepsilon_{r,0} - R_{B,s}\kappa_0 \tag{5.106}$$

$$N_{p,0} = R_{A,p}\varepsilon_{r,0} - R_{B,p}\kappa_0 + \sum_{i=1}^{m_p}\left(A_{p(i)}E_{p(i)}\varepsilon_{p(i),\text{init}}\right) \tag{5.107}$$

where A_c and B_c are the area and the first moment of area about the x-axis of the compressive concrete above the neutral axis (i.e. the section properties of the intact compressive concrete). The steel rigidities are:

$$R_{A,s} = \sum_{i=1}^{m_s}\left(A_{s(i)}E_{s(i)}\right) \tag{5.108}$$

$$R_{B,s} = \sum_{i=1}^{m_s}\left(y_{s(i)}A_{s(i)}E_{s(i)}\right) \tag{5.109}$$

$$R_{I,s} = \sum_{i=1}^{m_s}\left(y_{s(i)}^2 A_{s(i)}E_{s(i)}\right) \tag{5.110}$$

$$R_{A,p} = \sum_{i=1}^{m_p}\left(A_{p(i)}E_{p(i)}\right) \tag{5.111}$$

$$R_{B,p} = \sum_{i=1}^{m_p}\left(y_{p(i)}A_{p(i)}E_{p(i)}\right) \tag{5.112}$$

$$R_{I,p} = \sum_{i=1}^{m_p}\left(y_{p(i)}^2 A_{p(i)}E_{p(i)}\right) \tag{5.113}$$

Noting that:

$$P_{init} = \sum_{i=1}^{m_p} \left(A_{p(i)} E_{p(i)} \varepsilon_{p(i),init} \right) \tag{5.114}$$

and defining the internal moment caused by P_{init} about the reference axis as:

$$M_{init} = -\sum_{i=1}^{m_p} \left(y_{p(i)} A_{p(i)} E_{p(i)} \varepsilon_{p(i),init} \right) \tag{5.115}$$

Equation 5.104 can be re-written as:

$$N_{int,0} = \int_{A_c} E_{c,0}(\varepsilon_{r,0} - y\kappa_0)dA + (R_{A,s} + R_{A,p})\varepsilon_{r,0} - (R_{B,s} + R_{B,p})\kappa_0 + P_{init} \tag{5.116}$$

Remembering that for equilibrium $N_{ext,0} = N_{int,0}$, Equation 5.116 can be re-expressed as:

$$N_{ext,0} - P_{init} = \int_{A_c} E_{c,0}(\varepsilon_{r,0} - y\kappa_0)dA + (R_{A,s} + R_{A,p})\varepsilon_{r,0} - (R_{B,s} + R_{B,p})\kappa_0 \tag{5.117}$$

Similarly, the following expression based on moment equilibrium can be derived:

$$M_{ext,0} - M_{init} = -\int_{A_c} E_{c,0}(\varepsilon_{r,0} - y\kappa_0)ydA - (R_{B,s} + R_{B,p})\varepsilon_{r,0} + (R_{I,s} + R_{I,p})\kappa_0 \tag{5.118}$$

For a reinforced concrete section comprising rectangular components (e.g. rectangular flanges and webs) loaded in pure bending (i.e. $N_{ext,0} = N_{int,0} = 0$) and with no prestress, Equation 5.117 becomes a quadratic equation that can be solved to obtain the location of the neutral axis $y_{n,0}$.

If the cross-section is prestressed or the axial load $N_{ext,0}$ is not equal to zero (i.e. if $N_{ext,0} - P_{init} \neq 0$), dividing Equation 5.118 by Equation 5.117 gives:

$$\frac{M_{ext,0} - M_{init}}{N_{ext,0} - P_{init}} = \frac{-\int_{A_c} E_{c,0}(\varepsilon_{r,0} - y\kappa_0)y\,dA + (R_{B,s} + R_{B,p})\varepsilon_{r,0} + (R_{I,s} + R_{I,p})\kappa_0}{\int_{A_c} E_{c,0}(\varepsilon_{r,0} - y\kappa_0)dA + (R_{A,s} + R_{A,p})\varepsilon_{r,0} - (R_{B,s} + R_{B,p})\kappa_0}$$

and dividing the top and bottom of the right hand side by κ_0 and recognising that at the axis of zero strain $y = y_{n,0} = \varepsilon_{r,0}/\kappa_0$, the above expression becomes:

$$\frac{M_{ext,0} - M_{init}}{N_{ext,0} - P_{init}} = \frac{-\int_{A_c} E_{c,0}(y_{n,0} - y)y dA - (R_{B,s} + R_{B,p})y_{n,0} + (R_{I,s} + R_{I,p})}{\int_{A_c} E_{c,0}(y_{n,0} - y)dA + (R_{A,s} + R_{A,p})y_{n,0} - (R_{B,s} + R_{B,p})} \quad (5.119)$$

For a rectangular section of width b, Equation 5.119 becomes:

$$\frac{M_{ext,0} - M_{init}}{N_{ext,0} - P_{init}} = \frac{-\int_{y=y_{n,0}}^{y=d_{ref}}\left[E_{c,0}(y_{n,0} - y)by\right]dy - (R_{B,s} + R_{B,p})y_{n,0} + (R_{I,s} + R_{I,p})}{\int_{y=y_{n,0}}^{y=d_{ref}}\left[E_{c,0}(y_{n,0} - y)b\right]dy + (R_{A,s} + R_{A,p})y_{n,0} - (R_{B,s} + R_{B,p})}$$

$$(5.120)$$

Equation 5.120 may be solved for $y_{n,0}$ relatively quickly using a simple trial and error search.

When $y_{n,0}$ is determined, and the depth of the intact compressive concrete above the cracked tensile zone is known, the properties of the compressive concrete (A_c, B_c and I_c) with respect to the reference axis may be readily calculated. Similarly, the axial rigidity and the stiffness related to the first and second moments of area of the cracked section about the reference axis (i.e. $R_{A,0}$, $R_{B,0}$ and $R_{I,0}$) are calculated at time τ_0 using Equations 5.33, 5.34 and 5.37 and are re-expressed here as:

$$R_{A,0} = A_c E_{c,0} + R_{A,s} + R_{A,p} \quad (5.121)$$

$$R_{B,0} = B_c E_{c,0} + R_{B,s} + R_{B,p} \quad (5.122)$$

$$R_{I,0} = I_c E_{c,0} + R_{I,s} + R_{I,p} \quad (5.123)$$

Using the same solution procedure previously adopted for an uncracked cross-section, the system of equilibrium equations governing the problem (Equation 5.38) is rewritten here as:

$$\mathbf{r}_{ext,0} = \mathbf{D}_0 \boldsymbol{\varepsilon}_0 + \mathbf{f}_{p,init} \quad (5.38)$$

where:

$$\mathbf{r}_{\text{ext},0} = \begin{bmatrix} N_{\text{ext},0} \\ M_{\text{ext},0} \end{bmatrix} \tag{5.39}$$

$$\mathbf{D}_0 = \begin{bmatrix} R_{A,0} & -R_{B,0} \\ -R_{B,0} & R_{I,0} \end{bmatrix} \tag{5.40}$$

$$\boldsymbol{\varepsilon}_0 = \begin{bmatrix} \varepsilon_{r,0} \\ \kappa_0 \end{bmatrix} \tag{5.41}$$

$$\mathbf{f}_{p,\text{init}} = \begin{bmatrix} \sum P_{\text{init}(i)} \\ \sum -y_{p(i)} P_{\text{init}(i)} \end{bmatrix} = \begin{bmatrix} P_{\text{init}} \\ M_{\text{init}} \end{bmatrix} \tag{5.42}$$

The vector $\boldsymbol{\varepsilon}_0$ is readily obtained by solving the equilibrium equations (Equation 5.38) giving:

$$\boldsymbol{\varepsilon}_0 = \mathbf{D}_0^{-1}\left(\mathbf{r}_{\text{ext},0} - \mathbf{f}_{p,\text{init}}\right) = \mathbf{F}_0\left(\mathbf{r}_{\text{ext},0} - \mathbf{f}_{p,\text{init}}\right) \tag{5.43}$$

where:

$$\mathbf{F}_0 = \frac{1}{R_{A,0}R_{I,0} - R_{B,0}^2} \begin{bmatrix} R_{I,0} & R_{B,0} \\ R_{B,0} & R_{A,0} \end{bmatrix} \tag{5.44}$$

The stress distribution related to the concrete and reinforcement can then be calculated from the constitutive equations specified in Equations 5.100 through 5.103.

As an alternative approach, the solution may also be conveniently obtained using the cross-sectional properties of the transformed section. For example, for the cross-section of Figure 5.17, the transformed cross-section in equivalent areas of concrete for the short term analysis is shown in Figure 5.18.

The cross-sectional rigidities of the transformed section defined in Equations 5.121 through 5.123 can be re-calculated as:

$$R_{A,0} = A_0 E_{c,0} \tag{5.124}$$

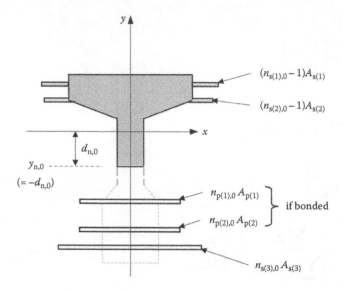

Figure 5.18 Transformed cracked section with bonded reinforcement transformed into equivalent areas of concrete.

$$R_{B,0} = B_0 E_{c,0} \tag{5.125}$$

$$R_{I,0} = I_0 E_{c,0} \tag{5.126}$$

where A_0 is the area of the transformed cracked concrete section, and B_0 and I_0 are the first and second moments of the transformed area about the reference x-axis at first loading.

Substituting Equations 5.124 through 5.126 into Equation 5.44, the matrix $\mathbf{F}_0$ becomes:

$$\mathbf{F}_0 = \frac{1}{E_{c,0}(A_0 I_0 - B_0^2)}\begin{bmatrix} I_0 & B_0 \\ B_0 & A_0 \end{bmatrix} \tag{5.127}$$

EXAMPLE 5.7

The depth of the concrete compression zone d_n and the short-term stress and strain distributions are to be calculated on the prestressed concrete beam cross-section shown in Figure 5.19, when $M_{ext,0} = 400$ kNm (and $N_{ext,0} = 0$). The section contains two layers of non-prestressed reinforcement as shown (each with $E_s = 2 \times 10^5$ MPa) and one layer of bonded prestressing steel ($E_p = 2 \times 10^5$ MPa). The prestressing force before transfer is

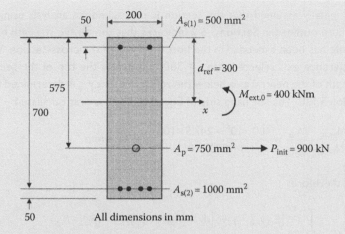

Figure 5.19 Cross-section (Example 5.7).

P_{init} = 900 kN (i.e. $\sigma_{p,init}$ = 1200 MPa). The tensile strength of the concrete is 3.5 MPa and the elastic modulus is $E_{c,0}$ = 30,000 MPa.

From Equations 5.108 through 5.113:

$$R_{A,s} = \left(A_{s(1)} + A_{s(2)}\right)E_s = \left(500 + 1,000\right) \times 200,000$$

$$= 300 \times 10^6 \text{ N}$$

$$R_{B,s} = \left(y_{s(1)}A_{s(1)} + y_{s(2)}A_{s(2)}\right)E_s$$

$$= (250 \times 500 - 400 \times 1,000) \times 200,000 = -55.0 \times 10^9 \text{ Nmm}$$

$$R_{I,s} = \left(y^2{}_{s(1)}A_{s(1)} + y^2{}_{s(2)}A_{s(2)}\right)E_s$$

$$= [250^2 \times 500 + (-400)^2 \times 1,000] \times 200,000 = 38.25 \times 10^{12} \text{ Nmm}^2$$

$$R_{A,p} = A_p E_p = 750 \times 200,000 = 150 \times 10^6 \text{ N}$$

$$R_{B,p} = y_p A_p E_s = -275 \times 750 \times 200,000 = -41.25 \times 10^9 \text{ Nmm}$$

$$R_{I,p} = y^2{}_p A_p E_s = (-275)^2 \times 750 \times 200,000 = 11.34 \times 10^{12} \text{ Nmm}^2$$

The vector of actions due to initial prestress is given by Equation 5.42:

$$\mathbf{f}_{p,init} = \begin{bmatrix} P_{init} \\ M_{init} \end{bmatrix} = \begin{bmatrix} 900 \times 10^3 \\ -(-275) \times 900 \times 10^3 \end{bmatrix} = \begin{bmatrix} 900 \times 10^3 \text{ N} \\ +247.5 \times 10^6 \text{ Nmm} \end{bmatrix}$$

If it is initially assumed that the section is uncracked, an analysis using the procedure outlined in Section 5.6.2 indicates that the tensile strength of the concrete has been exceeded in the bottom fibres of the cross-section. With the reference axis selected at $d_{ref} = 300$ mm below the top of the section, the depth of the neutral axis below the reference axis $y_{n,0}$ is determined from Equation 5.120. The left hand side of Equation 5.120 is first calculated:

$$\frac{M_{ext,0} - M_{init}}{N_{ext,0} - P_{init}} = \frac{400 \times 10^6 - 247.5 \times 10^6}{0 - 900 \times 10^3} = -169.\dot{4} \text{ mm}$$

and therefore:

$$-169.\dot{4} = \frac{-\int_{y=y_{n,0}}^{y=300} \left[E_c(y_{n,0}-y)yb\right]dy - (R_{B,s}+R_{B,p})y_{n,0} + (R_{I,s}+R_{I,p})}{\int_{y=y_{n,0}}^{y=300} \left[E_{c,0}(y_{n,0}-y)b\right]dy + (R_{A,s}+R_{A,p})y_{n,0} - (R_{B,s}+R_{B,p})}$$

$$= \frac{-\int_{y=y_{n,0}}^{y=300} 30,000 \times 200 \times (y_{n,0}-y)y \, dy}{}$$

$$= \frac{-(-55.0-41.25) \times 10^9 y_{n,0} + (38.25+11.34) \times 10^{12}}{\int_{y=y_{n,0}}^{y=300} 30,000 \times 200 \times (y_{n,0}-y) \, dy}$$

$$+ (300+150) \times 10^6 y_{n,0} + (55.0+41.25) \times 10^9$$

$$= \frac{-\left|6.0 \times 10^6 \times (0.5y_{n,0}y^2 - 0.3\dot{3}y^3)\right|_{y_{n,0}}^{300} + 96.25 \times 10^9 y_{n,0} + 49.59 \times 10^{12}}{\left|6.0 \times 10^6 \times (y_{n,0}y - 0.5y^2)\right|_{y_{n,0}}^{300} + 450 \times 10^6 y_{n,0} + 96.25 \times 10^9}$$

$$= \frac{y_{n,0}^3 - 173.7 \times 10^3 y_{n,0} + 103.6 \times 10^6}{-3y_{n,0}^2 + 2.25 \times 10^3 y_{n,0} - 173.75 \times 10^3}$$

Solving gives $y_{n,0} = -208.1$ mm and the depth of the neutral axis below the top surface is $d_n = d_{ref} - y_{n,0} = 508.1$ mm.

The properties of the compressive concrete (A_c, B_c and I_c) with respect to the reference axis are:

$$A_c = 508.1 \times 200 - 500 = 101,100 \text{ mm}^2$$

$$B_c = 508.1 \times 200 \times (300 - 254.05) - 500 \times (300 - 50) = +4.545 \times 10^6 \text{ mm}^3$$

$$I_c = 200 \times 508.1^3/12 + 508.1 \times 200 \times (300 - 254.0)^2 - 500 \times (300 - 50)^2$$

$$= 2370 \times 10^6 \text{ mm}^4$$

The cross-sectional rigidities $R_{A,0}$, $R_{B,0}$ and $R_{I,0}$ are obtained from Equations 5.121 through 5.123:

$$R_{A,0} = A_c E_{c,0} + R_{A,s} + R_{A,p} = 101,100 \times 30,000 + 300 \times 10^6 + 150 \times 10^6$$

$$= 3484 \times 10^6 \text{ mm}^4$$

$$R_{B,0} = B_c E_{c,0} + R_{B,s} + R_{B,p} = 4.545 \times 10^6 \times 30,000 - 55 \times 10^9 - 41.25 \times 10^9$$

$$= 40.08 \times 10^9 \text{ mm}^3$$

$$R_{I,0} = I_c E_{c,0} + R_{I,s} + R_{I,p} = 2370 \times 10^6 \times 30,000 + 38.25 \times 10^{12} + 11.34 \times 10^{12}$$

$$= 120.7 \times 10^{12} \text{ mm}^4$$

From Equation 5.44:

$$\mathbf{F}_0 = \frac{1}{R_{A,0}R_{I,0} - R_{B,0}^2}\begin{bmatrix} R_{I,0} & R_{B,0} \\ R_{B,0} & R_{A,0} \end{bmatrix} = \begin{bmatrix} 288.2 \times 10^{-12} & 95.71 \times 10^{-15} \\ 95.71 \times 10^{-15} & 8.318 \times 10^{-15} \end{bmatrix}$$

and the strain vector is obtained from Equation 5.43:

$$\varepsilon_0 = \mathbf{F}_0 \left(\mathbf{r}_{ext,0} - \mathbf{f}_{p,init} \right)$$

$$= \begin{bmatrix} 288.2 \times 10^{-12} & 95.71 \times 10^{-15} \\ 95.71 \times 10^{-15} & 8.318 \times 10^{-15} \end{bmatrix} \begin{bmatrix} 0 - 900 \times 10^3 \\ 400 \times 10^6 - 247.5 \times 10^6 \end{bmatrix}$$

$$= \begin{bmatrix} -244.8 \times 10^{-6} \\ 1.182 \times 10^{-6} \text{ mm}^{-1} \end{bmatrix}$$

The top (y = 300 mm) and bottom (y = −450 mm) fibre strains are:

$$\varepsilon_{0(top)} = \varepsilon_{r,0} - 300 \times \kappa_0 = (-244.8 - 300 \times 1.182) \times 10^{-6} = -599 \times 10^{-6}$$

$$\varepsilon_{0(bot)} = \varepsilon_{r,0} + 450 \times \kappa_0 = (-244.8 + 450 \times 1.182) \times 10^{-6} = +287 \times 10^{-6}$$

The distribution of strains is shown in Figure 5.20b.

The top fibre stress in the concrete and the stress in the non-prestressed reinforcement are (Equations 5.100 and 5.102):

$$\sigma_{c,0(top)} = E_{c,0}\varepsilon_{0(top)} = 30,000 \times (-599 \times 10^{-6}) = -18.0 \text{ MPa}$$

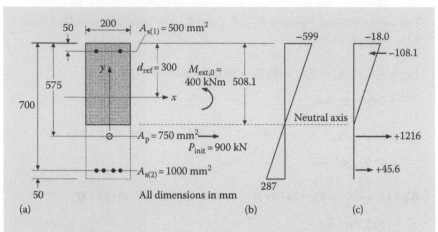

Figure 5.20 Stress and strain distributions for cracked section (Example 5.7). (a) Section. (b) Strain ($\times 10^{-6}$). (c) Stress (MPa).

$$\sigma_{s(1),0} = E_s(\varepsilon_{r,0} - y_{s(1)}\kappa_0) = 200,000 \times (-244.8 - 250 \times 1.182) \times 10^{-6}$$

$$= -108.1\,\text{MPa}$$

$$\sigma_{s(2),0} = E_s(\varepsilon_{r,0} - y_{s(2)}\kappa_0) = 200,000 \times (-244.8 + 400 \times 1.182) \times 10^{-6}$$

$$= +45.6\,\text{MPa}$$

and the stress in the prestressing steel is given by Equation 5.103:

$$\sigma_{p,0} = E_p(\varepsilon_{r,0} - y_p\kappa_0 + \varepsilon_{p,\text{init}})$$

$$= 200,000 \times (-244.8 + 275 \times 1.182 + 6,000) \times 10^{-6} = +1,216\ \text{MPa}$$

The stresses are plotted in Figure 5.20c.

5.9 TIME-DEPENDENT ANALYSIS OF CRACKED CROSS-SECTIONS

5.9.1 Simplifying Assumption

For a cracked cross-section under sustained actions, creep causes a gradual change of the position of the neutral axis under sustained loads and the size of the concrete compressive zone gradually increases with time. With the size of the cross-section gradually changing, and hence the sectional properties of the concrete changing, the principle of superposition does not apply. To accurately account for the increasing size of the compressive zone, a detailed numerical analysis is required in which time is discretised into many small steps.

However, if one assumes that the depth of the concrete compression zone d_n remains constant with time, the time analysis of a fully-cracked cross-section using the *Age-adjusted Effective Modulus Method* (AEMM) is essentially the same as that outlined in Section 5.7.3. This assumption greatly simplifies the analysis and usually results in relatively little error in the calculated deformations.

5.9.2 Long-term analysis of a cracked cross-section subjected to combined axial force and bending using the AEMM

Consider the fully-cracked cross-section shown in Figure 5.21a subjected to a sustained external bending moment $M_{ext,k}$ and axial force $N_{ext,k}$. Both the short-term and time-dependent strain distributions are shown in Figure 5.21b.

For the time analysis, the steel reinforcement and prestressing tendons (if any) are assumed to be linear-elastic (as for the short-term analysis) and the constitutive relationship for the concrete at τ_k is similar to Equation 5.57 as follows:

$$\sigma_{c,k} = \overline{E}_{e,k}\left(\varepsilon_k - \varepsilon_{cs,k}\right) + \overline{F}_{e,0}\sigma_{c,0} \quad \text{for } y \geq y_{n,0} \tag{5.128}$$

and

$$\sigma_{c,k} = 0 \quad \text{for } y < y_{n,0} \tag{5.129}$$

$$\sigma_{s(i),k} = E_{s(i)}\varepsilon_k \tag{5.130}$$

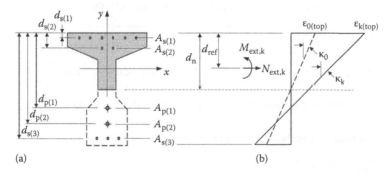

Figure 5.21 Fully-cracked cross-section – Time analysis (AEMM). (a) Cross-section. (b) Strain.

$$\sigma_{p(i),k} = E_{p(i)} \left(\varepsilon_k + \varepsilon_{p(i),init} - \varepsilon_{p.rel(i),k} \right) \tag{5.131}$$

where $\bar{E}_{e,k}$, $\bar{F}_{e,0}$ and $\varepsilon_{p.rel(i),k}$ are as defined previously (and given in Equations 5.55, 5.58 and 5.68).

At time τ_k, the internal axial force $N_{int,k}$ and moment $M_{int,k}$ on the cross-section are given by Equations 5.76 and 5.79, and the axial rigidity and the stiffness related to the first and second moments of area ($R_{A,k}$, $R_{B,k}$ and $R_{I,k}$, respectively) are given by Equations 5.77, 5.78 and 5.80 (ignoring the cracked concrete below the neutral axis). The equilibrium equations are expressed in Equation 5.82 and solving using Equation 5.89 gives the strain vector at time τ_k. The stresses in the concrete, steel reinforcement and tendons at time τ_k are then calculated from Equations 5.91, 5.92 and 5.93, respectively.

EXAMPLE 5.8

Calculate the change of stress and strain with time on the cracked prestressed cross-section of Example 5.7 using the AEMM. The cracked section and the initial strain distribution are shown in Figure 5.20. The actions on the section are assumed to be constant throughout the time period under consideration (i.e. τ_0 to τ_k) and equal to:

$$N_{ext,0} = N_{ext,k} = 0 \quad \text{and} \quad M_{ext,0} = M_{ext,k} = 400 \text{ kNm}$$

The relevant material properties are

$$E_{c,0} = 30 \text{ GPa} \quad E_s = E_p = 200 \text{ GPa} \quad \varphi_{cc}(\tau_k,\tau_0) = 2.5$$

$$\chi(\tau_k,\tau_0) = 0.65 \quad \varepsilon_{cs}(\tau_k) = -400 \times 10^{-6} \quad \varphi_p(\tau_k,\sigma_{p(i),0}) = 0.02$$

and the steel reinforcement is assumed to be linear-elastic.

From Example 5.7:

$$d_n = 508.1 \text{ mm} \quad \varepsilon_{r,0} = -244.8 \times 10^{-6} \quad \kappa_0 = 1.182 \times 10^{-6} \text{ mm}^{-1}$$

$$A_c = 101.1 \times 10^3 \text{ mm}^2 \quad B_c = +4.545 \times 10^6 \text{ mm}^3 \quad I_c = 2370 \times 10^6 \text{ mm}^4$$

and the rigidities of the steel reinforcement and tendons are:

$$R_{A,s} = 300 \times 10^6 \text{ N}$$

$$R_{B,s} = -55.0 \times 10^9 \text{ Nmm}$$

$R_{1,s} = 38.25 \times 10^{12} \text{ Nmm}^2$

$R_{A,p} = 150 \times 10^6 \text{ N}$

$R_{B,p} = -41.25 \times 10^9 \text{ Nmm}$

$R_{1,p} = 11.34 \times 10^{12} \text{ Nmm}^2$

From Equations 5.55 and 5.58:

$$\bar{E}_{e,k} = \frac{30,000}{1 + 0.65 \times 2.5} = 11,430 \text{ MPa} \quad \text{and} \quad \bar{F}_{e,0} = \frac{2.5 \times (0.65 - 1.0)}{1.0 + 0.65 \times 2.5} = -0.333$$

From Equation 5.85, the axial force and moment resisted by the concrete part of the cross-section at time τ_0 are:

$$N_{c,0} = A_c E_{c,0} \varepsilon_{r,0} - B_c E_{c,0} \kappa_0 = -903.7 \times 10^3 \text{ N}$$

$$M_{c,0} = -B_c E_{c,0} \varepsilon_{r,0} + I_c E_{c,0} \kappa_0 = +117.4 \times 10^6 \text{ Nmm}$$

and from Equations 5.77, 5.78 and 5.80, the cross-sectional rigidities $R_{A,k}$, $R_{B,k}$ and $R_{1,k}$ are:

$$R_{A,k} = A_c \bar{E}_{e,k} + R_{A,s} + R_{A,p} = 1606 \times 10^6 \text{ N}$$

$$R_{B,k} = B_c \bar{E}_{e,k} + R_{B,s} + R_{B,p} = -44.31 \times 10^9 \text{ Nmm}$$

$$R_{1,k} = I_c \bar{E}_{e,k} + R_{1,s} + R_{1,p} = 76.67 \times 10^{12} \text{ Nmm}^2$$

Equation 5.90 gives:

$$\mathbf{F}_k = \frac{1}{R_{A,k} R_{1,k} - R_{B,k}^2} \begin{bmatrix} R_{1,k} & R_{B,k} \\ R_{B,k} & R_{A,k} \end{bmatrix} = \begin{bmatrix} 632.9 \times 10^{-12} & -365.8 \times 10^{-15} \\ -365.8 \times 10^{-15} & 13.25 \times 10^{-15} \end{bmatrix}$$

From Equations 5.85 through 5.88:

$$\mathbf{f}_{cr,k} = \bar{F}_{e,0} \begin{bmatrix} N_{c,0} \\ M_{c,0} \end{bmatrix} = \begin{bmatrix} +301.2 \times 10^3 \\ -39.14 \times 10^6 \end{bmatrix}$$

$$\mathbf{f}_{cs,k} = \begin{bmatrix} A_c \\ -B_c \end{bmatrix} \bar{E}_{e,k} \varepsilon_{cs,k} = \begin{bmatrix} -462.3 \times 10^3 \\ +20.77 \times 10^6 \end{bmatrix}$$

$$\mathbf{f}_{p,init} = \sum_{i=1}^{m_p} \begin{bmatrix} A_{p(i)} E_{p(i)} \varepsilon_{p(i),init} \\ y_{p(i)} A_{p(i)} E_{p(i)} \varepsilon_{p(i),init} \end{bmatrix} = \begin{bmatrix} 900 \times 10^3 \\ 247.5 \times 10^6 \end{bmatrix} \quad \text{(from Example 5.7)}$$

$$f_{p,rel,k} = \sum_{i=1}^{m_p} \begin{bmatrix} A_{p(i)}E_{p(i)}\varepsilon_{p(i),0}\varphi_{p(i)} \\ y_{p(i)}A_{p(i)}E_{p(i)}\varepsilon_{p(i),0}\varphi_{p(i)} \end{bmatrix} = \begin{bmatrix} 18\times10^3 \\ 4.95\times10^6 \end{bmatrix}$$

The strain vector is obtained from Equation 5.89:

$$\varepsilon_k = F_k\left(r_{ext,k} - f_{cr,k} + f_{cs,k} - f_{p,init} + f_{p,rel,k}\right)$$

$$\varepsilon_k = \begin{bmatrix} 632.9\times10^{-12} & -365.8\times10^{-15} \\ -365.8\times10^{-15} & 13.25\times10^{-15} \end{bmatrix}\begin{bmatrix} (0-301.2-462.3-900+18)\times10^3 \\ (400+39.14+20.77-247.5+4.95)\times10^6 \end{bmatrix}$$

$$= \begin{bmatrix} -1121\times10^{-6} \\ 3.483\times10^{-6}\ \text{mm}^{-1} \end{bmatrix}$$

The top (y = 300 mm) and bottom (y = −450 mm) fibre strains are:

$\varepsilon_{k(top)} = \varepsilon_{r,k} - 300 \times \kappa_k = (-1121 - 300 \times 3.483) \times 10^{-6} = -2166 \times 10^{-6}$

$\varepsilon_{k(btm)} = \varepsilon_{r,k} + 450 \times \kappa_k = (-1121 + 450 \times 3.483) \times 10^{-6} = +446 \times 10^{-6}$

At the neutral axis depth at $y_{n,0}$ = −208.1 mm (below the reference axis):

$\varepsilon_{k(dn)} = \varepsilon_{r,k} - (-208.1) \times \kappa_k = (-1121 + 208.1 \times 3.483) \times 10^{-6}$

$= -396 \times 10^{-6}$

The concrete stresses at time τ_k at the top fibre (y = +300 mm) and at the bottom of the compressive concrete are obtained from Equation 5.91:
At top of section:

$\sigma_{c,k} = 11{,}430 \times (-2{,}166 + 400) \times 10^{-6} - 0.33\dot{3} \times (-18.0) = -14.2\ \text{MPa}$

At y = −208.1 mm:

$\sigma_{c,k} = 11{,}430 \times (-396 + 400) \times 10^{-6} - 0.33\dot{3} \times 0 = 0.0\ \text{MPa}$

The stress in the non-prestressed reinforcement are (Equation 5.92):

$\sigma_{s(1),k} = E_s(\varepsilon_{r,k} - y_{s(1)}\kappa_k) = -398\ \text{MPa}$

$\sigma_{s(2),k} = E_s(\varepsilon_{r,k} - y_{s(2)}\kappa_k) = +54.4\ \text{MPa}$

and the stress in the prestressing steel is given by (Equation 5.93):

$\sigma_{p,k} = E_p\left(\varepsilon_{r,k} - y_p\kappa_k + \varepsilon_{p.init,k} - \varepsilon_{p.rel,k}\right) = +1143\ \text{MPa}$

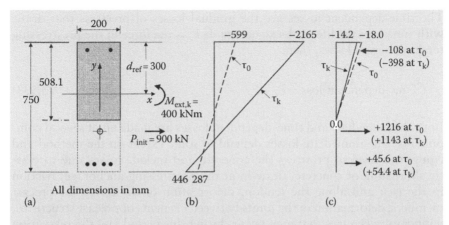

Figure 5.22 Initial and time-dependent strain and stress distributions. (a) Section. (b) Strain ($\times 10^{-6}$). (c) Stress (MPa).

The results are plotted in Figure 5.22. It can be seen that the loss of prestress in the tendon on this cross-section is only 73 MPa or only 6% of the initial prestress. With relaxation losses at 2%, creep and shrinkage have resulted in only 4% loss as the change in strain at the tendon level is relatively small after cracking.

5.10 LOSSES OF PRESTRESS

5.10.1 Definitions

The losses of prestress that occur in a tendon are categorised as either *immediate losses* or *time-dependent losses* and are illustrated in Figure 5.23.

The immediate losses occur when the prestress is transferred to the concrete and are the difference between the force imposed on the tendon by the hydraulic prestressing jack P_j and the force in the tendon immediately after transfer at a particular point along its length P_i and can be expressed as:

$$\text{Immediate loss} = P_j - P_i \qquad (5.132)$$

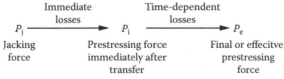

Figure 5.23 Losses of prestress in the tendons.

The time-dependent losses are the gradual losses of prestress that occur with time over the life of the structure. If P_e is the force in the prestressing tendon after all losses, then:

$$\text{Time-dependent loss} = P_i - P_e \qquad (5.133)$$

Both the immediate and time-dependent losses are made up of several components. The immediate losses depend to some extent on the method and equipment used to prestress the concrete and include losses due to elastic shortening of concrete, draw-in at the prestressing anchorage, friction in the jack and along the tendon, deformation of the forms for precast members, deformation in the joints between elements of precast structures, temperature changes that may occur during this period and the relaxation of the tendon in a pretensioned member between the time of casting the concrete and the time of transfer (particularly significant when the concrete is cured at elevated temperatures prior to transfer).

Time-dependent losses are the gradual losses of prestress that occur with time over the life of the structure. These include losses caused by the gradual shortening of concrete at the steel level due to creep and shrinkage, relaxation of the tendon after transfer and time-dependent deformation that may occur within the joints in segmental construction.

5.10.2 Immediate losses

The magnitude of the immediate losses is taken as the sum of the losses caused by each relevant phenomenon. Where appropriate, the effects of one type of immediate loss on the magnitude of other immediate losses should be considered. For example, in a pretensioned member, the loss caused by relaxation of the tendon prior to transfer will affect the magnitude of the immediate loss caused by elastic deformation of concrete.

5.10.2.1 Elastic deformation losses

Pretensioned members: The change in strain in a tendon in a pretensioned member immediately after transfer $\Delta\varepsilon_{p,0}$ caused by elastic shortening of the concrete is equal to the instantaneous strain in the concrete at the steel level $\varepsilon_{cp,0}$:

$$\varepsilon_{cp,0} = \frac{\sigma_{cp,0}}{E_{c,0}} = \Delta\varepsilon_{p,0} = \frac{\Delta\sigma_{p,0}}{E_p}$$

The corresponding loss of stress in the tendons at transfer is therefore, the product of the modular ratio $(E_p/E_{c,0})$ and the stress in the adjacent concrete at the tendon level $\sigma_{cp,0}$ and is given by:

$$\Delta\sigma_{p,0} = \frac{E_p}{E_{c,0}} \sigma_{cp,0} \qquad (5.134)$$

Post-tensioned members: For post-tensioned members with one tendon, or with two or more tendons stressed simultaneously, the elastic deformation of the concrete occurs during the stressing operation before the tendons are anchored. In this case, elastic shortening losses are zero. In a member containing more than one tendon and where the tendons are stressed sequentially, stressing of a tendon causes an elastic shortening loss in all previously stressed and anchored tendons. Consequently, the first tendon to be stressed suffers the largest elastic shortening losses and the last tendon to be stressed suffers no elastic shortening losses at all. Elastic shortening losses in the tendons stressed early in the prestressing sequence can be reduced by restressing the tendons (prior to grouting of the prestressing ducts).

It is relatively simple to calculate the elastic shortening losses in an individual tendon of a post-tensioned member provided the stressing sequence is known. For most cases, it is sufficient to determine the average loss of stress as:

$$\Delta\sigma_p = \frac{n-1}{2n} \frac{E_p}{E_{c,0}} \frac{P}{A} \qquad (5.135)$$

where n is the number of tendons and P/A is the average concrete compressive stress. In post-tensioned members, the tendons are not bonded to the concrete until grouting of the duct occurs some time after the stressing sequence is completed. It is the shortening of the member between the anchorage plates that leads to elastic shortening, and not the strain at the steel level, as is the case for pretensioned members.

5.10.2.2 Friction in the jack and anchorage

The loss caused by friction in the jack and anchorage depends on the jack pressure and the type of jack and anchorage system used. It is usually allowed for during the stressing operation and is generally relatively small.

5.10.2.3 Friction along the tendon

In post-tensioned members, friction losses occur along the tendon during the stressing operation. Friction between the tendon and the duct causes a gradual reduction in prestress with the distance along the tendon (L_{pa}) from the jacking end. The coefficient of friction between the tendon and the duct depends basically on the condition of the surfaces in contact, the profile of the duct, the nature of the tendon and its preparation. The magnitude of the friction loss depends on the tendon length L_{pa}, the total angular change of the tendon over that length, as well as the size and type of the duct containing the tendon. An estimation of the stress in tendon at any point a distance L_{pa} along the tendon from the jacking end may be made using [2]:

$$\sigma_{pa} = \sigma_{pj}e^{-\mu(\alpha_{tot}+\beta_p L_{pa})} \qquad (5.136)$$

where:

σ_{pj} is the stress in the tendon at the jacking end;

μ is the friction curvature coefficient that depends on the type of duct. In the absence of more specific data, and when all tendons in contact within the same duct are stressed simultaneously, AS3600-2009 [2] suggests that for greased and wrapped coating on wire or strand μ = 0.15; for bright and zinc-coated metal sheathing μ = 0.15–0.2 and for bright and zinc-coated flat ducts commonly used in slabs μ = 0.2. Higher values should be used if either the tendon or the duct is rusted. For tendons showing a high but still acceptable amount of rusting, the value of μ may increase by 20% for bright and zinc-coated metal sheathing. If the wires or strand in contact in the one duct are stressed separately, μ may be significantly greater than the values given above and should be checked by tests. For external tendons passing over machined cast-steel saddles, μ may increase markedly for large movements of tendons across the saddles;

α_{tot} is the sum in radians of the absolute values of successive angular deviations of the tendon over the length L_{pa}. Care should be taken during construction to achieve the same cable profile as assumed in the design;

β_p is an estimate of the angular deviation (in radians/m) due to wobble effects in the straight or curved parts of the tendon and depends on the rigidity of the sheaths, on the spacing and fixing of their supports, on the care taken in placing the prestressing tendons, on the clearance of tendons in the duct, on the stiffness of the tendons and on the precautions taken during concreting. In segmental construction, the angular deviation per metre (β_p) may be greater in the event of mismatching of ducts and the designer should allow for this possibility. The most important parameter affecting the rigidity of the sheaths is their diameter d_{duct}. AS3600-2009 [2] suggests the following:

• For sheathing containing wires or strands:

β_p = 0.024–0.016 rad/m   when $d_{duct} \leq 50$ mm

β_p = 0.016–0.012 rad/m   when $50 < d_{duct} \leq 90$ mm

$\beta_p = 0.012-0.008$ rad/m   when $d_{duct} > 90$ mm

- For flat metal ducts containing wires or strands:

  $\beta_p = 0.024-0.016$ rad/m

- For sheathing containing bars:

  $\beta_p = 0.016-0.008$ rad/m   when $d_{duct} \leq 50$ mm

- For bars with a greased-and-wrapped coating:

  $\beta_p = 0.008$ rad/m

EXAMPLE 5.9

Calculate the friction losses in the prestressing cable in the end-span of the post-tensioned girder of Figure 5.24. For this cable, $\mu = 0.2$ and $\beta_p = 0.01$.

From Equation 5.136:

At B: $\sigma_{pa} = \sigma_{pj}e^{-0.2(0.105+0.01\times9)} = 0.962\,\sigma_{pj}$ (i.e. 3.8% losses)

At C: $\sigma_{pa} = \sigma_{pj}e^{-0.2(0.210+0.01\times18)} = 0.925\sigma_{pj}$ (i.e. 7.5% losses)

At D: $\sigma_{pa} = \sigma_{pj}e^{-0.2(0.315+0.01\times25)} = 0.893\sigma_{pj}$ (i.e. 10.7% losses)

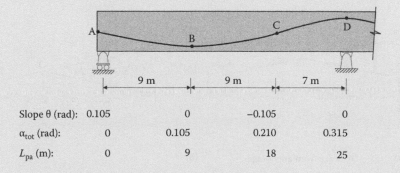

Slope θ (rad):	0.105		0		−0.105		0
α_{tot} (rad):	0		0.105		0.210		0.315
L_{pa} (m):	0		9		18		25

Figure 5.24 Tendon profile for end span (Example 5.9).

5.10.2.4 Anchorage losses

In post-tensioned members, some slip or draw-in occurs when the pre-stressing force is transferred from the jack to the anchorage. This causes an additional loss of prestress. The amount of slip depends on the type of anchorage. For wedge-type anchorages used for strand, the slip (Δ_{slip}) may be as high as 6 mm. The loss of prestress caused by Δ_{slip} decreases with distance from the anchorage owing to friction and, for longer tendons, may be negligible at the critical design section. However, for short tendons, this loss may be significant and should not be ignored in design.

The loss of tension in the tendon caused by slip is opposed by friction in the same way as the initial prestressing force is opposed by friction, but in the opposite direction, i.e. μ and β_p are the same. The variations of pre-stressing force along a member due to friction before anchoring the tendon (calculated using Equation 5.137) and after anchoring are shown in Figure 5.25, where the mirror image reduction in prestressing force in the vicinity of the anchorage is caused by slip at the anchorage. The slope of the draw-in line adjacent to the anchorage has the same magnitude as the friction loss line but the opposite sign. It follows that tendons with a small drape (and therefore small α) will suffer anchorage slip losses over a longer length of tendon than tendons with a large drape (larger α).

In order to calculate the draw-in loss at the anchorage ΔP_{di}, the length of the draw-in line L_{di} must be determined. By equating the anchorage slip Δ_{slip} with the integral of the change in strain in the steel tendon over the length of the draw-in line, L_{di} may be determined. If α is the slope of the friction loss line (i.e. the friction loss per unit length) as shown in Figure 5.25, the loss of prestress due to draw-in at a distance x from point O in Figure 5.25b is $(\Delta P_{di})_x = 2\alpha x$ and Δ_{slip} can be estimated as follows:

$$\Delta_{slip} = \int_0^{L_{di}} \frac{2\alpha x}{E_p A_p} dx = \frac{\alpha L_{di}^2}{E_p A_p} \tag{5.137}$$

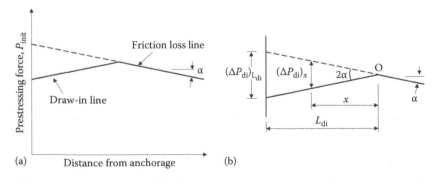

(a)   Distance from anchorage   (b)

Figure 5.25 Variation of prestress adjacent to the anchorage due to draw-in. (a) Prestressing force versus distance from anchorage. (b) Loss of prestress in vicinity of anchorage.

and rearranging gives:

$$L_{di} = \sqrt{\frac{E_p A_p \Delta_{slip}}{\alpha}} \tag{5.138}$$

The immediate loss of prestress at the anchorage caused by Δ_{slip} is:

$$(\Delta P_{di})_{L_{di}} = 2\alpha L_{di} \tag{5.139}$$

The immediate loss of prestress near an anchorage can be determined from geometry using Figure 5.25. At a distance of more than L_{di} from the live end anchorage, the immediate loss of prestress due to Δ_{slip} is zero.

The magnitude of the slip that should be anticipated in design is usually supplied by the anchorage manufacturer and should be checked on site. Cautious overstressing at the anchorage is often an effective means of compensating for slip.

5.10.2.5 Other causes of immediate losses

Additional immediate losses may occur due to deformation of the forms of precast members and deformation in the construction joints between the precast units in segmental construction and these losses must be assessed in design. Any change in temperature between the time of stressing the tendon and the time of casting the concrete in a pretensioned member will also cause immediate losses, as will any difference in the temperature between the stressed tendons and the concrete during heat treatment.

5.10.3 Time-dependent losses of prestress

5.10.3.1 Discussion

We have seen in Sections 5.7 and 5.9 that, in addition to causing time-dependent increases in deflection or camber, both compressive creep and shrinkage of the concrete cause gradual shortening of a concrete member and this, in turn, leads to time-dependent shortening of the prestressing tendons, and a consequent reduction in the prestressing force. These time-dependent losses of prestress are in addition to the losses caused by steel relaxation and may adversely affect the long-term serviceability of the structure and should be accounted for in design.

In Section 5.7, a time analysis was presented for determining the effects of creep and shrinkage of concrete, and relaxation of the tendon on the long-term stresses and deformations of a prestressed concrete cross-section of any shape and containing any layout of prestressed and non-prestressed reinforcement.

In this section, the approximate procedures specified in AS3600-2009 [2] for calculating time-dependent losses of prestress are outlined. These methods are of limited value and often give misleading results because they do not adequately account for the significant loss of pre-compression in the concrete that occurs when non-prestressed reinforcement is present. For a realistic estimate of the time-dependent losses of prestress in the tendon, and the redistribution of stresses between the bonded reinforcement and the concrete, the methods described in Sections 5.7 and 5.9 are recommended.

For members containing only tendons, the loss in tensile force in the tendons is simply equal to the loss in compressive force in the concrete. Where the member contains a significant amount of longitudinal non-prestressed reinforcement, there is a gradual transfer of the compressive prestressing force from the concrete into the bonded reinforcement. Shortening of the concrete, due to creep and shrinkage, causes a shortening of the bonded reinforcement, and therefore, an increase in compressive stress in the steel. The gradual increase in compressive force in the bonded reinforcement is accompanied by an equal and opposite decrease in the compressive force in the concrete. The loss of compressive force in the concrete is therefore, considerably greater than the loss of tensile force in the tendon. The redistribution of stresses with time was discussed in Section 5.7, and illustrated, for example, in Figures 5.14 and 5.15, where the immediate strain and stress distributions (at time τ_0 immediately after the application of both prestress and the applied moment $M_{ext.0}$) and the long-term strain and stress distributions (after creep and shrinkage at time τ_k) on prestressed concrete cross-sections are shown.

AS3600-2009 [2] suggests that the total time-dependent loss of prestress should be estimated by adding the calculated losses of prestress due to shrinkage, creep and relaxation. However, separate calculation of these losses is problematic as the time-dependent losses interact with each other, and this interaction should be considered when the sum of all the losses is determined. For example, the loss in tendon force due to creep and shrinkage of the concrete decreases the average force in the tendon with time, and this in turn reduces the relaxation loss. Restraint to shrinkage often substantially reduces the compressive stresses in the concrete at the steel level, and this may significantly affect the creep of the concrete at this level and reduce losses due to creep.

5.10.3.2 Shrinkage losses

If a concrete member of length L contains no bonded reinforcement (and no bonded tendons) and is unrestrained at its supports and along its length, the member will shorten due to shrinkage by an amount equal to $\varepsilon_{cs}L$. If the member contained an unbonded post-tensioned tendon with an anchorage at each end of the member, the tendon would shorten by the same amount and the change of stress in the tendon due to shrinkage (ignoring the effects of friction) would be constant along its length and equal to:

$$\Delta\sigma_{p.cs} = E_p\varepsilon_{cs} \tag{5.140}$$

where ε_{cs} is the shrinkage strain at the time under consideration and may be estimated using the procedures outlined in Section 4.2.5.4.

In concrete structures, unrestrained contraction is unusual. Reinforcement and bonded tendons embedded in the concrete provide restraint to shrinkage and reduce the shortening of the member. This, in turn, reduces the loss of prestress in any tendon within the member. If the reinforcement and bonded tendons are symmetrically placed on a cross-section so that the resultant restraining force is axial, the change in strain in a bonded tendon due to shrinkage and the corresponding change of stress in the tendon may be expressed as:

$$\Delta\varepsilon_{p.cs} = \frac{\varepsilon_{cs}}{1 + \bar{n}_e p} \tag{5.141}$$

and

$$\Delta\sigma_{p.cs} = \frac{E_p\varepsilon_{cs}}{1 + \bar{n}_e p} \tag{5.142}$$

where $p = A_s/A_g$, A_s is the total area of bonded steel (reinforcement plus tendons); A_g is the gross area of the cross section; $\bar{n}_e = E_s/\bar{E}_{e,k}$ is the age-adjusted effective modular ratio; and $\bar{E}_{e,k}$ is the age-adjusted effective modulus of concrete given by Equation 5.55.

Where the centroid of the bonded steel is at an eccentricity e_s below the centroidal axis of a rectangular concrete cross-section, the change of strain at the centroid of the bonded steel is approximated by Equation 5.143 and the corresponding change of stress in a tendon at this location is given by Equation 5.144:

$$\Delta\varepsilon_{p.cs} = \frac{\varepsilon_{cs}}{1 + \bar{n}_e p \left[1 + 12\left(\dfrac{e_s}{D}\right)^2\right]} \tag{5.143}$$

and

$$\Delta\sigma_{p.cs} = \frac{E_p\varepsilon_{cs}}{1 + \bar{n}_e p \left[1 + 12\left(\dfrac{e_s}{D}\right)^2\right]} \tag{5.144}$$

where D is the cross-section depth measured perpendicular to the centroidal axis.

In most practical situations, the bonded reinforcement and bonded tendons are not symmetrically located on the cross-section and the restraining force is not axial. In this case, the tensile restraining force will induce a curvature on the section and the change of strain and stress in a tendon will depend on its position on the cross-section. Equation 5.144 will usually provide a reasonable estimate of shrinkage induced change of stress in a tendon, particularly if the position of the tendon is close to the centroid of the bonded steel area (including both the non-prestressed reinforcement and bonded tendons).

5.10.3.3 Creep losses

Creep strain in the concrete at the level of the bonded tendon depends on the stress history of the concrete at that level. Because the concrete stress varies with time, a reliable estimate of creep losses requires a detailed time analysis of the cross-section (such as that presented in Section 5.7). An approximate estimate of creep losses can be made by assuming the concrete stress at the tendon level remains constant with time and equal to the short-term value $\sigma_{c0.p}$ calculated using the initial prestressing force (prior to any time-dependent losses) and the sustained portion of all the service loads. With this assumption, the creep strain that develops in the concrete at the tendon level $\varepsilon_{cc.p}$ and the corresponding change in stress in the tendon due to creep $\Delta\sigma_{p.cc}$ are given by:

$$\varepsilon_{cc.p} = \Delta\varepsilon_{cc.p} = \varphi_{cc}(\sigma_{c0.p}/E_{c.0}) \tag{5.145}$$

and

$$\Delta\sigma_{p.cc} = E_p\varphi_{cc}(\sigma_{c0.p}/E_{c.0}) \tag{5.146}$$

As mentioned previously, the restraint to creep (and shrinkage) caused by the bonded reinforcement reduces the compressive force on the concrete with time and the stress in the concrete at the tendon level is never constant. The above equations therefore, will always overestimate creep. In recognition of this and in the absence of a more refined analysis, AS3600-2009 [2] permits a reduction of 20% in the creep strain calculated using the above equation and the change in stress in the tendon due to creep may be taken as:

$$\Delta\sigma_{p.cc} = 0.8E_p\varphi_{cc}(\sigma_{c0.p}/E_{c.0}) \tag{5.147}$$

5.10.3.4 Relaxation losses

The loss of stress in a tendon due to relaxation depends on the sustained stress in the steel. Owing to creep and shrinkage in the concrete, the stress in the tendon decreases with time at a faster rate than would occur due to relaxation alone. Since the steel strain is reducing with time due to concrete creep and shrinkage, the relaxation losses are reduced from those that would occur in a constant strain relaxation test. With the design relaxation R determined from Equation 4.30, the percentage loss of prestress due to relaxation may be calculated from a detailed time analysis such as described in Section 5.7. In the absence of such an analysis, AS3600-2009 [2] permits the change of stress in the tendon due to relaxation to be approximated by:

$$\Delta\sigma_{p.rel} = -R\left(1 - \frac{|\Delta\sigma_{p.cs} + \Delta\sigma_{p.cc}|}{\sigma_{p.0}}\right)\sigma_{p.0} \tag{5.148}$$

where $\Delta\sigma_{p.cs}$ and $\Delta\sigma_{p.cc}$ are the changes in stress in the tendon caused by shrinkage and creep, respectively, and are usually compressive; $\sigma_{p.0}$ is the tendon stress just after transfer under the sustained service loads and the design relaxation R is expressed as a decimal (not as a percentage). The absolute values of $\Delta\sigma_{p.cs}$ and $\Delta\sigma_{p.cc}$ are used in Equation 5.148, to convert the negative changes of stress into positive losses.

EXAMPLE 5.10

For the post-tensioned concrete cross-section and material properties shown in Figure 5.26, the immediate strain and stress distributions at τ_0 when subjected to prestress and an external bending moment of $M_{ext,0} = 100$ kNm were calculated in Example 5.3 and, with the external moment remaining constant in time, the long-term strain and stress distributions at τ_k were calculated in Example 5.5 accounting for the effects of creep and shrinkage

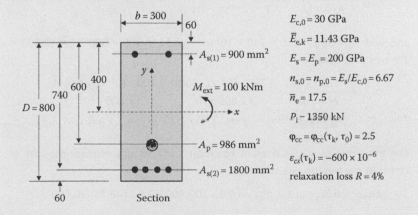

Figure 5.26 Cross-section and material properties (Example 5.10).

of the concrete and relaxation of the tendon. The immediate and long-term strain and stress distributions were shown in Figure 5.14, where we saw that the stress in the tendon decreased from 1369 MPa at τ_0 to 1100 MPa at τ_k.

Determine the time-dependent losses of prestress in the tendon using the approximate procedures specified in AS3600-2009 [2] and discussed in Section 5.10.3.

Shrinkage loss:

In this example, the gross area of the section is $A_g = b\,D = 240{,}000$ mm². The total area of bonded steel is $A_s = A_{s(1)} + A_{s(2)} + A_p = 3686$ mm² and its centroid is $e_s = 136.5$ mm below the centroidal axis of the gross section. The steel area to gross area ratio is $p = A_s/A_g = 0.0154$, and from Equation 5.144:

$$\Delta\sigma_{p.cs} = \frac{E_p \varepsilon_{cs}}{1+\bar{n}_e p\left[1+12\left(\dfrac{e_s}{D}\right)^2\right]} = \frac{200{,}000\times(-0.0006)}{1+17.5\times0.0154\left[1+12\left(\dfrac{136.5}{800}\right)^2\right]}$$

$$= -88.1\,\text{MPa}$$

The calculated stress determined using the time analysis of Section 5.7.3 caused by restrained shrinkage in an initially unloaded tendon in an initially unloaded cross-section is −86.4 MPa, and this is in close agreement with the approximation of Equation 5.144.

Creep loss:

A conservative approximation of the concrete stress at the tendon level based on gross section properties is:

$$\sigma_{c0.p} = \frac{-P_i}{A_g} - \frac{P_i y_p^2}{I_g} - \frac{M_{ext,0} y_p}{I_g}$$

and, with $A_g = 300 \times 800 = 240 \times 10^3$ mm² and $I_g = 300 \times 800^3/12 = 12{,}800 \times 10^6$ mm⁴, we have:

$$\sigma_{c0.p} = \frac{-1350\times10^3}{240\times10^3} - \frac{1{,}350\times10^3\times(-200)^2}{12{,}800\times10^6} - \frac{100\times10^6\times(-200)}{12{,}800\times10^6}$$

$$= -8.28\,\text{MPa}$$

and this value of $\sigma_{c0.p}$ is used in Equation 5.147 to approximate the loss of stress in the tendon due to creep:

$$\Delta\sigma_{p.cc} = 0.8 E_p \varphi_{cc}(\sigma_{c0.p}/E_{c.0}) = 0.8 \times 200{,}000 \times 2.5 \times (-8.28)/30{,}000$$

$$= -110.4\,\text{MPa}$$

Using the time analysis of Section 5.7.3, the loss of stress in the tendon due to creep is equal to -130.2 MPa and, for this cross-section with a relatively small sustained moment, the approximation of Equation 5.147 is in fact unconservative.

Relaxation loss:
With the stress in the tendon immediately after transfer $\sigma_{p.0} = P_i/A_p = 1369$ MPa, the loss of stress in the tendon due to relaxation is obtained from Equation 5.148:

$$\Delta\sigma_{p.rel} = -R\left(1 - \frac{|\Delta\sigma_{p.cs} + \Delta\sigma_{p.cc}|}{\sigma_{p.0}}\right)\sigma_{p.0}$$

$$= -0.04\left(1 - \frac{|-88.1 - 110.4|}{1369}\right) \times 1369 = -46.8\,\text{MPa}$$

Using the time analysis of Section 5.7.3, the loss of stress in the tendon due to relaxation is equal to -52.6 MPa and, for this cross-section, the approximation of Equation 5.148 is slightly unconservative.

Total time-dependent losses:
Summing the losses caused by shrinkage, creep and relaxation we get:

$$\Delta\sigma_{p.time} = \Delta\sigma_{p.cs} + \Delta\sigma_{p.cc} + \Delta\sigma_{p.rel} = -245.3\,\text{MPa}$$

This is 17.9% of the initial prestress in the tendon and, in this example, slightly underestimates (by 8.8%) the losses (-269 MPa) determined in Example 5.5 using the more rigorous procedure of Section 5.7.

5.11 DEFLECTION CALCULATIONS

5.11.1 General

If the axial strain and curvature are known at regular intervals along a member, it is a relatively simple task to determine the deformation of that member. Consider the statically determinate member AB of span L subjected to the axial and transverse loads shown in Figure 5.27a. The axial deformation of the member e_{AB} (either elongation or shortening) is obtained by integrating the axial strain at the centroid of the member $\varepsilon_a(z)$ over the length of the member, as shown:

$$e_{AB} = \int_0^L \varepsilon_a(z)\,dz \tag{5.149}$$

where z is measured along the member.

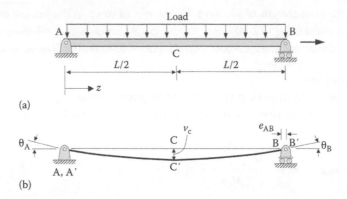

Figure 5.27 Deformation of a statically determinate member. (a) Original geometry. (b) Deformed shape.

Provided that deflections are small and that simple beam theory is applicable, the slope θ and deflection v at any point z along the member are obtained by integrating the curvature $\kappa(z)$ over the length of the member as follows:

$$\theta = \int \kappa(z) dz \qquad (5.150)$$

and

$$v = \iint \kappa(z) dz\, dz \qquad (5.151)$$

Equations 5.150 and 5.151 are quite general and apply to both elastic and inelastic material behaviour.

If the axial strain and curvature are calculated at any time after loading at a pre-selected number of points along the member shown in Figure 5.27a and, if a reasonable variation of strain and curvature is assumed between adjacent points, it is a simple matter of geometry to determine the deformation of the member. For convenience, some simple equations are given below for the determination of the deformation of a single span and of a cantilever. If the axial strain ε_a and the curvature κ are known at the mid-span and at each end of the member shown in Figure 5.27 (i.e. at supports A and B and at the mid-span C), the axial deformation e_{AB}, the slope at each support θ_A and θ_B and the deflection at mid-span v_C are given by Equations 5.152 to 5.159.

For a linear variation of strain and curvature:

$$e_{AB} = \frac{L}{4}\left(\varepsilon_{aA} + 2\varepsilon_{aC} + \varepsilon_{aB}\right) \tag{5.152}$$

$$v_{C} = \frac{L^{2}}{48}\left(\kappa_{A} + 4\,\kappa_{C} + \kappa_{B}\right) \tag{5.153}$$

$$\theta_{A} = \frac{L}{24}\left(5\,\kappa_{A} + 6\,\kappa_{C} + \kappa_{B}\right) \tag{5.154}$$

$$\theta_{B} = -\frac{L}{24}\left(\kappa_{A} + 6\,\kappa_{C} + 5\,\kappa_{B}\right) \tag{5.155}$$

For a parabolic variation of strain and curvature:

$$e_{AB} = \frac{L}{6}\left(\varepsilon_{aA} + 4\,\varepsilon_{aC} + \varepsilon_{aB}\right) \tag{5.156}$$

$$v_{C} = \frac{L^{2}}{96}\left(\kappa_{A} + 10\kappa_{C} + \kappa_{B}\right) \tag{5.157}$$

$$\theta_{A} = \frac{L}{6}\left(\kappa_{A} + 2\kappa_{C}\right) \tag{5.158}$$

$$\theta_{B} = -\frac{L}{6}\left(2\kappa_{C} + \kappa_{B}\right) \tag{5.159}$$

In addition to the simple span shown in Figure 5.27, Equations 5.152 through 5.159 also apply to any member in a statically indeterminate frame, provided the strain and curvature at each end and at mid-span are known.

Consider the fixed-end cantilever shown in Figure 5.28. If the curvatures at the fixed support at A and the free-end at B are known, then

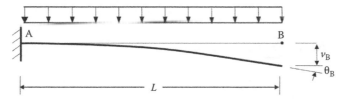

Figure 5.28 Deformation of a cantilever beam.

the slope and deflection at the free end of the member are given by Equations 5.160 through 5.163.

For a linear variation of curvature:

$$\theta_B = -\frac{L}{2}\left(\kappa_A + \kappa_B\right) \tag{5.160}$$

$$\upsilon_B = -\frac{L^2}{6}\left(2\kappa_A + \kappa_B\right) \tag{5.161}$$

For a parabolic variation of curvature (typical of what occurs in a uniformly loaded cantilever):

$$\theta_B = -\frac{L}{3}\left(\kappa_A + 2\kappa_B\right) \tag{5.162}$$

$$\upsilon_C = -\frac{L^2}{4}\left(\kappa_A + \kappa_B\right) \tag{5.163}$$

5.11.2 Short-term moment curvature relationship and tension stiffening

For any prestressed concrete section, the instantaneous moment-curvature relationship before cracking is linear elastic. For an uncracked cross-section, the instantaneous curvature may be calculated using the procedure of Section 5.6.2. In particular, from Equation 5.43, the instantaneous curvature is:

$$\kappa_0 = \frac{R_{B,0}N_{R,0} + R_{A,0}M_{R,0}}{R_{A,0}R_{I,0} - R_{B,0}^2} \tag{5.164}$$

where $R_{A,0}$, $R_{B,0}$ and $R_{I,0}$ are the cross-sectional rigidities given by Equations 5.33, 5.34 and 5.37; $N_{R,0}$ is the sum of the external axial force (if any) and the resultant compressive prestressing force exerted on the cross-section by the tendons; and $M_{R,0}$ is the sum of the external moment and the resultant moment about the centroidal axis caused by the compressive forces exerted by the tendons:

$$N_{R,0} = N_{ext,0} - \sum_{i=1}^{m_p} P_{init(i)} \tag{5.165}$$

$$M_{R,0} = M_{ext,0} + \sum_{i=1}^{m_p} y_{p(i)} P_{init(i)} \tag{5.166}$$

For uncracked, prestressed concrete cross-sections, if the reference axis is taken as the centroidal axis of the transformed section, the flexural rigidity $E_{c,0}I_{uncr}$ is in fact the rigidity $R_{I,0}$ calculated using Equation 5.37, where I_{uncr} is the second moment of area of the uncracked transformed cross-section about its centroidal axis. Codes of practice generally suggest that, for short-term deflection calculations, I_{uncr} may be approximated by the second moment of area of the gross cross-section about its centroidal axis. The initial curvature caused by the applied moment and prestress acting on any uncracked cross-section may therefore, be approximated by:

$$\kappa_0 = \frac{M_{R,0}}{E_{c,0}I_{uncr}} \tag{5.167}$$

After cracking, the instantaneous moment-curvature relationship can be determined using the analysis described in Section 5.8 (and illustrated in Example 5.7) for any level of applied moment greater than the cracking moment, provided the assumption of linear-elastic material behaviour remains valid for both the steel reinforcement/tendons and the concrete in compression. The analysis of the cross-section of Figure 5.19 after cracking in pure bending (i.e. when $M_{ext,0} = 400$ kNm and $N_{ext,0} = 0$) was illustrated in Example 5.7. If the analysis is repeated for different values of applied moment ($M_{ext,0}$ greater than the cracking moment M_{cr}), the instantaneous moment versus curvature ($M_{ext,0}$ vs. κ_0) relationship for the cross-section can be determined and is shown in Figure 5.29. In addition

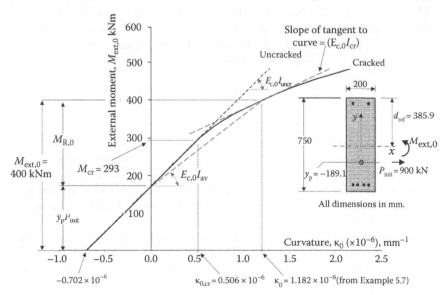

Figure 5.29 Short-term moment-curvature relationship for the prestressed concrete cross-section of Figure 5.19.

to the post-cracking relationship, the linear relationship prior to cracking (Equation 5.167) is also shown in Figure 5.29, with $E_{c,0}I_{uncr} = 242.4 \times 10^{12}$ Nmm^2 in this example.

It is noted that after cracking the neutral axis gradually rises as the applied moment increases. With the area of concrete above the crack becoming smaller, the second moment of area of the cracked section I_{cr} decreases as the applied moment increases. This is not the case for reinforced concrete sections where the depth to the neutral axis remains approximately constant with increasing moment and I_{cr} is constant in the post-cracking range.

For the cross-section shown in Figure 5.29, the x-axis has been taken to coincide with the centroidal axis of the transformed cross-section, and so the y coordinate of the prestressing steel (y_p) is numerically equal to the eccentricity of the prestressing force e.

In this case, from Equation 5.166, $M_{R,0} = M_{ext,0} + y_p P_{init} = M_{ext,0} - e P_{init}$. In Figure 5.29, at any moment $M_{ext,0}$ greater than the cracking moment (M_{cr}), the curvature is:

$$\kappa_0 = \frac{M_{R,0}}{E_{c,0}I_{av}} \tag{5.168}$$

where $E_{c,0}I_{av}$ is the secant stiffness. The secant stiffness $E_{c,0}I_{av}$ corresponding to the external moment of $M_{ext,0} = 400$ kNm is shown in Figure 5.29, with $y_p P_{init} = -170.2$ kNm and, therefore, $M_{R,0} = 229.8$ kNm. With the value of curvature determined in Example 5.7 at this moment equal to 1.182×10^{-6}, the stiffness $E_{c,0}I_{av}$ is obtained from Equation 5.168:

$$E_{c,0}I_{av} = \frac{M_{R,0}}{\kappa_0} = \frac{229.8 \times 10^6}{1.182 \times 10^{-6}} = 194.4 \times 10^{12} \text{ Nmm}^2$$

A conservative estimate of the instantaneous deflection of a prestressed concrete member is obtained if the value of $E_{c,0}I_{av}$ for the cross-section at the point of maximum moment is taken as the flexural rigidity of the member. In addition, when designing for crack control, variations in tensile steel stresses after cracking can be determined from the cracked section analysis.

The tangent stiffness $E_{c,0}I_{cr}$ is also shown in Figure 5.29. The second moment of area of the cracked section I_{cr} may be obtained using the cracked section analysis of Section 5.8. In that analysis, the tangent stiffness of the cracked cross-section was expressed as $R_{I,0}$ and was given by Equation 5.123 or 5.126. In our example, we can now calculate the tangent stiffness $E_{c,0}I_{cr}$ with respect to the centroidal axis when the external moment is $M_{ext,0} = 400$ kNm using the rigidities of the cracked section $R_{A,0}$, $R_{B,0}$ and $R_{I,0}$ as determined in Example 5.7:

$$E_{c,0}I_{cr} = \frac{R_{A,0}R_{I,0} - R_{B,0}^2}{R_{A,0}} = 120.2 \times 10^{12} \text{ Nmm}^2$$

If the reference axis had corresponded to the centroidal axis of the cracked section, then $R_{B,0}$ would equal zero and $E_{c,0}I_{cr} = R_{I,0}$.

For small variations in applied moment, curvature increments can be calculated using I_{cr}. In reinforced concrete construction, I_{cr} is constant and equal to I_{av}, but this is not so for prestressed concrete. For moment-curvature graph of Figure 5.29, at $M_{ext,0} = 400$ kNm, we have $I_{uncr} = 8080 \times 10^6$ mm^4, $I_{av} = 6480 \times 10^6$ mm^4 and $I_{cr} = 4008 \times 10^6$ mm^4. It is noted that for this 200 mm by 750 mm rectangular cross-section, $I_g = 7030 \times 10^6$ mm^4, which is 13% less than I_{uncr}.

For the case where $M_{ext,0} = 0$ in Figure 5.29, the internal moment caused by the resultant prestressing force about the centroidal axis of the uncracked section causes an initial negative curvature of $M_{R,0}/(E_{c,0}I_{uncr}) = -e_pP_{init}/ (E_{c,0}I_{uncr}) = -0.702 \times 10^{-6}$ mm^{-1}. If the beam remained unloaded for a period of time after transfer and, if shrinkage occurred during this period, the restraint provided by the bonded reinforcement to shrinkage would introduce a positive change of curvature $\kappa_{cs,0}$ provided the centroid of the bonded reinforcement is below the centroidal axis of the cross-section (as is the case in Figure 5.29). Shrinkage before loading causes the curve in Figure 5.29 to shift to the right. The restraint to shrinkage also causes tensile stresses in the bottom fibres of the cross-section, and this may significantly reduce the cracking moment. The effect of a modest early shrinkage on the moment-curvature relationship of Figure 5.29 is illustrated in Figure 5.30.

For cracked prestressed concrete members, the stiffness of the cracked cross-section, calculated using the procedure outlined in Section 5.8.3, may underestimate the actual stiffness of the member in the cracked region. The intact concrete between adjacent cracks carries tensile force, mainly in the direction of the reinforcement, due to the bond between the steel and the concrete. The average tensile stress in the concrete is therefore, not zero and may be a significant fraction of the tensile strength of concrete. The stiffening effect of the uncracked tensile concrete is known as *tension stiffening*. The moment curvature relationship of Figure 5.29 is reproduced in Figure 5.31. Also shown as the dashed line in Figure 5.31 is the moment versus *average curvature* relationship, with the average curvature being determined for a segment of beam containing two or more primary cracks. The hatched region between the curves at moments greater than the cracking moment M_{cr} represents the tension stiffening effect, i.e. the contribution of the tensile concrete between the primary cracks to the cross-sectional stiffness.

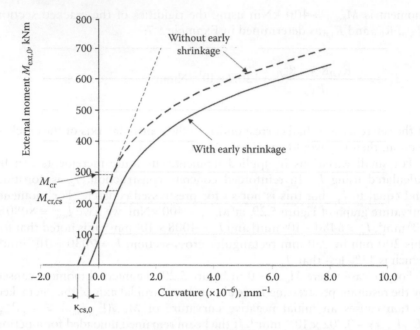

Figure 5.30 Effect of early shrinkage on the short-term moment-curvature relationship for a prestressed concrete cross-section.

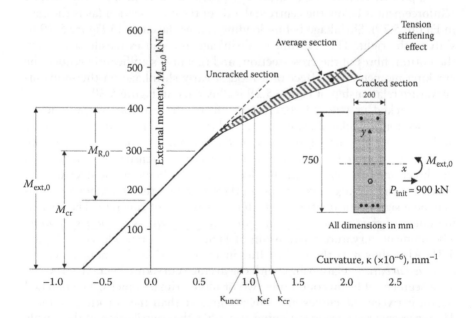

Figure 5.31 Short-term moment-average curvature relationship for the prestressed concrete cross-section of Figure 5.19.

For conventionally reinforced members, tension stiffening contributes significantly to the member stiffness, particularly when the maximum moment is not much greater than the cracking moment. However, as the moment level increases, the tension stiffening effect decreases owing to additional secondary cracking at the level of the bonded reinforcement.

For a prestressed member (or a reinforced member subjected to significant axial compression), the effect of tension stiffening is less pronounced because the loss of stiffness due to cracking is more gradual and significantly smaller.

Shrinkage-induced cracking and tensile creep cause a reduction of the tension stiffening effect with time. Repeated or cyclic loading also causes a gradual breakdown of tension stiffening.

Tension stiffening is usually accounted for in design by an empirical adjustment to the stiffness of the fully cracked cross-section as discussed in the following section.

5.11.3 Short-term deflection

If the initial curvature is determined at the mid-span and at each end of the span of a beam or slab, the short-term deflection can be estimated using Equations 5.153 or 5.157, whichever is appropriate.

For uncracked members, the initial curvature is given by Equation 5.167. For cracked prestressed members, the initial curvature may be determined from Equation 5.168, with $E_{c,0}I_{av}$ determined for the cracked section. This will be conservative unless an adjustment is made to include the tension stiffening effect. In codes of practice, this adjustment is often made using simplified techniques involving the determination of an effective second moment of area I_{ef} for the member. A number of empirical equations are available for estimating I_{ef}. Most have been developed specifically for reinforced concrete, where for a cracked member, I_{ef} lies between the second moments of area of the uncracked cross-section I_{uncr} and of the cracked transformed section I_{cr} about their centroidal axes. We have seen that for a prestressed concrete section, I_{cr} varies with the applied moment as the depth of the crack gradually changes and its value at any load level is usually considerably less than I_{av}, as illustrated in Figure 5.29. The equations used for estimating I_{ef} for a reinforced section are not therefore, directly applicable to prestressed concrete.

The following two well-known procedures for modelling tension stiffening may be applied to prestressed concrete provided I_{av} replaces I_{cr} in the original formulations:

1. The empirical equation for I_{ef} proposed by Branson [10] is adopted in many codes and specifications for reinforced concrete members, including AS3600-2009 [2]. For a prestressed concrete section, the following form of the equation can be used:

$$I_{ef} = I_{av} + (I_{uncr} - I_{av})(M_{cr}/M_{ext,0})^3 \leq I_{uncr} \tag{5.169}$$

where $M_{ext,0}$ is the maximum bending moment at the section, based on the short-term serviceability design load or the construction load and M_{cr} is the cracking moment. The cracking moment is best determined by undertaking a time analysis, as outlined in Section 5.7, to determine the effects of creep and shrinkage on the time-dependent redistribution of stresses between the concrete and the bonded reinforcement. In the absence of such an analysis, AS3600-2009 [2] suggests that the cracking moment may be taken as:

$$M_{cr} = Z \left(f'_{ct.f} - \sigma_{cs} + P/A \right) + Pe \geq 0.0 \tag{5.170}$$

where Z is the section modulus of the uncracked section, referred to the extreme fibre at which cracking occurs; $f'_{ct.f}$ is the characteristic flexural tensile strength of concrete (specified as $f'_{ct.f} = 0.6\sqrt{f'_c}$); P is the effective prestressing force (after all losses); e is the eccentricity of the effective prestressing force measured to the centroidal axis of the uncracked section; A is the area of the uncracked cross-section; and σ_{cs} is the maximum shrinkage-induced tensile stress on the uncracked section at the extreme fibre at which cracking occurs.

In the absence of more refined calculation, σ_{cs} may be taken as:

$$\sigma_{cs} = \left(\frac{2.5 p_w - 0.8 p_{cw}}{1 + 50 p_w} E_s \varepsilon_{cs}^* \right) \tag{5.171}$$

where p_w is the web reinforcement ratio for the tensile steel $(A_s + A_{pt})/(b_w d)$; p_{cw} is the web reinforcement ratio for the compressive steel, if any, $A_{sc}/(b_w d)$; A_{st} is the area on non-prestressed tensile reinforcement; A_{pt} is the area of prestressing steel in the tensile zone; A_{sc} is the area of non-prestressed compressive reinforcement; E_s is the elastic modulus of the steel in MPa; and ε_{cs}^* is the final design shrinkage strain (after 30 years).

2. An alternative approach for the inclusion of the tension stiffening effect is presented here and uses a procedure similar to that specified in Eurocode 2 [11]. The instantaneous curvature of a cracked prestressed section κ_{ef} accounting for tension stiffening (see Figure 5.31) may be calculated as a weighted average of the

values calculated on a cracked section (κ_{cr}) and on an uncracked section (κ_{uncr}) as follows:

$$\kappa_{ef} = \zeta\kappa_{cr} + (1 - \zeta)\kappa_{uncr} \qquad (5.172)$$

where ζ is a distribution coefficient that accounts for the moment level and the degree of cracking. For prestressed concrete flexural members, ζ may be taken as:

$$\zeta = 1 - \beta\left(\frac{M_{cr}}{M_{ext,0}}\right)^2 \qquad (5.173)$$

where β is a coefficient to account for the effects of duration of loading or repeated loading on the average deformation and equals 1.0 for a single, short-term load and 0.5 for sustained loading or many cycles of repeated loading; M_{cr} is the external moment at which cracking first occurs; and $M_{ext,0}$ is the external moment at which the instantaneous curvature is to be calculated (see Figure 5.31).

The introduction of $\beta = 0.5$ in Equation 5.173 for long-term loading reduces the cracking moment by about 30% and is a crude way of accounting for shrinkage-induced tension and time-dependent cracking.

If we express the curvatures in Equation 5.172 in terms of the flexural rigidities, i.e. $\kappa_{ef} = M_{R,0}/(E_c I_{ef})$, $\kappa_{cr} = M_{R,0}/(E_c I_{av})$ and $\kappa_{uncr} = M_{R,0}/(E_c I_{uncr})$, Equation 5.172 can be rearranged to give:

$$I_{ef} = \frac{I_{av}}{1 - \beta\left(1 - \dfrac{I_{av}}{I_{uncr}}\right)\left(\dfrac{M_{cr}}{M_{ext,0}}\right)^2} \leq I_{uncr} \qquad (5.174)$$

An expression similar to Equation 5.174 for reinforced concrete was first proposed by Bischoff [12].

Several other approaches have been developed for modelling the tension stiffening phenomenon. However, for most practical prestressed members, the maximum in-service moment is less than the cracking moment and cracking is not an issue. Even for those members that crack under service loads, the maximum moment is usually not much greater than the cracking moment and tension stiffening is not very significant. A conservative, but often a quite reasonable estimate of deflection can be obtained by ignoring tension stiffening and using $E_{c,0} I_{av}$ (from Equation 5.168) in the calculations.

EXAMPLE 5.11

Determine the *short-term or instantaneous* deflection of a uniformly loaded, simply-supported, post-tensioned beam of span 12 m immediately after first loading. An elevation of the member is shown in Figure 5.32, together with details of the cross-section at mid-span (which is identical with the

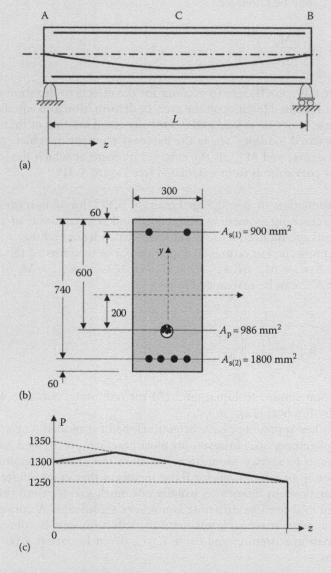

Figure 5.32 Beam details (Example 5.11). (a) Elevation. (b) Section at mid-span. (c) Prestressing force.

cross-section analysed in Example 5.3). The prestressing cable is parabolic with the depth of the tendon below the top fibre d_p at each support equal to 400 mm and at mid-span d_p = 600 mm, as shown. The non-prestressed reinforcement is uniform throughout the span.

Owing to friction and draw-in losses, the prestressing force at the left support is P = 1300 kN, at mid-span P = 1300 kN and at the right support P = 1250 kN, as shown in Figure 5.32c. The tendon is unbonded and housed inside a 60 mm diameter ungrouted duct.

Two service load cases are to be considered:

1. a uniformly distributed load of 6 kN/m (which is the self-weight of the member); and
2. a uniformly distributed load of 40 kN/m.

The material properties are: $E_{c,0}$ = 30 GPa, $E_s = E_p$ = 200 GPa, f_c' = 40 MPa and the flexural tensile strength is taken to be $f_{ct.f}' = 0.6\sqrt{f_c'} = 3.8$ MPa.

At support A: The applied moment at support A is zero for both load cases. The prestressing tendon is located at the mid-depth of the section (d_p = 400 mm) and the prestressing force P = 1300 kN. With the reference axis taken as the centroidal axis of the gross cross-section (as shown in Figure 5.32b) and using the cross-sectional analysis described in Section 5.6.2, the initial strain at the centroidal axis and the curvature are determined using Equation 5.43 as:

$$\varepsilon_{r,0} = -171.7 \times 10^{-6} \quad \text{and} \quad \kappa_0 = +0.0204 \times 10^{-6} \text{ mm}^{-1}$$

At support B: The prestressing force is 1250 kN and the tendon is located 400 mm below the top fibre. As at support A, $M_{ext,0}$ = 0 and, solving Equation 5.43, we get:

$$\varepsilon_{r,0} = -165.1 \times 10^{-6} \quad \text{and} \quad \kappa_0 = +0.0196 \times 10^{-6} \text{ mm}^{-1}$$

At mid-span C: The prestressing force is 1300 kN at a depth of 600 mm below the top fibre and, assuming no shrinkage has occurred prior to loading, the cracking moment may be estimated from Equation 5.170:

$$M_{cr} = Z\left(f_{ct.f}' + P/A\right) + Pe_p$$

$$= \frac{300 \times 800^2}{6}\left(3.8 + \frac{1300 \times 10^3}{300 \times 800}\right) + 1300 \times 10^3 \times 200$$

$$= 555 \times 10^6 \text{ Nmm} = 555 \text{ kNm}$$

Using the more accurate uncracked section analysis of Section 5.6.2, the second moment of area of the uncracked cross-section is I_{uncr} = 14,450 × 10^6 mm^4 and the cracking moment is determined to be M_{cr} = 581 kNm.

For load case (a): $M_{ext,0} = \dfrac{6 \times 12^2}{8} = 108$ kNm

The cross-section is uncracked and from Equation 5.43:

$\varepsilon_{r,0} = -170.1 \times 10^{-6}$ and $\kappa_0 = -0.337 \times 10^{-6}$ mm^{-1}

For load case (b): $M_{ext,0} = \dfrac{40 \times 12^2}{8} = 720$ kNm

The cross-section has cracked, and from Equation 5.120, the depth to the neutral axis is $d_n = 443.7$ mm. From Equation 5.43:

$\varepsilon_{r,0} = -70.5 \times 10^{-6}$ and $\kappa_0 = +1.610 \times 10^{-6}$ mm^{-1}

The value of I_{av} calculated from Equation 5.168 is:

$$I_{av} = \frac{M_{R,0}}{E_{c,0}\kappa_0} = \frac{M_{ext,0} + y_p P}{E_{c,0}\kappa_0} = \frac{720 \times 10^6 - 200 \times 1,300 \times 10^3}{30,000 \times 1.610 \times 10^{-6}}$$

$$= 9,530 \times 10^6 \text{ mm}^4$$

Deflection: With the initial curvature calculated at each end of the member and at mid-span, and with a parabolic variation of curvature along the beam, the short-term deflection at mid-span for each load case is determined using Equation 5.157.

For load case (a):

$$v_C = \frac{12,000^2}{96}\left[0.0204 + 10 \times (-0.337) + 0.0196\right] \times 10^{-6} = -5.0 \text{ mm} (-)$$

For load case (b):

$$v_C = \frac{12,000^2}{96}\left[0.0204 + 10 \times 1.610 + 0.0196\right] \times 10^{-6} = +24.2 \text{ mm} (\downarrow)$$

For load case (b), tension stiffening in the cracked region of the member near mid-span has been ignored. To include the effects of tension stiffening in the calculations, the effective second moment of area given by Equation 5.169 can be used instead of I_{av} for the estimation of curvature:

$$I_{ef} = I_{av} + (I_{uncr} - I_{av})(M_{cr} / M_{ext,0})^3$$

$$= [9,530 + (14,450 - 9,530) (581/720)^3] \times 10^6 = 12,110 \times 10^6 \text{ mm}^4$$

The revised curvature at mid-span for load case (b) is:

$$\kappa_0 = \frac{M_{ext,0} + y_p P}{E_{c,0}I_{ef}} = \frac{720 \times 10^6 - 200 \times 1,300 \times 10^3}{30,000 \times 12110 \times 10^6} = 1.266 \times 10^{-6} \text{ mm}^{-1}$$

and the revised mid-span deflection for load case (b) is:

$$v_C = \frac{12,000^2}{96}(0.0204 + 10 \times 1.266 + 0.0196) \times 10^{-6} = +19.0 \text{ mm } (\downarrow)$$

Alternatively, for the inclusion of the effects of tension stiffening in load case (b), Equation 5.174 could be used to estimate I_{ef}:

$$I_{ef} = \frac{9,530 \times 10^6}{1 - \left(1 - \frac{9,530 \times 10^6}{14,450 \times 10^6}\right)\left(\frac{581}{720}\right)^2} = 12,240 \times 10^6 \text{ mm}^4$$

The revised curvature at mid-span for load case (b) is:

$$\kappa_0 = \frac{M_{ext,0} + y_p P}{E_{c,0} I_{ef}} = 1.252 \times 10^{-6} \text{ mm}^{-1}$$

and the revised mid-span deflection for load case (b) is:

$$v_C = \frac{12,000^2}{96}(0.0204 + 10 \times 1.252 + 0.0196) \times 10^{-6} = +18.8 \text{ mm } (\downarrow)$$

5.11.4 Long-term deflection

Long-term deflections due to concrete creep and shrinkage are affected by many variables, including load intensity, mix proportions, member size, age at first loading, curing conditions, total quantity of compressive and tensile reinforcing steel, level of prestress, relative humidity and temperature. To account accurately for these parameters, a time analysis similar to that described in Sections 5.7.3 and 5.9.2 is required. The change in curvature during any period of sustained load may be calculated using Equation 5.89. Typical calculations are illustrated in Examples 5.5 and 5.6 for uncracked cross-sections and in Example 5.8 for a cracked cross-section.

When the final curvature has been determined at each end of the member and at mid-span, the long-term deflection can be calculated using either Equation 5.153 or 5.157.

In prestressed concrete construction, a large proportion of the sustained external load is often balanced by the transverse force exerted by the tendons. Under this balanced load, the short-term deflection may be zero, but the long-term deflection is not zero. The restraint to creep and shrinkage offered by non-symmetrically placed bonded reinforcement on a section can cause significant time-dependent curvature and, hence, significant deflection of the member. The use of a simple deflection multiplier to calculate long-term deflection from the short-term deflection is therefore, not satisfactory.

In this section, approximate procedures are presented, which allow a rough estimate of long-term deflections. In some situations, this is all that is required. However, for most applications, the procedures outlined in Sections 5.7.3 and 5.9.2 are recommended.

5.11.4.1 Creep-induced curvature

The creep-induced curvature $\kappa_{cc}(t)$ of a particular cross-section at any time t due to a sustained service load first applied at age τ_0 may be obtained from:

$$\kappa_{cc}(t) = \kappa_{sus,0} \frac{\varphi_{cc}(t, \tau_0)}{\alpha} \tag{5.175}$$

where $\kappa_{sus,0}$ is the instantaneous curvature due to the sustained service loads; $\varphi_{cc}(t,\tau_0)$ is the creep coefficient at time t due to load first applied at age τ_0; and α is a creep modification factor that accounts for the effects of cracking and the restraining action of the reinforcement on creep and may be estimated from either Equation 5.176, 5.177 or 5.178 [6,14].

For a cracked reinforced concrete section in pure bending ($I_{ef} < I_{uncr}$), $\alpha = \alpha_1$, where:

$$\alpha_1 = \left(0.48p^{-0.5}\right)\left(\frac{I_{cr}}{I_{ef}}\right)^{0.33}\left[1 + (125p + 0.1)\left(\frac{A_{sc}}{A_{st}}\right)^{1.2}\right] \tag{5.176}$$

For an uncracked reinforced or prestressed concrete section ($I_{ef} = I_{uncr}$), $\alpha = \alpha_2$, where:

$$\alpha_2 = 1.0 + \left(45p - 900p^2\right)\left(1 + \frac{A_{sc}}{A_{st}}\right) \tag{5.177}$$

and A_{st} is the equivalent area of bonded reinforcement in the tensile zone (including bonded tendons); A_{sc} is the area of the bonded reinforcement in the compressive zone between the neutral axis and the extreme compressive fibre; p is the tensile reinforcement ratio $A_{st}/(b\,d_o)$ and d_o is the depth from the extreme compressive fibre to the centroid of the outermost layer of tensile reinforcement. The area of any bonded reinforcement in the tensile zone (including bonded tendons) not contained in the outermost layer of tensile reinforcement (i.e. located at a depth d_1 less than d_o) should be included in the calculation of A_{st} by multiplying that area by d_1/d_o. For the purpose of calculating A_{st}, the tensile zone is that zone that would be in tension due to the applied moment acting in isolation.

For a cracked prestressed concrete section or for a cracked reinforced concrete section subjected to bending and axial compression, α may be taken as:

$$\alpha = \alpha_2 + (\alpha_1 - \alpha_2)\left(\frac{d_{n1}}{d_n}\right)^{2.4} \qquad (5.178)$$

where α_1 is determined from Equation 5.176; d_n is the depth of the intact compressive concrete on the cracked section; and d_{n1} is the depth of the intact compressive concrete on the cracked section ignoring the axial compression and/or the prestressing force (i.e. the value of d_n for an equivalent cracked reinforced concrete section in pure bending containing the same quantity of bonded reinforcement).

5.11.4.2 Shrinkage-induced curvature

The shrinkage-induced curvature on a reinforced or prestressed concrete section is approximated by:

$$\kappa_{cs}(t) = -\left[\frac{k_r \varepsilon_{cs}(t)}{D}\right] \qquad (5.179)$$

where D is the overall depth of the section; ε_{cs} is the shrinkage strain (note that ε_{cs} is a negative value); and k_r depends on the quantity and location of bonded reinforcement A_{st} and A_{sc} and may be estimated from Equation 5.180 through 5.183, as appropriate [6,14].

For a cracked reinforced concrete section in pure bending ($I_{ef} < I_{uncr}$), $k_r = k_{r1}$, where:

$$k_{r1} = 1.2\left(\frac{I_{cr}}{I_{ef}}\right)^{0.67}\left(1 - 0.5\frac{A_{sc}}{A_{st}}\right)\left(\frac{D}{d_o}\right) \qquad (5.180)$$

For an uncracked cross-section ($I_{ef} = I_{uncr}$), $k_r = k_{r2}$, where:

$$k_{r2} = (100p - 2500p^2)\left(\frac{d_o}{0.5D} - 1\right)\left(1 - \frac{A_{sc}}{A_{st}}\right)^{1.3} \quad \text{when } p = A_{st}/bd_o \leq 0.01 \qquad (5.181)$$

$$k_{r2} = (40p + 0.35)\left(\frac{d_o}{0.5D} - 1\right)\left(1 - \frac{A_{sc}}{A_{st}}\right)^{1.3} \quad \text{when } p = A_{st}/bd_o > 0.01 \quad (5.182)$$

In Equations 5.180 through 5.182, A_{sc} is defined as the area of the bonded reinforcement on the compressive side of the cross-section. This is a different definition to that provided under Equation 5.177. Whilst bonded steel near the compressive face of a cracked cross-section that is located at or below the neutral axis will not restrain compressive creep, it will provide restraint to shrinkage and it will be effective in reducing shrinkage-induced curvature on a cracked section.

For a cracked prestressed concrete section or for a cracked reinforced concrete section subjected to bending and axial compression, k_r may be taken as:

$$k_r = k_{r2} + (k_{r1} - k_{r2})\left(\frac{d_{n1}}{d_n}\right) \qquad (5.183)$$

where k_{r1} and k_{r2} are determined from Equations 5.180 through 5.182 by replacing A_{st} with $(A_{st} + A_{pt})$; and d_n and d_{n1} are as defined after Equation 5.178.

Equations 5.175 through 5.183 have been developed [14] as empirical fits to results obtained from a parametric study of the creep- and shrinkage-induced changes in curvature on reinforced and prestressed concrete cross-sections under constant sustained internal actions using the age-adjusted effective modulus method (AEMM) of analysis presented in Sections 5.7 and 5.9.

EXAMPLE 5.12

The final time-dependent deflection of the beam described in Example 5.11 and illustrated in Figure 5.32 is to be calculated. It is assumed that the duct is grouted soon after transfer and the tendon is effectively bonded to the concrete for the time period τ_0 to τ_k. As in Example 5.11, two load cases are to be considered:

(a) a uniformly distributed constant sustained load of 6 kN/m;
(b) a uniformly distributed constant sustained load of 40 kN/m.

For each load case, the time-dependent material properties are:

$\varphi_{cc}(\tau_k,\tau_0) = 2.5; \quad \chi(\tau_k,\tau_0) = 0.65; \quad \varepsilon_{cs}(\tau_k) = -450 \times 10^{-6};$
$\varphi_p(\tau_k,\sigma_{p(i),init}) = 0.03.$

All other material properties are as specified in Example 5.11.

(i) Calculation using the refined method (AEMM Analysis):

At support A: The sustained moment at support A is zero for both load cases and the prestressing force is $P = 1300$ kN at $d_p = 400$ mm. Using the procedure outlined in Section 5.7.3 and solving Equation 5.89, the strain at the reference axis and the curvature at time τ_k are:

$$\varepsilon_{r,k} = -843 \times 10^{-6} \quad \text{and} \quad \kappa_k = +0.260 \times 10^{-6} \text{ mm}^{-1}$$

At support B: As at support A, the sustained moment is zero, but the prestressing force is $P = 1250$ kN at $d_p = 400$ mm. Solving Equation 5.89, the strain at the reference axis and the curvature at time τ_k are:

$$\varepsilon_{r,k} = -824 \times 10^{-6} \quad \text{and} \quad \kappa_k = +0.254 \times 10^{-6} \text{ mm}^{-1}$$

At mid-span C: For load case (a), $M_{ext,k} = 108$ kNm and the prestressing force is 1300 kN at a depth of 600 mm below the top fibre. For this uncracked section, solving Equation 5.89 gives:

$$\varepsilon_{r,k} = -823 \times 10^{-6} \quad \text{and} \quad \kappa_k = -0.492 \times 10^{-6} \text{ mm}^{-1}$$

For load case (b), $M_{ext,0} = 720$ kNm and for this cracked cross-section with $d_n = 443.7$ mm, Equation 5.89 gives:

$$\varepsilon_{r,k} = -771 \times 10^{-6} \quad \text{and} \quad \kappa_k = +4.056 \times 10^{-6} \text{ mm}^{-1}$$

To include tension stiffening, we may use Equations 5.172 and 5.173. If the cross-section for load case (b) was considered to be uncracked, and the uncracked cross-section reanalysed, the final curvature is $(\kappa_k)_{uncr} = +3.263 \times 10^{-6} \text{ mm}^{-1}$. With the cracking moment determined in Example 5.11 to be $M_{cr} = 581$ kNm, Equation 5.173 gives:

$$\zeta = 1 - 0.5\left(\frac{581}{720}\right)^2 = 0.674$$

and from Equation 5.172:

$$\kappa_{ef,k} = 0.674 \times 4.056 \times 10^{-6} + (1 - 0.674) \times 3.263 \times 10^{-6}$$
$$= 3.798 \times 10^{-6} \text{ mm}^{-1}$$

Deflection: With the final curvature calculated at each end of the member and at mid-span, and with a parabolic variation of curvature along the beam, the long-term deflection at mid-span for each load case is determined using Equation 5.157.

For load case (a):

$$v_C = \frac{12,000^2}{96}\left[0.260 + 10 \times (-0.492) + 0.254\right] \times 10^{-6} = -6.6 \text{ mm } (\uparrow)$$

For load case (b), including the effects of tension stiffening:

$$v_C = \frac{12,000^2}{96}\left[0.260 + 10 \times 3.798 + 0.254\right] \times 10^{-6} = +57.7 \text{ mm } (\downarrow)$$

(ii) Calculation using the simplified method (Equations 5.175 through 5.183):

Load Case (a):

The creep and shrinkage induced curvatures at each support and at mid-span are estimated using Equation 5.175 and 5.179, respectively. At supports A and B, the cross-section is uncracked, with $A_{sc} = A_{s(1)} = 900 \text{ mm}^2$, $A_{st} = A_{s(2)} = 1800 \text{ mm}^2$ (noting that the tendons are at mid-depth and therefore, not in the tension zone) and $\rho = A_{st}/bd_o = 0.00811$. From Equation 5.177:

$$\alpha = \alpha_2 = 1.0 + \left(45 \times 0.00811 - 900 \times 0.00811^2\right)\left(1 + \frac{900}{1800}\right) = 1.46$$

and, from Equation 5.181:

$$k_r = k_{r2} = (100 \times 0.00811 - 2500 \times 0.00811^2)\left(\frac{740}{0.5 \times 800} - 1\right)\left(1 - \frac{900}{1800}\right)^{1.3}$$

$$= 0.223$$

From Equation 5.175, the creep induced curvatures are:

At support A : $\kappa_{cc}(t) = 0.0204 \times 10^{-6} \times \dfrac{2.5}{1.46} = 0.035 \times 10^{-6} \text{ mm}^{-1}$

At support B : $\kappa_{cc}(t) = 0.0196 \times 10^{-6} \times \dfrac{2.5}{1.46} = 0.034 \times 10^{-6} \text{ mm}^{-1}$

and, from Equation 5.179, the shrinkage induced curvature at each support is:

$$\kappa_{cs}(t) = -\left[\frac{0.223 \times (-450 \times 10^{-6})}{800}\right] = +0.125 \times 10^{-6} \text{ mm}^{-1}$$

At the mid-span C, the cross-section is uncracked, with $A_{sc} = 900$ mm², $A_{st} = A_{s(2)} + A_p(d_p/d_o) = 1800 + 986 \times (600/740) = 2599$ mm² and $p = A_{st}/bd_o = 0.0117$. From Equation 5.177 and Equation 5.182, we get, respectively:

$$\alpha = \alpha_2 = 1.0 + \left(45 \times 0.0117 - 900 \times 0.0117^2\right)\left(1 + \frac{900}{2599}\right) = 1.54$$

and

$$k_{r2} = (40 \times 0.0117 + 0.35)\left(\frac{740}{0.5 \times 800} - 1\right)\left(1 - \frac{900}{2599}\right)^{1.3} = 0.400$$

From Equation 5.175 and 5.179, the creep and shrinkage induced curvatures at mid-span are, respectively:

$$\kappa_{cc}(t) = -0.337 \times 10^{-6} \times \frac{2.5}{1.54} = -0.547 \times 10^{-6} \text{ mm}^{-1}$$

and

$$\kappa_{cs}(t) = -\left[\frac{0.400 \times (-450 \times 10^{-6})}{800}\right] = +0.225 \times 10^{-6} \text{ mm}^{-1}$$

The final curvature at each cross-section is the sum of the instantaneous, creep and shrinkage induced curvatures:

At support A : $\kappa(t) = (0.0204 + 0.035 + 0.125) \times 10^{-6}$

$$= +0.181 \times 10^{-6} \text{ mm}^{-1}$$

At support B : $\kappa(t) = (0.0196 + 0.034 + 0.125) \times 10^{-6}$

$$= +0.179 \times 10^{-6} \text{ mm}^{-1}$$

At mid-span C : $\kappa(t) = (-0.337 - 0.547 + 0.225) \times 10^{-6}$

$$= -0.659 \times 10^{-6} \text{ mm}^{-1}$$

From Equation 5.157, the long-term deflection at mid-span for load case (a) is:

$$v_C = \frac{12,000^2}{96}\left[0.181 + 10 \times (-0.659) + 0.179\right] \times 10^{-6} = -9.3 \text{ mm } (\uparrow)$$

For this load case, the simplified equations (Equations 5.175 through 5.183) overestimate the upward camber of the uncracked beam calculated using the more refined AEMM method (−6.6 mm).

Load Case (b):

As for load case (a), at supports A and B, the cross-sections are uncracked, with $A_{sc} = A_{s(1)} = 900$ mm^2, $A_{st} = A_{s(2)} = 1800$ mm^2 and p = $A_{st}/bd_o = 0.00811$. The final curvatures at the two supports are identical to those calculated for load case (a):

$$\text{At support A:} \quad \kappa(t) = (0.0204 + 0.035 + 0.125) \times 10^{-6}$$

$$= +0.181 \times 10^{-6} \text{ mm}^{-1}$$

$$\text{At support B:} \quad \kappa(t) = (0.0196 + 0.034 + 0.125) \times 10^{-6}$$

$$= +0.179 \times 10^{-6} \text{ mm}^{-1}$$

At mid-span C, the cross-section is cracked, with $A_{sc} = 900$ mm^2, $A_{st} = A_{s(2)} + A_p(d_p/d_o) = 1800 + 986 \times (600/740) = 2599$ mm^2 and $p = A_{st}/bd_o = 0.0117$. From Example 5.11, $d_n = 443.7$ mm. If the prestress is ignored and a cracked section analysis is performed on the equivalent reinforced concrete cross-section, we determine that $d_{nl} = 227.3$ mm (where d_{nl} is defined in the text under Equation 5.178) and $I_{cr} = 5360 \times 10^6$ mm^4. In Example 5.11, the value of I_{ef} determined using Equation 5.169 was calculated as $I_{ef} = 12,110 \times 10^6$ mm^4. From Equations 5.176 and 5.177, we get:

$$\alpha_1 = \left(0.48 \times 0.0117^{-0.5}\right)\left(\frac{5,360 \times 10^6}{12,110 \times 10^6}\right)^{0.33}\left[1+(125 \times 0.0117 + 0.1)\left(\frac{900}{2,599}\right)^{1.2}\right]$$

$$= 4.87$$

and

$$\alpha_2 = 1.0 + \left(45 \times 0.0117 - 900 \times 0.0117^2\right)\left(1 + \frac{900}{2599}\right) = 1.54$$

and from Equation 5.179, we have:

$$\alpha = 1.54 + (4.87 - 1.54)\left(\frac{227.3}{443.7}\right)^{2.4} = 2.21$$

The creep induced curvature at mid-span is given by Equation 5.175:

$$\kappa_{cc}(t) = 1.610 \times 10^{-6} \times \frac{2.5}{2.21} = 1.821 \times 10^{-6} \text{ mm}^{-1}$$

From Equations 5.180 and 5.182, we get:

$$k_{r1} = 1.2 \left(\frac{5,360 \times 10^6}{12,110 \times 10^6} \right)^{0.67} \left(1 - 0.5 \times \frac{900}{2,599} \right) \left(\frac{800}{740} \right) = 0.621$$

and

$$k_{r2} = (40 \times 0.0117 + 0.35) \left(\frac{740}{0.5 \times 800} - 1 \right) \left(1 - \frac{900}{2599} \right)^{1.3} = 0.400$$

Equation 5.183 gives:

$$k_r = 0.400 + (0.621 - 0.400) \left(\frac{227.3}{443.7} \right) = 0.513$$

and from Equation 5.179, the shrinkage induced curvature at mid-span is:

$$\kappa_{cs}(t) = -\left[\frac{0.514 \times (-450 \times 10^{-6})}{800} \right] = +0.289 \times 10^{-6} \text{ mm}^{-1}$$

The final curvature at mid-span is therefore:

$$\kappa(t) = (1.610 + 1.820 + 0.289) \times 10^{-6} = +3.720 \times 10^{-6} \text{ mm}^{-1}$$

From Equation 5.157, the long-term deflection at mid-span for load case (b) (including the effects of tension stiffening) is:

$$v_C = \frac{12,000^2}{96} (0.181 + 10 \times 3.720 + 0.179) \times 10^{-6} = +56.3 \text{ mm } (\downarrow)$$

For this load case, the deflection determined using the simplified equations (Equations 5.175 through 5.183) is in good agreement with the final long-term deflection of the cracked beam calculated using the more refined AEMM method (+57.7 mm).

5.12 CRACK CONTROL

5.12.1 Flexural crack control

When flexural cracking occurs in a prestressed concrete beam or slab, the axial prestressing force on the concrete controls the propagation of the crack and, unlike flexural cracking in a reinforced concrete member, the crack does not suddenly propagate to its full height (usually a large percentage of the depth of the cross-section). The height of a flexural crack gradually increases as the load increases and the loss of stiffness due to cracking is far more gradual than for a reinforced concrete member. The change in strain at the tensile steel level at first cracking is much less than that in a conventionally reinforced section with similar quantities of bonded reinforcement. After cracking, therefore, a prestressed beam generally suffers less deformation than the equivalent reinforced concrete beam, with finer, less extensive cracks. Flexural crack control in prestressed concrete beams and slabs is therefore not usually a critical design consideration provided bonded reinforcement is provided in the tensile zone.

AS3600-2009 [2] states that if the maximum tensile stress in the concrete due to short-term service loads is less than $0.25\sqrt{f_c'}$, the section may be considered to be uncracked and no further consideration needs to be given to crack control. This can be problematic, since time-dependent cracking may occur due to the loss of compressive stress in the concrete due to the restraint provided by the bonded reinforcement to creep and shrinkage deformations.

If the maximum tensile concrete stress due to short-term service loads is greater than $0.25\sqrt{f_c'}$, AS3600-2009 [2] requires that bonded reinforcement and/or bonded tendons be provided near the tensile face with a centre-to-centre spacing not exceeding 300 mm. In addition, one of the following alternatives must be satisfied:

1. the calculated maximum flexural tensile stress at the extreme concrete tensile fibre due to short-term service loads must be less than $0.6\sqrt{f_c'}$; or
2. the increment in the tensile stress in the steel near the tension face, as the applied load increases from its value when the extreme fibre is at zero stress (the decompression load) to the full short-term service load, must not exceed the maximum values given in Table 5.3.

Crack control is deemed to be provided in (1), if the maximum tensile stress calculated on the uncracked transformed cross-section does not exceed the lower characteristic flexural tensile strength of the concrete. In this case, cracking may still occur, but the change in tensile concrete and steel strains will not be great and crack control will not be a problem, provided some bonded steel at a spacing less than 300 mm is located in the tensile zone.

The alternative provision for crack control in (2) is to limit the change in stress that occurs in the tensile steel due to cracking to a maximum value that depends on the diameter of the bonded reinforcement or tendons. Although the values given in Table 5.3 are generally considered to be conservative, flexural crack control is rarely a problem in prestressed concrete structures containing bonded reinforcement in the tensile zone.

Table 5.3 Maximum increment of steel stress for flexural crack control
in prestressed members [2]

Beams		Slabs		
Nominal reinforcement bar diameter, d_b (mm)	Maximum increment of steel stress (MPa)	Nominal reinforcement bar diameter, d_b (mm)	Maximum increment of steel stress (MPa) when overall depth $D \leq 300$ mm	Maximum increment of steel stress (MPa) when overall depth $D > 300$ mm
≤12	330	≤10	320	360
16	280	12	300	330
20	240	16	265	280
24	210	20		240
≥28	200	24		210
		≥28		200
All bonded tendons	200	All bonded tendons		200

5.12.2 Crack control for restrained shrinkage and temperature effects

We saw in the previous section that flexural cracks are rarely a problem in prestressed concrete beams and slabs, provided of course that bonded reinforcement at reasonable spacing crosses the crack and that the member does not deflect excessively. In contrast, direct tension cracks due to restrained shrinkage and temperature changes may lead to serviceability problems, particularly in regions of low moment and in slab directions with little or no prestress. Such cracks usually extend completely through the member and are more parallel sided than flexural cracks. If uncontrolled, these cracks can become very wide and lead to waterproofing and corrosion problems. They can also disrupt the integrity and the structural action of the member.

Evidence of direct tension type cracks is common in concrete slab systems. For example, consider a typical one-way beam and slab floor system. The load is usually carried by the slab in the *primary direction* across the span to the supporting beams, while in the orthogonal direction (the *secondary direction*), the bending moment is small. Shrinkage is the same in both directions and restraint to shrinkage usually exists in both directions.

In the primary direction, prestress may eliminate flexural cracking, but if the level of prestress is such that flexural cracking does occur, shrinkage will cause small increases in the widths of the flexural cracks and may cause additional flexure type cracks in the previously uncracked regions. However, in the secondary direction, which is in effect a direct tension situation, there may be little or no prestress and shrinkage may cause a few widely spaced cracks that penetrate completely through the slab. Frequently, more reinforcement is required in the secondary direction to control these direct tension cracks than is required for bending in the primary direction. As far as cracking is concerned, it is not unreasonable to say that shrinkage is a greater problem when it is not accompanied by flexure and when the level of prestress is low.

When determining the amount of reinforcement required in a slab to control shrinkage- and temperature-induced cracking, AS3600-2009 [2] states that account should be taken of the influence of bending, the degree of restraint against in-plane movements and the exposure classification.

Where the ends of a slab are restrained and the slab is not free to expand or contract in the secondary direction, the minimum area of reinforcement required by AS3600-2009 [2] in the restrained direction is given by either Equation 5.184, 5.185 or 5.186, as appropriate.

For a slab fully enclosed within a building except for a brief period of weather exposure during construction:

1. where a strong degree of control over cracking is required for appearance or where cracks may reflect through finishes:

$$\left(A_s\right)_{\min} = (6.0 - 2.5\sigma_{cp})bD \times 10^{-3} \tag{5.184}$$

2. where a moderate degree of control over cracking is required and where cracks are inconsequential or hidden from view:

$$\left(A_s\right)_{\min} = (3.5 - 2.5\sigma_{cp})bD \times 10^{-3} \tag{5.185}$$

3. where a minor degree of control over cracking is required:

$$\left(A_s\right)_{\min} = (1.75 - 2.5\sigma_{cp})bD \times 10^{-3} \tag{5.186}$$

in which σ_{cp} is the average prestress P/A. For slabs greater than 500 mm in thickness, the reinforcement required near each surface may be determined assuming that $D = 250$ mm in Equations 5.184 through 5.186.

For all other surfaces and exposure environments in classification A1 and for exposure classification A2, either Equation 5.184 or 5.185 applies depending on the degree of crack control required. For the more severe exposure classifications B1, B2, C1 and C2, Equation 5.184 always applies. These exposure classifications are defined in Table 2.6.

In the primary direction of a one-way slab or in each direction of a two-way slab, the minimum quantity of reinforcement according to AS3600-2009 [2] is the greater of the minimum quantity required for the strength limit state or 75% of the minimum area required by Equations 5.184 through 5.186, as appropriate.

Equation 5.184 applies, for example, to a slab in a building where visible cracks could cause aesthetic problems or cracking to brittle floor finishes. Equation 5.185 would apply, for example, in the case of an interior slab where visible cracking could be tolerated or in the case of an interior slab that was later to be covered by a floor covering or a false ceiling. The area of steel specified in Equation 5.186 provides very little control, since the steel will yield at first cracking. It is therefore, difficult to imagine a situation where this value applies, except in an unrestrained direction or for a slab with closely spaced control joints that eliminate restraint.

5.12.3 Crack control at openings and discontinuities

Openings and discontinuities in slabs are the cause of stress concentrations that may result in diagonal cracks emanating from re-entrant corners. Additional reinforcing bars are generally required to trim the hole and to control the propagation of these cracks. A suitable method of estimating the number and size of the trimming bars is to postulate a possible crack and to provide reinforcement to carry a force at least equivalent to the area of the crack surface multiplied by the mean direct tensile strength of the concrete. For crack control, the maximum stress in the trimming bars should be limited to 200 MPa.

While this additional reinforcement is required for serviceability to control cracking at re-entrant corners, it should not be assumed that this same steel is satisfactory for strength. For a small hole through a slab, it is generally sufficient for bending to place additional steel on either side of the hole equivalent to the steel that must be terminated at the face of the opening. The effects of a large hole or opening should be determined by appropriate analysis accounting for the size, shape and position of the opening. Plastic methods of design, such as the yield line method (see Section 12.9.7) or the simplified strip method, are convenient ways of designing such slabs to meet requirements for strength.

REFERENCES

1. ACI318-11M 2011. (2011). *Building Code Requirements for Reinforced Concrete*. Detroit, MI: American Concrete Institute.
2. AS3600-2009. (2009). *Australian Standard for Concrete Structures*. Standards Association of Australia, New South Wales, Sydney.
3. Magnel, G. (1954). *Prestressed Concrete*, 3rd edn. London, U.K.: Concrete Publications Ltd.
4. Lin, T.Y. (1963). *Prestressed Concrete Structures*. New York: Wiley.
5. Warner, R.F. and Faulkes, K.A. (1979). *Prestressed Concrete*. Melbourne, Victoria, Australia: Pitman Australia.

6. Gilbert, R.I. and Ranzi, G. (2011). *Time-Dependent Behaviour of Concrete Structures*. London, U.K.: Spon Press.
7. Gilbert, R.I. (1988). *Time Effects in Concrete Structures*. Amsterdam, The Netherlands: Elsevier.
8. Ghali, A., Favre, R. and Elbadry, M. (2002). *Concrete Structures: Stresses and Deformations*, 3rd edn. London, U.K.: Spon Press.
9. AS5100.5-2004. (2006). Australian standard for bridge design – Part 5: Concrete. Standards Australia, Sydney, New South Wales, Australia, 2006, 164 pp.
10. Branson, D.E. (1963). Instantaneous and time-dependent deflection of simple and continuous reinforced concrete beams. Alabama Highway Research Report, No. 7, Bureau of Public Roads, Montgomery, AL.
11. Eurocode 2: (2004), Design of concrete structures – part 1–1: General rules and rules for buildings – BS EN 1992-1-1:2004. European Committee for Standardization, Brussels, Belgium.
12. Bischoff, P.H. (2005). Reevaluation of deflection prediction for concrete beams reinforced with steel and FRP bars. *Journal of Structural Engineering ASCE*, 131(5), 752–767.
13. Bazant, Z.P. (1972). Prediction of concrete creep effects using age-adjusted effective modulus method. *ACI Journal*, 69, 212–217.
14. Gilbert, R.I. (2001). Deflection calculation and control – Australian code amendments and improvements (Chapter 4). In *ACI International SP 203, Code Provisions for Deflection Control in Concrete Structures*. Farmington Hills, MI: American Concrete Institute, pp. 45–78.

Chapter 6

Ultimate flexural strength

6.1 INTRODUCTION

An essential design objective for a structure or a component of a structure is the provision of adequate strength. The consequences and costs of strength failures are high, and therefore, the probability of such failures must be very small.

The satisfaction of concrete and steel stress limits at service loads does not necessarily ensure adequate strength and does not provide a reliable indication of either the actual strength or the safety of a structural member. It is important to consider the nonlinear behaviour of the member in the overloaded range to ensure that it has an adequate structural capacity. Only by calculating the ultimate capacity of a member can a sufficient margin between the service load and the ultimate load be guaranteed.

The ultimate strength of a cross-section in bending M_u is calculated from a rational and well-established procedure involving consideration of the strength of both the concrete and the steel in the compressive and tensile parts of the cross-section. The prediction of ultimate flexural strength is described and illustrated in this chapter. When M_u is determined, the design requirements for the strength limit state (as discussed in Section 2.4) may be checked and satisfied.

In addition to calculating the strength of a section, a measure of the ductility of each section must also be established. Ductility is an important objective in structural design. Ductile members undergo large deformations prior to failure, thereby providing warning of failure and allowing indeterminate structures to establish alternative load paths. In fact, it is only with adequate ductility that the predicted strength of indeterminate members and structures can be achieved in practice.

6.2 FLEXURAL BEHAVIOUR AT OVERLOADS

The load at which collapse of a flexural member occurs is called the *ultimate load*. If the member has sustained large deformations prior to reaching the ultimate load, it is said to have ductile behaviour. If, on the other hand, it has only undergone relatively small deformations prior to failure, the member is said to have brittle behaviour. There is no defined deformation or curvature that distinguishes ductile behaviour from brittle behaviour. Codes of practice, however, usually impose a ductility requirement by limiting the curvature of a beam or slab at the ultimate load to some minimum value, thereby ensuring that significant deformation occurs in a flexural member prior to failure.

Since beam failures that result from a breakdown of bond between the concrete and the steel reinforcement, or from excessive shear or from failure of the anchorage zone, tend to be brittle in nature, every attempt should be made to ensure that, if a beam is overloaded, a ductile flexural failure would initiate the collapse. Therefore, the design philosophy should ensure that a flexural member does not fail before the required design moment capacity of the critical section is attained.

Consider the prestressed concrete cross-section shown in Figure 6.1. The section contains non-prestressed reinforcement in the compressive and tensile zones and bonded tensile prestressing steel. Typical strain and stress distributions for four different values of applied moment are also shown in Figure 6.1. As the applied moment M increases from typical in-service levels into the overload range, the neutral axis gradually rises and eventually, material behaviour becomes nonlinear. The non-prestressed tensile steel may yield (if its strain ε_{st} exceeds the yield strain ε_{sy}, where $\varepsilon_{sy} = f_{sy}/E_s$), the

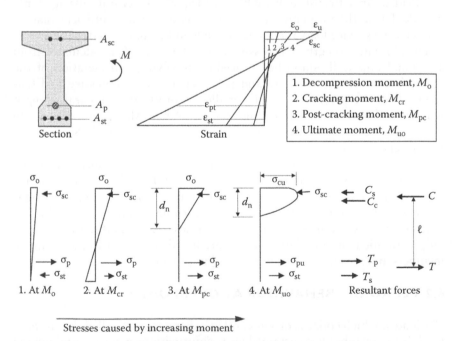

Figure 6.1 Stress and strain distributions caused by increasing moment.

prestressed steel may enter the nonlinear part of its stress–strain curve as ε_{pt} increases, the concrete compressive stress distribution becomes nonlinear when the extreme fibre stress exceeds about $0.5f_c'$ and the non-prestressed compressive steel may yield (if the magnitude of its strain ε_{sc} exceeds the yield strain ε_{sy}).

A flexural member that is designed to exhibit ductile behaviour usually has failure of the critical section preceded by yielding of the bonded tensile steel, i.e. by effectively exhausting the capacity of the tensile steel to carry any additional force. Such a member is said to be under-reinforced.

Because the stress–strain curve for the prestressing steel has no distinct yield point and the stress increases monotonically as the strain increases (see Figure 4.14), the capacity of the prestressing steel to carry additional force is never entirely used up until the steel actually fractures. When the steel strain ε_{pt} exceeds about 0.01 (for wire or strand), the stress–strain curve becomes relatively flat and the rate of increase of stress with strain is small. After yielding of the steel, the resultant internal tensile force (i.e. $T = T_s + T_p$ in Figure 6.1) remains approximately constant (as does the resultant internal compressive force C, which is equal and opposite to T). The moment capacity can be further increased slightly by an increase in the lever arm between C and T. Under increasing deformation, the neutral axis rises, the compressive zone becomes smaller and the compressive concrete stress increases. Eventually, after considerable deformation, a compressive failure of the concrete above the neutral axis occurs and the section reaches its ultimate capacity. It is, however, the strengths of the prestressing tendons and the non-prestressed reinforcement in the tensile zone that control the strength of a ductile section. In fact, the difference between the moment at first yielding of the tensile steel and the ultimate moment is usually relatively small.

A flexural member which is over-reinforced, on the other hand, does not have significant ductility at failure and fails without the prestressed or non-prestressed tensile reinforcement reaching yield or deforming significantly after yield. At the ultimate load condition, both the tensile strain at the steel level and the section curvature are relatively small and, consequently, there is little deformation or warning of failure.

Because it is the deformation at failure that defines ductility, it is both usual and reasonable in design to define a minimum ultimate curvature to ensure the ductility of a cross-section. This is often achieved by placing a maximum limit on the depth to the neutral axis at the ultimate load condition. Ductility can be increased by the inclusion of non-prestressed reinforcing steel in the compression zone of the beam. With compressive steel included, the internal compressive force C is shared between the concrete and the steel. The volume of the concrete stress block above the neutral axis is therefore reduced and, consequently, the depth to the neutral axis is decreased. Some compressive reinforcement is normally included in beams to provide anchorage for transverse shear reinforcement.

Ductility is desirable in prestressed (and reinforced) concrete flexural members. In continuous or statically indeterminate members, ductility is particularly necessary. Large curvatures are required at the peak moment regions in order to permit the inelastic moment redistribution that must occur if the moment diagram assumed in design is to be realised in practice. Consider the stress distribution caused by the ultimate moment on the section in Figure 6.1. The resultant compressive force of magnitude C equals the resultant tensile force T, and the ultimate moment capacity M_{uo} is calculated from the internal couple:

$$M_{uo} = C\ell = T\ell \tag{6.1}$$

The lever arm ℓ between the internal compressive and tensile resultants (C and T) is usually about $0.9d$, where d is the *effective depth* of the section and may be defined as the distance from the extreme compressive fibre to the position of the resultant tensile force in all the steel on the tensile side of the neutral axis.

To find the lever arm ℓ more accurately, the location of the resultant compressive force in the concrete C needs to be determined by considering the actual stress–strain relationship for concrete in the compression zone and locating the position of its centroid.

6.3 ULTIMATE FLEXURAL STRENGTH

6.3.1 Assumptions

In the analysis of a cross-section to determine its ultimate bending strength M_{uo}, the following assumptions are usually made:

1. the variation of strain on the cross-section is linear, i.e. strains in the concrete and the bonded steel are calculated on the assumption that plane sections remain plane;
2. perfect bond exists between the concrete and the bonded reinforcement and the bonded tendons;
3. concrete carries no tensile stress, i.e. the tensile strength of the concrete is ignored; and
4. the stress in the compressive concrete and in the steel reinforcement (both prestressed and non-prestressed) is obtained from actual or idealised stress–strain relationships for the respective materials.

6.3.2 Idealised rectangular compressive stress block for concrete

In order to simplify numerical calculations for ultimate flexural strength, codes of practice usually specify idealised rectangular stress blocks for the compressive concrete above the neutral axis. The dimensions of the stress block are calibrated such that the volume of the stress block and the position of its centroid are approximately the same as in the real curvilinear stress block.

In Figure 6.2a, an under-reinforced section at the ultimate moment is shown. The section has a single layer of bonded prestressing steel. The strain diagram and the actual concrete stress distribution at ultimate are also shown. In Figure 6.2b, the idealised rectangular stress block specified in AS3600-2009 [1] to model the compressive stress distribution in the concrete above the neutral axis is shown.

At the ultimate moment, the extreme fibre compressive strain ε_{cu} is taken to be 0.003 in AS3600-2009 [1]. In reality, the actual extreme fibre strain measured in experiments at ultimate conditions may not even be close to 0.003. However, for under-reinforced members, with the flexural strength very much controlled by the strength of the tensile steel (both prestressed and non-prestressed), variation in the assumed value of ε_{cu} does not have a significant effect on M_{uo}.

The depth of the rectangular stress block (in Figure 6.2b) is γd_n and the uniform stress intensity is $\alpha_2 f_c'$. For the rectangular section of Figure 6.2b, the hatched area A' ($=\gamma d_n b$) is therefore assumed to be subjected to a

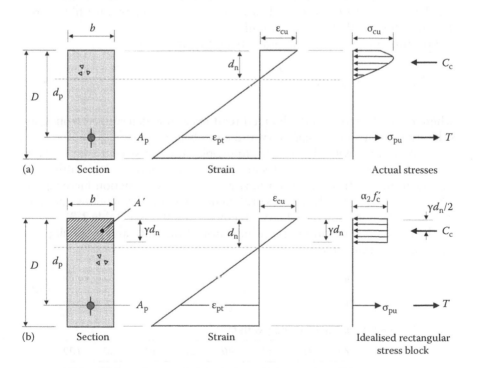

Figure 6.2 Ultimate moment conditions and rectangular stress block. (a) Actual curvilinear stress block. (b) Idealised stress block [1].

uniform stress of $\alpha_2 f_c'$. In AS3600-2009 [1], α_2 depends on the compressive strength of concrete and is given by:

$$\alpha_2 = 1.0 - 0.003 f_c' \quad \text{within the limits } 0.67 \leq \alpha_2 \leq 0.85 \tag{6.2}$$

The parameter γ depends on the concrete strength and is specified in AS3600-2009 [1] as:

$$\gamma = 1.05 - 0.007 f_c' \quad \text{within the limits } 0.67 \leq \gamma \leq 0.85 \tag{6.3}$$

Values of α_2 and γ for the standard concrete strength grades in AS3600-2009 [1] are given in Table 6.1.

For the rectangular section of Figure 6.2b, the hatched area A' $(=\gamma d_n b)$ is therefore assumed to be subjected to a uniform stress of $\alpha_2 f_c'$ and the resultant compressive force C_c is the volume of the rectangular stress block given by:

$$C_c = \alpha_2 f_c' A' = \alpha_2 f_c' \gamma d_n b \tag{6.4}$$

and the line of action of C_c passes through the centroid of the hatched area A', that is at a depth of $\gamma d_n/2$ below the extreme compressive fibre (provided, of course, that A' is rectangular).

The ultimate moment is obtained from Equation 6.1:

$$M_{uo} = T\ell = \sigma_{pu} A_p \left(d_p - \frac{\gamma d_n}{2} \right) \tag{6.5}$$

where σ_{pu} is the stress in the bonded tendons and is determined from considerations of strain compatibility and equilibrium.

In accordance with the design philosophy for the strength limit states outlined in Section 2.4, the design strength is obtained by multiplying M_{uo} (determined from Equation 6.5) by a capacity reduction factor ϕ. In AS3600-2009 [1], for ductile flexural members with steel tendons and normal ductility non-prestressed reinforcement: $\phi = 0.8$ (see Table 2.3).

The ultimate curvature κ_u is an indicator of ductility and is the slope of the strain diagram at failure:

$$\kappa_u = \frac{\varepsilon_{cu}}{d_n} = \frac{0.003}{d_n} \tag{6.6}$$

Table 6.1 Variation of α_2 and γ with f_c'

f_c' (MPa)	20	25	32	40	50	65	80	100
α_2	0.85	0.85	0.85	0.85	0.85	0.805	0.76	0.70
γ	0.85	0.85	0.826	0.77	0.70	0.67	0.67	0.67

Large deformations at ultimate are associated with ductile failures. To ensure ductility of a section at failure, AS3600-2009 [1] suggests that the depth to the neutral axis d_n at the ultimate moment should not exceed $0.36d_o$, where d_o is the depth from the extreme compressive fibre to the centroid of the outermost layer of tensile reinforcement or tendons (but not less than 0.8 times the overall depth of the cross-section). For the cross-section shown in Figure 6.2, provided the depth to the prestressing steel d_p is greater than $0.8D$, $d_o = d_p$ and the minimum curvature to ensure ductile failure is:

$$(\kappa_u)_{min} = \frac{0.003}{0.36d_p} = \frac{0.00833}{d_p} \tag{6.7}$$

6.3.3 Prestressed steel strain components (for bonded tendons)

For reinforced concrete sections, the strain in the reinforcing steel and in the concrete at the steel level is the same at every stage of loading, while for the tendons on a prestressed concrete section, this is not the case. The strain in the bonded prestressing steel at any stage of loading is equal to the strain caused by the initial prestress plus the change in strain in the concrete at the steel level caused by the applied load. To calculate accurately the ultimate flexural strength of a section, an accurate estimate of the final strain in the prestressed and non-prestressed steel is required. The tensile strain in the prestressing steel at ultimate ε_{pu} is much larger than the tensile strain in the concrete at the steel level, owing to the large initial prestress. For a bonded tendon, ε_{pu} is usually considered to be the sum of several sub-components. Figure 6.3 shows the instantaneous strain distributions on a prestressed section at three stages of loading.

Stage (a) shows the elastic instantaneous concrete strain caused by the effective prestress P_e, when the externally applied moment is zero. The

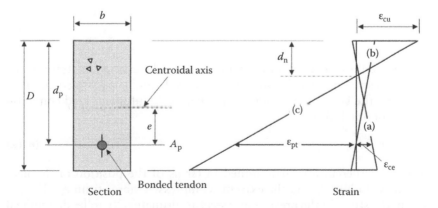

Figure 6.3 Instantaneous strain distributions at three stages of loading.

instantaneous strain in the concrete at the steel level is compressive, with magnitude approximately equal to:

$$\varepsilon_{ce} = \frac{1}{E_c}\left(\frac{P_e}{A} + \frac{P_e e^2}{I}\right) \tag{6.8}$$

where A is the area of the section; I is the second moment of area of the section about its centroidal axis; and e is the eccentricity of the prestressing force (as shown in Figure 6.3).

The stress and strain in the prestressing steel at stage (a) are:

$$\sigma_{pe} = \frac{P_e}{A_p} \tag{6.9}$$

and

$$\varepsilon_{pe} = \frac{\sigma_{pe}}{E_p} \tag{6.10}$$

provided that the steel stress is within the elastic range.

Stage (b) is the concrete strain distribution when the applied moment is sufficient to decompress the concrete at the steel level. Provided that there is bond between the steel and the concrete, the change in strain in the prestressing steel is equal to the change in concrete strain at the steel level. The strain in the prestressing steel at stage (b) is therefore equal to the value at stage (a) plus a tensile increment of strain equal in magnitude to ε_{ce} (from Equation 6.8).

Strain diagram (c) in Figure 6.3 corresponds to the ultimate load condition. The concrete strain at the steel level ε_{pt} can be expressed in terms of the extreme compressive fibre strain ε_{cu} and the depth to the neutral axis at ultimate d_n, and is given by:

$$\varepsilon_{pt} = \varepsilon_{cu}\frac{d_p - d_n}{d_n} \tag{6.11}$$

From the requirements of strain compatibility, the change in strain in the bonded prestressing steel between load stages (b) and (c) is also equal to ε_{pt}. Therefore, the strain in the bonded tendon at the ultimate load condition may be obtained from:

$$\varepsilon_{pu} = \varepsilon_{pe} + \varepsilon_{ce} + \varepsilon_{pt} \tag{6.12}$$

and ε_{pu} can therefore be determined in terms of the position of the neutral axis at failure d_n and the extreme compressive fibre strain ε_{cu}. If ε_{pu} is known, the stress in the prestressing steel at ultimate σ_{pu} can be determined from the stress–strain diagram for the prestressing steel. With the area of prestressing steel known, the tensile force at ultimate T_p can be calculated. In general, however, the steel stress is not known at failure and it is necessary to equate the tensile force in the steel tendon (plus the tensile force in any non-prestressed tensile steel) with the concrete compressive force (plus the compressive force in any non-prestressed compressive steel) in order to locate the neutral axis depth and hence find ε_{pu}.

In general, the magnitude of ε_{ce} in Equation 6.12 is very much smaller than either ε_{pe} or ε_{pt}, and may often be ignored without introducing serious errors.

6.3.4 Determination of M_{uo} for a singly reinforced section with bonded tendons

Consider the section shown in Figure 6.2a and the idealised compressive stress block shown in Figure 6.2b. In order to calculate the ultimate bending strength using Equation 6.5, the depth to the neutral axis d_n and the final stress in the prestressing steel σ_{pu} must first be determined.

An iterative trial and error procedure is usually used to determine the value of d_n for a given section. The depth to the neutral axis is adjusted until horizontal equilibrium is satisfied, i.e. $C = T$ in which both C and T are functions of d_n. For this singly reinforced cross-section, C is the volume of the compressive stress block given by Equation 6.4 and T depends on the strain in the prestressing steel ε_{pu}. For any value of d_n, the strain in the prestressing steel is calculated using Equation 6.12 (and Equations 6.8 through 6.11). The steel stress at ultimate σ_{pu}, which corresponds to the calculated value of strain ε_{pu}, can be obtained from the stress–strain curve for the prestressing steel and the corresponding tensile force is $T = T_p = \sigma_{pu} A_p$.

When the correct value of d_n is found (i.e. when $C = T$), the ultimate flexural strength M_{uo} may be calculated from Equation 6.5.

A suitable iterative procedure is outlined below and illustrated in Example 6.1. About three iterations are usually required to determine a good estimate of d_n and hence, M_{uo}.

1. With ε_{cu} taken to be 0.003, select an *appropriate* trial value of d_n and determine ε_{pu} from Equation 6.12 and C_c from Equation 6.4. By equating the tensile force in the steel to the compressive force in the concrete, the stress in the tendon may be determined:

$$T_p = \sigma_{pu} A_p = C_c = \alpha_2 f_c' \gamma d_n b \quad \therefore \sigma_{pu} = \frac{\alpha_2 f_c' \gamma d_n b}{A_p}$$

2. Plot the point ε_{pu} and σ_{pu} on the graph containing the stress–strain curve for the prestressing steel (as illustrated in Figure 6.5). If the point falls on the curve, then the value of d_n selected in step 1 is correct. If the point is not on the curve, then the stress–strain relationship for the prestressing steel is not satisfied and the value of d_n is not correct.

3. If the point ε_{pu} and σ_{pu} obtained in step 2 is not sufficiently close to the stress–strain curve for the steel, repeat steps 1 and 2 with a new estimate of d_n. A larger value for d_n is required if the point plotted in step 2 is below the stress–strain curve and a smaller value is required if the point is above the curve.
4. Interpolate between the plots from steps 2 and 3 to obtain a close estimate for ε_{pu} and σ_{pu} and the corresponding value for d_n.
5. With the values of σ_{pu} and d_n determined in step 4, calculate the ultimate moment M_{uo}. If the area above the neutral axis is rectangular, M_{uo} is obtained from Equation 6.5. Non-rectangular-shaped cross-sections are considered in Section 6.6.

EXAMPLE 6.1

The ultimate flexural strength M_{uo} of the rectangular section of Figure 6.4a is to be calculated. The steel tendon consists of ten 12.7 mm diameter strands (A_p = 1000 mm²) with an effective prestress P_e = 1200 kN. The stress–strain relationship for prestressing steel is as shown in Figure 6.5, and its elastic modulus is E_p = 195 × 10³ MPa. The concrete properties are f_c' = 40 MPa and E_c = 32,750 MPa.

From Table 6.1: α_2 = 0.85 and γ = 0.77 for f_c' = 40 MPa

The initial strain in the tendons due to the effective prestress is given by Equation 6.10:

$$\varepsilon_{pe} = \frac{P_e}{E_p A_p} = \frac{1,200 \times 10^3}{195,000 \times 1,000} = 0.00615$$

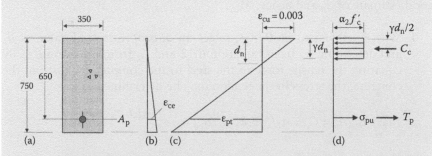

Figure 6.4 Section details and stress and strain distributions at ultimate (Example 6.1). (a) Section. (b) Strain due to P_e. (c) Strain at ultimate. (d) Concrete stress block at ultimate.

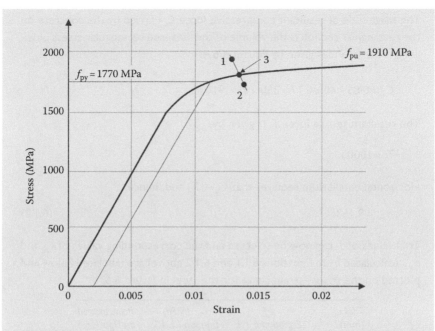

Figure 6.5 Stress–strain curve for strand (Example 6.1).

The strain in the concrete caused by the effective prestress at the level of the prestressing steel (ε_{ce} in Figure 6.4b) is calculated using Equation 6.8. Because ε_{ce} is small compared with ε_{pe}, it is usually acceptable to use the properties of the gross cross-section for its determination:

$$\varepsilon_{ce} = \frac{1}{32,750}\left(\frac{1,200\times10^3}{750\times350} + \frac{1,200\times10^3\times275^2}{350\times750^3/12}\right) = 0.000365$$

The concrete strain at the prestressed steel level at failure is obtained from Equation 6.11:

$$\varepsilon_{pt} = 0.003\times\left(\frac{650-d_n}{d_n}\right)$$

The final strain in the prestressing steel is given by Equation 6.12:

$$\varepsilon_{pu} = 0.00615 + 0.000365 + 0.003\times\left(\frac{650-d_n}{d_n}\right) \tag{6.1.1}$$

The magnitude of resultant compressive force C_c carried by the concrete on the rectangular section is the volume of the idealised rectangular stress block in Figure 6.4d and is given by Equation 6.4:

$$C_c = 0.85 \times 40 \times 0.77 \times 350 \times d_n = 9163 d_n$$

The resultant tensile force T_p is given by:

$$T_p = 1000 \times \sigma_{pu}$$

Horizontal equilibrium requires that $C_c = T_p$, and, hence:

$$\sigma_{pu} = 9.163 d_n \qquad\qquad (6.1.2)$$

Trial values of d_n can now be selected and the corresponding values of ε_{pu} and σ_{pu}, (calculated from Equations 6.1.1 and 6.1.2 above) are tabulated below and plotted on the stress–strain curve for the steel in Figure 6.5.

Trial d_n (mm)	ε_{pu} Equation 6.1.1	σ_{pu} (MPa) Equation 6.1.2	Point plotted on Figure 6.5
210	0.0128	1924	1
190	0.0138	1741	2
197	0.0134	1805	3

Point 3 lies sufficiently close to the stress–strain curve for the tendon, and therefore, the correct value for d_n is close to 197 mm ($0.303 d_p$).

From Equation 6.6, the curvature at ultimate is:

$$\kappa_u = \frac{0.003}{197} = 15.2 \times 10^{-6} \text{ mm}^{-1}$$

which is greater than the minimum value required for ductility given by Equation 6.7:

$$(\kappa_u)_{min} = \frac{0.00833}{650} = 12.8 \times 10^{-6} \text{ mm}^{-1}$$

The ultimate moment is found using Equation 6.5:

$$M_{uo} = 1805 \times 1000 \left(650 - \frac{0.77 \times 197}{2} \right) = 1036 \times 10^6 \text{ Nmm} = 1036 \text{ kNm}$$

The design strength of the section in flexure is ϕM_{uo}, where the value of ϕ for bending is obtained from Table 2.3 (i.e. $\phi = 0.8$ when $k_{uo} = d_n/d_p \leq 0.36$). In structural design, the moment M^* caused by the most severe factored load combination for strength (see Section 2.3.2) must be less than or equal to ϕM_{uo}.

6.3.5 Determination of M_{uo} for sections containing non-prestressed reinforcement and bonded tendons

Frequently, in addition to the prestressing reinforcement, prestressed concrete beams contain non-prestressed longitudinal reinforcement in both the compressive and tensile zones. This reinforcement may be included for a variety of reasons. For example, non-prestressed reinforcement is included in the tensile zone to provide additional flexural strength when the strength provided by the prestressing steel is not adequate. Non-prestressed tensile steel is also included to improve crack control when cracking is anticipated at service loads. Non-prestressed compressive reinforcement may be used to strengthen the compressive zone in beams that might otherwise be over-reinforced. In such beams, the inclusion of compression reinforcement not only increases the ultimate strength, but also increases the curvature at failure and, therefore, improves ductility.

The use of compressive reinforcement also reduces long-term deflections caused by creep and shrinkage and, therefore, improves serviceability. If for no other reason, compression reinforcement may be included to provide anchorage and bearing for the transverse reinforcement (stirrups) in beams.

When compressive reinforcement is included, closely spaced transverse ties should be used to laterally brace the highly stressed bars in compression and prevent them from buckling outward. In general, the spacing of these ties should not exceed about 16 times the diameter of the compressive bar.

Consider the *doubly reinforced* section shown in Figure 6.6a. The resultant compressive force consists of a steel component C_s ($=\sigma_{sc} A_{sc}$) and a concrete component C_c ($=\alpha_2 f_c' \gamma d_n b$). The stress in the compressive reinforcement is determined from the geometry of the linear strain diagram shown in Figure 6.6b. The magnitude of strain in the compressive steel is:

$$\varepsilon_{sc} = \frac{0.003(d_n - d_{sc})}{d_n} \tag{6.13}$$

If ε_{sc} is less than or equal to the yield strain of the non-prestressed steel ($\varepsilon_{sy} = f_{sy}/E_s$), then the stress in the compressive steel is $\sigma_{sc} = \varepsilon_{sc} E_s$. If ε_{sc} exceeds the yield strain, then $\sigma_{sc} = f_{sy}$.

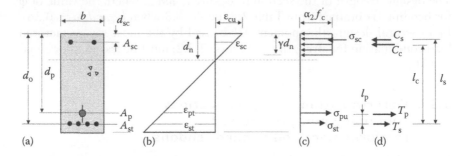

Figure 6.6 Doubly reinforced rectangular cross-section at the ultimate moment. (a) Section. (b) Strain. (c) Stresses. (d) Forces.

The resultant tensile force in Figure 6.6d consists of a prestressed component T_p (=$\sigma_{pu} A_p$) and a non-prestressed component T_s (=$\sigma_{st} A_{st}$). The stress in the non-prestressed tensile steel is determined from the strain at ultimate ε_{st}, given by:

$$\varepsilon_{st} = \frac{0.003(d_o - d_n)}{d_n} \tag{6.14}$$

If $\varepsilon_{st} \leq \varepsilon_{sy}$, then $\sigma_{st} = \varepsilon_{st} E_s$. If $\varepsilon_{st} > \varepsilon_{sy}$, then $\sigma_{st} = f_{sy}$.

In order to calculate the depth to the neutral axis d_n at ultimate, a trial and error procedure similar to that outlined in Section 6.3.4 can be employed. Successive values of d_n are tried until the value that satisfies the following horizontal equilibrium equation is determined:

$$T_p + T_s = C_c + C_s \tag{6.15}$$

Since one of the reasons for the inclusion of compressive reinforcement is to improve ductility, most doubly reinforced beams are, or should be, under-reinforced, that is the non-prestressed tensile steel is at yield at ultimate. Whether or not the compressive steel has yielded depends on its depth from the top compressive surface of the section d_{sc} and on the depth to the neutral axis d_n.

For any value of d_n, with the stresses in the compressive and tensile reinforcement determined from the strains ε_{sc} and ε_{st} (given by Equations 6.13 and 6.14, respectively), Equation 6.15 can be expanded as:

$$C_c = \alpha_2 f'_c \gamma d_n b$$

$$= T_p + T_s - C_s = \sigma_{pu} A_p + \sigma_{st} A_{st} - \sigma_{sc} A_{sc}$$

which can be rearranged to give:

$$\sigma_{pu} = \frac{\alpha_2 f_c' \gamma d_n b - \sigma_{st} A_{st} + \sigma_{sc} A_{sc}}{A_p} \tag{6.16}$$

When the value of σ_{pu} (calculated from Equation 6.16) and the value of ε_{pu} (calculated from Equation 6.12) together satisfy the stress–strain relationship of the prestressing steel, the correct value of d_n has been found. If it has been assumed that the non-prestressed steel has yielded in the calculations, the corresponding steel strains should be checked to ensure that the steel has, in fact, yielded. If the compressive steel is not at yield, then the compressive force C_s has been over-estimated and the correct value of d_n is slightly greater than the calculated value. The compressive steel stress σ_{sc} in Equation 6.16 should be taken as $\varepsilon_{sc} E_s$ instead of f_{sy}. Further iteration may be required to determine the correct value of d_n and the corresponding internal forces C_c, C_s, T_p and T_s.

With horizontal equilibrium satisfied, the ultimate moment of the section may be determined by taking moments of the internal forces about any convenient point on the cross-section. Taking moments about the non-prestressed tensile reinforcement level gives:

$$M_{uo} = C_c \, l_c + C_s \, l_s - T_p \, l_p \tag{6.17}$$

For the rectangular section shown in Figure 6.6, the lever arms to each of the internal forces in Equation 6.17 are:

$$l_c = d_o - \frac{\gamma d_n}{2} \quad l_s = d_o - d_{sc} \quad l_p = d_o - d_p$$

In the aforementioned equations, C_s and C_c are the *magnitudes* of the compressive forces in the steel and concrete, respectively, and are therefore considered to be positive.

The ultimate curvature is obtained from Equation 6.6, and the minimum curvature required for ductility is given by:

$$(\kappa_u)_{min} = \frac{0.00833}{d_o} \tag{6.18}$$

which is the same as Equation 6.7, except that the depth to the prestressing steel d_p is replaced by d_o (the depth to the bottom layer of the tensile steel reinforcement).

EXAMPLE 6.2

To the cross-section shown in Figure 6.4a and analysed in Example 6.1, three 24 mm diameter non-prestressed reinforcing bars (A_{st} = 1350 mm²) are added in the tensile zone at a depth d_o = 690 mm. Calculate the ultimate flexural strength M_{uo} of the section. The yield stress and elastic modulus of the non-prestressed steel are f_{sy} = 500 MPa and E_s = 200,000 MPa, respectively, and all other material properties and cross-sectional details are as specified in Example 6.1.

The strain in the prestressing steel at ultimate is as calculated in Example 6.1:

$$\varepsilon_{pu} = 0.00652 + 0.003 \times \left(\frac{650 - d_n}{d_n} \right) \tag{6.2.1}$$

and the magnitude of the compressive force C_c carried by the concrete above the neutral axis is:

$$C_c = \alpha_2 f_c' \gamma d_n b = 0.85 \times 40 \times 0.77 \times 350 \times d_n = 9163 d_n$$

From Equation 6.14, the non-prestressed steel is at yield, that is $\varepsilon_{st} \geq \varepsilon_{sy}$ ($=f_{sy}/E_s$ = 0.0025), provided that the depth to the neutral axis d_n is less than or equal to $0.5455d_o$ (=376 mm). If σ_{st} is assumed to equal f_{sy}, the resultant tensile force T ($=T_p + T_s$) is given by:

$$T = \sigma_{pu} A_p + f_{sy} A_{st} = 1000\sigma_{pu} + (500 \times 1350)$$

$$= 1000 \times (\sigma_{pu} + 675)$$

and, enforcing horizontal equilibrium (i.e. $C_c = T$), we get:

$$\sigma_{pu} = 9.163 d_n - 675 \tag{6.2.2}$$

Trial values of d_n are now selected and the respective values of ε_{pu} and σ_{pu} are tabulated below and plotted on the stress–strain curve in Figure 6.7.

Trial d_n (mm)	ε_{pu} Equation 6.2.1	σ_{pu} (MPa) Equation 6.2.2	Point plotted on Figure 6.7
280	0.0105	1891	4
260	0.0110	1707	5
264	0.0109	1744	6

Since point 6 lies sufficiently close to the stress–strain curve for the tendon, the value for d_n is taken as 264 mm.

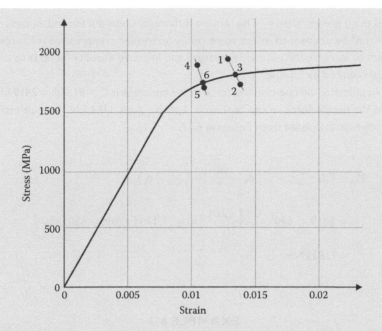

Figure 6.7 Stress–strain curve for strand (Example 6.2).

It is apparent in Figure 6.7 that the strain in the prestressing steel at ultimate is decreased by the introduction of tensile reinforcement (from point 3 to point 6) and the depth to the neutral axis is increased. From Equation 6.6, the ultimate curvature is:

$$\kappa_u = \frac{0.003}{264} = 11.4 \times 10^{-6} \text{ mm}^{-1}$$

and this is 25.4% less than that obtained in Example 6.1 (where $A_{st} = 0$).

The depth to the neutral axis d_n is much less than $0.5455d_o$ (=376 mm) and therefore, the non-prestressed steel has yielded, as previously assumed. The depth from the top surface to the resultant force in the tensile steel at ultimate is:

$$d = \frac{\sigma_{pu}A_p d_p + f_{sy}A_{st}d_o}{\sigma_{pu}A_p + f_{sy}A_{st}} - 661 \text{ mm}$$

The minimum curvature required to ensure some measure of ductility is obtained from Equation 6.18:

$$(\kappa_u)_{min} = \frac{0.00833}{690} = 12.08 \times 10^{-6} \text{ mm}^{-1}$$

and this is greater than κ_u. The section is therefore non-ductile and, in design, it would be prudent to insert some non-prestressed compressive reinforcement to increase the ultimate curvature and improve ductility (at least to the level required by Equation 6.18).

At ultimate, the compressive force in the concrete is $C_c = 9163d_n = 2419$ kN and the tensile force in the tendon is $T_p = \sigma_{pu} A_p = 1744$ kN. The ultimate moment is calculated from Equation 6.17:

$$M_{uo} = C_c l_c - T_p l_p = C_c \left(d_o - \frac{\gamma d_n}{2} \right) - T_p(d_o - d_p)$$

$$= 2419 \times \left(690 - \frac{0.77 \times 264}{2} \right) \times 10^{-3} - 1744 \times (690 - 650) \times 10^{-3}$$

$$= 1353 \text{ kNm}$$

EXAMPLE 6.3

Consider the effect on both strength and ductility of the cross-section of Example 6.2 if two 24 mm diameter bars are included in the compression zone. Details of the cross-section are shown in Figure 6.8, together with the stress and strain distributions at ultimate. All data are as specified in Examples 6.1 and 6.2.

From Examples 6.1 and 6.2, the ultimate strain in the tendons is:

$$\varepsilon_{pu} = 0.00652 + 0.003 \times \left(\frac{650 - d_n}{d_n} \right) \qquad (6.3.1)$$

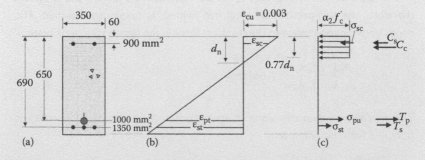

Figure 6.8 Section details and stress and strain distributions at ultimate (Example 6.3). (a) Section. (b) Strain. (c) Stress block and resultant forces.

and the strain in the non-prestressed tensile reinforcement in Example 6.2 is greater than ε_{sy} and, hence, $\sigma_{st} = f_{sy}$.

The magnitude of the compressive steel strain at ultimate is given by Equation 6.13:

$$\varepsilon_{sc} = \frac{0.003(d_n - 60)}{d_n} \tag{6.3.2}$$

and the stress in the compression steel can be readily obtained from ε_{sc} for any value of d_n.

By equating $C = T$, the expression for σ_{pu} given by Equation 6.16 becomes:

$$\sigma_{pu} = \frac{0.85 \times 40 \times 0.77 \times 350 \times d_n - 500 \times 1350 + \sigma_{sc} \times 900}{1000}$$

$$= 9.163d_n - 675 + 0.9\sigma_{sc} \tag{6.3.3}$$

Values of ε_{pu}, ε_{sc}, σ_{sc} and σ_{pu} for trial values of d_n are tabulated below and plotted as points 7 to 9 in Figure 6.9.

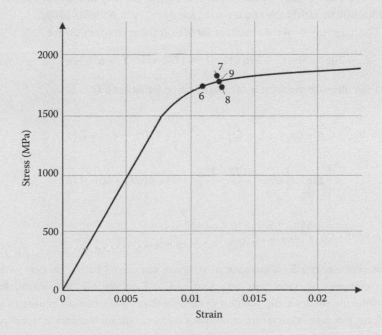

Figure 6.9 Stress–strain curve for strand (Example 6.3).

Trial d_n (mm)	ε_{pu} Equation 6.3.1	ε_{sc} Equation 6.3.2	σ_{sc} (MPa)	σ_{pu} (MPa) Equation 6.3.3	Point plotted on Figure 6.9
230	0.0120	0.00222	443.5	1832	7
220	0.0124	0.00218	436.4	1734	8
225	0.0122	0.00220	440.0	1783	9

From Figure 6.9, point 9 lies on the actual stress–strain curve and therefore, the neutral axis depth is taken as $d_n = 225$ mm.

It is apparent from Figure 6.9 that the strain in the prestressing steel at ultimate is increased by the introduction of compressive reinforcement (from point 6 to point 9) and the depth to the neutral axis is decreased. The ultimate curvature is obtained from Equation 6.6:

$$\kappa_u = \frac{0.003}{225} = 13.33 \times 10^{-6} \text{ mm}^{-1}$$

which represents a 17.3% increase in final curvature caused by the introduction of the compressive reinforcement. The curvature κ_u is greater than the minimum value $(\kappa_u)_{min}$ specified in AS3600-2009 [1] $(=12.08 \times 10^{-6}$ mm^{-1} obtained in Example 6.2 from Equation 6.18). The introduction of the two bars in the compressive zone has improved the ductility of the cross-section sufficiently to satisfy the requirement for ductility in AS3600-2009.

The magnitudes of the resultant forces on the cross-section are:

$$C_c = 2062 \text{ kN} \quad C_s = 396 \text{ kN} \quad T_p = 1783 \text{ kN} \quad T_s = 675 \text{ kN}$$

and the ultimate moment is calculated using Equation 6.17:

$$M_{uo} = C_c l_c + C_s l_s - T_p l_p = C_c \left(d_o - \frac{\gamma d_n}{2} \right) + C_s (d_o - d_{sc}) - T_p (d_o - d_p)$$

$$= \left[2062 \times \left(690 - \frac{0.77 \times 225}{2} \right) + 396 \times (690 - 60) - 1783 \right.$$

$$\left. \times (690 - 650) \right] \times 10^{-3} = 1422 \text{ kN} \cdot \text{m}$$

This represents a 5.1% increase in strength compared to the section without compressive steel that was analysed in Example 6.2. In general, for non-ductile sections, the addition of compressive reinforcement causes a significant increase in curvature at ultimate (i.e. a significant increase in ductility) and a less significant, but nevertheless appreciable, increase in strength.

6.4 APPROXIMATE PROCEDURE IN AS3600-2009

6.4.1 Bonded tendons

An approximate equation is specified in AS3600-2009 [1] to estimate the stress in a bonded tendon at the ultimate moment. The procedure is generally conservative and may be used in lieu of the more accurate determination of σ_{pu} based on strain compatibility (as outlined in the previous sections). For example, when the effective prestress σ_{pe} ($=P_e/A_p$) is not less than $0.5f_{pb}$, AS3600-2009 [1] specifies that the stress in the bonded steel at ultimate may be taken as:

$$\sigma_{pu} = f_{pb}\left(1 - \frac{k_1 k_2}{\gamma}\right) \tag{6.19}$$

where the parameter γ is defined in Equation 6.3 and illustrated in Figure 6.2b. The term k_1 depends on the particular type of prestressing steel, with $k_1 = 0.4$ when $f_{py}/f_{pb} < 0.9$ and $k_1 = 0.28$ when $f_{py}/f_{pb} \geq 0.9$. The term k_2 is given by:

$$k_2 = \frac{1}{b_{ef}d_pf'_c}\left[A_{pt}f_{pb} + (A_{st} - A_{sc})f_{sy}\right] \tag{6.20}$$

where b_{ef} is the width of the compressive face of the cross-section; A_{pt} is defined as the cross-sectional area of the tendons in that part of the cross-section that will be tensile under ultimate load conditions; and d_p is the distance from the extreme compressive fibre to the centroid of A_{pt}.

According to AS3600-2009 [1], when compression reinforcement is present, k_2 should be taken not less than 0.17. In addition, if the depth to the compressive steel d_{sc} exceeds $0.15\ d_p$, then A_{sc} should be set to zero in Equation 6.20.

Other more accurate procedures are available for calculating σ_{pu} based on the strain compatibility approach of the previous section, together with recognised mathematical expressions for the shape of the stress–strain curve of the prestressing steel, such as that proposed by Loov [2].

Consider a rectangular section, such as that shown in Figure 6.6, containing tensile prestressing steel and both tensile and compressive non-prestressed reinforcement of areas (A_p, A_{st} and A_{sc}, respectively). Assuming the non-prestressed tensile steel is at yield, the total tensile force in the steel at ultimate is:

$$T = T_p + T_s = A_p\,\sigma_{pu} + A_{st}\,f_{sy}$$

The magnitude of the total compressive force C consists of a concrete component C_c (given by Equation 6.4) and a steel component $C_s = A_{sc}\sigma_{sc}$:

$$C = C_c + C_s = \alpha_2 f_c' \gamma d_n b + A_{sc}\sigma_{sc}$$

If the stress in the prestressing steel at ultimate (σ_{pu}) is obtained from Equation 6.19, the depth to the neutral axis is obtained by equating C and T:

$$d_n = \frac{A_p\sigma_{pu} + A_{st}f_{sy} - A_{sc}\sigma_{sc}}{\alpha_2 f_c' \gamma b} \tag{6.21}$$

If we initially assume the compressive reinforcement is at yield (i.e. $\sigma_{sc} = f_{sy}$), the value of d_n calculated from Equation 6.21 can be used to check that the compressive steel has in fact yielded. If the steel has not yielded, a revised estimate of σ_{sc} ($=E_s\varepsilon_{sc}$) should be used to calculate a new value of d_n. Relatively few iterations are required for convergence.

By taking moments about the level of the tensile steel, the ultimate moment M_{uo} is given by:

$$M_{uo} = C_c\left(d_o - \frac{\gamma d_n}{2}\right) + C_s\left(d_o - d_{sc}\right) - T_p(d_o - d_p) \tag{6.22}$$

where d_o is the distance from the extreme compressive surface to the non-prestressed tensile reinforcement.

EXAMPLE 6.4

Recalculate the ultimate strength M_{uo} for the cross-section shown in Figure 6.4a (determined using the actual stress–strain relationship of the prestressing steel in Example 6.1) using the approximate estimate of σ_{pu} obtained from Equation 6.19. For this example, the properties of the prestressing steel are f_{pb} = 1910 MPa and f_{py} = 1770 MPa (obtained from Figure 6.5).

With f_{py}/f_{pb} = 0.93, k_1 = 0.28 and from Equation 6.20:

$$k_2 = \frac{1}{350 \times 650 \times 40}(1000 \times 1910) = 0.21$$

From Equation 6.19, the stress in the tendon at ultimate is approximated by:

$$\sigma_{pu} = 1910 \times \left(1 - \frac{0.28 \times 0.21}{0.77}\right) = 1764 \text{ MPa}$$

and this is 2.3% less than the more accurate value of 1805 MPa obtained by trial and error in Example 6.1. From Equation 6.21:

$$d_n = \frac{1000 \times 1764}{0.85 \times 40 \times 0.77 \times 350} = 192.5 \text{ mm}$$

which compares with d_n = 197 mm in Example 6.1. The ultimate strength is calculated using Equation 6.22:

$$M_{uo} = 1764 \times 1000 \times \left(650 - \frac{0.77 \times 192.5}{2}\right) \times 10^{-6} = 1016 \text{ kNm}$$

This is about 2% more conservative than the value obtained in Example 6.1. As expected, the simplified empirical procedure predicts a reasonable and conservative estimate of strength. However, the ultimate curvature is less conservative, because the depth to the neutral axis is underestimated. The ultimate curvature is obtained from Equation 6.6:

$$\kappa_u = \frac{0.003}{192.5} = 15.58 \times 10^{-6} \text{ mm}^{-1}$$

and is 2.3% greater than the value determined in Example 6.1. Nevertheless, for practical purposes, the simplified method is a useful design alternative.

EXAMPLE 6.5

Recalculate the flexural strength and ductility of the cross-section of Example 6.3 using the approximate formula for σ_{pu} given in Equation 6.19.

As in Example 6.4, k_1 = 0.28. As the depth to the compressive steel d_{sc} does not exceed $0.15d_p$, A_{sc} may be included in Equation 6.20:

$$k_2 = \frac{1}{350 \times 650 \times 40}\left[1000 \times 1910 + (1350 - 900) \times 500\right] = 0.235$$

and

$$\sigma_{pu} = 1910 \times \left(1 - \frac{0.28 \times 0.235}{0.77}\right) = 1747 \text{ MPa}$$

Assuming all non-prestressed steel is at yield, the depth to the neutral axis is obtained from Equation 6.21:

$$d_n = \frac{1000 \times 1747 + 1350 \times 500 - 900 \times 500}{0.85 \times 40 \times 0.77 \times 350} = 215 \, mm$$

Checking the strain in the non-prestressed reinforcement, Equation 6.14 gives:

$$\varepsilon_{st} = \frac{0.003(690 - 215)}{215} = 0.0066 \gg \varepsilon_{sy}(=0.0025)$$

and Equation 6.13 gives:

$$\varepsilon_{sc} = \frac{0.003(215 - 60)}{215} = 0.00216 < \varepsilon_{sy}$$

and so the assumption that the compressive steel has yielded is incorrect. The value of d_n determined earlier has been slightly underestimated and so too has the above value of ε_{sc}. Revising the estimate of the compressive steel stress to $\sigma_{sc} = E_s \varepsilon_{sc} = 200,000 \times 0.00216 = 432$ MPa, Equation 6.21 is used to determine a revised depth to the neutral axis:

$$d_n = \frac{1000 \times 1747 + 1350 \times 500 - 900 \times 432}{0.85 \times 40 \times 0.77 \times 350} = 222 \, mm$$

and the revised top steel strain is:

$$\varepsilon_{sc} = \frac{0.003(222 - 60)}{222} = 0.00219$$

The corresponding steel stress is $\sigma_{sc} = 200,000 \times 0.00219 = 438$ MPa. One more iteration gives $d_n = 221.3$ mm, $\varepsilon_{sc} = 0.002187$ and $\sigma_{sc} = 200,000 \times 0.002187 = 437.3$ MPa.

The magnitudes of the resultant forces on the cross-section are:

$$C_c = 2028 \, kN, \quad C_s = 394 \, kN, \quad T_p = 1747 \, kN \quad and \quad T_s = 675 \, kN$$

and the ultimate moment is calculated using Equation 6.22:

$$M_{uo} = C_c \left(d_o - \frac{\gamma d_n}{2} \right) + C_s \left(d_o - d_{sc} \right) - T_p (d_o - d_p)$$

$$= \left[2028 \times \left(690 - \frac{0.77 \times 221.3}{2} \right) + 394 \times (690 - 60) - 1747 \right.$$

$$\left. \times (690 - 650) \right] \times 10^{-3} = 1405 \text{ kNm}$$

which is 1.2% less than the more accurate value calculated in Example 6.3.

6.4.2 Unbonded tendons

In post-tensioned concrete, where the prestressing steel is not bonded to the concrete, the stress in the tendon at ultimate σ_{pu} is significantly less than that predicted by Equation 6.19 and accurate determination of the ultimate flexural strength is more difficult than for a section containing bonded tendons. This is because the final strain in the tendon is more difficult to determine accurately. The ultimate strength of a section containing unbonded tendons may be as low as 75% of the strength of an equivalent section containing bonded tendons. Hence, from a strength point of view, bonded construction is to be preferred. Indeed, in AS3600-2009 [1], bonded construction is mandatory (except for slabs on ground).

An unbonded tendon is not restrained by the concrete along its length, and slip between the tendon and the duct takes place as the external loads are applied and the member deforms. The steel strain is more uniform along the length of the member and tends to be lower in regions of maximum moment than would be the case for a bonded tendon. The ultimate strength of the section may be reached before the stress in the unbonded tendon reaches its yield stress f_{py}. For members not containing any bonded reinforcement, crack control may be a problem if cracking occurs in the member for any reason. If flexural cracking occurs, the number of cracks in the tensile zone is fewer than in a beam containing bonded reinforcement, but the cracks are wider and less serviceable.

AS3600-2009 [1] specifies Equations 6.23 and 6.24 for the ultimate stress (σ_{pu}) in a post-tensioned tendon that is not yet bonded to the concrete. In no case shall σ_{pu} be taken greater than f_{py}.

1. If the span to depth ratio of the member is 35 or less:

$$\sigma_{pu} = \sigma_{pe} + 70 + \frac{f_c' b_{ef} d_p}{100 A_{pt}} \leq \sigma_{pe} + 400 \tag{6.23}$$

2. If the span to depth ratio of the member is greater than 35:

$$\sigma_{pu} = \sigma_{pe} + 70 + \frac{f_c' b_{ef} d_p}{300 A_{pt}} \leq \sigma_{pe} + 200 \qquad (6.24)$$

where b_{ef} is the effective width of the compression face or flange of the member; and σ_{pe} is the effective stress in the tendon under service loads after allowing for all losses (i.e. $\sigma_{pe} = P_e/A_p$).

To ensure robustness and some measure of crack control, it is a good practice to include non-prestressed bonded tensile reinforcement in members where the post-tensioned tendons are to remain unbonded for a significant period during and after construction. However, it is the practice in most post-tensioning applications in Australia to grout the tendons within the duct after (or during) construction, thereby ensuring that the tendon is bonded to the surrounding concrete. In many parts of the world, however, including North America, unbonded construction is permitted and, indeed, is commonly adopted.

EXAMPLE 6.6

The ultimate flexural strength of a simply-supported post-tensioned beam containing a single unbonded cable is to be calculated. The beam spans 12 m and its cross-section at mid-span is shown in Figure 6.4a. Material properties and prestressing arrangement are as specified in Example 6.1.

The stress in the tendon caused by the effective prestressing force P_e = 1200 kN is:

$$\sigma_{pe} = \frac{P_e}{A_p} = \frac{1200 \times 10^3}{1000} = 1200 \text{ MPa}$$

With the span-to-depth ratio equal to 16, the stress in the unbonded tendon at ultimate, according to AS3600-2009 [1], is given by Equation 6.23:

$$\sigma_{pu} = 1200 + 70 + \frac{40 \times 350 \times 650}{100 \times 1000} = 1361 \text{ MPa}$$

and therefore the tensile force in the steel is T_p = 1361 kN (=C_c). This is almost 25% lower than the value determined in Example 6.1 where the tendon was bonded to the concrete. The depth to the neutral axis is calculated using Equation 6.21:

$$d_n = \frac{1000 \times 1361}{0.85 \times 40 \times 0.77 \times 350} = 148.5 \text{ mm}$$

and Equation 6.5 gives:

$$M_{uo} = 1361 \times 1000 \times \left(650 - \frac{0.77 \times 148.5}{2} \right) \times 10^{-6} = 807 \text{ kNm}$$

In Example 6.1, the ultimate bending strength of the same cross-section with a bonded tendon was calculated to be 1036 kNm. Clearly, the strength afforded by a post-tensioned tendon is significantly reduced if it remains unbonded.

6.5 DESIGN CALCULATIONS

6.5.1 Discussion

The magnitude of the prestressing force P_e and the quantity of the prestressing steel A_p are usually selected to satisfy the serviceability requirements of the member, that is to control deflection or to reduce or eliminate cracking. With serviceability satisfied, the member is then checked for adequate strength. The ultimate moment M_{uo} for the section containing the prestressing steel (plus any non-prestressed steel added for crack control or deflection control) is calculated and the design strength is compared with the design action, in accordance with the design requirements outlined in Section 2.4. For example, in AS3600-2009 [1], the flexural strength is ϕM_{uo}. The design action M^* is the moment caused by the most severe factored load combination specified for the strength limit state (see Section 2.3.2). The design requirement is expressed by $\phi M_{uo} \geq M^*$.

The prestressing steel needed for the satisfaction of serviceability requirements may not be enough to provide adequate strength. When this is the case, the ultimate moment capacity can be increased by the inclusion of additional non-prestressed tensile reinforcement. Additional compressive reinforcement may also be required to improve ductility.

6.5.2 Calculation of additional non-prestressed tensile reinforcement

Consider the singly-reinforced cross-section shown in Figure 6.10a. It is assumed that the effective prestress P_e, the area of the prestressing steel A_p and the cross-sectional dimensions have been designed to satisfy the serviceability requirements of the member. The idealised strain and stress distributions specified in AS3600-2009 [1] for the ultimate limit state are

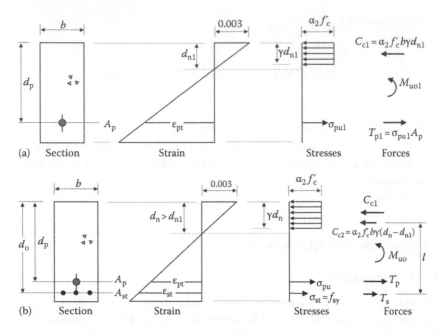

Figure 6.10 Cross-section containing tensile reinforcement – ultimate limit state. (a) Singly reinforced prestressed cross-section. (b) Cross-section containing both prestressed and non-prestressed tensile.

also shown in Figure 6.10a. The ultimate moment for the section, denoted as M_{uo1}, is calculated as follows:

$$M_{uo1} = \sigma_{pu1}A_p\left(d_p - \frac{\gamma d_{n1}}{2}\right)$$ (6.25)

where the tendon stress at ultimate σ_{pu1} can be calculated from the actual stress–strain curve for the steel (as illustrated in Example 6.1) or from the approximation of Equation 6.19 (as illustrated in Example 6.4).

If the design strength ϕM_{uo1} is greater than or equal to M^*, then no additional tensile steel is necessary, and the cross-section has adequate strength. If ϕM_{uo1} is less than M^*, the section is not adequate and additional tensile reinforcement is required.

In addition to providing adequate strength, it is important also to ensure that the section is ductile. To ensure that the curvature at ultimate κ_u is large enough to provide sufficient ductility, an upper limit for the depth to the neutral axis of about $0.36d_o$ is specified in AS3600-2009 [1]. However, to ensure ductility, a more satisfactory range for the depth to the neutral axis at failure is $d_n \leq 0.3\, d_o$. If the value of d_{n1} in Figure 6.10a is outside this range, some additional non-prestressed compressive reinforcement is required to relieve the concrete compressive zone and reduce the depth to the neutral axis. The design procedure outlined in Section 6.5.3 for doubly reinforced cross-sections is recommended in such a situation.

For the cross-section shown in Figure 6.10a, if ϕM_{uo1} is less than M^* and if d_{n1} is small so that ductility is not a problem, the aim in design is to calculate the minimum area of non-prestressed tensile reinforcement A_{st} that must be added to the section to satisfy strength requirements (i.e. the value of A_{st} such that $\phi M_{uo} = M^*$). In Figure 6.10b, the cross-section containing A_{st} is shown, together with the revised strain and stress distributions at ultimate. With d_n small enough to ensure ductility, the tensile steel strain ε_{st} is greater than the yield strain ε_{sy} $(=f_{sy}/E_s)$, so that $\sigma_{st} = f_{sy}$. The addition of A_{st} to the cross-section causes an increase in the resultant tension at ultimate $(T_p + T_s)$ and hence an increase in the resultant compression $C_c (=C_{c1} + C_{c2})$. To accommodate this additional compression, the depth of the compressive stress block in Figure 6.10b must be greater than the depth of the stress block in Figure 6.10a (i.e. $\gamma d_n > \gamma d_{n1}$). The increased value of d_n results in a reduction in the ultimate curvature (i.e. a decrease in ductility), a reduction in the strain in the prestressing steel and a consequent decrease in σ_{pu}. While the decrease in σ_{pu} is relatively small, it needs to be verified that the modified cross-section possesses adequate ductility (i.e. that the value of d_n remains less than about $0.3\, d_o$).

If σ_{pu} is assumed to remain constant, a first estimate of the magnitude of the area of non-prestressed steel A_{st} required to increase the ultimate strength from M_{uo1} (the strength of the section prior to the inclusion of the additional steel) to M_{uo} (the required strength of the section) may be obtained from:

$$A_{st} = \frac{M_{uo} - M_{uo1}}{f_{sy}\ell} \tag{6.26}$$

where ℓ is the lever arm between the tension force in the additional steel T_s and the equal and opposite compressive force C_{c2} which results from the increase in the depth of the compressive stress block. The lever arm ℓ may be approximated initially as:

$$\ell = 0.9(d_o - \gamma d_{n1}) \tag{6.27}$$

where d_{n1} is the depth to the neutral axis corresponding to M_{uo1}.

EXAMPLE 6.7

The ultimate strength of the singly reinforced cross section shown in Figure 6.11 is M_{uo1} = 1154 kNm. The stress and strain distributions corresponding to M_{uo1} are also shown in Figure 6.11, and the material properties are f_c' = 40 MPa (α_2 = 0.85 and γ = 0.77) and f_{pb} = 1910 MPa. Calculate the additional amount of non-prestressed tensile reinforcement located at d_o = 840 mm (f_{sy} = 500 MPa) required to increase the ultimate strength of the section to M_{uo} = 1500 kNm.

For the section in Figure 6.11, d_{n1} = 160 mm = $0.191d_o$ and the section is ductile. If the additional tensile steel is to be added at d_o = 840 mm, then the lever arm ℓ in Equation 6.27 may be approximated by:

$$\ell = 0.9(d_o - \gamma d_{n1}) = 0.9 \times (840 - 0.77 \times 160) = 645 \text{ mm}$$

and the required area of non-prestressed steel is estimated using Equation 6.26:

$$A_{st} = \frac{(1500 - 1154) \times 10^6}{500 \times 645} = 1073 \text{ mm}^2$$

Choose four 20 mm diameter bars (A_{st} = 1240 mm²) located at a depth d_o = 840 mm.

A check of this section to verify that $M_{uo} \geq 1500$ kNm, and also that the section is ductile, can now be made using the trial and error procedure illustrated in Example 6.2.

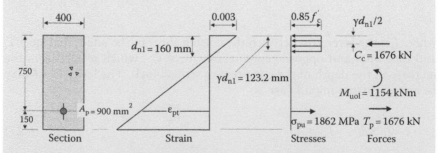

Figure 6.11 Singly reinforced cross-section at ultimate (Example 6.7).

6.5.3 Design of a doubly reinforced cross-section

For a singly reinforced section (such as that shown in Figure 6.12a) in which d_{nl} is greater than about $0.3d_o$, the inclusion of additional tensile reinforcement may cause ductility problems. In such cases, the ultimate strength may be increased by the inclusion of suitable quantities of both tensile and compressive non-prestressed reinforcement without causing any reduction in curvature, i.e. without increasing d_n. If the depth to the neutral axis is held constant at d_{nl}, the values of both C_c (the compressive force carried by the concrete) and T_p (the tensile force in the prestressing steel) in Figure 6.12a and b are the same. In each figure, C_c is equal to T_p. With the strain diagram in Figure 6.12b known, the strains at the levels of the non-prestressed steel may be calculated using Equations 6.13 and 6.14, and hence, the non-prestressed steel stresses σ_{st} and σ_{sc} may be determined. The equal and opposite forces which result from the inclusion of the non-prestressed steel are:

$$T_s = A_{st}\sigma_{st} \tag{6.28}$$

and

$$C_s = A_{sc}\sigma_{sc} \tag{6.29}$$

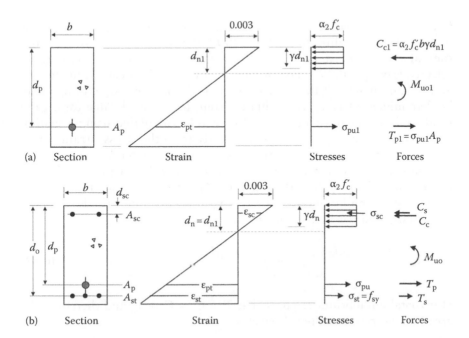

Figure 6.12 Doubly reinforced section at ultimate. (a) Cross-section containing prestressed steel only. (b) Cross-section containing top and bottom non-prestressed reinforcement.

When the depth to the compressive reinforcement is less than γd_n, the compressive force C_s could be calculated as $C_s = A_{sc}(\sigma_{sc} - \alpha_2 f_c')$, in order to account for the voids in the compressive concrete created by the compressive reinforcement.

If M_{uo1} is the strength of the singly reinforced section in Figure 6.12a (calculated using Equation 6.25) and M_{uo} is the required strength of the doubly reinforced cross-section, the minimum area of the tensile reinforcement is given by:

$$A_{st} = \frac{M_{uo} - M_{uo1}}{\sigma_{st}(d_o - d_{sc})} \qquad (6.30)$$

For conventional non-prestressed steel, σ_{st} is at yield (i.e. $\sigma_{st} = f_{sy}$) provided that $\varepsilon_{st} \geq \varepsilon_y$, and the depth to the neutral axis d_n satisfies the stated ductility requirements. For equilibrium, the forces in the top and bottom non-prestressed steel are equal and opposite, i.e. $C_s = T_s$, since $C_c = T_p$. From Equations 6.28 and 6.29:

$$A_{sc} = \frac{A_{st}\sigma_{st}}{\sigma_{sc}} \qquad (6.31)$$

If the depth to the neutral axis in Figure 6.12b is greater than about $0.36d_o$, then according to AS3600-2009, the section is non-ductile and the value of d_n must be reduced. An appropriate value of d_n may be selected (say $d_n = 0.3d_o$). For this value of d_n, all the steel strains (ε_{sc}, ε_{st} and ε_{pt}) and hence all the steel stresses at ultimate (σ_{sc}, σ_{st} and σ_{pu}) may be determined. Once ε_{pt} is calculated from the assumed value for d_n, the total strain in the prestressing steel ε_{pu} can be calculated using Equation 6.12 and the stress σ_{pu} can be read directly from the stress–strain curve. In this way, the magnitude of the tensile force in the tendon ($T_p = A_p\sigma_{pu}$) and the compressive force in the concrete $C_c = \alpha_2 f_c' b\gamma d_n$ can be evaluated. If the required strength of the section is M_{uo}, the minimum area of compressive steel can be obtained by taking moments about the level of the non-prestressed tensile reinforcement:

$$A_{sc} = \frac{M_{uo} + T_p(d_o - d_p) - C_c(d_o - 0.5\gamma d_n)}{\sigma_{sc}(d_o - d_{sc})} \qquad (6.32)$$

Horizontal equilibrium requires that $T_s = C_c + C_s - T_p$ and therefore, the area of non-prestressed tensile steel is:

$$A_{st} = \frac{\alpha_2 f_c'\gamma d_n b + A_{sc}\sigma_{sc} - A_p\sigma_{pu}}{\sigma_{st}} \qquad (6.33)$$

EXAMPLE 6.8

Determine the additional non-prestressed steel required to increase the ultimate flexural strength of the section in Figure 6.4 (and analysed in Example 6.1) to M_{uo} = 1400 kNm. Take the depth to the additional tensile steel as d_o = 690 mm and the reinforcement yield stress and elastic modulus as f_{sy} = 500 MPa and E_s = 200,000 MPa, respectively.

From Example 6.1, M_{uo1} = 1036 kNm and d_{n1} = 197 mm. If only non-prestressed tensile steel were to be added, the lever arm ℓ in Equation 6.27 would be:

$$\ell = 0.9 \times (690 - 0.77 \times 197) = 485 \text{ mm}$$

and from Equation 6.26:

$$A_{st} = \frac{(1400 - 1036) \times 10^6}{500 \times 485} = 1501 \text{ mm}^2$$

This corresponds to the addition of five 20 mm diameter bars (1550 mm²) in the bottom of the section shown in Figure 6.4 at a depth d_o = 690 mm.

A check of the section to verify that $M_{uo} \geq$ 1400 kNm can next be made using the trial and error procedure illustrated in Example 6.2. In this example, however, the neutral axis depth increases above 0.36 d_o and the curvature at ultimate is less than the minimum value recommended in AS3600-2009 [1]. For this cross-section, it is appropriate to supply the additional moment capacity via both tensile and compressive non-prestressed reinforcement.

If the depth to the neutral axis is held constant at the value determined in Example 6.1, i.e. d_{n1} = 197 mm, then the stress and strain in the prestressed steel remain as previously calculated, i.e. ε_{pu} = 0.0134 and σ_{pu} = 1805 MPa.

If the depth to the compressive reinforcement is d_{sc} = 60 mm, then from Equation 6.13:

$$\varepsilon_{sc} = \frac{0.003(197 - 60)}{197} = 0.00209 < \varepsilon_{sy} \quad \text{and} \quad \sigma_{sc} = E_s \varepsilon_{sc} = 417 \text{ MPa}$$

From Equation 6.14:

$$\varepsilon_{st} = \frac{0.003(690 - 197)}{197} = 0.00751 > \varepsilon_{sy} \quad \text{and} \quad \sigma_{st} = f_{sy} = 500 \text{ MPa}$$

The areas of additional tensile and compressive steel are obtained using Equations 6.30 and 6.31, respectively:

$$A_{st} = \frac{(1500 - 1036) \times 10^6}{500 \times (690 - 60)} = 1473 \text{ mm}^2$$

$$A_{sc} = \frac{1473 \times 500}{417} = 1766 \text{ mm}^2$$

A suitable solution is to include four 24 mm diameter reinforcing bars in the top of the section (at d_{sc} = 60 mm) and four 24 mm diameter bars in the bottom of the section (at d_o = 690 mm).

6.6 FLANGED SECTIONS

Flanged sections such as those shown in Figure 6.13a are commonly used in prestressed concrete construction, where the bending efficiency of I-, T- and box-shaped sections can be effectively utilised. Frequently, in the construction of prestressed floor systems, beams or wide bands are poured monolithically with the slabs. In such cases, a portion of slab acts as either a top or a bottom flange of the beam, as shown in Figure 6.13b.

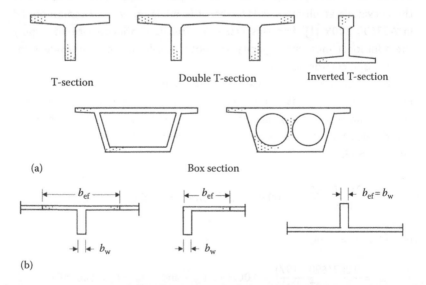

Figure 6.13 Typical flanged sections. (a) Precast sections. (b) Monolithic sections.

AS3600-2009 [1] specifies the width of the slab that may be assumed to be part of the beam cross-section (i.e. the *effective width* of the flange, b_{ef}) as follows:

For T-sections: $b_{ef} = b_w + 0.2a$ (6.34)

For L-sections: $b_{ef} = b_w + 0.1a$ (6.35)

except that the overhanging part of the effective flange should not exceed half the clear distance to the next parallel beam. The term b_w is the width of the web of the cross-section and a is the distance along the beam between the points of zero bending moment and may be taken as the actual span for simply-supported members and 0.7 times the actual span for continuous members.

It is recommended in ACI 318M-11 [3] that the effective width of the flange of a T-beam should not exceed one quarter of the span length of the beam, and the effective overhanging flange width on each side of the web should not exceed eight times the slab thickness. For L-beams with a slab on one side only, the effective overhanging flange width should not exceed the lesser of one-twelfth of the span of the beam and six times the slab thickness. Although these are not formal requirements of AS3600-2009 [1], their satisfaction is recommended here.

The flexural strength calculations discussed in Section 6.3 can also be used to determine the flexural strength of non-rectangular sections. The equations developed earlier for rectangular sections are directly applicable provided the depth of the idealised rectangular stress block is less than the thickness of the compression flange, i.e. provided the portion of the cross-section subjected to the uniform compressive stress is rectangular (b_{ef} wide and γd_n deep). The ultimate strength M_{uo} is unaffected by the shape of the concrete section below the compressive stress block and depends only on the area and position of the steel reinforcement and tendons in the tensile zone. If the compressive stress block acts on a non-rectangular portion of the cross-section, some modifications to the formulae are necessary to calculate the resulting concrete compressive force and its line of action.

Consider the T-sections shown in Figure 6.14, together with the idealised rectangular stress blocks (previously defined in Figure 6.2b). If $\gamma d_n \le t$ (as in Figure 6.14a), the area of the concrete in compression A' is rectangular and the strength of the section is identical with that of a rectangular section of width b_{ef} containing the same tensile steel at the same effective depth. Equation 6.22 may therefore be used to calculate the strength of such a section. The depth of the neutral axis d_n may be calculated using Equation 6.21, except that b_{ef} replaces b in the denominator.

If $\gamma d_n > t$, the area of concrete in compression A' is T-shaped, as shown in Figure 6.14b. Although not strictly applicable, the idealised stress block

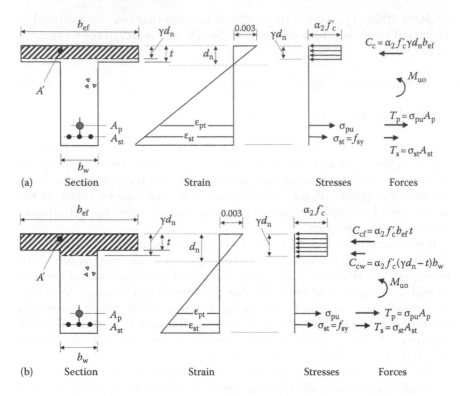

Figure 6.14 Flanged sections subjected to the ultimate moment. (a) Compressive stress block in the flange. (b) Compressive stress block in the flange and web.

may still be used on this non-rectangular compressive zone. A uniform stress of $\alpha_2 f_c'$ may therefore be considered to act over the area A'.

It is convenient to separate the resultant compressive force in the concrete into a force in the flange C_{cf} and a force in the web C_{cw} as follows (and shown in Figure 6.14b):

$$C_{cf} = \alpha_2 f_c' t b_{ef} \qquad (6.36)$$

and

$$C_{cw} = \alpha_2 f_c' (\gamma d_n - t) b_w \qquad (6.37)$$

By equating the tensile and compressive forces, the depth to the neutral axis d_n can be determined by trial and error, and the ultimate moment M_{uo} can be obtained by taking moments of the internal forces about any convenient point on the cross-section.

EXAMPLE 6.9

Evaluate the ultimate flexural strength of the standardised double tee section shown in Figure 6.15. The cross-section contains a total of 26 12.7 mm diameter strands (13 in each cable) placed at an eccentricity of 408 mm to the centroidal axis. The effective prestressing force P_e is 3250 kN. The stress–strain relationship for the prestressing steel is shown in Figure 6.16, and its elastic modulus and tensile strength are E_p = 195,000 MPa

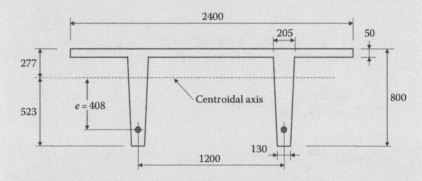

Figure 6.15 Double tee cross-section (Example 6.9). All dimensions in the figure are in mm.

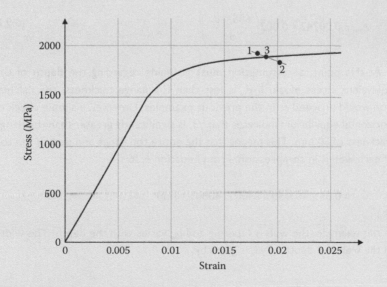

Figure 6.16 Stress–strain curve for strand (Example 6.9).

and f_{pb} = 1910 MPa, respectively. The properties of the section and other relevant material data are as follows:

$A = 371 \times 10^3$ mm²; $I = 22.8 \times 10^9$ mm⁴; $Z_b = 43.7 \times 10^6$ mm³;
$Z_t = 82.5 \times 10^6$ mm³; $A_p = 26 \times 100 = 2{,}600$ mm²; $E_c = 31{,}900$ MPa;
$f'_c = 40$ MPa; $\alpha_2 = 0.85$; $\gamma = 0.77$

Using the same procedure as was illustrated in Example 6.1, the strain components in the prestressing steel are obtained from Equations 6.8 through 6.11:

$$\varepsilon_{ce} = \frac{1}{31{,}900}\left(\frac{3{,}250\times10^3}{371\times10^3} + \frac{3{,}250\times10^3\times408^2}{22.8\times10^9}\right) = 0.00102$$

$$\varepsilon_{pe} = \frac{3{,}250\times10^3}{2{,}600\times195{,}000} = 0.00641$$

$$\varepsilon_{pt} = 0.003\left(\frac{685-d_n}{d_n}\right)$$

and from Equation 6.12:

$$\varepsilon_{pu} = 0.00743 + 0.003\left(\frac{685-d_n}{d_n}\right) \tag{6.9.1}$$

At this point, an assumption must be made regarding the depth of the equivalent stress block. If d_n is less than the flange thickness, the calculation would proceed as in the previous examples. However, a simple check of horizontal equilibrium indicates that γd_n is significantly greater than the flange thickness of 50 mm. This means that the entire top flange and part of the top of each web is in compression. From Equation 6.36:

$$C_{cf} = 0.85\times40\times50\times2400 = 4080\times10^3 \text{ N}$$

In this example, the web is tapering and b_w varies with the depth. The width of the web at a depth of γd_n is given by:

$$b_{w1} = 210 - \frac{\gamma d_n}{10}$$

The compressive force in the web is therefore:

$$C_{cw} = 0.85 \times 40 \times (0.77d_n - 50) \times \left(\frac{205 + b_{wl}}{2}\right) \times 2$$

$$= -2.016d_n^2 + 10,996d_n - 705,500$$

The resultant compression force is the sum of the flange and web compressive forces:

$$C_c = C_{cf} + C_{cw} = -2.016d_n^2 + 10,996d_n + 3,375,000$$

and the resultant tensile force in the tendons is:

$$T_p = 2600\,\sigma_{pu}$$

Equating C_c and T_p gives:

$$\sigma_{pu} = -0.000775d_n^2 + 4.229d_n + 1298 \tag{6.9.2}$$

Trial values of d_n may now be used to determine ε_{pu} and σ_{pu} from the above expressions and the resulting points are tabulated below and plotted on the stress–strain diagram of Figure 6.16.

Trial d_n(mm)	ε_{pu} Equation 6.9.1	σ_{pu} (MPa) Equation 6.9.2	Point plotted on Figure 6.16
150	0.0181	1915	1
130	0.0202	1834	2
142	0.0189	1883	3

Since point 3 lies sufficiently close to the stress–strain curve for the tendon, the value taken for d_n is 142 mm.

The depth of the stress block is $\gamma d_n = 109.3$ mm, which is greater than the flange thickness (as was earlier assumed). The resultant forces on the cross-section are:

$$C_c = T_p = 2600 \times 1883 \times 10^{-3} = 4896 \text{ kN}$$

For this section, $d_n = 0.207d_p < 0.36d_p$ and therefore, according to AS3600-2009 [1], the failure is ductile. The compressive force in the flange

C_{cf} = 4080 kN acts 25 mm below the top surface and the compressive force in the web C_{cw} = 816 kN acts at the centroid of the trapezoidal areas of the webs above γd_n, i.e. 79.5 mm below the top surface.

By taking moments of these internal compressive forces about the level of the tendons, we get:

$$M_{uo} = 4080 \times (685 - 25) \times 10^{-3} + 816 \times (685 - 79.5) \times 10^{-3}$$

$$= 3187 \text{ kNm}$$

6.7 DUCTILITY AND ROBUSTNESS OF PRESTRESSED CONCRETE BEAMS

6.7.1 Introductory remarks

Ductility is the ability of a structure or structural member to undergo large plastic deformations without a significant loss of load-carrying capacity. Ductility is important for many reasons. It provides indeterminate structures with alternative load paths and the ability to redistribute internal actions as the collapse load is approached. After the onset of cracking, concrete structures are nonlinear and inelastic. The stiffness varies from location to location, depending on the extent of cracking and the reinforcement/tendon layout. In addition, the stiffness of a particular cross-section or region is time-dependent, with the distribution of internal actions changing under service loads due to creep and shrinkage, as well as other imposed deformations such as support settlements and temperature changes and gradients. All these factors cause the actual distribution of internal actions in an indeterminate structure to deviate from that assumed in an elastic analysis. Despite these difficulties, codes of practice, including AS3600-2009 [1], permit the design of concrete structures based on elastic analysis. This is quite reasonable provided the critical regions possess sufficient ductility (plastic rotational capacity) to enable the actions to redistribute as the collapse load is approached. If critical regions have little ductility (such as in over-reinforced elements), the member may not be able to undergo the necessary plastic deformation and the safety of the structure could be compromised.

Ductility is also important to resist impact and cyclic loading, and to provide robustness. With proper detailing, ductile structures can absorb the energy associated with sudden impact (as may occur in an accident or a blast) or cyclic loading (such as a seismic event) without collapse of the structure. With proper detailing, ductile structures can also be designed to resist progressive collapse.

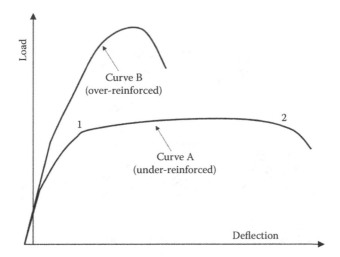

Figure 6.17 Ductile and non-ductile load–deflection curves.

Figure 6.17 shows the load–deflection curves for two prestressed concrete beams, one under-reinforced (Curve A) and one over-reinforced (Curve B). Curve A indicates ductile behaviour with large plastic deformations developing as the peak load is approached. The relatively flat post-yield plateau (1–2) in Curve A, where the structure deforms while maintaining its full load-carrying capacity (or close to it) is characteristic of ductile behaviour. Curve B indicates non-ductile or brittle behaviour, with relatively little plastic deformation before the peak load. There is little or no evidence of a flat *plastic plateau* as the peak load is approached and the beam immediately begins to unload when the peak load is reached.

Structures with load–deflection relationships similar to Curve B in Figure 6.17 are simply too brittle to perform adequately under significant impact or seismic loading, and they cannot resist progressive collapse. Prestressed concrete beams can be designed to be robust and not to suddenly collapse when overloaded, but ductility is the key and a ductile load–displacement relationship such as that shown as Curve A in Figure 6.17 is an essential requirement. AS3600-2009 [1] states that structures should be designed to be robust, but qualitative statements rather than quantitative recommendations are made. To design a prestressed structure for robustness, some quantitative measure of robustness is required.

Beeby [4] stated that a structure is robust, if it is able to absorb damage resulting from unforeseen events without collapse. He also argued that this could form the basis of a design approach to quantify robustness. Explosions or impacts are clearly inputs of energy. Beeby [4] suggests that accidents or even design mistakes could also be considered as inputs of energy and that robustness requirements could be quantified in terms of a structure's ability to absorb energy.

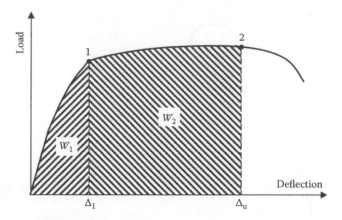

Figure 6.18 Typical under-reinforced load–deflection curve.

The area under the load–deflection response of a member or structure is a measure of the energy absorbed by the structure in undergoing
that deformation. Consider the load–deflection response of a simply-
supported under-reinforced prestressed beam shown in Figure 6.18. The
area under the curve up to point 1 (before the tensile steel yields) is
W_1 and represents the elastic energy. The area under the curve between
points 1 and 2 (when the plastic hinge has developed and the peak load
is reached) is W_2, which represents the plastic energy. A minimum value
of the ratio W_2/W_1 could be specified to ensure an acceptable level of
ductility and, if all members and connections were similarly ductile and
appropriately detailed, then an acceptable level of robustness (or resistance to collapse) could be achieved.

A ductile simply-supported member is one for which W_2/W_1 exceeds about
3.0, but for statically indeterminate structures where significant redistribution of internal actions may be required as the peak load is approached,
satisfaction of the following is recommended:

$$\frac{W_2}{W_1} \geq 4.0 \qquad\qquad (6.38)$$

6.7.2 Calculation of hinge rotations

A typical moment-curvature relationship for an under-reinforced prestressed concrete cross-section was shown in Figure 1.15 and an idealised
elastic–plastic moment–curvature curve is shown in Figure 6.19. A plastic
hinge is assumed to develop at a point in a beam or slab when the peak
(ultimate) moment is reached at a curvature of κ_y and rotation of the plastic
hinge occurs as the curvature increases from κ_y to κ_u. The rotation at the
plastic hinge θ_h is the change in curvature multiplied by the length of the

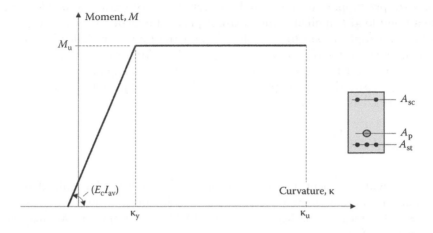

Figure 6.19 Idealised elastic–plastic moment–curvature relationship.

plastic hinge l_h in the direction of the member axis. For under-reinforced cross-sections with ductile tensile reinforcement and tendons, the length of the plastic hinge l_h is usually taken to be equal to the effective depth of the beam or slab d. The maximum rotation available at a plastic hinge may therefore be approximated by:

$$\theta_h = l_h(\kappa_u - \kappa_y) \approx d(\kappa_u - \kappa_y) \tag{6.39}$$

6.7.3 Quantifying ductility and robustness of beams and slabs

To investigate the ductility of a prestressed concrete beam, it is convenient to idealise the load–deflection curve as elastic-perfectly plastic. As an example, consider the idealised load–deflection response shown in Figure 6.20

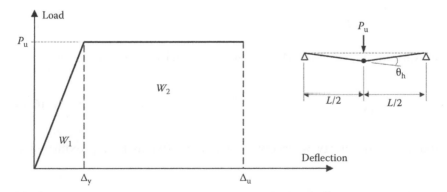

Figure 6.20 Idealised load–deflection curve.

of a simply-supported prestressed concrete beam of span L and subjected to a point load P applied at mid-span. A plastic hinge develops at mid-span when the applied load first reaches P_u and the mid-span deflection is Δ_y. The moment at the plastic hinge is $M_u = P_u L/4$. After the formation of the plastic hinge, it is assumed that the deflection at mid-span increases from Δ_y to Δ_u by rotation of the plastic hinge through an angle θ_h and this can be described as follows:

$$(\Delta_u - \Delta_y) = \frac{L\theta_h}{4} \tag{6.40}$$

The ultimate moment at the plastic hinge M_u may be determined from Equation 6.17. For example, if we assume the beam contains $p = (A_p + A_{st})/bd = 0.01$, with $f_c' = 32$ MPa, $E_c = 28{,}600$ MPa, the moment M_u may be determined from:

$$M_u = \sigma_{pu} A_p \left(d_p - \frac{\gamma d_n}{2} \right) + f_{sy} A_{st} \left(d_{st} - \frac{\gamma d_n}{2} \right) \approx \beta b d^2 \tag{6.41}$$

where b is the section width; and d is the effective depth to the resultant of the tensile forces in the prestressed and non-prestressed steel.

The term β depends on the area, position and strength of the reinforcement and tendons. In our example, β is about 4.5. For the centrally loaded simply-supported beam, the deflection Δ_y may be approximated as:

$$\Delta_y = \frac{P_u L^3}{48 E_c I_{av}} = \frac{M_u L^2}{12 E_c I_{av}} \tag{6.42}$$

and, for the stated material properties and steel quantities, the average moment of inertia of the cracked cross-section is approximated by:

$$I_{av} = 0.045 b d^3 \tag{6.43}$$

Substituting Equations 6.41 and 6.43 into Equation 6.42 gives:

$$\Delta_y = \frac{L^2}{3430 d} \tag{6.44}$$

and the elastic energy W_1 (shown in Figure 6.20) may be approximated as:

$$W_1 = \frac{P_u \Delta_y}{2} = 0.00262 \, b d L \tag{6.45}$$

If we assume that satisfaction of Equation 6.38 is required for robustness, the minimum internal plastic energy W_2 that must be absorbed during the hinge rotation is $4W_1 = 0.0105\,bdL$ and this must equal the external work:

$$P_u(\Delta_u - \Delta_y) = W_2 = 0.0105\,bdL \tag{6.46}$$

and substituting Equations 6.40 and 6.41 into Equation 6.46, we get:

$$\left(4\frac{M_u}{L}\right)\left(\frac{L\theta_h}{4}\right) = 4.5bd^2\theta_h = 0.0105\,bdL$$

$$\therefore\ \theta_h = 2.33\times10^{-3}\frac{L}{d} \tag{6.47}$$

It is evident that the plastic rotation required at the hinge at mid-span depends on the span to effective depth ratio. To achieve a ductility corresponding to $W_2/W_1 = 4.0$, the rotation required at the hinge at mid-span and the span to final deflection ratio (L/Δ_u) are determined from Equations 6.47, 6.40 and 6.44 and given in the following table.

L/d	$\theta_h(rad)$	(L/Δ_u)
10	0.023	114
14	0.033	82
18	0.042	64
22	0.051	52
26	0.061	44

REFERENCES

1. AS3600-2009. *Australian Standard for Concrete Structures*. Standards Australia, Sydney, New South Wales, Australia.
2. Loov, R.E. (1988). A general equation for the steel stress for bonded prestressed concrete members. *Journal of the Prestressed Concrete Institute*, 33, 108–137.
3. ACI 318M-11 2011. Building code requirements for reinforced concrete. American Concrete Institute, Detroit, MI.
4. Beeby, A.W. 1999. Safety of structures, a new approach to robustness. *The Structural Engineer*, 77(4), 16–21.

Chapter 7

Ultimate strength in shear and torsion

7.1 INTRODUCTION

In Chapter 5, methods were presented for the determination of the strains and stresses normal to a cross-section caused by the longitudinal prestress and the bending moment acting at the cross-section. Procedures for calculating the flexural strength of beams were discussed in Chapter 6. In structural design, shear failure must also be guarded against. Shear failure is sudden and difficult to predict with accuracy. It results from diagonal tension in the web of a concrete member produced by shear stress in combination with the longitudinal normal stress. Torsion, or twisting of the member about its longitudinal axis, also causes shear stresses, which lead to diagonal tension in the concrete and consequential inclined cracking.

Conventional reinforcement in the form of transverse stirrups is used to carry the tensile forces in the webs of prestressed concrete beams after the formation of diagonal cracks. This reinforcement should be provided in sufficient quantities to ensure that flexural failure, which can be predicted accurately and is usually preceded by extensive cracking and large deformation, will occur before diagonal tension failure.

In slabs and footings, a local shear failure at columns or under concentrated loads may also occur. This so-called *punching shear* type of failure often controls the thickness of flat slabs and plates in the regions above the supporting columns. In this chapter, the design for adequate strength of prestressed concrete beams in shear and in combined shear and torsion is described. Procedures for determining the punching shear strength of slabs and footings are also presented.

7.2 SHEAR IN BEAMS

7.2.1 Inclined cracking

Cracking in prestressed concrete beams subjected to overloads, as shown in Figure 7.1, depends on the local magnitudes of moment and shear. In regions where the moment is large and the shear is small, vertical flexural

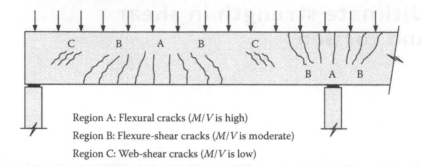

Region A: Flexural cracks (M/V is high)
Region B: Flexure-shear cracks (M/V is moderate)
Region C: Web-shear cracks (M/V is low)

Figure 7.1 Types of cracking at overload.

cracks appear after the normal tensile stress in the extreme concrete fibres exceeds the tensile strength of concrete. These are the crack referred to in Sections 5.8.1 and 5.12.1 and are shown in Figure 7.1 as crack type A.

Where both the moment and shear force are relatively large, flexural cracks that are vertical at the extreme fibres become inclined as they extend deeper into the beam owing to the presence of shear stresses in the beam web. These inclined cracks, which are often quite flat in a prestressed beam, are called *flexure-shear* cracks and are designated crack type B in Figure 7.1. If adequate shear reinforcement is not provided, a flexure-shear crack may lead to a so-called shear-compression failure, in which the area of concrete in compression above the advancing inclined crack is so reduced as to be no longer adequate to carry the compression force resulting from flexure.

A second type of inclined crack sometimes occurs in the web of a prestressed beam in the regions where moment is small and shear is large, such as the cracks designated type C adjacent to the discontinuous support and near the point of contraflexure in Figure 7.1. In such locations, high principal tensile stress may cause inclined cracking in the mid-depth region of the beam before flexural cracking occurs in the extreme fibres. These cracks are known as *web-shear* cracks and occur most often in beams with relatively thin webs.

7.2.2 Effect of prestress

The longitudinal compression introduced by prestress delays the formation of each of the crack types shown in Figure 7.1. The effect of prestress on the formation and direction of inclined cracks can be seen by examining the stresses acting on a small element located at the centroidal axis of the uncracked beam shown in Figure 7.2. Using a simple Mohr's circle construction, the principal stresses and their directions are readily found.

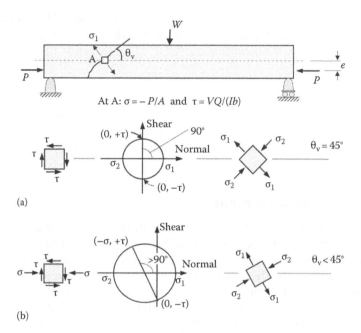

At A: $\sigma = -P/A$ and $\tau = VQ/(Ib)$

(a)

(b)

Figure 7.2 **Effect of prestress on the principal stresses in a beam web. (a) At P = 0. (b) At P > 0.**

When the principal tensile stress σ_1 reaches the tensile strength of concrete, cracking occurs and the cracks form in the direction perpendicular to the direction of σ_1.

When the prestress is zero, σ_1 is equal to the shear stress τ and acts at 45° to the beam axis, as shown in Figure 7.2a. If diagonal cracking occurs, it will be perpendicular to the principal tensile stress, that is at 45° to the beam axis. When the prestress is not zero, the normal compressive stress σ ($=P/A$) reduces the principal tension σ_1, as illustrated in Figure 7.2b. The angle between the principal stress direction and the beam axis increases, and consequently, if cracking occurs, the inclined crack is flatter. Prestress therefore improves the effectiveness of any transverse reinforcement (stirrups) that may be used to increase the shear strength of a beam. With prestress causing the inclined crack to be flatter, a larger number of vertical stirrup legs are crossed by the crack and, consequently, a larger tensile force can be carried across the crack.

In the case of I-beams, the maximum principal tension may not occur at the centroidal axis of the uncracked beam where the shear stress is greatest, but may occur at the flange–web junction where shear stresses are still high and the longitudinal compression is reduced by external bending.

If the prestressing tendon is inclined at an angle θ_p, the vertical component of prestress P_v ($=P\sin\theta_p \approx P\theta_p$) usually acts in the opposite direction to the load-induced shear. The force P_v may therefore be included as a significant part of the shear strength of the cross-section. Alternatively, P_v may be treated as an applied load and the net shear force V to be resisted by the section may be taken as:

$$V = V_{\text{loads}} - P_v \tag{7.1}$$

In summary, the introduction of prestress increases the shear strength of a reinforced concrete beam. Nevertheless, prestressed sections often have thin webs, and the thickness of the web may be governed by shear strength considerations.

7.2.3 Web reinforcement

In a beam containing no shear reinforcement, the shear strength is reached when inclined cracking occurs. The inclusion of shear reinforcement, usually in the form of vertical stirrups, increases the shear strength. After inclined cracking, the shear reinforcement carries tension across the cracks and resists widening of the cracks. Adjacent inclined cracks form in a regular pattern as shown in Figure 7.3a. The behaviour of the beam after cracking is explained conveniently in terms of an analogous truss, first described by Ritter [1] and shown in Figure 7.3b.

The web members of the analogous truss resist the applied shear and consist of vertical tension members (which represent the vertical legs of the closely spaced steel stirrups) and inclined compression members (which

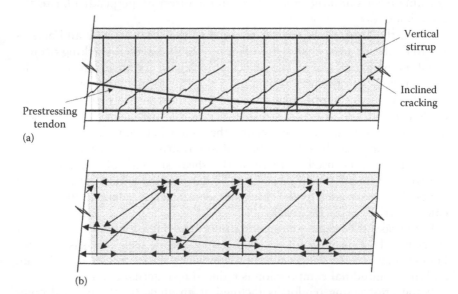

Figure 7.3 The analogous truss used to model a beam with shear reinforcement. (a) Beam elevation after inclined cracking. (b) The truss analogy.

model the concrete segments between the inclined cracks). In reality, there exists a continuous field of diagonal compression in the concrete between the diagonal cracks. This is idealised in the analogous truss by the discrete diagonal compression struts. In a similar manner, the vertical members of the analogous truss may represent a number of more closely spaced vertical stirrups. The top compressive chord of the analogous truss represents the concrete compressive zone plus any longitudinal compressive reinforcement, and the bottom chord models the longitudinal prestressed and nonprestressed reinforcement in the tensile zone. At each panel point along the bottom chord of the analogous truss, the vertical component of the compressive force in the inclined concrete strut must equal the tension in the vertical steel member, and the horizontal component must equal the change in the tensile force in the bottom chord (i.e. the change in force in the prestressing tendon and any other longitudinal non-prestressed reinforcement).

The analogous truss can be used to visualise the flow of forces in a beam after inclined cracking, but it is at best a simple model of a rather complex situation. The angle of the inclined compressive strut θ_v has traditionally been taken as 45°, although in practical beams, it is usually less. The stirrup stresses predicted by a 45° analogous truss are considerably higher than those measured in real beams [2], because the truss is based on the assumption that the entire shear force is carried by the vertical stirrups. In fact, part of the shear is carried by dowel action of the longitudinal tensile steel and part by friction on the mating surfaces of the inclined cracks (known as *aggregate interlock*). Some shear is also carried by the uncracked concrete compressive zone. In addition, the truss model neglects the tension carried by the concrete between the inclined cracks. The stress in the vertical leg of a stirrup in a real concrete beam is therefore a maximum at the inclined crack and is significantly lower away from the crack.

At the ultimate limit state, shear failure may be initiated by yielding of the stirrups or, if large amounts of web reinforcement are present, crushing of the concrete compressive strut. The latter is known as *web-crushing* and is usually avoided by placing upper limits on the quantity of web reinforcement. Not infrequently, premature shear failure occurs because of inadequately anchored stirrups. The truss analogy shows that the stirrup needs to be able to carry the full tensile force from the bottom panel point (where the inclined compressive force is resolved both vertically and horizontally) to the top panel point. To achieve this, care must be taken to detail the stirrup anchorages adequately to ensure that the full tensile capacity of the stirrup can be developed at any point along the vertical leg. After all, an inclined crack may cross the vertical leg of the stirrup at any point.

Larger diameter longitudinal bars should be included in the corners of the stirrup to form a rigid cage and to improve the resistance to pull-out of the hooks at the stirrup anchorage. These longitudinal bars also disperse the concentrated force from the stirrup and reduce the likelihood of splitting in the plane of the stirrup anchorage. Stirrup hooks should be located on the compression side of the beam where anchorage conditions are most favourable and the clamping action of the transverse compression greatly increases the resistance to pull-out. If the stirrup hooks are located on the tensile side of the beam, anchorage may be lost if flexural cracks form in the plane of the stirrup. In current practice, stirrup anchorages are most often located at the top of a beam. In the negative moment regions of such beams, adjacent to the internal supports, for example where shear and moment are relatively large, the shear capacity may be significantly reduced owing to the loss of stirrup anchorages after flexural cracking.

AS3600-2009 [3] requires that shear reinforcement, of area not less than that calculated as being necessary at any cross-section, must be provided for a distance D from that cross-section in the direction of decreasing shear, where D is the overall depth of the member. The first stirrup at each end of a span should be located within 50 mm from the face of the support. AS3600-2009 also requires that shear reinforcement extends as close to the compression face and the tension face of the member as cover requirements and the proximity of other reinforcement and tendons will permit. In addition, AS3600-2009 states that bends in bars used as stirrups shall enclose, and be in contact with, a longitudinal bar with a diameter not less than the diameter of the fitment (stirrup) bar.

In Figure 7.4, some satisfactory and some unsatisfactory stirrup arrangements are shown. Stirrup hooks should be bent through an angle of at least 135°. A 90° bend (a cog) will become ineffective should the cover be lost, for any reason, and will not provide adequate anchorage. AS3600-2009 states that fitment cogs of 90° are not to be used when the cog is located within 50 mm of any concrete surface.

In addition to carrying diagonal tension produced by shear, and controlling inclined web cracks, closed stirrups also provide increased ductility to a beam by confining the compressive concrete. The *open stirrups* shown in Figure 7.4b are commonly used, particularly in post-tensioned beams

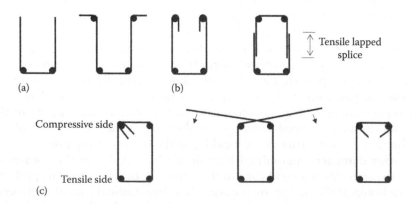

Figure 7.4 Stirrup shapes. (a) Incorrect. (b) Undesirable (but satisfactory). (c) Satisfactory.

where the opening at the top of the stirrup facilitates the placement and positioning of the post-tensioning duct along the member. This form of stirrup does not provide confinement for the concrete in the compression zone and is undesirable in heavily reinforced beams where confinement of the compressive concrete may be required to improve ductility of the member.

It is good practice to use adequately anchored stirrups, even in areas of low shear, particularly when tensile steel quantities are relatively high and cross-section ductility is an issue. Notwithstanding, AS3600-2009 [3] suggests that no shear reinforcement is required in regions of low shear in a shallow beam (where $D \leq 750$ mm). Although permitted by the Standard [1], it is not recommended here.

7.2.4 Shear strength

From the point of view of structural design, the shear strength of a beam containing no shear reinforcement V_u is the load required to cause the first inclined crack. In a beam containing web reinforcement, the ultimate strength in shear is usually calculated as the sum of the strength provided by the stirrups V_{us} and the strength provided by the concrete V_{uc}:

$$V_u = V_{uc} + V_{us} \tag{7.2}$$

In Figure 7.5, the transfer of shear force across a diagonal crack is shown. The part of the shear force carried by shear stresses in the uncracked concrete compression zone is V_c, the part carried by bearing and friction between the two surfaces of the inclined crack is V_a and the part carried by dowel action in the longitudinal steel crossing the crack is V_d. Because it is difficult to determine the magnitude of the force associated with each of these load-carrying mechanisms, they are usually lumped together and represented by a single empirical term for the shear strength contributed by the concrete V_{uc}.

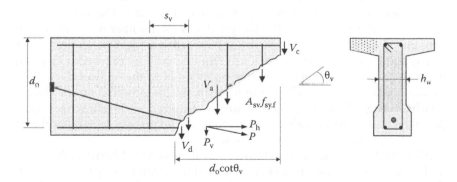

Figure 7.5 Transfer of shear at an inclined crack.

The contribution of stirrups to the shear strength of the beam V_{us} depends on the area of the vertical legs of each stirrup A_{sv}, the yield stress of the transverse steel reinforcement $f_{sy.f}$ and the number of stirrups that cross the inclined crack. As already stated, the longitudinal prestress causes the slope of the inclined crack θ_v to be less than 45°. It is reasonable to take the length of the horizontal projection of the inclined crack to be $d_o \cot \theta_v$, where d_o is the distance from the extreme compressive fibre to the centroid of the out-ermost layer of tensile steel (but need not be taken less than 0.8 times the overall depth of the member). The number of stirrups crossing the diagonal crack is therefore $d_o \cot \theta_v / s_v$, where s_v is the spacing of the stirrups required for shear in the direction of the member axis.

The total contribution of the stirrups to the shear strength of the section is the capacity of a stirrup times the number of stirrups crossing the inclined crack. AS3600-2009 [3] states that:

- for shear reinforcement perpendicular to the axis of the beam:

$$V_{us} = \frac{A_{sv} f_{sy.f} d_o \cot \theta_v}{s_v} \tag{7.3}$$

- for inclined shear reinforcement:

$$V_{us} = \frac{A_{sv} f_{sy.f} d_o}{s_v} \left(\sin \alpha_v \cot \theta_v + \cos \alpha_v \right) \tag{7.4}$$

where α_v is the angle between the inclined shear reinforcement and the lon-gitudinal tensile reinforcement.

In some building codes, the angle of the inclined crack θ_v is taken to be constant and equal to 45° (i.e. $\cot \theta_v = 1$ in Equations 7.3 and 7.4). However, in other codes, θ_v is not constant and may be varied between specified lim-its. It is evident from Equations 7.3 and 7.4 that the contribution of stirrups to the shear strength of a beam depends on θ_v. The flatter the inclined crack (i.e. the smaller the value of θ_v), the greater is the number of effective stir-rups and the greater is the value of V_{us}. In order to achieve the desired shear strength in design, fewer stirrups are required as θ_v is reduced. However, if the slope of the diagonal compression member in the analogous truss of Figure 7.3b is small, the change in force in the longitudinal tensile steel is relatively large. More longitudinal steel is required in the shear span near the support than would otherwise be the case and greater demand is placed on the anchorage requirements of these bars.

AS3600-2009 [3] states that θ_v may be taken as 45°. Alternatively, a lim-ited variable angle truss model is permitted in which θ_v may be selected in the range 30° and 60°, except that the minimum value for θ_v depends on the magnitude of the factored design shear force V^* and should be taken to vary linearly from 30°, when $V^* = \phi V_{u.min}$, to 45°, when $V^* = \phi V_{u.max}$. That is the minimum value for θ_v permitted by AS3600-2009 [3] is given by:

$$(\theta_v)_{min} = 30 + \frac{15(V^* - \phi V_{u.min})}{\phi V_{u.max} - \phi V_{u.min}} \tag{7.5}$$

and any value for θ_v that is greater than or equal to $(\theta_v)_{min}$ may be chosen up to a maximum value of 60°.

In Equation 7.5, $V_{u.min}$ is the ultimate shear strength of a beam containing the minimum area of shear reinforcement $A_{sv.min}$ and should be taken as:

$$V_{u.min} = V_{uc} + 0.10\sqrt{f_c'}\, b_v d_o \geq V_{uc} + 0.6 b_v d_o \tag{7.6}$$

The term b_v is the effective width of the web for shear and is taken as $b_v = b_w - 0.5\Sigma d_d$, where b_w is the width of the web and Σd_d is the sum of the diameters of the grouted ducts, if any, in a horizontal plane across the web. The corresponding minimum area of shear reinforcement is specified as:

$$A_{sv.min} = \frac{0.06\sqrt{f_c'}\, b_v s_v}{f_{sy.f}} \geq \frac{0.35 b_v s_v}{f_{sy.f}} \tag{7.7}$$

Equation 7.6 is obtained by substituting Equation 7.7 into Equations 7.2 and 7.3, with $\theta_v = 30°$.

In Equation 7.5, $V_{u.max}$ is the maximum allowable shear strength for a section and is limited by web-crushing, that is failure of the diagonal concrete compression strut in the analogous truss. The maximum shear strength is given by:

$$V_{u.max} = 0.2 f_c'\, b_v d_o + P_v \tag{7.8}$$

where P_v is the vertical component of the prestressing force at the section under consideration. The vertical force P_v is also accounted for in Equation 7.6 when calculating $V_{u.min}$, in which case it is included in the concrete contribution V_{uc} (as indicated in Equations 7.9 and 7.11).

The strength reduction factor for shear in Equation 7.5 is specified in AS3600-2009 [3] to be $\phi = 0.7$.

With θ_v limited to a minimum value of 30°, the following requirements are imposed on the anchorage of the longitudinal reinforcement at a support:

(1) at a simple support:
    Sufficient positive moment reinforcement must be anchored past the anchor point, such that the anchored reinforcement can develop a tensile force of $V^* \cot\theta_v/\phi$ plus any other tensile forces that may be carried by the longitudinal tensile reinforcement at that point. V^* is the design shear force at a distance $d_o \cot\theta_v$ from the anchor point. The *anchor point* is taken to be halfway along the length of bearing at a support

or determined by calculating the width of the compressive strut entering the truss node at the support (see Figure 8.26 for clarification).
In addition, not less than 50% of the positive moment reinforcement
required at mid-span should extend past the face of a simple support
for a length of 12 bar diameters or an equivalent anchorage.

(2) At a continuous of flexurally restrained support:

Not less than 25% of the total positive reinforcement required at
mid-span must continue past the near face of the support.

AS3600-2009 [3] also requires that the steel necessary for flexure at any
particular section must be provided and developed at a section a distance
D along the beam in the direction of increasing shear.

From much research and many laboratory tests, it appears that the contribution of the concrete to the shear strength is not less than the shear
force that initially caused the diagonal crack to form. For this reason, in
AS3600-2009 [3], the contribution of the concrete to the shear strength of
a section V_{uc} is taken to be the smaller of the shear forces required to produce either flexure-shear or web-shear cracking, as outlined in the following. However, if the cross-section under consideration is already cracked in
flexure, only flexure-shear cracking need be considered.

7.2.4.1 Flexure-shear cracking

The shear force required to produce an inclined flexure-shear crack is taken
as the sum of the shear force that exists when the flexural crack first develops, the additional shear force required to produce the inclined portion
of the crack (which extends a distance of about $d_o \cot \theta_v$ along the beam
in the direction of increasing moment) and the vertical component of the
prestressing force. The first and third of these shear force components are
easily calculated. The second is usually determined using empirical expressions developed from test data.

AS3600-2009 suggests that for flexure-shear cracking, the value of V_{uc}
(in N) may be taken as:

$$V_{uc} = \beta_1 \beta_2 \beta_3 b_v d_o f_{cv} \left[\frac{(A_{st} + A_{pt})}{b_v d_o} \right]^{1/3} + V_o + P_v \qquad (7.9)$$

where:

- β_1 is a size effect factor and for members where the cross-sectional
  area of shear reinforcement provided (A_{sv}) is equal to or greater than
  the area $A_{sv.min}$ specified in Equation 7.7:

$$\beta_1 = 1.1 \left(1.6 - \frac{d_o}{1000} \right) \geq 1.1$$

otherwise, $\beta_1 = 1.1(1.6 - d_o/1000) \geq 0.8$

- β_2 depends on the axial force present at the cross-section:
  $\beta_2 = 1$ for members in pure bending
  $\beta_2 = 1 - (N^*/3.5A_g) \geq 0$ for members with axial tension
  $\beta_2 = 1 + (N^*/14A_g) \geq 0$ for members with axial compression
  where N^* is the absolute value of the design axial force on the section excluding prestress.
- $\beta_3 = 1$, except when concentrated loads are applied close to the support, β_3 may be taken as $2d_o/a_v \leq 2$, provided the applied loads and the support are oriented so as to create diagonal compression over the length a_v. The dimension a_v is the distance from the section at which shear is being considered to the face of the nearest support.
- $f_{cv} = (f'_c)^{1/3} \leq 4$ MPa
- A_{st} and A_{pt} are, respectively, the cross-sectional areas of the longitudinal non-prestressed and prestressed tensile steel provided in the tensile zone and fully anchored at the cross-section under consideration.
- V_o is the shear force which exists at the section when the bending moment at that section equals the decompression moment M_o (i.e. the moment which causes zero stress in the extreme tensile fibre and may be taken as $Z\sigma_{cp.f}$, where Z is the section modulus and $\sigma_{cp.f}$ is the compressive stress caused by prestress at the extreme fibre where cracking occurs). For statically determinate members:

$$V_o = \frac{M_o}{\left| M^*/V^* \right|} \tag{7.10}$$

where M^* and V^* are the factored design moment and shear force, respectively, at the section under consideration. When the prestress and the applied moment both produce tension on the same extreme fibre of the member, V_o shall be taken as zero. For statically indeterminate members, secondary shear forces and bending moments due to prestress should be taken into account when determining V_o and M_o.
- P_v is the vertical component of prestress at the section under consideration. If the vertical component of prestress is adding to, rather than resisting, the transverse component of the compressive force in the diagonal strut, P_v shall be taken as negative.

7.2.4.2 Web-shear cracking

If a cross-section is uncracked in flexure, the shear force required to produce web-shear cracking is given by:

$$V_{uc} = V_t + P_v \tag{7.11}$$

where V_t is the shear force, which, when combined with the normal stresses caused by the prestress and the external loads, would produce a principal tensile stress of f'_{ct} (see Equation 4.4) at either the centroidal axis, the level of a prestressing duct or at the intersection of the web and the flange (if any), whichever is the more critical.

V_t may be found analytically, or graphically using a Mohr's circle construction, by setting $\sigma_1 = f'_{ct} = 0.36\sqrt{f'_c}$ in the following equation:

$$\sigma_1 = \sqrt{\left(\frac{\sigma}{2}\right)^2 + \tau^2} + \left(\frac{\sigma}{2}\right) \tag{7.12}$$

where the normal stress σ and the shear stress τ are given by:

$$\sigma = -\frac{P_e}{A} + \frac{P_e e y}{I} - \frac{M_t y}{I} = a_{Vt0} + a_{Vt1}V_t \tag{7.13}$$

and

$$\tau = \frac{V_t Q}{Ib} = b_{Vt1}V_t \tag{7.14}$$

in which M_t is the moment at the cross-section when the shear is V_t (and may be calculated as $M_t = V_t\,M^*/V^*$) and the remaining terms are given by:

$$a_{Vt0} = -\frac{P_e}{A} + \frac{P_e e y}{I} \tag{7.15}$$

$$a_{Vt1} = -\frac{M^*}{V^*}\frac{y}{I} \tag{7.16}$$

and

$$b_{Vt1} = \frac{Q}{Ib} \tag{7.17}$$

b is the appropriate width of the web, and is equal to b_v at the level of any prestressing duct and b_w at points remote from the duct, and Q is the first moment about the centroidal axis of that part of the area of the cross-section between the level under consideration and the extreme fibre.

Using Equations 7.12 through 7.17, the value of V_t (that produces $\sigma_1 = f'_{ct}$) is determined as follows:

$$V_t = \frac{-a_{Vt1}f'_{ct} + \sqrt{(a_{Vt1}f'_{ct})^2 - 4b^2_{Vt1}a_{Vt0}f'_{ct} + 4(b_{Vt1}f'_{ct})^2}}{2b^2_{Vt1}} \tag{7.18}$$

7.2.5 Summary of design requirements for shear

The design requirements for shear specified in AS3600-2009 [3] are summarised below.

1. The design shear strength of a section is ϕV_u, where $V_u = V_{uc} + V_{us}$, as stated in Equation 7.2.
2. The shear strength contributed by the concrete V_{uc} is the lesser of the values obtained from Equations 7.9 and 7.11.
3. The contribution of the shear reinforcement to the ultimate shear strength V_{us} is given by the following equation:

$$V_{us} = \frac{A_{sv} f_{sy.f} d_o}{s_v} \left(\sin \alpha_v \cot \theta_v + \cos \alpha_v \right) \tag{7.4}$$

where s_v is the centre-to-centre spacing of the shear reinforcement measured parallel to the axis of the member; θ_v is the angle of the concrete compression strut to the horizontal and may be conservatively taken as 45° or, alternatively, may be taken as any value between 30° and 60°, but greater than $(\theta_v)_{min}$ given by Equation 7.5. $(\theta_v)_{min}$ is assumed to vary linearly from 30°, when $V^* = \phi V_{u.min}$, to 45°, when $V^* = \phi V_{u.max}$; V^* is the factored design shear force; $V_{u.min}$ and $V_{u.max}$ are defined in Equations 7.6 and 7.8, respectively; and α_v is the angle between the inclined shear reinforcement and the longitudinal tensile reinforcement.

4. The maximum transverse shear to be designed for near a support is taken as the shear at
   a. the face of the support; or
   b. a distance d_o from the face of the support, provided:
      i. diagonal cracking cannot take place at the support or extend into the support;
      ii. there are no concentrated loads closer than $2d_o$ from the face of the support;
      iii. the value of β_3 in Equation 7.9 is taken to be equal to one; and
      iv. the transverse shear reinforcement required at d_o from the support is continued unchanged to the face of the support.

   AS3600-2009 [3] also requires that the longitudinal tensile reinforcement required at d_o from the face of the support shall be continued into the support and shall be fully anchored past that face.

   Note that when the support is above the beam, diagonal cracking can take place at, and extend into, the support and item (a) applies.

5. Where the factored design shear force $V^* \leq 0.5\phi V_{uc}$, no shear reinforcement is required, except that where the overall depth of the beam exceeds 750 mm, minimum shear reinforcement should be provided as given by Equation 7.7.

Where $0.5\phi V_{uc} < V^* \leq \phi V_{u.min}$, minimum shear reinforcement should be provided as given by Equation 7.7.

These minimum shear reinforcement requirements may be waived for shallow beams, where $V^* \leq \phi V_{uc}$ and D does not exceed the greater of 250 mm and half the width of the web, and for slabs, where $V^* \leq \phi V_{uc}$.

Where $V^* > \phi V_{u.min}$, shear reinforcement should be provided in accordance with Equation 7.3 (for vertical stirrups) or Equation 7.4 (for inclined stirrups).

6. In no case should the ultimate shear strength V_u exceed $V_{u.max}$ (as defined in Equation 7.8).

7. The maximum spacing between stirrups measured in the direction of the beam axis should not exceed the lesser of 0.5D or 300 mm, except that where $V^* \leq \phi V_{u.min}$, the spacing may be increased to 0.75D or 500 mm, whichever is smaller. The maximum transverse spacing between the vertical legs of a stirrup measured across the web of a beam should not exceed the lesser of 600 mm and the overall depth of the cross-section D.

8. The quantity of shear reinforcement calculated as being necessary at any cross-section should be provided for a distance D from the section in the direction of decreasing shear.

The first stirrup at each end of a span should be positioned no more than 50 mm from the face of the adjacent support.

9. Stirrups should be anchored on the compression side of the beam using standard hooks bent through an angle of at least 135° around a larger diameter longitudinal bar. It is important that the stirrup anchorage be located as close to the compression face of the beam as is permitted by concrete cover requirements and the proximity of other reinforcement and tendons.

7.2.5.1 Design equation

The factored design shear force must be less than or equal to the design strength:

$$V^* \leq \phi V_u = \phi(V_{uc} + V_{us}) \tag{7.19}$$

where the capacity reduction factor ϕ for shear is 0.7 in AS3600-2009 [3] (see Table 2.3).

Substituting Equation 7.3 into Equation 7.19 gives:

$$V^* \leq \phi V_{uc} + \frac{\phi A_{sv} f_{sy.f} d_o \cot\theta_v}{s_v}$$

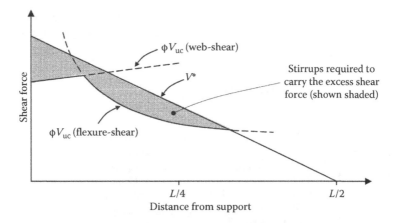

Figure 7.6 Web steel requirements for a uniformly loaded beam.

and the design equation for vertical stirrups becomes:

$$\frac{A_{sv}}{s_v} \geq \frac{1}{\phi}\left(\frac{V^* - \phi V_{uc}}{f_{sy.f}d_o \cot\theta_v}\right) \qquad (7.20)$$

The use of Equation 7.20 for the design of web reinforcement is tedious (see Example 7.1) even for straightforward or ordinary cases. The critical value for V_{uc} (controlled by the onset of either flexure-shear or web-shear cracking) must be determined at each section along the beam and varies from section to section. However, the process is repetitive and may be easily implemented in a spreadsheet or computer program.

The shear reinforcement requirements can be visualised by plotting the variation of both the applied shear (V^*) and the shear strength provided by the concrete (ϕV_{uc}) along the span of a uniformly loaded member, as shown in Figure 7.6. The hatched area represents the design strength that must be supplied by the shear reinforcement (ϕV_{us}).

EXAMPLE 7.1

The shear reinforcement for the post-tensioned beam shown in Figure 7.7 is to be designed. The beam is simply-supported over a span of 30 m and carries a uniformly distributed load, consisting of an imposed load $w_Q = 25$ kN/m and a permanent load $w_G = 40$ kN/m (which includes the beam self-weight). The beam is prestressed by a bonded parabolic cable with an eccentricity of 700 mm at mid-span and zero at each support. The area of the prestressing steel is $A_p = 3800$ mm^2 and the duct diameter is 120 mm. The prestressing

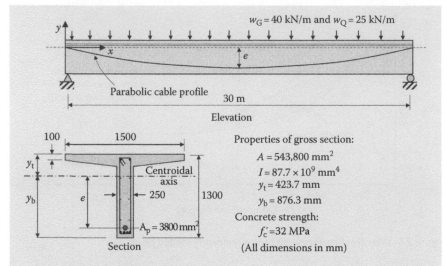

Figure 7.7 Beam details (Example 7.1).

force at each support is 4500 kN and at mid-span is 4200 kN and, elsewhere, is assumed to vary linearly along the beam length. The concrete cover to the reinforcement is taken to be 40 mm.

The factored design load combination for the strength limit state (see Section 2.3.2) is given by Equation 2.2:

$$w^* = 1.2\,w_G + 1.5\,w_Q = 1.2 \times 40 + 1.5 \times 25 = 85.5 \text{ kN/m}$$

At x m from support A:

$$V^* = 1282.5 - 85.5x \quad \text{and} \quad M^* = 1282.5x - 42.75x^2$$

Using Equations 1.3 and 1.4, the distance of the parabolic prestressing cable below the centroidal axis of the section at x m from A and the slope of the cable at that point are:

$$y = -2.8\left[\frac{x}{30} - \left(\frac{x}{30}\right)^2\right] \quad \text{and} \quad y' = -\frac{2.8}{30}\left(1 - \frac{x}{15}\right)$$

Assuming 40 mm concrete cover, 12 mm diameter stirrups and 28 mm diameter longitudinal bars in the corners of the stirrups, the depth d_o is:

$$d_o = 1300 - 40 - 12 - 14 = 1234 \text{ mm}$$

In Table 7.1, a summary of the calculations and reinforcement requirements at a number of sections along the beam is presented. In the following, sample calculations are provided for the sections at 1 m from the support and at 2 m from the support.

At x = 1.0 m:

From the above equations:

$V* = 1197$ kN, $M* = 1240$ kNm, $y = -0.0902$ m,

$e = 90.2$ mm, $y' = 0.0871$ rad,

$\theta = 0.0871$ rad, $P_e = 4480$ kN, $b_v = b_w - 0.5\Sigma d_d = 190$ mm,

$f_{cv} = 32^{1/3} = 3.17$ MPa, and $P_v = P_e\theta = 390$ kN (P_v is positive, because it is resisting the applied shear forces).

Flexure-shear cracking: The decompression moment is:

$$M_o = \frac{I}{y_b}\left(\frac{P_e}{A} + \frac{P_e e y_b}{I}\right) = P_e\left(e + \frac{I}{y_b A}\right)$$

$$= 4,480 \times 10^3\left(90.2 + \frac{87.7 \times 10^9}{876.3 \times 543,800}\right) = 1,229 \times 10^6 \text{ Nmm}$$

and the corresponding shear force V_o is calculated using Equation 7.10:

$$V_o = \frac{1229}{|1240/1197|} = 1186 \text{ kN}$$

Assuming 4–28 mm diameter longitudinal tensile reinforcing bars in the bottom of the cross-section at d_o from the top surface ($A_{st} = 2480$ mm²), including one 28 mm bar in each corner of the stirrups, the shear force required to produce a flexure-shear crack is obtained from Equation 7.9:

$$V_{uc} = 1.1 \times 1.0 \times 1.0 \times 190 \times 1234 \times 3.17 \times\left(\frac{2480 + 3800}{190 \times 1234}\right)^{1/3} \times 10^{-3}$$

$$+ 1186 + 390 = 1821 \text{ kN}$$

Web-shear cracking: Since $M*$ is only just greater than the decompression moment M_o, this section is unlikely to crack in flexure, so web-shear cracking may be critical. Checks should be made at the centroidal axis, at the tendon level and at the flange–web intersection.

At the centroidal axis: The first moment of the area below the centroidal axis is:

$$Q = 0.5 \times 250 \times 876.3^2 = 96.0 \times 10^6 \text{ mm}^3$$

The effective web width at the centroidal axis is $b = b_w = 250$ mm. From Equations 7.15 through 7.17:

$$a_{Vt0} = -\frac{P_e}{A} = -\frac{4,480 \times 10^3}{543,800} = -8.24 \text{ MPa}$$

$$a_{Vt1} = 0 \quad \text{and} \quad b_{Vt1} = \frac{Q}{Ib} = 4.378 \times 10^{-6} \text{ mm}^{-2}$$

and, with $\sigma_1 = f'_{ct} = 0.36\sqrt{f'_c} = 2.04$ MPa, Equation 7.13 gives $V_t = 1045$ kN.

At the level of the prestressing tendon (i.e. 90.2 mm below the centroidal axis):

$$Q = 0.5 \times 250 \times (876.3 - 90.2) \times (876.3 + 90.2) = 95.0 \times 10^6 \text{ mm}^3$$

and the effective web width is $b = b_v = 190$ mm. From Equations 7.15 through 7.17:

$$a_{Vt0} = -\frac{4,480 \times 10^3}{543,800} + \frac{4,480 \times 10^3 \times 90.2 \times (-90.2)}{87.7 \times 10^9} = -8.65 \text{ MPa}$$

$$a_{Vt1} = -\frac{1240 \times 10^6}{1197 \times 10^3} \frac{(-90.2)}{87.7 \times 10^9} = 1.07 \times 10^{-6} \text{ mm}^{-3}$$

$$b_{Vt1} = \frac{95.0 \times 10^6}{87.7 \times 10^9 \times 190} = 5.70 \times 10^{-6} \text{ mm}^{-2}$$

and, with $\sigma_1 = f'_{ct} = 0.36\sqrt{f'_c} = 2.04$ MPa, Equation 7.18 gives $V_t = 786$ kN which is less than the value at the centroidal axis. It also is less than the value at the flange–web intersection (not calculated here).

With $V_t = 786$ kN, Equation 7.11 gives: $V_{uc} = 786 + 390 = 1176$ kN.

Clearly, at this section, web-shear cracking occurs at a lower load than flexure-shear cracking and is therefore critical. Thus:

$$\phi V_{uc} = 0.7 \times 1176 = 823 \text{ kN}$$

This is less than the design shear force V^*, and therefore, shear reinforcement is required.

Stirrup design: In this example, 12 mm diameter single stirrups (two vertical legs) with $A_{sv} = 220$ mm² and $f_{sy.f} = 250$ MPa are to be used. To find the inclination of the diagonal compressive strut, $(\theta_v)_{min}$ is required from Equation 7.5, and the maximum and minimum shear strengths, $V_{u.max}$ and $V_{u.min}$, must first be calculated. From Equation 7.8, the maximum shear strength (limited by web-crushing) is:

$$V_{u.max} = 0.2 \times 32 \times 190 \times 1234 \times 10^{-3} + 390 = 1891\,kN$$

$$\phi V_{u.max} = 1324\,kN$$

From Equation 7.6:

$$V_{u.min} = 1176 + 0.6 \times 190 \times 1234 \times 10^{-3} = 1317\,kN$$

$$\phi V_{u.min} = 922\,kN$$

With $V^* = 1197$ kN, Equation 7.5 gives:

$$(\theta_v)_{min} = 30 + \frac{15(1197 - 922)}{1324 - 922} = 40.3^\circ$$

Taking $\theta_v = (\theta_v)_{min}$, and the design equation for vertical stirrups, (Equation 7.20) gives:

$$\frac{A_{sv}}{s_v} \geq \frac{1}{0.7} \left(\frac{(1197 - 823) \times 10^3}{250 \times 1234 \times \cot 40.3} \right) = 1.469$$

and with $A_{sv} = 220$ mm^2, the stirrup spacing must satisfy: $s_v \leq 150$ mm.

Using 12 mm diameter mild steel stirrups at 150 mm centres satisfies both the minimum steel and maximum spacing (300 mm) requirements of AS3600-2009.

At x = 2.0 m:

$V^* = 1111.5$ kN, $M^* = 2394$ kN·m, $y = -0.174$ m, $e = 174$ mm, $y' = 0.0809$ rad, $\theta = 0.0809$ rad, $P_e = 4460$ kN,   and $P_v = P_e\theta = 361$ kN.

As before:

$$b_v = 190\,mm \quad \text{and} \quad f_{cv} = 32^{1/3} = 3.17\,MPa.$$

Flexure-shear cracking: The decompression moment is:

$$M_o = \frac{I}{y_b} \left(\frac{P_e}{A} + \frac{P_e e y_b}{I} \right) = P_e \left(e + \frac{I}{y_b A} \right)$$

$$= 4,460 \times 10^3 \left(174 + \frac{87.7 \times 10^9}{876.3 \times 543,800} \right) = 1,598 \times 10^6\,Nmm$$

and the corresponding shear force V_o is calculated using Equation 7.10:

$$V_o = \frac{1936}{|3463/1026|} = 742 \text{ kN}$$

As previously assumed, A_{st} = 2480 mm^2 and the shear force required to produce a flexure-shear crack is therefore obtained from Equation 7.9:

$$V_{uc} = 1.1 \times 1.0 \times 1.0 \times 190 \times 1234 \times 3.17 \times \left(\frac{2480 + 3800}{190 \times 1234}\right)^{1/3} \times 10^{-3} + 742 + 361$$

$$= 1348 \text{ kN}$$

Web-shear cracking: Flexural cracking is deemed to occur at the cracking moment M_{cr} when the extreme tensile fibre reaches the flexural tensile strength of concrete (here taken as $f_{ct.f} = 0.6\sqrt{f_c'} = 3.39 \text{ MPa}$). The cracking moment is:

$$M_{cr} = \frac{I}{y_b}\left(f_{ct.f} + \frac{P_e}{A} + \frac{P_e e y_b}{I}\right)$$

$$= \frac{87.7 \times 10^9}{876.3}\left(3.39 + \frac{4,460 \times 10^3}{543,800} + \frac{4,460 \times 10^3 \times 174 \times 876.3}{87.7 \times 10^9}\right)$$

$$= 1,937 \times 10^6 \text{ Nmm}$$

Since M^* is greater than the cracking moment M_{cr}, this section will crack in flexure, so web-shear cracking need not be considered.

For this cross-section, therefore, ϕV_{uc} = 0.7 × 1348 = 944 kN.

Stirrup design: At this section, the maximum shear strength (Equation 7.8) and the shear strength with minimum reinforcement (Equation 7.6) are:

$$V_{u.max} = 0.2 \times 32 \times 190 \times 1234 \times 10^{-3} + 361 = 1862 \text{ kN}$$

$$\phi V_{u.max} = 1303 \text{ kN}$$

and

$$V_{u.min} = 1348 + 0.6 \times 190 \times 1234 \times 10^{-3} = 1489 \text{ kN}$$

$$\phi V_{u.min} = 1042 \text{ kN}$$

With V^* = 1111.5 kN, Equation 7.5 gives:

$$(\theta_v)_{min} = 30 + \frac{15 \times (1111.5 - 1042)}{1303 - 1042} = 34.0°$$

Table 7.1 Summary of results – Example 7.1

x(m)	V* (kN)	M* (kNm)	Web-shear ϕV_{uc} (kN)	Flexure-shear ϕV_{uc} (kN)	$\phi V_{u,min}$ (kN)	$\phi V_{u,max}$ (kN)	$(\theta_v)_{min}$ (kN)	Specified spacing of 12 mm mild steel stirrups Equation 7.20 (mm)
1	1197	1240	823	1275	922	1324	40.3	150 (150)
2	1111.5	2394	—	944	1042	1303	34.0	420 (300)
3	1026	3463	—	805	904	1282	34.8	309 (300)
4	940.5	4446	—	715	814	1262	34.2	310 (300)
5	855	5344	—	645	744	1242	33.3	345 (300)
6	769.5	6156	—	585	684	1222	32.4	407 (300)
7	684	6883	—	531	630	1202	31.4	509 (300)
8	598.5	7524	—	481	579	1183	30.5	685 (300)
9	513	8080	—	433	531	1163	30	1024 (500)
10	427.5	8550	—	386	485	1144	30	2006 (500)

Taking $\theta_v = (\theta_v)_{min}$, and the design equation for vertical stirrups (Equation 7.20) gives:

$$\frac{A_{sv}}{s_v} \geq \frac{1}{0.7}\left(\frac{(1111.5 - 944)\times 10^3}{250 \times 1234 \times \cot(34.0)}\right) = 0.523$$

and with $A_{sv} = 220$ mm², the stirrup spacing must satisfy: $s_v \leq 420$ mm.

We will adopt 12 mm diameter mild steel stirrups at 300 mm spacing. This corresponds to the maximum stirrup spacing permitted by AS3600-2009 [3] when V^* exceeds $\phi V_{u.min}$.

For other cross-sections, results are shown in Table 7.1. When x exceeds 8 m, the design shear V^* is less than $\phi V_{u.min}$ and the minimum amount of shear reinforcement is required. However, for much of the span, the maximum spacing requirements of 300 mm (when $V^* \geq \phi V_{u\,min}$) and 500 mm (when $V^* < \phi V_{u.min}$) specified in AS3600-2009 govern the design.

7.3 TORSION IN BEAMS

7.3.1 Compatibility torsion and equilibrium torsion

In addition to bending and shear, some members are subjected to twisting about their longitudinal axes. A common example is a spandrel beam supporting the edge of a monolithic floor, as shown in Figure 7.8a. The floor loading causes torsion to be applied along the length of the beam. A second

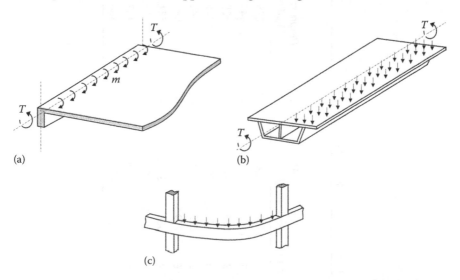

(a)   (b)

(c)

Figure 7.8 Members subjected to torsion. (a) Compatibility torsion. (b) Equilibrium torsion. (c) Equilibrium and compatibility torsion.

example is a box girder bridge carrying a load in one eccentric traffic lane, as shown in Figure 7.8b. Members which are curved in plan such as the beam in Figure 7.8c may also carry significant torsion.

For the design of spandrel beams, designers often disregard torsion and rely on the redistribution of internal forces to find an alternative load path. This may or may not lead to a satisfactory design. When torsional cracking occurs in the spandrel, its torsional stiffness is reduced and, therefore, the restraint provided to the slab edge is reduced. Additional rotation of the slab edge occurs and the torsion in the spandrel decreases.

Torsion that may be reduced by redistribution, such as the torsion in the spandrel beam, is often called *compatibility torsion*. Whereas indeterminate structures generally tend to behave in accordance with the design assumptions, full redistribution will occur only if the structure possesses adequate ductility and may be accompanied by excessive cracking and large local deformations. Ductile reinforcement is essential. For some statically indeterminate members (and for statically determinate members) twisted about their longitudinal axes, torsion is required for equilibrium and cannot be ignored. In the case of the box girder bridge of Figure 7.8b, for example torsion cannot be disregarded and will not be redistributed, as there is no alternative load path. This is *equilibrium torsion* and must be considered in design.

The behaviour of beams carrying combined bending, shear and torsion is complex. Most current design recommendations rely heavily on gross simplifications and empirical estimates derived from experimental observations and the provisions of AS3600-2009 are no exception.

7.3.2 Effects of torsion

Prior to cracking, the torsional stiffness of a member may be calculated using elastic theory. The contribution of reinforcement to the torsional stiffness before cracking is insignificant and may be ignored. When cracking occurs, the torsional stiffness decreases significantly and is very dependent on the quantity of steel reinforcement. In addition to causing a large reduction of stiffness and a consequential increase in deformation (twisting), torsional cracks tend to propagate rapidly and tend to be wider and more unsightly than flexural cracks.

Torsion causes additional longitudinal stresses in the concrete and the steel, and additional transverse shear stresses. Large torsion results in a significant reduction in the load-carrying capacity in bending and shear. To resist torsion after the formation of torsional cracks, additional longitudinal reinforcement and closely spaced closed stirrups are required. Cracks caused by pure torsion form a spiral pattern around the beam, hence the need for closed ties with transverse reinforcement near the top and bottom surfaces of a beam as well as the side faces. Many such cracks usually develop at relatively close centres and failure eventually occurs on a warped failure surface. The angles between the crack and the beam axis on each face of the beam are approximately the same and are here denoted θ_t. In AS3600-2009 [3], θ_t is taken to be the same as the angle of the inclined compressive strut θ_v when designing for shear.

After torsional cracking, the contribution of concrete to the torsional resistance of a reinforced or prestressed concrete member drops significantly. Any additional torque must be carried by the transverse reinforcement. Tests show that prestress increases the torsional stiffness of a member significantly, but does not greatly affect the strength in torsion. The introduction of prestress delays the onset of torsional cracking, thereby improving the member stiffness and increasing the cracking torque. The strength contribution of the concrete after cracking, however, is only marginally increased by prestress and the contribution of the transverse reinforcement is unchanged by prestress.

For a beam in pure torsion, the behaviour after cracking can be described in terms of the three-dimensional analogous truss shown in Figure 7.9. The closed stirrups act as transverse tensile web members (both vertical and horizontal), the longitudinal reinforcement in each corner of the stirrups acts as the longitudinal chords of the truss and the compressive web members inclined at an angle θ_t on each face of the truss represent the concrete between the inclined cracks on each face of the beam and carry the inclined compressive forces.

The three-dimensional truss analogy ignores the contribution of the interior concrete to the post-cracking torsional strength of the member. The diagonal compressive struts are located on each face of the truss and, in the actual beam, diagonal compressive stress is assumed to be located close to each surface of the member. The beam is therefore assumed to behave similarly to a hollow thin-walled section. Tests of members in pure torsion tend to support these assumptions.

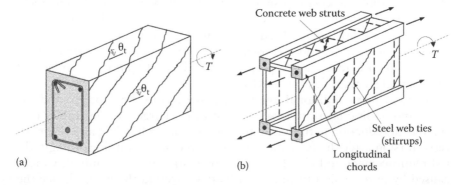

Figure 7.9 Three-dimensional truss analogy for a beam in pure torsion. (a) Beam segment. (b) Analogous truss.

Design models for reinforced and prestressed concrete beams in torsion are usually based on a simple model such as that described earlier.

7.3.3 Design provisions for torsion

The provisions in the Australian Standard for the design of beams subjected to torsion, and to torsion combined with flexure and shear, are outlined in this section and represent a simple and efficient design approach. Several conservative assumptions are combined with a variable angle truss model and the design procedure for shear presented in Sections 7.2.4 and 7.2.5.

7.3.3.1 Compatibility torsion

Provided certain detailing requirements are satisfied, AS3600-2009 [3] allows compatibility torsion (called *secondary torsion* in the Standard) to be ignored. The standard states that where torsional strength is not required for the equilibrium of the structure and the torsion in a member is induced solely by the angular rotation of adjoining members (such as the spandrel beam shown in Figure 7.8a), it is permissible to disregard the torsional stiffness of the member in the analysis and to disregard torsion in the member provided the minimum torsion reinforcement provisions given in (1), (2) and (3) in the following are satisfied.

1. Transverse reinforcement in the form of fully anchored closed stirrups must be provided with an area A_{sv} that provides a torsional capacity of at least $0.25T_{uc}$, but not less than $A_{sv.min}$ specified in Equation 7.7. The term T_{uc} is the ultimate strength in pure torsion of a beam without any stirrups (i.e. the torsion required to first produce torsional cracks) and is defined subsequently in Equation 7.23. The spacing of the closed fitments should not be greater than the lesser if $0.12u_t$ and 300 mm. The dimension u_t is the perimeter of the polygon with vertices at the centres of the longitudinal bars at the corners of the closed stirrups.
2. Areas of longitudinal reinforcement shall be provided to resist the following design tensile forces, taken as additional to any other design tensile forces:
   a. In the flexural tensile zone:

$$0.5f_{sy.f}\left(\frac{A_{sw}}{s_t}\right)u_t\cot^2\theta_v \tag{7.21}$$

b. In the flexural compressive zone:

$$0.5f_{\text{sy.f}}\left(\frac{A_{\text{sw}}}{s_t}\right)u_t\cot^2\theta_v - F_c^* \ (\geq 0) \tag{7.22}$$

where θ_v is the angle between the axis of the compressive strut and the longitudinal axis of the member (as shown in Figure 7.5); A_{sw} is the cross-sectional area of the bar forming the closed stirrup; s_t is the stirrup spacing along the longitudinal axis of the member; and F_c^* is the absolute value of the design force in the compressive zone due to flexure.

3. The longitudinal reinforcement required in (2) should be placed as close as practicable to the corners of the cross-section and, in all cases, at least one longitudinal bar shall be provided at each corner of the closed stirrups.

7.3.3.2 Equilibrium torsion

Where torsion is required for equilibrium, it must be considered in design and the ultimate torsional strength of each cross-section T_u must be calculated. When analysing a structure to determine the design torsion and shear for the strength limit state (i.e. T^* and V^*, respectively), the elastic uncracked stiffness may be used.

For a beam without torsional reinforcement, T_u is taken as the torsional strength of the concrete T_{uc} and is given by:

$$T_{uc} = J_t(0.3\sqrt{f_c'})\sqrt{\left(1+\frac{10\sigma_{cp}}{f_c'}\right)} \tag{7.23}$$

In this equation, T_{uc} is an estimate of the torsion required to cause first cracking in an otherwise unloaded beam. The torsional constant J_t may be taken as:

$J_t = 0.33x^2y$ for solid rectangular sections

$= 0.33\Sigma x^2y$ for solid T-shaped, L-shaped or I-shaped sections

$= 2A_m b_w$ for thin-walled hollow sections

where A_m is the area enclosed by the median lines of the walls of a single closed cell; and b_w is the minimum thickness of the wall of the hollow section.

The terms x and y are, respectively, the shorter and longer overall dimensions of the rectangular part(s) of the solid section. The term $0.3\sqrt{f_c'}$ in Equation 7.23 represents the tensile strength of concrete, the beneficial effect of the prestress on T_{uc} is accounted for by the term $\sqrt{(1 + 10\sigma_{cp}/f_c')}$ and σ_{cp} is the average effective prestress P_e/A.

When T^* and V^* are determined using the elastic uncracked stiffness of the structure, torsional reinforcement is required in a member if:

(1) $T^* \geq 0.25\, \phi\, T_{uc}$ \hfill (7.24)

or

(2) $\dfrac{T^*}{\phi T_{uc}} + \dfrac{V^*}{\phi V_{uc}} > 0.5$ \hfill (7.25)

or if the overall depth does not exceed the greater of 250 mm or half the width of the web:

$$\frac{T^*}{\phi T_{uc}} + \frac{V^*}{\phi V_{uc}} > 1.0 \tag{7.26}$$

where T_{uc} is determined from Equation 7.23; and V_{uc} is determined as the smaller of the values obtained from Equations 7.9 and 7.11.

When Equations 7.24 through 7.26 indicate that torsional reinforcement is required, additional closed stirrups must be specified and additional longitudinal reinforcement is required to satisfy Equations 7.21 and 7.22. The quantity of additional torsional reinforcement must satisfy the following requirement:

$$T^* \leq \phi T_{us} \tag{7.27}$$

where:

$$T_{us} = f_{sy.f} \left(\frac{A_{sw}}{s_t} \right) 2 A_t \cot \theta_v \tag{7.28}$$

and A_{sw} is the cross-sectional area of the bar forming the closed stirrup, s_t is the spacing between the additional stirrups required for torsion and A_t is the area of the polygon with vertices at the centres of the longitudinal bars at the corners of the closed stirrups.

Substituting Equation 7.28 into Equation 7.27 and rearranging gives the following design equation for determining the quantity of additional transverse reinforcement required for torsion:

$$\frac{A_{sw}}{s_t} \geq \frac{T^*}{\phi f_{sy.f} 2A_t \cot\theta_v} \tag{7.29}$$

As Equation 7.27 indicates, AS3600-2009 [3] makes the conservative assumption that the contribution of the concrete to the torsional strength of a beam cracked in torsion is zero.

The additional area of the longitudinal reinforcement that should be provided to resist the tensile forces outlined in Equations 7.21 and 7.22 is given by:

1. in the flexural tensile zone:

$$A_{st}^+ \geq 0.5 \frac{A_{sw}}{s_t} u_t \cot^2\theta_v \frac{f_{sy.f}}{f_{sy}} \tag{7.30}$$

2. in the flexural compressive zone:

$$A_{sc}^+ \geq 0.5 \frac{A_{sw}}{s_t} u_t \cot^2\theta_v \frac{f_{sy.f}}{f_{sy}} - \frac{F_c^*}{f_{sy}} \tag{7.31}$$

where the term (A_{sw}/s_t) is taken as the minimum value given by Equation 7.29.

Where torsional reinforcement is required, the area of transverse reinforcement A_{sv} should not be less than the larger of that required to provide a torsional capacity of $0.25T_{uc}$ and $A_{sv.min}$ specified in Equation 7.7 (in the form of closed ties or stirrups).

For a beam subjected to combined torsion and shear, care must be taken to avoid crushing of the compressive web struts before yielding of the torsional reinforcement. To this end, AS3600-2009 [3] requires the satisfaction of the following requirement:

$$\frac{T^*}{\phi T_{u.max}} + \frac{V^*}{\phi V_{u.max}} \leq 1.0 \tag{7.32}$$

where $V_{u.max}$ is obtained from Equation 7.8 and $T_{u.max}$ is calculated as:

$$T_{u.max} = 0.2f_c' J_t \tag{7.33}$$

As mentioned previously, when detailing the torsional reinforcement, the closed stirrups must be continuous around all sides of the section and be anchored so that the full strength of the bar can be developed at any point. The maximum spacing of the stirrups measured parallel to the longitudinal axis of the member is $0.12u_t$ or 300 mm, whichever is smaller. The additional longitudinal reinforcement must be enclosed within the stirrup and be as close to the corners of the section as possible. In all cases, at least one bar should be provided at each corner of the closed stirrup.

In Section 7.3.2, the desirability of avoiding torsional cracking at service loads was discussed. Equation 7.23 provides an estimate of the pure torsion required to cause first cracking, T_{uc}. When torsion is combined with shear, the torque required to cause first cracking is reduced. An estimate of the torque necessary to cause torsional cracking at a section can be obtained from the following interaction equation:

$$\frac{T_{cr}}{T_{uc}} + \frac{V_{cr}}{V_{uc}} = 1.0 \tag{7.34}$$

where T_{cr} and V_{cr} are the actual twisting moment and shear force, respectively, acting together at first cracking; and V_{uc} is the shear force required to cause inclined cracking when bending and shear are acting alone (and is the lesser of the values calculated using Equations 7.9 and 7.11).

If V_{cr} is expressed as $V_{cr} = T_{cr} (V^*/T^*) = T_{cr}/e_{T/V}$, then Equation 7.34 can be rearranged to give:

$$T_{cr} = \frac{T_{uc}\, e_{T/V} V_{uc}}{T_{uc} + e_{T/V} V_{uc}} \tag{7.35}$$

where $e_{T/V}$ is the ratio of design torsion to design shear (T^*/V^*). In design, it is often advisable to check that the applied torque T, under in-service conditions, is less than T_{cr}.

When torsion is accompanied by shear, the shear reinforcement requirements are determined in accordance with Sections 7.2.4 and 7.2.5. However, when torsion is present and when calculating the transverse shear reinforcement using Equation 7.20, AS3600-2009 [3] cautions that when load reversals can cause cracking in a zone usually in compression, the values of V_{uc} given by Equations 7.9 and 7.11 may not apply and that "V_{uc} shall be assessed or be taken as zero."

EXAMPLE 7.2

A prestressed concrete beam has a rectangular cross-section 400 mm wide and 550 mm deep. At a particular cross-section, the beam must resist the following factored design actions: $M^* = 300$ kNm, $V^* = 150$ kN and $T^* = 60$ kNm.

The dimensions and properties of the cross-section are shown in Figure 7.10. An effective prestress of 700 kN is applied at a depth of 375 mm by a

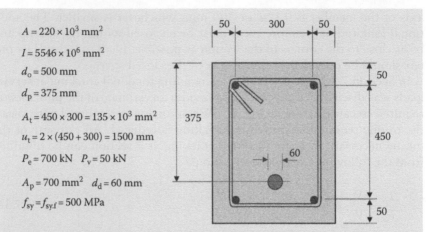

$A = 220 \times 10^3 \text{ mm}^2$

$I = 5546 \times 10^6 \text{ mm}^2$

$d_o = 500 \text{ mm}$

$d_p = 375 \text{ mm}$

$A_t = 450 \times 300 = 135 \times 10^3 \text{ mm}^2$

$u_t = 2 \times (450 + 300) = 1500 \text{ mm}$

$P_e = 700 \text{ kN} \quad P_v = 50 \text{ kN}$

$A_p = 700 \text{ mm}^2 \quad d_d = 60 \text{ mm}$

$f_{sy} = f_{sy.f} = 500 \text{ MPa}$

Figure 7.10 Cross-Section details (Example 7.2).

single cable consisting of seven 12.7 mm diameter strands in a grouted duct of 60 mm diameter. The area of the strands is A_p = 700 mm² with an ultimate tensile strength of f_{pb} = 1870 MPa. The vertical component of the prestressing force at the section under consideration is 50 kN, and the concrete strength is f_c' = 32 MPa.

Determine the longitudinal and transverse reinforcement requirements.

1. Initially, the cross-section should be checked for web-crushing. The effective width of the web for shear is:

$$b_v = b_w - 0.5d_d = 400 - 0.5 \times 60 = 370 \text{ mm}$$

and from Equation 7.8:

$$V_{u.max} = (0.2 \times 32 \times 370 \times 500 \times 10^{-3}) + 50 = 1234 \text{ kN}$$

The torsional constant J_t is:

$$J_t = 0.33x^2y = 0.33 \times 400^2 \times 550 = 29.0 \times 10^6 \text{ mm}^3$$

and Equation 7.33 gives:

$$T_{u.max} = (0.2 \times 32 \times 29.0 \times 10^6) \times 10^{-6} = 186 \text{ kNm}$$

The interaction equation for web-crushing (Equation 7.32) gives:

$$\frac{60}{0.7 \times 186} + \frac{150}{0.7 \times 1234} = 0.63 < 1.0$$

Therefore, web-crushing will not occur and the size of the cross-section is acceptable.

2. The longitudinal reinforcement required for bending must next be calculated. From the procedures outlined in Chapter 6, the ultimate bending strength provided by the seven prestressing strands (ignoring the longitudinal non-prestressed reinforcement) is $\phi M_{uo} = 319$ kNm (where $\phi = 0.8$ for bending as indicated in Table 2.3). With $\phi M_{uo} > M^*$, no non-prestressed steel is required for flexural strength.

3. To check whether torsional reinforcement is required, V_{uc} and T_{uc} must be calculated. Because bending is significant on this section, flexure-shear cracking will control and V_{uc} is obtained from Equation 7.9. The decompression moment M_o is given by:

$$M_o = Z_b\left(\frac{P_e}{A} + \frac{P_e e}{Z_b}\right) = 134\ \text{kNm}$$

and the corresponding shear force is calculated using Equation 7.10:

$$V_o = \frac{M_o V^*}{M^*} = 67\ \text{kN}$$

From Equation 7.9:

$$V_{uc} = 1.21 \times 1 \times 1 \times 370 \times 500 \times 32^{1/3} \times \left(\frac{700}{370 \times 500}\right)^{1/3} \times 10^{-3} + 67 + 50$$

$$= 228\ \text{kN}$$

With $\sigma_{cp} = P_e/A = 3.18$ MPa, from Equation 7.23:

$$T_{uc} = [29.0 \times 10^6 \times (0.3\sqrt{32})\sqrt{(1 + 10 \times 3.18/32)}] \times 10^{-6} = 69.5\ \text{kNm}$$

Both of the inequalities of Equations 7.24 and 7.25 are satisfied. That is:

$$T^* \ge 0.25\phi T_{uc} \quad \text{and} \quad \frac{T^*}{\phi I_{uc}} + \frac{V^*}{\phi V_{uc}} > 0.5$$

and, therefore, closed stirrups are required.

4. The shear strength of the section containing the minimum quantity of web reinforcement is obtained from Equation 7.6:

$$V_{u.min} = V_{uc} + 0.6\,b_v d_o = 228 + 0.6 \times 370 \times 500 \times 10^{-3} = 339\ \text{kN}$$

Since $V^* < \phi V_{u.min}$, $\theta_v = 30°$ and minimum transverse reinforcement is required for shear. Adopting 12 mm diameter closed stirrups ($A_{sv} = 220 \text{ mm}^2$), Equation 7.7 can be rearranged to give:

$$s_v \leq \frac{A_{sv}f_{sy.f}}{0.35b_v} = \frac{220 \times 500}{0.35 \times 370} = 849 \text{ mm}$$

That is at least 1.18 stirrups are required for shear per metre length along the beam.

The quantity of additional 12 mm closed stirrups required for torsion ($A_{sw} = 110 \text{ mm}^2$) is obtained from Equation 7.29:

$$\frac{A_{sw}}{s_t} \geq \frac{60 \times 10^6}{0.7 \times 500 \times 2 \times 135 \times 10^3 \times \cot(30)} = 0.367 \quad \therefore \, s_t \leq 300.1 \text{ mm}$$

That is an additional 3.33 stirrups are required for torsion per metre length of beam.

Summing the transverse steel requirements for shear and torsion, we have at least 4.51 stirrups required per metre, i.e. $s \leq 1000/4.51 = 222$ mm. This is more than the maximum permit permitted spacing of $0.12u_t = 0.12 \times 1500 = 180$ mm.

use 12 mm diameter stirrups at 180 mm centres ($f_{sy.f} = 500$ MPa).
Note: The required maximum spacing of transverse stirrups could have been calculated directly from s_v and s_t using:

$$\frac{1}{s} \geq \frac{1}{s_v} + \frac{1}{s_t} \quad \text{that is} \quad s \leq \frac{s_v s_t}{s_v + s_t}$$

5. With the prestressing steel providing sufficient longitudinal tensile steel to carry M^*, the additional longitudinal non-prestressed reinforcement required in the tension zone to resist torsion (with $f_{sy} = 500$ MPa) is obtained from Equation 7.30:

$$A_{st}^+ = 0.5 \times 0.367 \times 1500 \cot^2(30) \frac{500}{500} = 826 \text{ mm}^2$$

use 3–20 mm diameter deformed longitudinal bars in the bottom of the section, with two of these bars located in the corners of the stirrups.

The force in the compressive zone due to flexure F_c^* is approximately:

$$F_c^* \approx \frac{M^*}{0.9 \times d_p} = \frac{300 \times 10^6}{0.9 \times 375} \times 10^{-3} = 889 \text{ kN}$$

and from Equation 7.31:

$$A_{sc}^+ = 0.5 \times 0.367 \times 1500 \cot^2(30) \times \frac{500}{500} - \frac{889 \times 10^3}{500} < 0$$

Therefore, no additional longitudinal reinforcement is theoretically required in the compressive zone. We require two longitudinal bars in the top corners of the stirrups. Therefore:

use two 20 mm diameter top bars (one in each top corner of the stirrup).

7.4 SHEAR IN SLABS AND FOOTINGS

7.4.1 Punching shear

In the design of slabs and footings, strength in shear frequently controls the thickness of the member, particularly in the vicinity of a concentrated load or a column. Consider the pad footing shown in Figure 7.11. Shear failure may occur on one of two critical sections. The footing may act essentially as a wide beam and shear failure may occur across the entire width of the member, as illustrated in Figure 7.11a. This is *beam-type shear* (or one-way shear), and the shear strength of the critical section is calculated as for a beam. The critical section for this type of shear failure is usually assumed to be located at a distance d_o from the face of the column or concentrated load. Beam-type shear is often critical for footings but will rarely cause concern in the design of floor slabs.

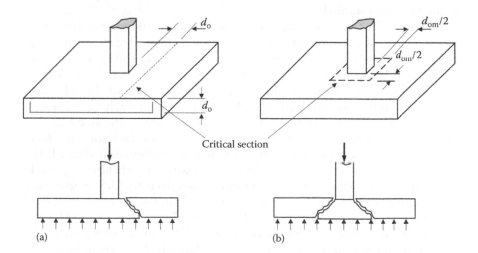

Figure 7.11 Shear failure surfaces in a footing or slab. (a) Beam-type shear. (b) Punching shear.

An alternative type of shear failure may occur in the vicinity of a concentrated load or column and is illustrated in Figure 7.11b. Failure may occur on a surface that forms a truncated cone or pyramid around the loaded area, as shown. This is known as *punching shear failure* (or two-way shear failure) and is often a critical consideration when determining the thickness of pad footings and flat slabs at the intersection of slab and column. The critical section for punching shear is usually taken to be geometrically similar to the loaded area and located at a distance $d_{om}/2$ from the face of the loaded area, where d_{om} is the mean value of d_o averaged around the critical shear perimeter. The critical section (or surface) is assumed to be perpendicular to the plane of the footing or slab. The remainder of this chapter is concerned with this type of shear failure.

The provisions for punching shear in the Australian Standard AS3600-2009 were developed from the results of a series of laboratory tests conducted on large-scale reinforced concrete edge column-slab specimens by Rangan and Hall [4,5] and Rangan [6,7]. The extension of their proposals to cover prestressed concrete slabs is both logical and simple. The design rules contained in AS3600-2009 [3] are outlined in Section 7.4.2. The rules are based on a simple model of the slab-column connection. Rangan and Hall [4,5] suggested that in order to determine the punching shear strength of a slab at a slab-column connection, the forces acting on the column and the capacity of the slab at each face of the column should be evaluated. Ideally in design, the column support should be large enough for the concrete to carry satisfactorily the moments and shears being transferred to the column without the need for any shear reinforcement. However, if this is not possible, procedures for the design of an adequate quantity of properly detailed reinforcement must be established.

In Figure 7.12, the way in which the moments and shears are transferred to an edge column in a flat plate floor is illustrated. Some of these forces are transferred at the front face of the column (M_b, V_b) and the remainder through the side faces as bending, torsion and shear (M_s, T_s, V_s). The front face must be able to carry M_b and V_b, and the side faces must have enough strength to carry M_s, T_s and V_s. A punching shear failure is initiated by the failure of a slab strip at either of the side faces, in combined bending, shear and torsion, or at the front face (or back face, in the case of an interior column), in combined bending and shear. If the concrete alone is unable to carry the imposed torsion and shear in the side faces, then transverse reinforcement in the spandrels (or side strips) must be designed. The provisions for the design of beam sections in combined torsion, bending and shear (outlined in Section 7.3.3) may be used for the design of the spandrel strips. Account should be taken, however, of the longitudinal restraint offered by the floor slab which prevents longitudinal expansion of the strip and substantially increases its torsional strength.

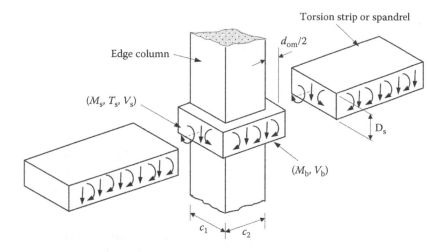

Figure 7.12 Forces at an edge column of a flat plate floor [4].

In laboratory tests, Rangan [6] observed that the longitudinal restraint provided by the floor slab increased the torsional strength of the spandrels by a factor of between four and six.

7.4.2 Design for punching shear

7.4.2.1 Introduction and definitions

Where shear failure can occur locally around a support or concentrated load, the design shear strength of the slab, in accordance with AS3600-2009 [3], is ϕV_u and must be greater than or equal to the design shear V^* acting on the *critical shear perimeter*. The *critical shear perimeter* is defined in AS3600-2009 [3] as being geometrically similar to the boundary of the effective area of a support or concentrated load and located at a distance $d_{om}/2$ therefrom. The *effective area* of a support or concentrated load is the area totally enclosing the actual support or load, for which the perimeter is a minimum. Both the critical shear perimeter and the effective area of a support are illustrated in Figure 7.13. Also shown in Figure 7.13 is the reduction of the shear perimeter caused by an opening through the thickness of the slab that is located within a distance of $2.5b_o$ from the critical perimeter. The term b_o is the dimension of the critical opening, as illustrated in the figure.

The punching shear strength V_u of a slab depends on the magnitude of the bending moment (M_v^*) being transferred from the slab to the support or loaded area. Accordingly, the design procedures with and without moment transfer are considered separately.

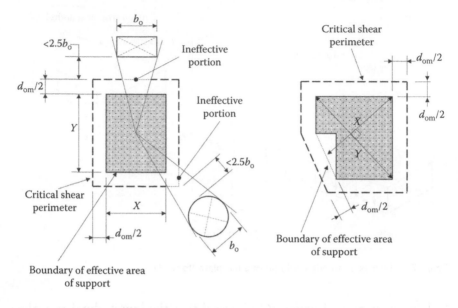

Figure 7.13 Critical shear perimeter [3].

7.4.2.2 Shear strength with no moment transfer

When no moment is transferred from the slab or footing to the column support (i.e. $M_v^* = 0$), or when the slab is subjected to a concentrated load, the punching shear strength of the slab is ϕV_{uo}, given by:

$$V_{uo} = u d_{om}(f_{cv} + 0.3\sigma_{cp}) \tag{7.36}$$

where u is the length of the critical shear perimeter (with account taken of the ineffective portions of the perimeter caused by adjacent openings); d_{om} is the average distance from the extreme compressive fibre to the outside layer of tensile flexural reinforcement (i.e. the average value of d_o around the critical shear perimeter); σ_{cp} is the average of the intensity of the effective prestress in the concrete (P_e/A) in each direction; and f_{cv} is a limiting concrete shear stress on the critical perimeter given by:

$$f_{cv} = 0.17\left(1 + \frac{2}{\beta_h}\right)\sqrt{f_c'} \leq 0.34\sqrt{f_c'} \tag{7.37}$$

The term β_h is the ratio of the longest overall dimension Y of the effective loaded area to the overall dimension X measured perpendicular to Y, as illustrated in Figure 7.13. It is noted that $f_{cv} = 0.34\sqrt{f_c'}$ for all cases in which $\beta_h \leq 2$.

If a properly designed, fabricated shear head is used to increase the shear strength, the upper limit for V_{uo} specified in the following equation may be used:

$$V_{uo} = ud_{om}(0.5\sqrt{f_c'} + 0.3\sigma_{cp}) \le 0.2ud_{om}f_c' \tag{7.38}$$

7.4.2.3 Shear strength with moment transfer

In the following, reference is made to the *torsion strips* associated with a particular slab-column connection. A torsion strip is a strip of slab, of width a, that frames into the side face of a column, as shown in Figure 7.14 (and also in Figure 7.12). In addition to the strip of slab, a torsion strip includes any beam that frames into the side face of the column. The longitudinal axis of a torsion strip is perpendicular to the direction of the spans used to calculate M_v^*.

AS3600-2009 [3] considers three cases for the determination of the punching shear strength of slab-column connection where an unbalanced moment M_v^* is transferred from the slab to the column:

1. where the torsion strip contains no beams and no closed ties;
2. where the torsion strip contains the minimum quantity of closed stirrups; and
3. where the torsion strips contain more than the minimum closed stirrups.

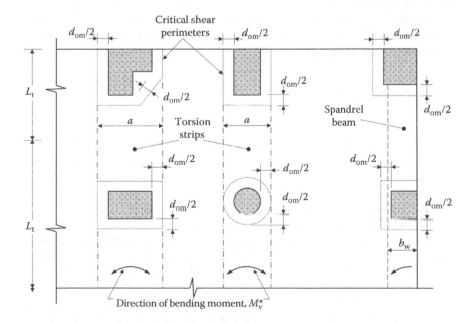

Figure 7.14 Torsion strips and spandrel beams [3].

Where the torsion strip contains no beams and no closed ties:

Consider a slab-column connection that is required to carry a factored design shear force V^* and an unbalanced moment M_v^*. The shear and torsion carried by the torsional strip at each side face of a column may be conservatively taken to be $V_s = (a/u)\, V^*$ and $T_s = 0.4 M_v^*$ (from Ref. 6). The design shear strength of the torsion strip is $V_{uc} = (a/u) V_{uo}$ (where V_{uo} is given by Equation 7.36), and the torsional strength may be obtained using Equation 7.23. If the width of the torsion strip a is greater than the overall slab depth D_s, and if a factor α is included to account for the restraint provided by the slab, then Equation 7.23 is modified as follows:

$$T_{uc} = 0.33 a D_s^2 (0.3\sqrt{f_c'})\, k\alpha$$

where $k = \sqrt{(1 + 10\sigma_{cp} / f_c')}$. By substituting these expressions into Equation 7.26 and by taking $D_s = 1.15 d_{om}$ and $k\alpha = 6$ (as recommended by Rangan in Reference 6), the following expression can be derived:

$$M_v^* \le 8 d_{om} \frac{a}{u}(\phi V_{uo} - V^*) \tag{7.39}$$

That is the strength of the unreinforced torsion strip is adequate, provided the combination of V^* and M_v^* satisfies the inequality of Equation 7.39. For design purposes, Equation 7.39 may be rearranged to give $V^* \le \phi V_u$, where, in AS3600-2009, $\phi = 0.7$ and the strength of the critical section V_u is obtained from Equation 7.39 as:

$$V_u \le \frac{V_{uo}}{\left[1 + (u M_v^* / 8 V^* a d_{om})\right]} \tag{7.40}$$

According to AS3600-2009 [3], Equation 7.40 is applicable to both reinforced and prestressed concrete slab-column connections, the only difference being the inclusion of the average prestress σ_{cp} in the estimate of V_{uo} in Equation 7.36.

If V^* is not less than ϕV_u, then the critical section must be either increased in size or strengthened by the inclusion of closed stirrups in the torsion strips. In practice, it is prudent to ensure that M_v^* satisfies Equation 7.39 (and hence $V^* \le \phi V_u$) so that no shear reinforcement is required. The introduction of a drop panel to increase the slab depth locally over the column support or the introduction of a column capital to increase the effective support area, and hence the critical shear perimeter, are the measures that are often adopted to increase V_{uo}, and hence V_u, to its required value.

Where the torsion strip contains the minimum quantity of closed stirrups:

This section applies to slab-column connections that may or may not have a transverse spandrel beam within the torsion strip. AS3600-2009 [3] specifies that if reinforcement for shear and torsion is required in the torsion strips, it shall be in the form of closed stirrups that extend for a distance not less than $L_t/4$ from the face of the support or concentrated load, where L_t is defined in Figure 7.14. The minimum cross-sectional area of the closed stirrup is taken as:

$$A_{\text{sw.min}} = 0.2y_1\, s_t/f_{\text{sy.f}} \tag{7.41}$$

where y_1 is the larger dimension of the closed rectangular stirrup; and s_t is the spacing of the stirrups in the direction of L_t.

The first stirrup should be located at not more than $s_t/2$ from the face of the support and the stirrup spacing s_t should not exceed the greater of 300 mm and D_b or D_s, as applicable. At least one longitudinal bar should be provided at each corner of the stirrup. Reinforcement details and dimensions are illustrated in Figure 7.15.

Using a similar derivation to that described for Equation 7.39 and with several conservative assumptions, Rangan [6] showed that, if a torsion strip contains the minimum quantity of closed stirrups (as specified in Equation 7.41), the strength is adequate, provided that:

$$M_v^* \le 2b_w\, \frac{a}{u}\left[\phi 1.2\frac{D_b}{D_s}V_{\text{uo}} - V^*\right] \tag{7.42}$$

where b_w and D_b are the web width and overall depth of the beam in the torsion strip, as shown in Figure 7.15. If the torsion strip contains no beam, then $b_w = a$ and $D_b = D_s$ (the slab thickness).

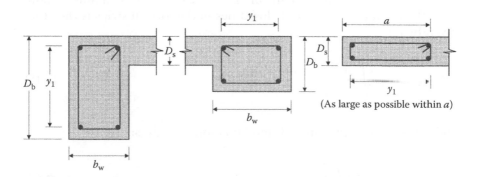

Figure 7.15 Shear reinforcement details and dimensions for slabs [3].

When M_v^* satisfies Equation 7.42, the shear strength of the critical section, with the minimum quantity of closed stirrups in the torsion strips, is given by:

$$V_{u.min} = \frac{1.2(D_b/D_s)V_{uo}}{\left[1 + uM_v^*/(2V^*ab_w)\right]} \tag{7.43}$$

In the case of a slab-column connection without any beams framing into the side face of the column, Equation 7.43 becomes:

$$V_{u.min} = \frac{1.2V_{uo}}{\left[1 + (uM_v^*/2V^*a^2)\right]} \tag{7.44}$$

When M_v^* does not satisfy Equation 7.42, i.e. when $V^* > \phi V_{u.min}$, the critical section must be increased in size or the side faces must be reinforced with more than the minimum quantity of closed stirrups.

Where the torsion strips contain more than the minimum closed stirrups:

Frequently, architectural considerations prevent the introduction of spandrel beams, column capitals, drop panels (or other slab thickenings) or the use of larger columns. In such cases, when M_v^* is greater than the limits specified in Equation 7.39 or 7.42 (as applicable), it is necessary to design shear reinforcement to increase the shear strength of the critical section. This may be the case for some edge or corner columns where the moment transferred from the slab to the column is relatively large and restrictions are placed on the size of the spandrel beams.

When closed stirrups are included in the torsion strips at the side faces of the critical section, the punching shear strength is proportional to $\sqrt{A_{sw}/s_t}$ and the shear strength of the critical section containing more than the minimum amount of closed ties in the torsion strips is therefore given by [6]:

$$V_u = V_{u.min}\sqrt{\frac{A_{sw}}{A_{sw.min}}} \tag{7.45}$$

and substituting Equation 7.41 into Equation 7.45 gives:

$$V_u = V_{u.min}\sqrt{\frac{A_{sw}f_{sy.f}}{0.2y_1s_t}} \tag{7.46}$$

To avoid web-crushing of the side faces of the critical section, AS3600-2009 [3] requires that the maximum shear strength be limited to:

$$V_{u.max} = 3V_{u.min}\sqrt{\frac{x}{y}} \qquad (7.47)$$

where x and y are the smaller and larger dimensions, respectively, of the cross-section of the torsion strip or spandrel beam.

By rearranging Equations 7.46, the amount of closed stirrups required in the torsion strip at the side face of the critical section must satisfy:

$$\frac{A_{sw}}{s_t} \geq \frac{0.2y_1}{f_{sy.f}} \left(\frac{V^*}{\phi V_{u.min}}\right)^2 \qquad (7.48)$$

It is emphasised that Equation 7.48 should only be used where size restrictions are such that the slab thickness and support sizes are too small to satisfy Equations 7.39 and 7.42. In general, it is more efficient and economical to provide column capitals and/or drop panels to overcome punching shear than it is to try to design and detail stirrups within the slab thickness.

EXAMPLE 7.3 INTERIOR COLUMN (CASE I)

The interior columns of a prestressed concrete flat plate are 100 by 400 mm in section and are located on a regular rectangular grid at 8 m centres in one direction and 6 m centres in the other. Check the adequacy of the punching shear strength of the critical shear perimeter of a typical interior column. The slab thickness is $D_s = 200$ mm, and the average depth to the bottom layer of tensile steel is $d_{om} = 160$ mm. The following data apply:

$\sigma_{cp} = 2.5$ MPa, $f_c' = 32$ MPa, $V^* = 520$ kN and $M_v^* = 40$ kNm.

For a square interior column, $\beta_h = 1$, and from Equation 7.37:

$$f_{cv} = 0.34\sqrt{32} = 1.92 \text{ MPa}$$

and the critical shear perimeter u and the width of the torsion strip a are:

$$u = 4 \times (400 + 160) = 2240 \text{ mm} \text{ and } a = 400 + 160 = 560 \text{ mm}$$

The shear strength of the critical section without moment transfer is calculated using Equation 7.36:

$$V_{uo} = 2240 \times 160 \times (1.92 + 0.3 \times 2.5) \times 10^{-3} = 957 \text{ kN}$$

Provided that M_v^* satisfies Equation 7.39, the shear strength of the critical section is given by Equation 7.40 and no shear reinforcement is necessary. In this case:

$$8d_{om}\frac{a}{u}(\phi V_{uo} - V^*) = 8 \times 160 \times \frac{560}{2240}(0.7 \times 957 - 520) \times 10^{-3} = 48.0 \text{ kNm}$$

which is greater than M_v^* and, therefore, the critical section is adequate without any shear reinforcement. The shear strength of the slab is obtained from Equation 7.40:

$$V_u \le \frac{957 \times 10^3}{1 + (2240 \times 40 \times 10^6)/(8 \times 520 \times 10^3 \times 560 \times 160)} = 772 \text{ kN}$$

and

$$\phi V_u = 541 \text{ kN} > V^*$$

Therefore, the punching shear strength of the critical section is adequate.

EXAMPLE 7.4   INTERIOR COLUMN (CASE 2)

The slab-column connection analysed in Example 7.3 is to be rechecked for the case when $V^* = 720$ kN and $M_v^* = 80$ kNm.

As in Example 7.3, $V_{uo} = 957$ kN and $\phi V_{uo} = 670$ kN and this is less than V^*, even without considering the unbalanced moment M_v^*. The critical shear perimeter is clearly not adequate. Shear and torsional reinforcement could be designed to increase the shear strength. However, successfully anchoring and locating stirrups within a 200 mm thick slab is difficult. An alternative solution is to use a fabricated steel shear head to improve resistance to punching shear. The most economical and structurally efficient solution, however, is to increase the size of the critical section. The slab thickness can often be increased locally by the introduction of a drop panel, or alternatively, the critical shear perimeter may be increased by introducing a column capital or simply by increasing the column dimensions. In general, provided such dimensional changes are architecturally acceptable, they represent the best structural solution.

Let the slab thickness be increased to 250 mm by the introduction of a 50 mm thick drop panel over the column in question (so that $d_{om} = 210$ mm):

$$u = 4 \times (400 + 210) = 2440 \text{ mm} \quad \text{and} \quad a = 400 + 210 = 610 \text{ mm}$$

From Equation 7.36:

$$V_{uo} = 2440 \times 210 \times (1.92 + 0.3 \times 2.5) \times 10^{-3} = 1368 \text{ kN}$$

and checking Equation 7.39 gives:

$$8d_{om} \frac{a}{u} (\phi V_{uo} - V^*) = 8 \times 210 \times \frac{610}{2440} (0.7 \times 1368 - 720) \times 10^{-3}$$

$$= 99.8 \text{ kNm} > M_v^*$$

Therefore, the shear strength of the critical section will be adequate and no shear reinforcement is required.

From Equation 7.40, $\phi V_u = 757$ kN, which is greater than V^*, as expected.

EXAMPLE 7.5   EDGE COLUMN

Consider the edge column-slab connection with critical shear perimeter as shown in Figure 7.16. The design shear and unbalanced moment are $V^* = 300$ kN and $M_v^* = 160$ kNm. The slab thickness is 220 mm, with no spandrel beams along the free slab edge. The average effective depth for the top tensile steel in the slab is $d_{om} = 180$ mm and $f_c' = 32$ MPa. The width of the torsion strip is 490 mm.

When designing a slab for punching shear at an edge (or corner) column, the average prestress σ_{cp} perpendicular to the free edge across the critical section (i.e. across the width b_t in Figure 7.16) should be taken as zero,

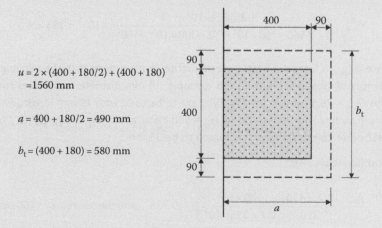

$u = 2 \times (400 + 180/2) + (400 + 180)$
$= 1560$ mm

$a = 400 + 180/2 = 490$ mm

$b_t = (400 + 180) = 580$ mm

Figure 7.16 Plan view of critical shear perimeter (Example 7.5).

unless care is taken to ensure that the slab tendons are positioned so that this part of the critical section is subjected to prestress. Often this is not physically possible, as discussed in Section 12.2 and illustrated in Figure 12.6. In this example, it is assumed that $\sigma_{cp} = 0$ perpendicular to the free edge and $\sigma_{cp} = 2.5$ MPa parallel to the edge.

As in Example 7.3, $f_{cv} = 1.92$ MPa and using Equation 7.36:

$$V_{uo} = [2 \times 490 \times 180 \times (1.92 + 0.3 \times 2.5) + 580 \times 180 \times (1.92 + 0.3 \times 0)] \times 10^{-3}$$

$$= 671 \, kN$$

Checking Equation 7.39 shows that:

$$8d_{om} \frac{a}{u} (\phi V_{uo} - V^*) = 8 \times 180 \times \frac{490}{1560} (0.7 \times 671 - 300) \times 10^{-3}$$

$$= 76.8 \, kNm < M_v^*$$

Therefore, the unreinforced critical shear perimeter is not adequate.

As mentioned in the previous examples, a local increase in the slab thickness or the introduction of a spandrel beam or a column capital may prove to be the best solution. For the purposes of this example, however, shear and torsional reinforcement will be designed in the torsion strip of width $a = 490$ mm and depth $D_s = 220$ mm at the side faces of the critical section.

From Equation 7.44, the strength of the critical section when the side faces contain the specified minimum quantity of closed stirrups is:

$$V_{u.min} = \frac{1.2 \times 671 \times 10^3}{1 + (1560 \times 160 \times 10^6) / (2 \times 300 \times 10^3 \times 490^2)} \times 10^{-3} = 295 \, kN$$

Since $\phi V_{u.min}$ is less than V^*, the torsion strips require more than the minimum quantity of closed stirrups. In this example, 10 mm diameter mild steel stirrups ($A_{sw} = 78.5$ mm^2, $f_{sy.f} = 250$ MPa) are to be used with 16 mm longitudinal bars in each corner of the stirrup ($f_{sy} = 500$ MPa), as shown in Figure 7.17. The clear cover to the stirrups is assumed to be 25 mm.

From Equation 7.48:

$$\frac{A_{sw}}{s_t} \geq \frac{0.2 \times 430}{250} \left(\frac{300 \times 10^3}{0.7 \times 295 \times 10^3} \right)^2 = 0.726 \quad \text{and therefore: } s_t \leq 108 \, mm$$

Use 10 mm stirrups at 100 mm centres in the torsion strips.

$A_{sw} = 78.5$ mm^2, $f_{sy.f} = 250$ MPa,
$f_{sy} = 500$ MPa
$y_1 = 490 - 2 \times 25 - 10 = 430$ mm
$x_1 = 220 - 2 \times 25 - 10 = 160$ mm
$u_t = 2 \times [(430 - 26) + (160 - 26)]$
$\quad = 1076$ mm

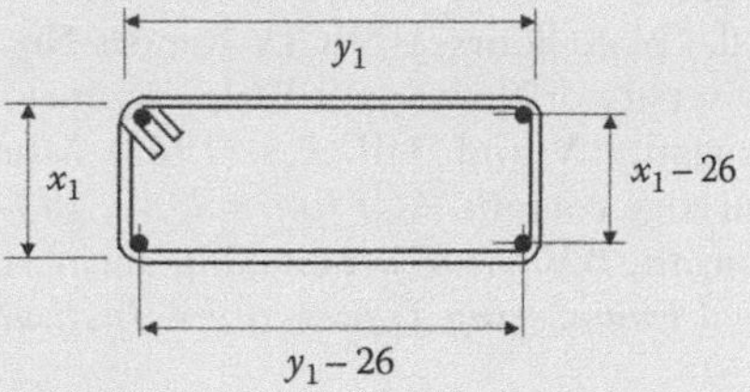

Figure 7.17 Details of closed stirrups in the torsion strips (Example 7.5).

From Equation 7.47:

$$V_{u.max} = 3 \times 295 \times \sqrt{\frac{220}{490}} = 593 \text{ kN}$$

and the angle of the inclined compression is obtained from Equation 7.5:

$$\theta_v = (\theta_v)_{min} = 30 + \frac{15\,(300 - 0.7 \times 295)}{0.7 \times 593 - 0.7 \times 295} = 36.7°$$

The minimum area of longitudinal tensile steel within the closed stirrups is obtained from Equation 7.30:

$$A_{st}^+ \geq 0.5 \times 0.726 \times 1076 \times \cot^2 36.7 \times \frac{250}{500} = 352 \text{ mm}^2$$

The 4–16 mm diameter longitudinal bars, one in each corner of the stirrup as shown in Figure 7.17, provide $A_{st}^+ = A_{sc}^+ = 400$ mm^2.

The shear strength of the critical section is given by Equations 7.46:

$$V_u = V_{u.min} \sqrt{\frac{A_{sw} f_{sy.f}}{0.2 y_1 s_t}} = 295 \times \sqrt{\frac{78.5 \times 250}{0.2 \times 430 \times 100}} = 446 \text{ kN}$$

and therefore $\phi V_u = 312$ kN $> V^*$. The cross-section of the proposed torsion strip is therefore adequate.

REFERENCES

1. Ritter, W. (1899). *Die Bauweise Hennebique (Construction Methods of Hennebique)*. Zurich: Schweizerische Bauzeitung.
2. Hognestad, E. (1952). What do we know about diagonal tension and web reinforcement in concrete? University of Illinois Engineering Experiment Station, Circular Series No. 64, Urbana.
3. AS3600-2009 (2009). *Australian Standard for Concrete Structures*. Standards Australia, Sydney, New South Wales, Australia.

4. Rangan, B.V. and Hall, A.S. (1983). Forces in the vicinity of edge-columns in flat plate floors. UNICIV Report No. R-203. School of Civil Engineering, University of New South Wales, Sydney, New South Wales, Australia.
5. Rangan, B.V. and Hall, A.S. (1983). Moment and shear transfer between slab and edge column. *ACI Journal*, 80, 183–191.
6. Rangan, B.V. (1987). Punching shear strength of reinforced concrete slabs. *Civil Engineering Transactions, Institution of Engineers, Australia*, CE 29, 71–78.
7. Rangan, B.V. (1987). Shear and torsion design in the new *Australian Standard for Concrete Structures. Civil Engineering Transactions, Institution of Engineers, Australia*, CE 29, 148–156.

Chapter 8

Anchorage zones

8.1 INTRODUCTION

In prestressed concrete structural members, the prestressing force is usually transferred from the prestressing steel to the concrete in one of two different ways. In post-tensioned construction, relatively small anchorage plates transfer the force from the tendon to the concrete immediately behind the anchorage by bearing at each end of the tendon. For pretensioned members, the force is transferred by bond between the steel and the concrete. In either case, the prestressing force is transferred in a relatively concentrated fashion, usually at the end of the member, and involves high local pressures and forces. A finite length of the member is required for the concentrated forces to disperse to form the linear compressive stress distribution usually assumed in design [1].

The length of member over which this dispersion of stress takes place is called the *transfer or transmission length* (in the case of pretensioned members) and the *anchorage length* (for post-tensioned members). Within these so-called *anchorage zones,* a complex stress condition exists. Transverse tension is produced by the dispersion of the longitudinal compressive stress trajectories and may lead to longitudinal cracking within the anchorage zone. Similar zones of stress exist in the immediate vicinity of any concentrated force, including the concentrated reaction forces at the supports of a member.

The anchorage length in a post-tensioned member and the magnitude of the transverse forces (both tensile and compressive), that act perpendicular to the longitudinal prestressing force, depend on the magnitude of the prestressing force and on the size and position of the anchorage plate or plates. Both single and multiple anchorages are commonly used in post-tensioned construction. A careful selection of the number, size and location of the anchorage plates can often minimise the transverse tension and hence minimise the transverse reinforcement requirements within the anchorage zone.

The stress concentrations within the anchorage zone in a pretensioned member are not usually as severe as in a post-tensioned anchorage zone. There is a more gradual transfer of prestress. The prestress is transmitted by bond over a significant length of the tendon, and there are usually numerous individual tendons that are well distributed throughout the anchorage zone. In addition, the high concrete bearing stresses behind the anchorage plates in post-tensioned members do not occur in pretensioned construction.

8.2 PRETENSIONED CONCRETE – FORCE TRANSFER BY BOND

In pretensioned concrete, the tendons are usually tensioned within a casting bed. The concrete is cast around the tendons and, after the concrete has gained sufficient strength, the pretensioning force is released. The extent of the anchorage zone and the distribution of stresses within that zone depend on the quality of bond between the tendon and the concrete. The transfer of prestress usually occurs only at the end of the member, with the steel stress varying from zero at the end of the tendon to the prescribed amount (full prestress) at some distance from the end. Over the *transfer length* (or *transmission length*) L_{pt}, bond stresses are high. The better the quality of the steel–concrete bond, the more efficient is the force transfer and the shorter is the transfer length. Outside the transfer length, bond stresses at transfer are small and the prestressing force in the tendon is approximately constant. Bond stresses and localised bond failures may occur outside the transfer length after the development of flexural cracks and under overloads, but a bond failure of the entire member involves failure of the anchorage zone at the ends of the tendons.

The main mechanisms that contribute to the strength of the steel–concrete bond are chemical adhesion of steel to concrete, friction at the steel–concrete interface and mechanical interlocking of concrete and steel (associated primarily with deformed or twisted strands). When the tendon is released from its anchorage within the casting bed and the force is transferred to the concrete, there is a small amount of tendon slip at the end of the member. This slippage destroys the bond for a short distance into the member at the released end, after which adhesion, friction and mechanical interlock combine to transfer the tendon force to the concrete.

During the stressing operation, there is a reduction in the diameter of the tendon due to the Poisson's ratio effect. The concrete is then cast around the highly tensioned tendon. When the tendon is released, the unstressed portion of the tendon at the end of the member returns to its original diameter, whilst at some distance into the member, where the tensile stress in the tendon is still high, the tendon remains at its reduced diameter. Within the transfer length, the tendon diameter varies as shown in Figure 8.1 and there is a radial pressure exerted on the surrounding concrete. This pressure produces a frictional component which assists in the transferring of force from the steel to the concrete. The wedging action due to this radial strain is known as the Hoyer effect [2].

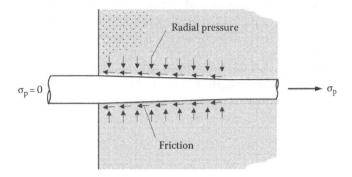

Figure 8.1 Hoyer effect [2].

The transfer length and the rate of development of the steel stress along the tendon depend on many factors, including the size of the strand (i.e. the surface area in contact with the concrete), the surface conditions of the tendon, the type of tendon, the degree of concrete compaction within the anchorage zone, the degree of cracking in the concrete within the anchorage zone, the method of release of the prestressing force into the member and the compressive strength of the concrete.

The factors of size and surface condition of a tendon affect bond capacity in the same way as they do for non-prestressed reinforcement. A light coating of rust on a tendon will provide greater bond than for steel that is clean and bright. The surface profile has a marked effect on transfer length. Stranded cables have a shorter transfer length than crimped, indented or plain steel wires of equivalent area owing to the interlocking between the helices forming the strand. The strength of concrete, within the range of strengths used in prestressed concrete members, does not greatly affect the transfer length. However, with increased concrete strength, there is greater shear strength of the concrete embedded between the individual wires in the strand.

An important factor in force transfer is the quality and degree of concrete compaction. The transfer length in poorly compacted concrete is significantly longer than that in well-compacted concrete. A prestressing tendon anchored at the top of a member generally has a greater transfer length than a tendon located near the bottom of the member. This is because the concrete at the top of a member is subject to increased sedimentation and is generally less well compacted than the concrete at the bottom of a member. When the tendon is released suddenly and the force is transferred to the concrete with impact, the transfer length is greater than for the case when the force in the steel is gradually imparted to the concrete.

Depending on the aforementioned factors, transfer lengths are generally within the range 50–150 times the tendon diameter. The force transfer is not linear, with about 50% of the force transferred in the first quarter of the transfer length and about 80% within the first half of the length. For design purposes, however, it is reasonable and generally conservative to assume a linear variation of steel stress over the entire transfer length.

AS3600-2001 [1] does not provide guidance on the distribution of bond stresses along the transfer length. Eurocode 2 [3] states that at release of the tendons, the prestress may be assumed to be transferred to the concrete by a constant bond stress f_{bpt} given by:

$$f_{bpt} = \eta_{p1}\eta_1 f_{ctd}(t) \tag{8.1}$$

where η_{p1} is a coefficient that takes into account the type of tendon and equals 2.7 for indented wire and 3.2 for strand; η_1 depends on the bond conditions and equals 1.0 for good bond conditions and 0.7 otherwise; and $f_{ctd}(t)$ is the design tensile strength of concrete at the time of release (i.e. the characteristic tensile strength value divided by a partial material factor of 1.5).

The *basic* value of the transmission length L_{pt} is specified in Eurocode 2 [3] as:

$$L_{pt} = \frac{\alpha_1 \alpha_2 d_b \sigma_{p0}}{f_{pbt}} \tag{8.2}$$

where α_1 depends on the method of release of the tendon and equals 1.0 for gradual release and 1.25 for sudden release; α_2 depends on the type of tendon and equals 0.25 for round wire and 0.19 for seven-wire strand; d_b is the nominal diameter of the tendon; and σ_{p0} is the tendon stress just after release.

The *design* value of the transmission length is taken as the least favourable of two alternative values $L_{pt1} = 0.8 L_{pt}$ or $L_{pt2} = 1.2 L_{pt}$ depending on the design situation. When local stresses are being checked at release, L_{pt1} is appropriate. When the ultimate limit state of the anchorage and the anchorage zone is being checked, L_{pt2} is appropriate.

AS3600-2009 [1] specifies the minimum values given in Table 8.1 for the transmission length depending on the type of tendon and the concrete strength at transfer. These values are independent of the level of initial prestress and ignore the level of cracking and degree of compaction within the anchorage zone. The Standard [1] assumes that no change in the position of the inner end of the transmission length occurs with time and also specifies that, within a distance of $0.1L_{pt}$ from the end of the tendon, the tendon should be assumed to be unstressed.

For seven-wire strand, ACI318M-11 [4] specifies that the transmission length is given by $L_{pt} = (\sigma_{pe}/21) \times d_b$. This corresponds to the value of $60d_b$

Table 8.1 Minimum transmission length for pretensioned tendons [1]

Types of tendon	L_{pt} for gradual release	
	$f_{cp} \geq 32$ MPa	$f_{cp} < 32$ MPa
Indented wire	100 d_b	175 d_b
Crimped wire	70 d_b	100 d_b
Ordinary and compact strand	60 d_b	60 d_b

specified in Table 8.1 when the effective stress in the tendon after all losses σ_{pe} is 1260 MPa.

More information about the transmission length may be obtained from specialist literature, including Refs [5–8].

The value of stress in the tendon, in regions outside the transmission length, remains approximately constant under service loads or whilst the member remains uncracked, and hence the transfer length remains approximately constant. After cracking in a flexural member, however, the behaviour becomes more like that of a reinforced concrete member and the steel stress increases with increasing moment. If the critical moment location occurs at or near the end of a member, such as may occur in a short-span beam or a cantilever, the required development length for the tendon is much greater than the transfer length. In such cases, the bond capacity of the tendons needs to be carefully considered.

The length of a pretensioned tendon from its end to the critical cross-section, where the ultimate stress in the tendon σ_{pu} is required, must be greater than the minimum development length L_p given in AS3600-2009 [1] as:

$$L_p = 0.145 \, (\sigma_{pu} - 0.67\sigma_{pe}) \, d_b \geq 60 \, d_b \tag{8.3}$$

and is illustrated in Figure 8.2. L_p is the sum of the transmission length L_{pt} plus the additional bonded length necessary to develop the increase of steel stress from σ_{pe} to σ_{pu}.

Where debonding of a strand is specified near the end of a member, and the design allows for tension at service loads within the development length, the minimum development length of the debonded strand is $2L_p$.

Where a prestressing tendon is not initially stressed, i.e. it is used in a member as non-prestressed reinforcement, and the tendon is required to develop its full characteristic breaking strength f_{pb}, the minimum development length required on either side of the critical cross-section is 2.5 times the minimum transmission length specified in Table 8.1. Care should be taken in situations where a sudden change in the effective depth of the tendon occurs due to an abrupt change in the member depth. In these locations, it may not be possible to develop the full strength of an initially untensioned tendon. Local bond failure may occur in the vicinity of the step, limiting the stress that can be developed in the tendon. Such a

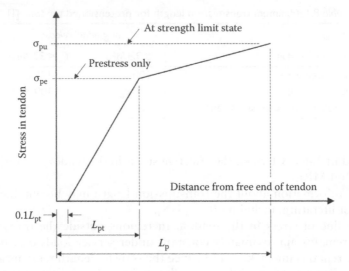

Figure 8.2 Variation of steel stress near the free end of a pretensioned tendon [1].

situation may develop if the calculated stress change in the strand required in the region of high local bond stresses exceeds about 500 MPa [9].

From their test results, Marshall and Mattock [10] proposed the following simple equation for determining the amount of transverse reinforcement A_{sb} (in the form of stirrups) in the end zone of a pretensioned member:

$$A_{sb} = 0.021 \frac{D}{L_{pt}} \frac{P}{\sigma_{sb}} \qquad (8.4)$$

where D is the overall depth of the member; P is the prestressing force immediately after transfer; and σ_{sb} is the permissible steel stress required for crack control and may be taken conservatively as 150 MPa.

The transverse steel A_{sb} should be equally spaced within $0.2D$ from the end face of the member.

8.3 POST-TENSIONED CONCRETE ANCHORAGE ZONES

8.3.1 Introduction

In post-tensioned concrete structures, failure of the anchorage zone is perhaps the most common cause of problems arising during construction. Such failures are difficult and expensive to repair, and usually necessitate replacement of the entire structural member. Anchorage zones may fail owing to uncontrolled cracking or splitting of the concrete resulting from insufficient, well-anchored transverse reinforcement. Bearing failures immediately behind the anchorage plate are also relatively common and may be caused by inadequately dimensioned bearing plates or poor workmanship resulting in poorly compacted concrete in the heavily reinforced region behind the bearing plate. Great care should therefore be taken in both the design and construction of post-tensioned anchorage zones.

Consider the case shown in Figure 8.3 of a single square anchorage plate ($h \times h$) centrally positioned at the end of a prismatic member of depth D and width b. In the disturbed region of length L_a immediately behind the anchorage plate (i.e. the anchorage zone), plane sections do not remain plane and simple beam theory does not apply. High bearing stresses at the anchorage plate disperse throughout the anchorage zone, creating high transverse stresses, until at a distance L_a from the anchorage plate the linear stress and strain distributions predicted by simple beam theory are produced. The dispersion of stress that occurs within the anchorage zone is illustrated in Figure 8.3b.

The stress trajectories directly behind the anchorage are convex to the centreline of the member, as shown, and therefore produce a transverse component of compressive stress normal to the member axis. Further from the anchorage, the compressive stress trajectories become concave to the member axis and, as a consequence, produce transverse tensile stress components. The stress trajectories are closely spaced directly behind the bearing plate where compressive stress is high, and become more widely spaced as the distance from the anchorage plate increases. St Venant's principle suggests that the length of the disturbed region, for the single centrally

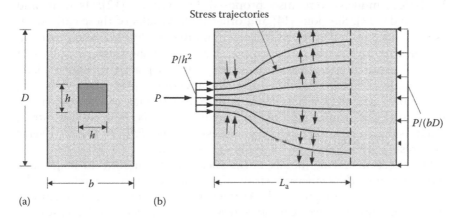

Figure 8.3 Stress trajectories for a centrally placed anchorage plate. (a) End elevation. (b) Side elevation.

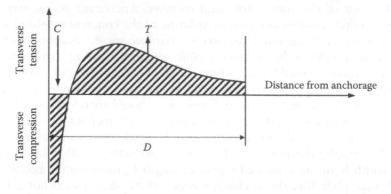

Figure 8.4 Distribution of transverse stress behind a single central anchorage.

located anchorage shown in Figure 8.3, is approximately equal to the depth of the member D. The variation of the transverse stresses along the centre-line of the member and normal to it, is represented in Figure 8.4.

The degree of curvature of the stress trajectories is dependent on the size of the bearing plate. The smaller the bearing plate, the larger are both the curvature and concentration of the stress trajectories, and hence the larger are the transverse tensile and compressive forces in the anchorage zone. The transverse tensile forces (often called *bursting* or *splitting* forces) need to be estimated accurately so that transverse reinforcement within the anchorage zone can be designed to resist them.

Elastic analysis can be used to analyse anchorage zones prior to the commencement of cracking. Early studies using photo-elastic methods [11] demonstrated the distribution of stresses within the anchorage zone. Analytical models were also proposed by Iyengar [12], Iyengar and Yogananda [13], Sargious [14] and others. The results of these early studies have since been confirmed by Foster and Rogowsky [15] (and others) in nonlinear finite element investigations. Figure 8.5a shows stress isobars of $|\sigma_y/\sigma_x|$ in an anchorage zone with a single centrally placed anchorage plate. Results are presented for three different anchorage plate sizes: $h/D = 0$, $h/D = 0.25$ and $h/D = 0.5$. These isobars are similar to those obtained in photo-elastic studies reported by Guyon [11]. σ_y is the transverse stress and σ_x is the average longitudinal compressive stress $P/(bD)$. The transverse compressive stress region in Figure 8.5a is shaded.

The effect of varying the size of the anchor plate on both the magnitude and position of the transverse stress along the axis of the member can be more clearly seen in Figure 8.5b. As the plate size increases, the magnitude of the maximum transverse tensile stress on the member axis decreases and its position moves further along the member (i.e. away from the anchorage plate). Tensile stresses also exist at the end surface of the anchorage zone in the corners adjacent to the bearing plate. Although these stresses are

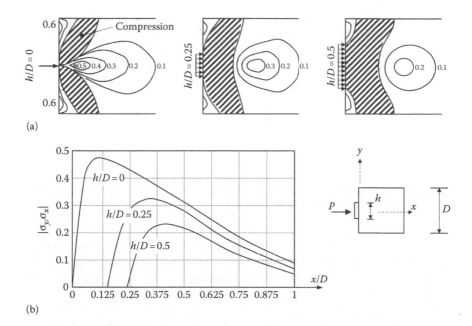

Figure 8.5 Transverse stress distributions for central anchorage [11]. (a) Stress isobars $|\sigma_y/\sigma_x|$. (b) Transverse stress along member axis.

relatively high, they act over a small area and the resulting tensile force is small. Guyon [11] suggested that a tensile force of about 3% of the longitudinal prestressing force is located near the end surface of a centrally loaded anchorage zone when h/D is greater than 0.10.

The position of the line of action of the prestressing force with respect to the member axis has a considerable influence on the magnitude and distribution of stress within the anchorage zone. As the distance of the applied force from the axis of the member increases, the tensile stress at the loaded face adjacent to the anchorage also increases.

Figure 8.6a illustrates the stress trajectories in the anchorage zone of a prismatic member containing an eccentrically positioned anchorage plate. At a length L_a from the loaded face, the concentrated bearing stresses disperse to the asymmetric stress distribution shown. The stress trajectories, which indicate the general flow of forces, are therefore unequally spaced, but will produce transverse tension and compression along the anchorage axis in a manner similar to that for the single centrally placed anchorage. Isobars of $|\sigma_y/\sigma_x|$ are shown in Figure 8.6b. High bursting forces exist along the axis of the anchorage plate and, away from the axis of the anchorage, tensile stresses are induced on the end surface. These end tensile stresses, or spalling stresses, are typical of an eccentrically loaded anchorage zone.

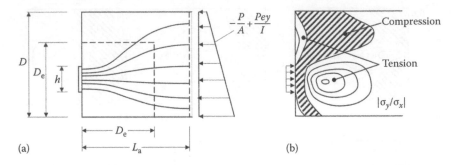

Figure 8.6 Diagrammatic stress trajectories and isobars for an eccentric anchorage [11].
(a) Side elevation. (b) Stress isobars.

Transverse stress isobars in the anchorage zones of members containing multiple anchorage plates are shown in Figure 8.7. The length of the member over which significant transverse stress exists (L_a) reduces with the number of symmetrically placed anchorages. The zone directly behind each anchorage contains bursting stresses and the stress isobars resemble those in a single anchorage centrally placed in a much smaller end zone, as indicated in Figure 8.7. Tension also exists at the end face between adjacent anchorage plates. Guyon [11] suggested that the tensile force near the end face between any two adjacent bearing plates is about 4% of the sum of the longitudinal prestressing forces at the two anchorages.

The isobars presented in this section are intended only as a means of visualising the structural behaviour. Concrete is not a linear-elastic material, and a cracked prestressed concrete anchorage zone does not behave exactly as depicted by the isobars in Figures 8.5 through 8.7. However, the linear-elastic analyses indicate the areas of high tension, both behind each anchorage plate and on the end face of the member, where cracking of the concrete can be expected during the stressing operation. The formation of such cracks reduces the stiffness in the transverse direction and leads to a significant redistribution of forces within the anchorage zone.

8.3.2 Methods of analysis

The design of the anchorage zone of a post-tensioned member involves both the arrangement of the anchorage plates to minimise transverse stresses and the determination of the amount and distribution of reinforcement to carry the transverse tension after cracking of the concrete. Relatively large amounts of transverse reinforcement, usually in the form of stirrups, are often required within the anchorage zone and careful detailing of the steel is essential to permit the satisfactory placement and compaction of the concrete. In thin-webbed members, the anchorage zone is often enlarged to form an *end-block* which is sufficient to accommodate the anchorage devices.

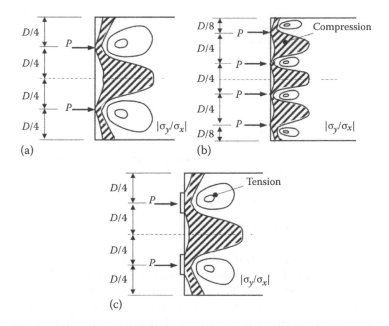

Figure 8.7 Transverse stress isobars for end zones with multiple anchorages [11].
(a) Two symmetrically placed anchorages – $h/D = 0$. (b) Four symmetrically
placed anchorages – $h/D = 0$. (c) Two symmetrically placed anchorage plates.

This also facilitates the detailing and fixing of the reinforcement and the
subsequent placement of concrete.

The anchorages usually used in post-tensioned concrete are patented by the
manufacturer and prestressing companies for each of the types and arrangements of tendons. The units are usually recessed into the end of the member,
and have bearing areas which are sufficient to prevent bearing problems in
well-compacted concrete. Often the anchorages are manufactured with *fins*
that are embedded in the concrete to assist in distributing the large concentrated force. Spiral reinforcement often forms part of the anchorage system
and is located immediately behind the anchorage plate to confine the concrete and thus significantly improve its bearing capacity.

As discussed in Section 8.3.1, the curvature of the stress trajectories determines the magnitude of the transverse stresses. In general, the dispersal of
the prestressing forces occurs through both the depth and the width of the
anchorage zone and therefore, transverse reinforcement must be provided
within the end zone in two orthogonal directions (usually, vertically and
horizontally on sections through the anchorage zone). The reinforcement
quantities required in each direction are usually obtained from separate
two-dimensional analyses, i.e. the vertical transverse tension is calculated
by considering the vertical dispersion of forces and the horizontal tension is
obtained by considering the horizontal dispersion of forces.

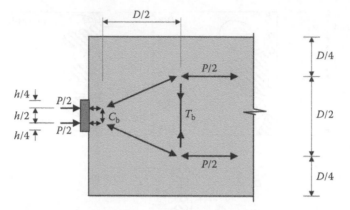

Figure 8.8 Strut-and-tie model of an anchorage zone.

The internal flow of forces in each direction can be visualised in several ways. A simple model is to consider truss action within the anchorage zone. For the anchorage zone of the beam of rectangular cross-section shown in Figure 8.8, a simple strut-and-tie model shows that transverse compression exists directly behind the bearing plate, with transverse tension, often called the bursting force (T_b), at some distance along the member. Design using strut-and-tie modelling is outlined in more detail in Section 8.4.

Consider the anchorage zone of the T-beam shown in Figure 8.9. The strut-and-tie arrangement shown is suitable for calculating both the vertical tension in the web and the horizontal tension across the flange.

An alternative model for estimating the internal tensile forces is to consider the anchorage zone as a deep beam loaded on one side by the bearing stresses immediately under the anchorage plate and resisted on the

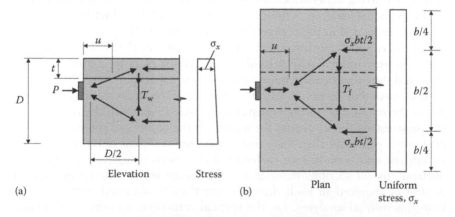

Figure 8.9 Vertical and horizontal tension in the anchorage zone of a post-tensioned T-beam. (a) Vertical tension in web. (b) Horizontal tension across flange.

other side by the statically equivalent, linearly distributed stresses in the beam. The depth of the deep beam is taken as the anchorage length L_a. This approach was proposed by Magnel [16] and was further developed by Gergely and Sozen [17] and Warner and Faulkes [18].

8.3.2.1 Single central anchorage

The beam analogy model is illustrated in Figure 8.10 for a single central anchorage, together with the bending moment diagram for the idealised beam. Since the maximum moment tends to cause bursting along the axis of the anchorage, it is usually denoted by M_b and called the *bursting moment*.

By considering one-half of the end-block as a free-body diagram, as shown in Figure 8.11, the bursting moment M_b required for rotational equilibrium is obtained from statics. Taking moments about any point on the member axis gives:

$$M_b = \frac{P}{2}\left(\frac{D}{4} - \frac{h}{4}\right) = \frac{P}{8}(D - h) \tag{8.5}$$

where M_b is resisted by the couple formed by the transverse forces C_b and T_b, as shown.

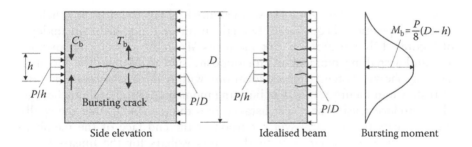

Figure 8.10 Beam analogy for a single centrally placed anchorage.

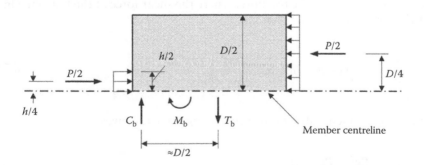

Figure 8.11 Free-body diagram of the top half of the anchorage zone in Figure 8.10.

As has already been established, the position of the resulting transverse (vertical) tensile force T_b in Figures 8.10 and 8.11 is located at some distance from the anchorage plate, as shown. For a linear-elastic anchorage zone, the exact position of T_b is the centroid of the area under the appropriate transverse tensile stress curve in Figure 8.5b. For the single, centrally placed anchorage of Figures 8.5, 8.10 and 8.11, the lever arm between C_b and T_b is approximately equal to $D/2$. This approximation also proves to be a reasonable one for a cracked concrete anchorage zone. Therefore, using Equation 8.5, we get:

$$T_b \approx \frac{M_b}{D/2} = \frac{P}{4}\left(1 - \frac{b}{D}\right) \qquad (8.6)$$

Expressions for the bursting moment and the horizontal transverse tension resulting from the lateral dispersion of bearing stresses across the width b of the section are obtained by replacing the depth D in Equations 8.5 and 8.6 with the width b.

8.3.2.2 Two symmetrically placed anchorages

Consider the anchorage zone shown in Figure 8.12a containing two anchorages, each positioned equidistant from the member axis. The beam analogy of Figure 8.12b indicates bursting moments M_b on the axis of each anchorage and a spalling moment M_s (of opposite sign to M_b) on the member axis, as shown. Potential crack locations within the anchorage zone are also shown in Figure 8.12a. The bursting moments behind each anchorage plate produce tension at some distance into the member, while the spalling moments produce transverse tension at the end face of the member. This simple analysis agrees with the stress isobars for the linear-elastic end-block of Figure 8.7c. Consider the free-body diagram shown in Figure 8.12c. The maximum bursting moment behind the top anchorage occurs at the distance x below the top fibre, where the shear force at the bottom edge of the free-body is zero. That is:

$$\frac{P}{D}x = \frac{P}{2h}(x-a) \quad \text{or} \quad x = \frac{aD}{(D-2h)} \qquad (8.7)$$

Summing moments about any point in Figure 8.12c gives:

$$M_b = \frac{Px^2}{2D} - \frac{P(x-a)^2}{4h} \qquad (8.8)$$

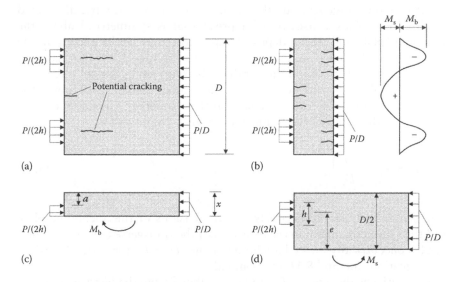

Figure 8.12 Beam analogy for a two symmetrically placed anchorages. (a) Side elevation. (b) Idealised beam with bursting and spalling moments. (c) Free-body
diagram of top portion of beam of depth x. (d) Free-body diagram of the top
half of the beam.

The maximum spalling moment M_s occurs at the member axis, where the
shear is also zero, and may be obtained by taking moments about any point
on the member axis in the free-body diagram of Figure 8.12d:

$$M_s = \frac{P}{2}\left(e - \frac{D}{4}\right)$$
(8.9)

After the maximum bursting and spalling moments have been determined, the resultant internal compressive and tensile forces can be estimated, provided that the lever arm between them is known. The internal
tension T_b produced by the maximum bursting moment M_b behind each
anchorage may be calculated from:

$$T_b = \frac{M_b}{l_b}$$
(8.10)

By examining the stress contours in Figure 8.7, the distance between the
resultant transverse tensile and compressive forces behind each anchorage
l_b depends on the size of the anchorage plate and the distance between the
plate and the nearest adjacent plate or free edge of the section. Guyon [11]
suggested an approximate method which involves the use of an *idealised
symmetric prism* for computing the transverse tension behind an eccentrically positioned anchorage. The assumption is that the transverse stresses

in the real anchorage zone are the same as those in a concentrically loaded idealised end-block consisting of a prism that is symmetrical about the anchorage plate and with a depth D_e equal to twice the distance from the axis of the anchorage plate to the nearest concrete edge. If the internal lever arm l_b is assumed to be half the depth of the symmetrical prism (i.e. $D_e/2$), then the resultant transverse tension induced along the line of action of the anchorage is obtained from an equation that is identical to Equation 8.6, except that the depth of the symmetric prism D_e replaces D:

$$T_b = \frac{P}{4}\left(1 - \frac{h}{D_e}\right)$$
(8.11)

where h and D_e are, respectively, the dimensions of the anchorage plate and the symmetric prism in the direction of the transverse tension T_b. For a single concentrically located anchorage plate, $D_e = D$ (for vertical tension) and Equations 8.6 and 8.11 are identical.

Alternatively, the tension T_b can be calculated from the bursting moment obtained from the statics of the real anchorage zone using a lever arm $l_b = D_e/2$. Guyon's symmetric prism concept is well accepted as a useful design procedure and has been incorporated into a number of building codes, including AS3600-2009 [1]. However, Equation 8.11 is an approximation that underestimates the transverse tension. Guyon [11] suggested that a conservative estimate of T_b will always result if the bursting tension calculated by Equation 8.11 is multiplied by D/D_e, but this may be very conservative.

For anchorage zones containing multiple bearing plates, the bursting tension behind each anchorage, for the case where all anchorages are stressed, may be calculated using Guyon's symmetric prisms. The depth of the symmetric prism D_e associated with a particular anchorage may be taken as the smaller of the following:

1. the distance in the direction of the transverse tension from the centre of the anchorage to the centre of the nearest adjacent anchorage; and
2. twice the distance in the direction of the transverse tension from the centre of the anchorage to the nearest edge of the anchorage zone.

For each symmetric prism, the lever arm l_b between the resultant transverse tension and compression may be taken as $D_e/2$.

The anchorage zone shown in Figure 8.13 contains two symmetrically placed anchorage plates located close together near the axis of the member. The stress contours show the bulb of tension immediately behind each anchorage plate. Also shown in Figure 8.13 is the symmetric prism to be used to calculate the resultant tension and the transverse reinforcement required in this region. Tension also exists further along the axis of the

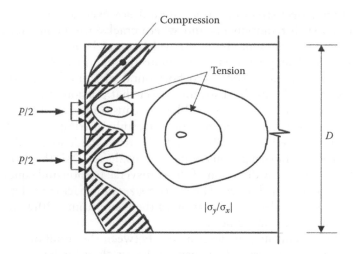

Figure 8.13 Two closely spaced symmetric anchorage plates.

member with a pattern of stress isobars similar to the pattern that occurs behind a single concentrically placed anchorage. AS3600-2009 [1] suggests that where the distance between two anchorages is less than 0.3 times the total depth of a member, consideration must also be given to the effects of the pair of anchorages acting in a manner similar to a single anchorage subject to the combined forces.

AS3600-2009 [1] also specifies that the loading cases to be considered in the design of a post-tensioned anchorage zone with multiple anchorage plates are: (1) all anchorages loaded; and (2) critical loading cases during the stressing operation.

8.3.3 Reinforcement requirements

In general, reinforcement should be provided to carry all the transverse tension in an anchorage zone. It is unwise to assume that the concrete will be able to carry any tension or that the concrete in the anchorage zone will not crack. The quantity of transverse reinforcement A_{sb} required to carry the transverse tension caused by bursting can be obtained by dividing the appropriate tensile force, calculated using Equation 8.6 or 8.11, by the permissible steel stress σ_{sb}, as follows:

$$A_{sb} = \frac{T_b}{\sigma_{sb}} \tag{8.12}$$

AS3600-2009 [1] suggests that for crack control in all anchorage zones and in other non-flexural members where a strong degree of cracking is required for appearance or where cracks may reflect through finishes, the maximum steel stress in service should not exceed 150 MPa. Where there are no such requirements and where cracks are inconsequential or hidden from view, an upper limit of 200 MPa may be used.

Equation 8.12 may be used to calculate the quantity of bursting reinforcement in both the vertical and horizontal directions. The transverse steel so determined must be distributed over that portion of the anchorage zone where the transverse tension is likely to cause cracking of the concrete. Therefore, the steel area A_{sb} should be uniformly distributed over the portion of beam located from $0.20D_e$ to $1.0D_e$ from the loaded end face [1]. For the particular bursting moment being considered, D_e is the depth of the symmetric prism in the direction of the transverse tension and equals D for a single concentric anchorage. The stirrup size and spacing so determined should also be provided in the portion of the beam from $0.20D_e$ to as near as practicable to the loaded face.

For spalling moments, the lever arm l_s between the resultant transverse tension T_s and compression C_s is usually larger than for bursting, as can be seen from the isobars in Figure 8.7. AS3600-2009 [1] suggests that for a single eccentric anchorage, the transverse tension at the loaded face remote from the anchorage may be calculated by assuming that l_s is half the overall depth of the member. Between two widely spaced anchorages, the transverse tension at the loaded face may be obtained by taking l_s equal to 0.6 times the spacing of the anchorages. The reinforcement required to resist the transverse tension at the loaded face A_{ss} is obtained from:

$$A_{ss} = \frac{T_s}{\sigma_{sb}} = \frac{M_s}{\sigma_{sb} l_s} \tag{8.13}$$

where σ_{sb} = 150 MPa. The steel area A_{ss} should be located as close to the loaded face as is permitted by concrete cover and compaction requirements.

8.3.4 Bearing stresses behind anchorages

Local concrete bearing failures can occur in post-tensioned members immediately behind the anchorage plates if the bearing area is inadequate or the concrete strength is too low. The design bearing strength F_b for unconfined concrete is specified in AS3600-2009 [1] as:

$$F_b = \phi 0.9 f_c' \sqrt{\frac{A_2}{A_1}} \quad (\leq \phi 1.8 f_c') \tag{8.14}$$

where f_c' is the compressive strength of the concrete at the time of first loading (i.e. at transfer); A_1 is the net bearing area; and A_2 is the largest area of the concrete supporting surface that is geometrically similar to and concentric with A_1 and $\phi = 0.6$ (see Table 2.3).

In commercial post-tensioned anchorages, the concrete immediately behind the anchorage is confined by spiral reinforcement (see Figure 3.6e), in addition to the transverse bursting and spalling reinforcement (often in the form of closed stirrups). In addition, the transverse compression at the loaded face immediately behind the anchorage plate significantly improves the bearing capacity of such anchorages. Therefore, provided the concrete behind the anchorage is well compacted, the bearing stress given by Equation 8.14 can usually be exceeded. Commercial anchorages are typically designed for bearing stresses of about 40 MPa, and bearing strength is specified by the manufacturer and is usually based on satisfactory test performance. For post-tensioned anchorage zones containing transverse confining reinforcement, the design bearing stress given by Equation 8.14 can usually be increased by at least 50%, but a maximum value of $\phi 2.5f_c'$ is recommended.

EXAMPLE 8.1   A SINGLE CONCENTRIC ANCHORAGE ON A RECTANGULAR SECTION

The anchorage zone of a flexural member with the dimensions shown in Figure 8.14 is to be designed. The size of the bearing plate is 315 mm square with a duct diameter of 106 mm, as shown. The jacking force is $P_j = 3000$ kN and the concrete strength when the full jacking force is applied is 50 MPa.

First consider the bearing stress immediately behind the anchorage plate. For bearing strength calculations, we use the strength load factor given in Section 2.3.2, i.e. the design load is $1.15\,P_j = 3450$ kN, and the capacity reduction factor given in Table 2.3 is $\phi = 0.6$.

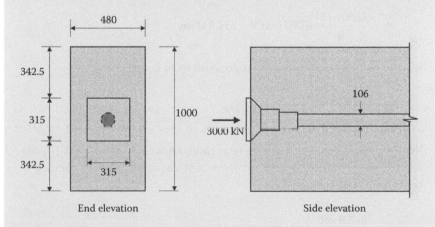

Figure 8.14 Details of anchorage zone (Example 8.1).

The net bearing area A_1 is the area of the bearing plate minus the area of the hollow duct:

$$A_1 = 315 \times 315 - \frac{\pi \times 106^2}{4} = 90.4 \times 10^3 \text{ mm}^2$$

and, for this anchorage:

$$A_2 = 480 \times 480 = 230 \times 10^3 \text{ mm}^2$$

The design bearing stress is:

$$\sigma_b = \frac{1.15 P_j}{A_1} = \frac{1.15 \times 3000 \times 10^3}{90.4 \times 10^3} = 38.2 \text{ MPa}$$

From Equation 8.14, the design strength in bearing is:

$$F_b = 0.6 \times 0.9 \times 50 \times \sqrt{\frac{230 \times 10^3}{90.4 \times 10^3}} = 43.1 \text{ MPa}$$

which is greater than σ_b and therefore acceptable. In practice, confinement reinforcement included behind the anchorage will significantly increase the design bearing strength.

Consider moments in the vertical plane:
The forces and bursting moments in the vertical plane are illustrated in Figure 8.15a. From Equation 8.5:

$$M_b = \frac{3000 \times 10^3}{8} (1000 - 315) = 256.9 \text{ kNm}$$

and the vertical bursting tension is obtained from Equation 8.6:

$$T_b = \frac{256.9 \times 10^6}{1000 / 2} \times 10^{-3} = 513.8 \text{ kN}$$

With σ_{sb} taken equal to 150 MPa in accordance with AS3600-2009 [1], the amount of vertical transverse reinforcement required to resist bursting is calculated from Equation 8.12:

$$A_{sb} = \frac{513.8 \times 10^3}{150} = 3425 \text{ mm}^2$$

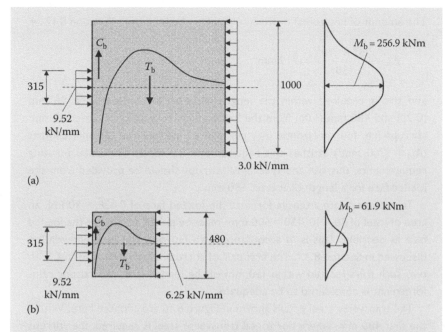

Figure 8.15 Bursting force and moment diagrams (Example 8.1). (a) Bursting in the vertical plane. (b) Bursting in the horizontal plane.

This area of transverse steel must be provided within the length of beam located from 0.20D to 1.0D from the loaded end face, i.e. over a length of 0.8D = 800 mm. Two 12 mm diameter stirrups (four vertical legs) are required every 100 mm along the 800 mm length (i.e. eight sets of stirrups gives A_{sb} = 8 × 4 × 110 = 3520 mm² within the 800 mm length). This size and spacing of stirrups must be provided over the entire anchorage zone, i.e. for a distance of 1000 mm from the loaded face.

Consider moments in the horizontal plane:
The forces and bursting moments in the horizontal plane are illustrated in Figure 8.15b. With b = 480 mm replacing D in Equations 8.5 and 8.6, the bursting moment and horizontal tension are:

$$M_b = \frac{3000 \times 10^3}{8}(480 - 315) = 61.9 \text{ kNm}$$

$$T_b = \frac{61.9 \times 10^6}{480/2} \times 10^{-3} = 257.8 \text{ kN}$$

The amount of horizontal transverse steel is obtained from Equation 8.12 as:

$$A_{sb} = \frac{257.8 \times 10^3}{150} = 1719 \text{ mm}^2$$

and this is required within the length of beam located between 96 mm ($0.2b$) and 480 mm ($1.0b$) from the loaded face. Four pairs of closed 12 mm stirrups (i.e. four horizontal legs per pair of stirrups) at 100 mm centres ($A_{sb} = 1760 \text{ mm}^2$) satisfies this requirement. To satisfy horizontal bursting requirements, this size and spacing of stirrups should be provided from the loaded face for a length of at least 480 mm.

To accommodate a tensile force at the loaded face of $0.03P_i = 90$ kN, an area of steel of $90 \times 10^3/150 = 600 \text{ mm}^2$ must be placed as close to the loaded face as possible. This is in accordance with Guyon's [11] recommendation discussed in Section 8.3.1. The first pair of stirrups supply 440 mm² and, with two such pairs located within 150 mm of the loaded face, the existing rein-forcement is considered to be adequate.

The transverse steel details shown in Figure 8.16 are adopted here. Within the first 480 mm, where horizontal transverse steel is required, the stirrups are closed at the top, as indicated, but for the remainder of the anchorage zone, between 480 and 1000 mm from the loaded face, open stirrups may be used to facilitate placement of the concrete. The first stirrup is placed as close as possible to the loaded face, as shown.

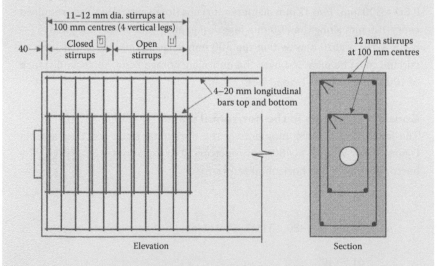

Figure 8.16 Reinforcement details (Example 8.1).

EXAMPLE 8.2 TWIN ECCENTRIC ANCHORAGES ON A RECTANGULAR SECTION

The anchorage shown in Figure 8.17 is to be designed. The jacking force at each of the two anchorages is P_j = 2000 kN and the concrete strength is f_c' = 40 MPa.

Check bearing stresses behind each anchorage:
As in Example 8.1, the design strength in bearing F_b is calculated using Equation 8.14:

$$A_1 = 265^2 - \frac{\pi \times 92^2}{4} = 63.6 \times 10^3 \text{ mm}^2;$$

$$A_2 = 450^2 = 202.5 \times 10^3 \text{ mm}^2$$

$$F_b = 0.6 \times 0.9 \times 40 \times \sqrt{\frac{202.5 \times 10^3}{63.6 \times 10^3}} = 38.5 \text{ MPa}$$

Using a load factor of 1.15 for prestress [1], the design bearing stress is:

$$\sigma_b = \frac{1.15 \times 2000 \times 10^3}{63.6 \times 10^3} = 36.2 \text{ MPa}$$

which is less than F_b and is therefore satisfactory.

Case (a): Consider the lower cable only stressed:
It is necessary first to examine the anchorage zone after just one of the tendons has been stressed. The stresses, forces and corresponding

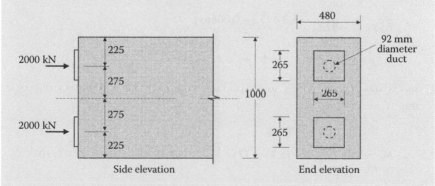

Figure 8.17 Twin anchorage arrangement (Example 8.2).

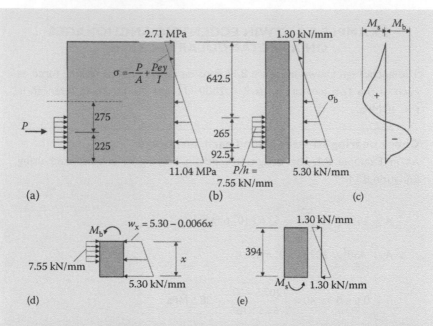

(a)   (b)   (c)

(d)   (e)

Figure 8.18 Actions on anchorage zone when the lower cable only is stressed (Example 8.2). (a) Side elevation and stress. (b) Forces. (c) Moments. (d) Free-body of analogous beam at M_b. (e) Free-body of analogous beam at M_s.

moments acting on the eccentrically loaded anchorage zone are shown in Figure 8.18a through c.

The maximum bursting moment M_b occurs at a distance x from the bottom surface at the point of zero shear in the free-body diagram of Figure 8.18d. From statics:

$$7.55 \times (x - 92.5) = \frac{5.3 + (5.3 - 0.0066x)}{2} x$$

$$\therefore x = 231.8 \text{ mm} \quad \text{and} \quad w_x = 3.77 \text{ kN/mm}$$

and

$$M_b = \left[5.3 \times \frac{231.8^2}{2} - (5.3 - 3.77) \times \frac{231.8^2}{6} - 7.55 \times \frac{(231.8 - 92.5)^2}{2} \right] \times 10^{-3}$$

$$= 55.5 \text{ kNm}$$

The maximum spalling moment M_s occurs at 394 mm below the top surface where the shear is also zero, as shown in Figure 8.18e, and from equilibrium:

$$M_s = 1.3 \times \frac{394^2}{6} \times 10^{-3} = 33.6 \text{ kNm}$$

Design for M_b:

The *symmetric prism* which is concentric with and directly behind the lower anchorage plate has a depth of $D_e = 450$ mm and is shown in Figure 8.19. From Equation 8.10:

$$T_b = \frac{M_b}{I_b} = \frac{55.5 \times 10^3}{450/2} = 246.5 \text{ kN}$$

By contrast, Equation 8.11 gives:

$$T_b = \frac{2000}{4} \left(1 - \frac{265}{450} \right) = 206 \text{ kN}$$

which is considerably less conservative in this case. Adopting the value of T_b obtained from the actual bursting moment, Equation 8.12 gives:

$$A_{sb} = \frac{246.5 \times 10^3}{150} = 1640 \text{ mm}^2$$

This area of steel must be distributed over a distance of $0.8D_e = 360$ mm.

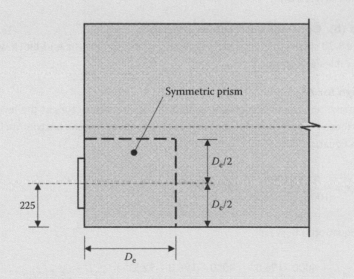

Figure 8.19 Symmetric prism for one eccentric anchorage (Example 8.2).

For the steel arrangement illustrated in Figure 8.21, 16 mm diameter and 12 mm diameter stirrups are used at the spacings indicated, i.e. a total of four vertical legs of area 620 mm² per stirrup location is used behind each anchorage. The number of such stirrups required in the 360 mm length of the anchorage zone is 1640/620 = 2.65, and therefore the maximum spacing of the stirrups is 360/2.65 = 135 mm. This size and spacing of stirrups is required from the loaded face to 450 mm therefrom. The spacing of the stirrups in Figure 8.21 is less than that calculated here because the horizontal bursting moment and spalling moment requirements are more severe. These are examined subsequently.

Design for M_s:
The lever arm l_s between the resultant transverse compression and tension forces that resist M_s is taken as $0.5D$ = 500 mm. The area of transverse steel required within $0.2D$ = 200 mm from the front face is given by Equation 8.13:

$$A_{ss} = \frac{33.6 \times 10^6}{150 \times 500} = 448 \text{ mm}^2$$

The equivalent of about four vertical 12 mm diameter steel legs is required close to the loaded face of the member to carry the resultant tension caused by spalling. This requirement is easily met by the three full depth 16 mm diameter stirrups (six vertical legs) located within $0.2D$ of the loaded face, as shown in Figure 8.21.

Case (b): Consider both cables stressed:
Figure 8.20 shows the force and moment distribution for the end-block when both cables are stressed.

Design for M_b
The maximum bursting moment behind the anchorage occurs at the level of zero shear, x mm below the top surface and x mm above the bottom surface. From Equation 8.7:

$$x = \frac{92.5 \times 1000}{(1000 - 2 \times 265)} = 196.8 \text{ mm}$$

and Equation 8.8 gives:

$$M_b = \left[\frac{4000 \times 196.8^2}{2 \times 1000} - \frac{4000 \times (196.8 - 92.5)^2}{4 \times 265} \right] \times 10^{-3} = 36.4 \text{ kNm}$$

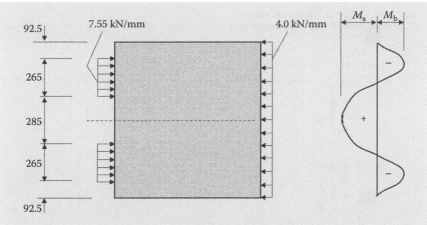

Figure 8.20 Forces and moments when both cables are stressed.

This is less than the value for M_b when only the single anchorage was stressed. Since the same symmetric prism is applicable here, the reinforcement requirements for bursting determined in case (a) are more than sufficient.

Design for M_s:
The spalling moment at the mid-depth of the anchorage zone (on the member axis) is obtained from Equation 8.9:

$$M_s = \frac{4000 \times 10^3}{2}\left(275 - \frac{1000}{4}\right) \times 10^{-6} = 50 \text{ kNm}$$

With the lever arm l_s taken as 0.6 times the spacing of the bearing plates, i.e. $l_s = 0.6 \times 550 = 330$ mm, the area of transverse steel required within $0.2D = 200$ mm of the loaded face is given by Equation 8.13:

$$A_{ss} = \frac{50 \times 10^6}{150 \times 330} = 1010 \text{ mm}^2$$

To avoid steel congestion, 16 mm diameter stirrups will be used close to the loaded face. Use six vertical legs of 16 mm diameter (1200 mm²) across the member axis within 200 mm of the loaded face, as shown in Figure 8.21.

Case (c): Consider horizontal bursting
Horizontal transverse steel must also be provided to carry the transverse tension caused by the horizontal dispersion of the total prestressing force ($P = 4000$ kN) from a 265 mm wide anchorage plate into a 480 mm wide section. With $b = 480$ mm used instead of D, Equations 8.5 and 8.6 give:

$$M_b = 107.5 \text{ kNm} \quad \text{and} \quad T_b = 448 \text{ kN}$$

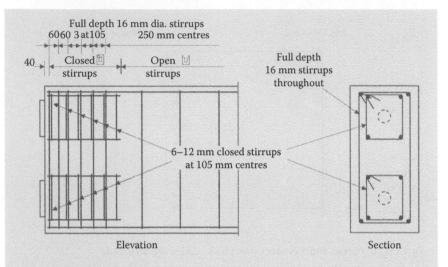

Elevation   Section

Figure 8.21 Reinforcement details (Example 8.2).

and the amount of horizontal steel is obtained from Equation 8.12:

$$A_{sb} = 2987 \text{ mm}^2$$

With the steel arrangement shown in Figure 8.21, six horizontal bars exist at each stirrup location (2–16 mm diameter bars and 4–12 mm diameter bars, i.e. 840 mm² at each stirrup location). The required stirrup spacing within the length $0.8b$ (=384 mm) is 108 mm. Therefore, within 480 mm from the end face of the beam, all available horizontal stirrup legs are required and, therefore, all stirrups in this region must be closed.

The reinforcement details shown in Figure 8.21 are adopted.

EXAMPLE 8.3   SINGLE CONCENTRIC ANCHORAGE IN A T-BEAM

The anchorage zone of the T-beam shown in Figure 8.22a is to be designed. The member is prestressed by strands located within a single 92 mm diameter duct, with a 265 mm square anchorage plate located at the centroidal axis of the cross-section. The jacking force is $P_j = 2000$ kN, and the concrete strength at transfer is 50 MPa. The distributions of forces on the anchorage zone in elevation and in plan are shown in Figure 8.22b and c, respectively.

The design bearing stress and the design strength in bearing are calculated as for the previous examples, with $\sigma_b = 36.2$ MPa and $F_b = 37.5$ MPa.

Consider moments in the vertical plane:
The maximum bursting moment occurs at the level of zero shear at x mm above the bottom of the cross-section. From Figure 8.22d:

$$2.044 \times x = 7.547 \times (x - 295.8) \quad \therefore x = 405.7 \text{ mm}$$

and

$$M_b = \left[\frac{2.044 \times 405.7^2}{2} - \frac{7.547 \times (405.7 - 295.8)^2}{2} \right] \times 10^{-3}$$

$$= 122.6 \text{ kNm}$$

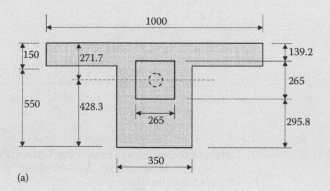

(a)

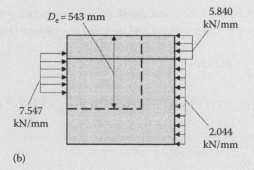

(b)

Figure 8.22 Details of the anchorage zone of the T-beam (Example 8.3). (a) End elevation. (b) Side elevation. (c) Plan. (d) Part side elevation. (Continued)

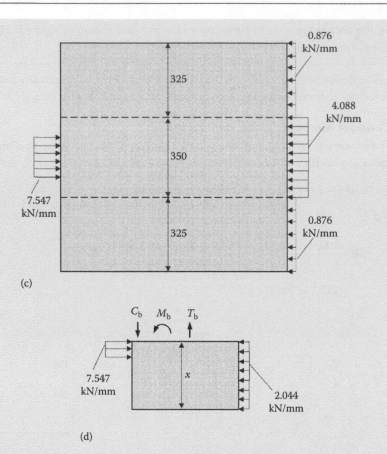

(c)

(d)

Figure 8.22 (Continued) Details of the anchorage zone of the T-beam (Example 8.3). (a) End elevation. (b) Side elevation. (c) Plan. (d) Part side elevation.

As indicated in Figure 8.22b, the depth of the symmetric prism associated with M_b is $D_e = 2 \times 139.2 + 265 = 543$ mm and the vertical tension is:

$$T_b = \frac{M_b}{D_e / 2} = 451 \text{kN}$$

The vertical transverse reinforcement required in the web is obtained from Equation 8.12:

$$A_{sb} = \frac{451 \times 10^3}{150} = 3007 \text{ mm}^2$$

This area of steel must be located within the length of the beam between $0.2D_e = 109$ mm and $D_e = 543$ mm from the loaded face.

By using 16 mm stirrups over the full depth of the web and 12 mm stirrups immediately behind the anchorage, as shown in Figure 8.23 (i.e. $A_{sb} = (2 \times 200) + (2 \times 110) = 620$ mm^2 per stirrup location), the number of double stirrups required is 3007/620 = 4.85 and the required spacing is (543 − 109)/4.85 = 90 mm, as shown. With two such pairs of stirrups located within 130 mm of the loaded face, Guyon's recommendation that steel be provided near the loaded face to carry $0.03P_i$ is satisfied.

Consider moments in the horizontal plane:

Significant lateral dispersion of prestress occurs in plan in the anchorage zone as the concentrated prestressing force finds its way out into the flange of the T-section. By taking moments of the forces shown in Figure 8.22c about a point on the axis of the anchorage, the horizontal bursting moment is:

$$M_b = \left(0.876 \times 325 \times 337.5 + 4.088 \times 175 \times 87.5 - 1000 \times 66.25\right) \times 10^{-3}$$

$$= 92.4 \text{ kNm}$$

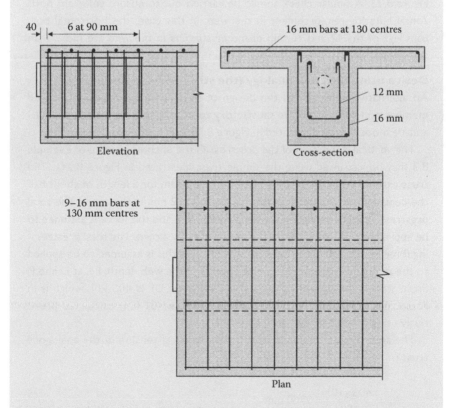

Elevation   Cross-section

Plan

Figure 8.23 Reinforcement details for anchorage zone of T-beam (Example 8.3).

Much of this bursting moment must be resisted by horizontal transverse tension and compression in the flange. Taking D_e equal to the flange width, the lever arm between the transverse tension and compression is $l_b = D_e/2 = 500$ mm and the transverse tension is calculated using Equation 8.10:

$$T_b = \frac{92.4 \times 10^3}{500} = 185 \text{ kN}$$

The area of horizontal transverse reinforcement required in the flange is therefore:

$$A_{sb} = \frac{185 \times 10^3}{150} = 1234 \text{ mm}^2$$

and this quantity should be provided within the flange and located between 200 and 1000 mm from the loaded face. Adopt 16 mm bars across the flange at 130 mm centres from the loaded face to 1000 mm therefrom, as shown in Figure 8.23. A similar check should be carried out to ensure sufficient horizontal bursting reinforcement in the web. In this case, the horizontal bottom legs of the 12 and 16 mm diameter stirrups in the web are more than sufficient.

Design using the truss analogy (the strut-and-tie method)

An alternative approach to the design of the anchorage zone in a flanged member, and perhaps a more satisfactory approach, involves the use of strut-and-tie modelling, as illustrated in Figure 8.24.

The vertical dispersion of the prestress in the anchorage zone of Example 8.3 may be visualised using the simple truss illustrated in Figure 8.24a. The truss extends from the bearing plate into the beam for a length of about half the depth of the symmetric prism (i.e. $D_e/2 = 272$ mm in this case). The total prestressing force carried in the flange is 876 kN, and this force is assumed to be applied to the analogous truss at A and at B, as shown. The total prestressing force in the web of the beam is 1124 kN, and this is assumed to be applied to the analogous truss at the quarter points of the web depth, i.e. at D and F. From statics, the tension force in the vertical tie DF is 405 kN, which is in reasonable agreement with the bursting tension (451 kN) calculated previously using the deep beam analogy.

The area of steel required to carry the vertical tension in the analogous truss is:

$$A_{sb} = \frac{405 \times 10^3}{150} = 2700 \text{ mm}^2$$

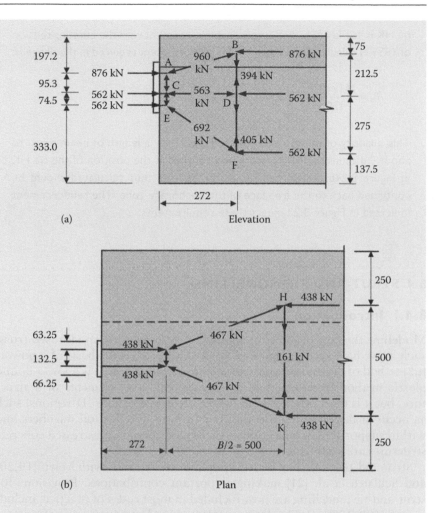

Figure 8.24 Truss analogy of the anchorage zone of T-beam (Example 8.3). (a) Vertical dispersion of prestress. (b) Horizontal dispersion of prestress.

and this should be located between $0.2D_e$ and D_e from the loaded face. According to the truss analogy, therefore, the vertical steel spacing of 90 mm in Figure 8.23 may be increased to 100 mm.

The horizontal dispersion of prestress into the flange is illustrated using the truss analogy of Figure 8.24b. After the prestressing force has dispersed vertically to point B in Figure 8.24a (i.e. at 272 mm from the anchorage plate), the flange force then disperses horizontally. The total flange force (876 kN) is applied to the horizontal truss at the quarter points across the flange, that is at points H and K in Figure 8.24b. From statics, the horizontal tension in the tie HK is 161 kN (which is in reasonable agreement with the bursting tension of 185 kN calculated previously). The reinforcement required in the flange is:

$$A_{sb} = \frac{161 \times 10^3}{150} = 1073 \text{ mm}^2$$

This quantity of reinforcement is required over a length of beam equal to about 0.8 times the flange width and centred at the position of the tie HK in Figure 8.24b. Reinforcement at the spacing thus calculated should be continued back to the free face of the anchorage zone. The reinforcement indicated in Figure 8.23 meets these requirements.

8.4 STRUT-AND-TIE MODELLING

8.4.1 Introduction

Modelling the flow of forces in an anchorage zone using an idealised truss, such as we have seen in Figures 8.8, 8.9 and 8.24, is the basis of a powerful method of design known as *strut-and-tie modelling*. It is a lower bound plastic method of design that can be applied to all elements of a structure, but it is most often used to design disturbed regions (D-regions) such as occur at discontinuities in the structure, in non-flexural members and within supports and connections. Anchorage zones in prestressed concrete structures are such regions.

Strut-and-tie modelling became popular in the 1980s with Marti [19,20] and Schlaich et al. [21] making important contributions. Provisions for strut-and-tie modelling are now included in most codes of practice, including AS3600-2009 [1]. The designer selects a load path consisting of internal concrete struts and steel ties connected at nodes. The internal forces carried by the struts and ties must be in equilibrium with the external loads. Each element of the strut-and-tie model (i.e. the concrete struts, the steel ties and the nodes connecting them) must then be designed and detailed so that the load path is everywhere sufficiently strong to carry the applied loads through the structure and into the supports. Care must be taken to ensure that strut-and-tie model selected is compatible with the applied loads and the supports, and that both the struts and the ties possess sufficient ductility to accommodate the redistribution of internal forces necessary to achieve the desired load path.

AS3600-2009 [1] permits the use of strut-and-tie modelling as a basis for strength design (and for evaluating strength) in non-flexural regions of members and specifies a range of requirements that must be satisfied, including: (1) loads are applied only at nodes with struts and ties carrying only axial force; (2) the model must be in equilibrium; (3) when determining the geometry of the model, the dimensions of the struts, ties and nodes must be accounted for; (4) if required, ties may cross struts; (5) struts are permitted to cross or intersect only at nodes; and (6) the angle between the axis of any strut and any tie at a node point shall not be less than 30° for a reinforced concrete tie or 20° in a prestressed concrete member when a tendon is acting as the tie.

8.4.2 Concrete struts

8.4.2.1 Types of struts

Depending on the geometry of the member and its supports and loading points, the struts in a strut-and-tie model can be either fan shaped, bottle shaped or prismatic, as shown in Figure 8.25. If unimpeded by the edges of a member or any penetrations through the member, compressive stress fields diverge. A prismatic strut, such as shown in Figure 8.25c, can only develop if the stress field is physically unable to diverge because of the geometry of the structure. When the compressive stress field can diverge without interruption and is not constrained at its ends, so that the stress trajectories remain straight, a fan-shaped strut results, as shown in Figure 8.25a. When the compressive stress field is free to diverge laterally along its length, but is constrained at either end, a bottle-shaped strut develops with curved stress trajectories similar to that shown in Figure 8.25b. Such curved compressive stress trajectories create bursting forces at right angles to the strut axis. AS3600-2009 [1] suggests that these bursting forces T_b can be determined using the polygon of forces shown

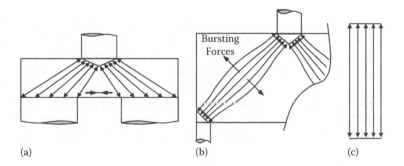

Figure 8.25 Types of concrete struts. (a) Fan-shaped struts. (b) Bottle-shaped strut. (c) Prismatic strut.

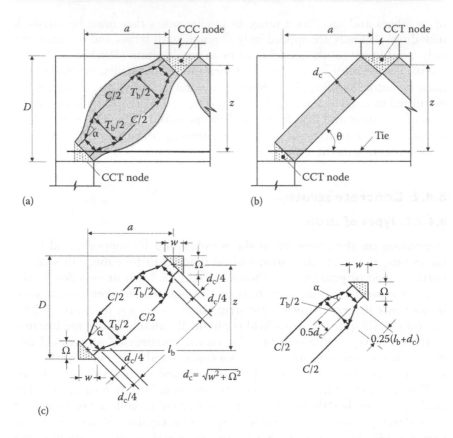

Figure 8.26 Model of internal forces in a bottle-shaped strut [1]. (a) Bottle-shaped strut and internal forces. (b) Idealised parallel sided strut. (c) Geometry of the polygon of forces in a bottle-shaped strut.

in Figure 8.26, where the angle α is known as the *divergence angle* of the bottle-shaped strut and l_b is the length of the bursting zone parallel to the axis of the strut. A transverse force is generated, wherever the direction of the compressive force in the strut changes. When the angle change, viewed from the axis of the strut, is convex (such as occurs at $d_c/4$ from the ends of the strut in Figure 8.26c), the transverse force is compressive and, when the angle change, viewed from the axis of the strut, is concave (such as occurs at $0.25\,(l_b + d_c)$ further along the strut in Figure 8.26c), the transverse force is tensile.

From considerations of equilibrium:

$$T_b = C \tan \alpha \tag{8.15}$$

8.4.2.2 Strength of struts

According to AS3600-2009 [1], the design strength of a strut is the product of the smallest cross-sectional area of the concrete strut at any point along its length A_c, the strength reduction factor ($\phi_{st} = 0.6$), a strut efficiency factor β_s and the in-situ strength of the concrete (taken as $0.9f_c'$):

$$\text{Design strength} = \phi_{st}\beta_s 0.9f_c' A_c \tag{8.16}$$

Properly detailed longitudinal reinforcement placed parallel to the axis of the strut and located within the strut may be used to increase the strength of a strut. The longitudinal reinforcement should be enclosed by suitably detailed ties or spiral reinforcement (see Chapter 14). The strength of a strut containing longitudinal reinforcement may be calculated as for a prismatic, pin-ended short column of cross-sectional area A_c and the same length as the strut (see Chapter 13).

The strut efficiency factor reduces the design strength of the strut to account for the weakening effects of the transverse tension T_b crossing the unconfined concrete. For prismatic struts, there is no transverse tension and the efficiency factor β_s may be taken as 1.0. For fan-shaped or bottle-shaped struts that are unconfined, AS3600-2009 [1] specifies the efficiency factor proposed by Foster and Malik [22], which was developed from the expression proposed by Collins and Mitchell [23]:

$$\beta_s = \frac{1}{1.0 + 0.66\cot^2\theta} \quad \text{(within the limits } 0.3 \leq \beta_s \leq 1.0\text{)} \tag{8.17}$$

where θ is defined as the angle between the axis of the strut and the axis of a tie passing though a common node, as shown in Figure 8.26b. Where more than one tie passes through a node at the end of the strut, the smallest value of θ should be used in Equation 8.17.

8.4.2.3 Bursting reinforcement in bottle-shaped struts

For the calculation of T_b (from Equation 8.15), AS3600-2009 [1] specifies different minimum values of the divergence angle α for the serviceability limit state and for the ultimate limit state, as follows:

For serviceability: $\tan\alpha = 0.5$, i.e. $\alpha_{min} = 26.6°$ $\qquad$ (8.18)

For ultimate strength: $\tan\alpha = 0.2$, i.e. $\alpha_{min} = 11.3°$ $\qquad$ (8.19)

The internal bursting tension T_b in a bottle-shaped strut reduces the compressive strength of the strut and, if T_b is significant, transverse reinforcement is required. Without adequate transverse reinforcement, splitting along the strut can initiate a sudden brittle failure of the strut. The bursting force required to cause first cracking is specified in AS3600-2009 [1] as:

$$T_{b.cr} = 0.7 \, b \, l_b \, f_{ct}' \tag{8.20}$$

where b is the width of the member; and l_b is the length of the bursting zone, as shown in Figure 8.26c.

AS3600-2009 [1] further specifies that if the internal tensile force T_b is greater than one-half of the tensile strength of the concrete over the length l_b of the bursting zone (i.e. if $T_b > 0.5T_{b.cr}$), adequate reinforcement is required to carry the entire bursting tension T_b at both the strength and the serviceability limit states and to ensure that in-service cracking is controlled.

For the control of cracking along the strut at service loads, AS3600-2009 [1] limits the maximum stress f_s in the transverse reinforcement to 150 MPa, wherever a strong degree of cracking is required for appearance or where cracks may reflect through finishes. Where cracks are inconsequential or hidden from view, the maximum stress in the transverse reinforcement at service loads may be increased to 200 MPa. In most design situations, the serviceability conditions will determine the amount of bursting steel required.

The transverse reinforcement requirements can be met by including reinforcement of areas A_{s1} and A_{s2} in two orthogonal directions γ_1 and γ_2 to the axis of the strut. Alternatively, transverse reinforcement of area A_{s1} may be provided in one direction only, provided the angle γ_1 between the axis of the strut and the reinforcement is not less than 40°. In the latter case, if the direction of the transverse reinforcement is not perpendicular to the plane of cracking, the component of the bursting force orthogonal to the reinforcement must be resisted by dowel action and aggregate interlock on the crack surface and, for this reason, the minimum value is placed on γ_1. The area(s) of steel required must satisfy the following conditions:

$$\text{For serviceability :} \quad \sum A_{si} f_{si} \sin \gamma_i \geq \max(T_{b.s}^*, T_{b.cr}) \tag{8.21}$$

$$\text{For strength:} \quad \phi_{st} \sum A_{si} f_{sy} \sin \gamma_i \geq T_b^* \tag{8.22}$$

where $T_{b.s}^*$ is the bursting tension caused by the design loads at the serviceability limit state; and T_b^* is the bursting tension caused by the factored design loads at the strength limit state.

AS3600-2009 [1] requires that the transverse steel quantities determined from Equations 8.21 and 8.22 should be uniformly distributed along the length of the bursting zone l_b:

$$l_b = \sqrt{z^2 + a^2} - d_c \tag{8.23}$$

and z, a and d_c are defined in Figure 8.26.

8.4.3 Steel ties

The ties in a strut-and-tie model consist of reinforcement, prestressing tendons or any combination thereof running uninterruptedly along the full length of the tie and adequately anchored within (or beyond) the node at each end of the tie. The reinforcement and/or tendons should be evenly distributed across the end nodes and arranged so that the resultant tension in the steel coincides with the axis of the tie in the strut-and-tie model.

The design strength of the tie is given by:

$$\phi_{st}T_u = \phi_{st}[A_{st}f_{sy} + A_p(\sigma_{pe} + \Delta\sigma_p)] \tag{8.24}$$

where the strength reduction factor $\phi_{st} = 0.8$ (see Section 2.4.3); σ_{pe} is the effective prestress force in tendons after all the losses; and $\Delta\sigma_p$ is the incremental force in the tendons due to the external loads. The sum $\sigma_{pe} + \Delta\sigma_p$ should not be taken to be greater than yield strength f_{py}.

AS3600-2009 [1] requires that for adequate anchorage at each end of the tie, all reinforcement shall be fully anchored in accordance with the procedures outlined in Section 14.3.2 and at least 50% of the steel should extend beyond the node. Alternatively, anchorage can be provided by a welded or mechanical anchorage entirely located beyond the node.

8.4.4 Nodes

At a node connecting struts and ties, at least three forces must be acting to satisfy equilibrium. The strength of the concrete within a node must also be checked. AS3600-2009 [1] identifies three types of nodes depending on the arrangement of the struts and ties entering the node. A CCC node is one with only struts (or compressive loading points or reactions) entering the node. For example, the node at the top of the strut in Figure 8.26a (and Figure 8.26b) is subjected to three compressive forces and is a CCC node. A CCT node is one with two or more struts and single tension tie entering the node. For example, the node at the bottom of the strut immediately above the reaction in Figure 8.26a is a CCT node. Finally, a CTT node is one with two or more tension ties entering the node.

When the strut-and-tie model is constructed so that all the strut forces entering a node are perpendicular to the node faces, the node is hydrostatic. The lengths of the node faces are proportional to the strut forces. The node faces are subjected to normal stress, without any shear component, and the compressive stress on each node face is identical. Although the design of hydrostatic nodes is straightforward, the forces entering the node may not be concurrent and hydrostatic nodes are often not possible. Non-hydrostatic nodes are commonly adopted when truss analysis software is used to determine member forces. The design of non-hydrostatic nodes, where the face of the node is not perpendicular to the strut force, requires the inclusion of the shear components in the node design.

The design strength of an unconfined node in compression depends on the number of ties entering the node. AS3600-2009 [1] states that the design strength of a node is adequate, provided the principal compressive stress σ_2^* on any nodal face, determined from the normal and shear stresses on that face, satisfies:

$$\sigma_2^* \leq \phi_{st}\beta_n 0.9f_c' \tag{8.25}$$

where $\phi_{st} = 0.6$; and β_n accounts for the high level of strain incompatibility between the ties and struts entering the node and equals 1.0 for CCC nodes, 0.8 for CCT nodes and 0.6 for CTT nodes.

Where the node is confined, the strength increases. AS3600-2009 [1] suggests that the increase in strength may be determined by test or calculation, but σ_2^* should not exceed $\phi_{st}1.8f_c'$.

REFERENCES

1. AS3600-2009 (2009). *Australian Standard for Concrete Structures.* Standards Australia, Sydney, New South Wales, Australia.
2. Hoyer, E. (1939). *Der Stahlsaitenbeton.* Berlin, Germany: Elsner.
3. Eurocode 2 (2004): Design of concrete structures – Part 1-1: General rules and rules for buildings – BS EN 1992-1-1:2004. British Standards Institution – European Committee for Standardization.
4. ACI 318-M-11 (2011). Building code requirements for reinforced concrete. Detroit, MI: American Concrete Institute.
5. Logan, D.R. (1997). Acceptance criteria for bond quality of strand for pretensioned concrete applications. *PCI Journal*, 42(2), 52–90.
6. Rose, D.R. and Russell, B.W. (1997). Investigation of standardized tests to measure the bond performance of prestressing strands. *PCI Journal*, 42(4), 56–80.
7. Martin, L. and Korkosz, W. (1995). Strength of prestressed members at sections where strands are not fully developed. *PCI Journal*, 40(5), 58–66.
8. Martin, L.D. and Perry, C.J. (2004). *PCI Design Handbook: Precast and Prestressed Concrete*, 6th edn. Chicago, IL: Precast/Prestressed Concrete Institute.
9. Gilbert, R.I. (2012). Unanticipated bond failure over supporting band beams in grouted post-tensioned slab tendons with little or no prestress. *Bond in Concrete, Fourth International Symposium*, June 17–20, Brescia, Italy.
10. Marshall, W.T. and Mattock A.H. (1962). Control of horizontal cracking in the ends of pretensioned prestressed concrete girders. *Journal of the Prestressed Concrete Institute*, 7(5), 56–74.
11. Guyon, Y. (1953). *Prestressed Concrete*, English edn. London, U.K.: Contractors Record and Municipal Engineering.

12. Iyengar, K.T.S.R. (1962). Two-dimensional theories of anchorage zone stresses in post-tensioned concrete beams. *Journal of the American Concrete Institute*, 59, 1443–1446.
13. Iyengar, K.T.S.R. and Yogananda, C.V. (1966). A three dimensional stress distribution problem in the end zones of prestressed beams. *Magazine of Concrete Research*, 18, 75–84.
14. Sargious, M. (1960). Beitrag zur Ermittlung der Hauptzugspannungen am Endauflager vorgespannter Betonbalken. Phd Dissertation. Stuttgart, Germany: Technische Hochschule.
15. Foster, S.J. and Rogowsky, D.M. (1997). Bursting forces in concrete members resulting from in-plane concentrated loads. *Magazine of Concrete Research*, 49(180), 231–240.
16. Magnel, G. (1954). *Prestressed Concrete*, 3rd edn. New York: McGraw-Hill.
17. Gergely, P. and Sozen, M.A. (1967). Design of anchorage zone reinforcement in prestressed concrete beams. *Journal of the Prestressed Concrete Institute*, 12(2), 63–75.
18. Warner, R.F. and Faulkes, K.A. (1979). *Prestressed Concrete*, 1st edn. Melbourne, Australia: Pitman Australia.
19. Marti, P. (1985). Truss models in detailing. *Concrete International – American Concrete Institute*, 7(1), 46–56.
20. Marti, P. (1985). Basic tools of reinforced concrete beam design. *Concrete International – American Concrete Institute*, 7(12), 66–73.
21. Schlaich, J., Schäfer, K., and Jennewein, M. (1987). Towards a consistent design of structural concrete, Special Report. *PCI Journal*, 32(3), 74–150.
22. Foster, S.J. and Malik, A.R. (2002). Evaluation of efficiency factor models used in strut-and-tie modelling of non-flexural members. *Journal of Structural Engineering, ASCE*, 128(5), 569–577.
23. Collins, M.P. and Mitchell, D. 1986. A rational approach to shear design—The 1984 Canadian Code Provisions. *ACI Structural Journal*, 83(6), 925–933.

Chapter 9

Composite members

9.1 TYPES AND ADVANTAGES OF COMPOSITE CONSTRUCTION

Composite construction in prestressed concrete usually consists of precast prestressed members acting in combination with a cast in-situ concrete component. The composite member is formed in at least two separate stages with some or all of the prestressing normally applied before the completion of the final stage. The precast and the cast in-situ elements are mechanically bonded to each other to ensure that the separate components act together as a single composite member.

Composite members can take a variety of forms. In building construction, the precast elements are often pretensioned slabs (which may be either solid or voided), or single or double tee-beams. The cast in-situ element is a thin, lightly reinforced topping slab placed on top of the precast units after the units have been erected to their final position in the structure. Single or double tee precast units are used extensively in building and bridge structures because of the economies afforded by this type of construction.

Composite prestressed concrete beams are widely used in the construction of highway bridges. For short- and medium-span bridges, standardised I-shaped or trough-shaped girders (which may be either pretensioned or post-tensioned) are erected between the piers and a reinforced concrete slab is cast onto and across the top flange of the girders. The precast girders and the in-situ slab are bonded together to form a stiff and strong composite bridge deck.

The two concrete elements, which together form the composite structure, may have different concrete strengths, different elastic moduli and different creep and shrinkage characteristics. The concrete in the precast element is generally of better quality than the concrete in the cast in-situ element, because it usually has a higher specified target strength and experiences better quality control and better curing conditions. With the concrete in the precast element being older and of better quality than the in-situ concrete, restraining actions will develop in the composite

355

structure with time owing to differential creep and shrinkage movements. These effects should be carefully considered in design.

Prestressed concrete composite construction has many advantages over non-composite construction. In some situations, a significant reduction in construction costs can be achieved. The use of precast elements can greatly speed up construction time. When the precast elements are standardised and factory produced, the cost of long-line pretensioning may be considerably less than the cost of post-tensioning on-site. Of course, the cost of transporting precast elements to the site must be included in these comparisons and it is often transportation difficulties that limit the size of the precast elements and the range of application of this type of construction. In addition, it is easier and more economical to manufacture concrete elements with high mechanical properties in a controlled prestressing plant rather than on a building or bridge site.

During construction, the precast elements can support the forms for the cast in-situ concrete, thereby reducing falsework and shoring costs. The elimination of scaffolding and falsework is often a major advantage over other forms of construction, and permits the construction to proceed without interruption to the work or traffic beneath. Apart from providing significant increases to both the strength and stiffness of the precast girders, the in-situ concrete can perform other useful structural functions. It can provide continuity at the ends of precast elements over adjacent spans. In addition, it provides lateral stability to the girders and also provides a means for carrying lateral loads back to the supports. Stage stressing can be used to advantage in some composite structures. A composite member consisting of a pretensioned, precast element and an in-situ slab may be subsequently post-tensioned to achieve additional economies of section. This situation may arise, for example when a relatively large load is to be applied at some time after composite action has been achieved.

Cross-sections of some typical composite prestressed concrete members commonly used in buildings and bridges are shown in Figure 9.1.

9.2 BEHAVIOUR OF COMPOSITE MEMBERS

The essential requirement for a composite member is that the precast and cast in-situ elements act together as one unit. To achieve this, it is necessary to have good bond between the two elements.

When a composite member is subjected to bending, a horizontal shear force develops at the interface between the precast and the in-situ elements. This results in a tendency for horizontal slip on the mating surfaces, if the bond is inadequate. Resistance to slip is provided by the naturally achieved adhesion and friction that occurs between the two elements. Often the top surface of the precast element is deliberately roughened during manufacture to improve its bonding characteristics and facilitate the transfer of horizontal shear through mechanical interlock. Where the contact surface

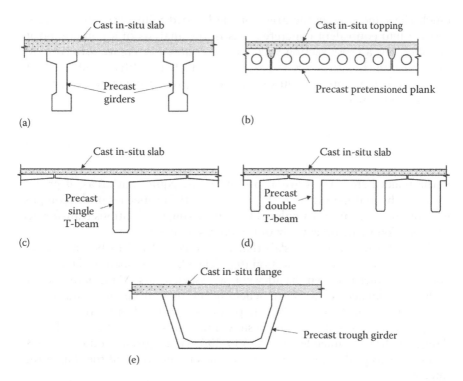

Figure 9.1 Typical composite prestressed concrete cross-sections. (a) Slab and girder. (b) Pretensioned plank plus topping. (c) Single T-sections. (d) Double T-sections. (e) Trough girder.

between the two elements is broad (such as in Figure 9.1b through d), natural adhesion and friction are usually sufficient to resist the horizontal shear. Where the contact area is small (such as between the slab and girders in Figure 9.1a and e), other provisions are necessary. Frequently, the web reinforcement in the precast girder is continued through the contact surface and anchored in the cast in-situ slab. This reinforcement resists horizontal shear primarily by dowel action, but assistance is also gained by clamping the mating surfaces together and increasing the frictional resistance.

If the horizontal shear on the element interface is resisted without slip (or with small slip only), the response of the composite member can be determined in a similar manner to that of a monolithic member. Stresses and strains on the composite cross-section due to service loads applied after the in-situ slab has been placed (and has hardened) may be calculated using the properties of the combined cross-section calculated using the transformed area method. If the elastic modulus of the concrete in the in-situ part of the cross-section E_{c2} is different to that in the precast element E_{c1}, it is convenient to transform the cross-sectional area of the in-situ element to an equivalent area of the precast concrete. This is achieved in much the same way as the areas of the bonded reinforcement are transformed into equivalent concrete areas in the analysis of a non-composite member. For a cross-section such as those shown in Figure 9.1a or e, for example if the in-situ concrete slab has an effective width b_{ef} and depth D_s, it is transformed into an equivalent area of precast concrete of depth D_s and width b_{tr}, where:

$$b_{tr} = \frac{E_{c2}}{E_{c1}} b_{ef} = n_c b_{ef} \tag{9.1}$$

If the bonded steel areas are also replaced by equivalent areas of precast concrete (by multiplying by E_s/E_{c1} or E_p/E_{c1}), the properties of the composite cross-section can be calculated by considering the fictitious transformed cross-section made up entirely of the precast concrete.

The width of the in-situ slab that can be considered to be an effective part of the composite cross-section (b_{ef}) depends on the span of the member and the distance between the adjacent precast elements. Maximum effective widths for flanged sections are generally specified in building codes, with the provisions of AS3600-2009 (1) previously outlined in Section 6.6. For composite members such as those shown in Figure 9.1a and e, the effective flange widths recommended by AS3600-2009 (1) are given in Equations 6.34 and 6.35, except that the term b_w now refers to the width of the slab–girder interface.

The design of prestressed concrete composite members is essentially the same as that of non-composite members, provided that certain behavioural differences are recognised and taken into account. It is important to appreciate that part of the applied load is resisted by the precast element(s) prior to the establishment of composite action. Care must be taken, therefore, when designing for serviceability to ensure that behaviour of the cross-section and its response to various load stages are accurately modelled. It is also necessary in design to ensure adequate horizontal shear capacity at the element interface. The design procedures for flexural, shear and torsional strengths are similar to that of a non-composite member.

9.3 STAGES OF LOADING

As mentioned in the previous section, the precast part of a composite member is required to carry loads prior to the establishment of composite action. When loads are applied during construction, before the cast in-situ slab has set, flexural stresses are produced on the precast element. After the in-situ concrete has been placed and cured, the properties of the cross-section are substantially altered for all subsequent loadings. Moments due to service live loads, for example modify the stress distribution in the precast element and introduce stresses into the cast in-situ slab. Creep and shrinkage of the concrete also cause a substantial redistribution of stress with time between the precast and the in-situ elements, and between the concrete and the bonded reinforcement in each element.

In the design of a prestressed concrete composite member, most of the following load stages usually need to be considered:

1. *Initial prestress at transfer in the precast element*: This normally involves calculation of elastic stresses due to both the initial prestress P_i and the self-weight of the precast member. This load stage frequently occurs off-site in a precasting plant.

2. *Period before casting in-situ slab*: This involves a time analysis to determine the stress redistribution and change in curvature caused by creep and shrinkage of the concrete in the precast element during the period after the precast element is prestressed and prior to casting the in-situ concrete. The only loads acting are the prestress (after initial losses) and the self-weight of the precast element. A reasonably accurate time analysis can be performed using the analysis described in Section 5.7.3. Typical concrete stresses at load stages 1 and 2 at the mid-span of the precast element are illustrated in Figure 9.2a.

3. *Immediately after casting the in-situ concrete and before composite action*: This load stage involves a short-term analysis of the precast element to calculate the instantaneous effects of the additional superimposed dead loads prior to composite action. If the precast element is unshored (i.e. not temporarily supported by props during construction), the superimposed dead load mentioned here includes the weight of the wet in-situ concrete. The additional increments of stress and instantaneous strain in the precast element are added to the stresses and strains obtained at the end of load stage 2. Typical concrete stresses at the critical section of an unshored member at load stage 3 are shown in Figure 9.2b.

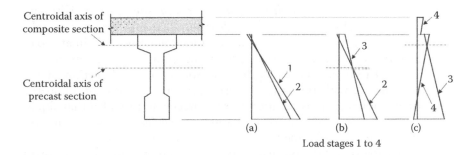

Load stages 1 to 4

Figure 9.2 Concrete stresses at the various load stages.

If the precast member is shored prior to placement of the cast in-situ slab, the applied loads do not produce internal actions or deformations in the member and the imposed loads are carried by the shoring. Therefore, no additional stresses or strains occur in a fully shored precast element at this load stage. When curing of the cast in-situ component has been completed and the shoring is removed, the self-weight of the cast in-situ concrete, together with any other loading applied at this time, produces deformations and flexural stresses and is considered in load stage 4.

4. *Immediately after the establishment of composite action*: This involves a short-term analysis of the composite cross-section (see Section 9.5.2) to determine the change of stresses and deformations on the composite cross-section as all the remaining loads are applied. The instantaneous effect of any dead load or service live load and any additional prestressing not previously considered (i.e. not applied previously to the non-composite precast element) are considered here. If cracking occurs, a cracked section analysis is required. Additional prestress may be applied to the composite member by re-stressing existing post-tensioned tendons or tensioning previously unstressed tendons.

   If the composite section remains uncracked, the increments of stress and strain calculated at this load stage on the precast part of the composite cross-section are added to the stresses and strains calculated in stage 3 prior to the establishment of composite action. Typical concrete stresses at the end of load stage 4 are shown in Figure 9.2c.

5. *Period after the establishment of composite action*: A time analysis of the composite cross-section is required (see Section 9.5.3) for the period beginning at the time the sustained load is first applied (usually soon after the in-situ concrete is poured) and ending after all creep and shrinkage deformations have taken place. The long-term effects of creep and shrinkage of concrete and relaxation of the prestressing steel on the behaviour of the composite section subjected to the sustained service loads are determined.

6. *The ultimate load condition for the composite section*: Ultimate strength checks are required for flexure, shear and torsion (if applicable) to ensure an adequate factor of safety. Under ultimate load conditions, the flexural strength of the composite section can be assumed to equal the strength of a monolithic cross-section of the same shape, with the same material properties, and containing the same amount and distribution of reinforcement, provided that slip at the interface between the precast and in-situ elements is small and full shear transfer is obtained. The stress discontinuity at the interface at service loads and the inelastic effects of creep and shrinkage have an insignificant effect on the ultimate strength and can be ignored at the ultimate load condition.

9.4 DETERMINATION OF PRESTRESS

In practice, the initial prestress and the eccentricity of prestress at the critical section in the precast element (P_i and e_{pc}, respectively) are calculated to satisfy preselected stress limits at transfer. In general, cracking is avoided at transfer by limiting the tensile stress to about $F_{tp} = 0.25\sqrt{f'_{cp}}$. In addition, in order to avoid unnecessarily large creep deformations, it is prudent to ensure that the initial compressive stresses do not exceed $F_{cp} = -0.5f'_{cp}$. In the case of trough girders, as shown in Figure 9.1e, the centroidal axis of the precast element is often not far above the bottom flange, so that loads applied to the precast element prior to or during placement of the in-situ slab may cause unacceptably large compressive stresses in the top fibres of the precast girder.

Satisfaction of stress limits in the precast element at transfer and immediately prior to the establishment of composite action (at the end of load stage 3) can be achieved using the procedure discussed in Section 5.4.1. For the case of a precast girder, Equations 5.1 through 5.4 become:

$$P_i \leq \frac{A_{pc}F_{tp} + \alpha_{t.pc}M_1}{\alpha_{t.pc}e_{pc} - 1} \tag{9.2}$$

$$P_i \leq \frac{-A_{pc}F_{cp} + \alpha_{b.pc}M_1}{\alpha_{b.pc}e_{pc} + 1} \tag{9.3}$$

$$P_i \geq \frac{-A_{pc}F_t + \alpha_{b.pc}M_3}{\Omega_3(\alpha_{b.pc}e_{pc} + 1)} \tag{9.4}$$

$$P_i \geq \frac{A_{pc}F_c + \alpha_{t.pc}M_3}{\Omega_3(\alpha_{t.pc}e_{pc} - 1)} \tag{9.5}$$

where e_{pc} is the eccentricity of prestress from the centroidal axis of the precast section; $\alpha_{t.pc} = A_{pc}/Z_{t.pc}$; $\alpha_{b.pc} = A_{pc}/Z_{b.pc}$; A_{pc} is the cross-sectional area of the precast member, and $Z_{t.pc}$ and $Z_{b.pc}$ are the top and bottom section moduli of the precast element, respectively. The moment M_1 is the moment applied at load stage 1 (usually resulting from the self-weight of the precast member), M_3 is the maximum in-service moment applied to the precast element prior to composite action (in load stage 3) and $\Omega_3 P_i$ is the prestressing force at load stage 3.

An estimate of the losses of prestress between transfer and the placement of the in-situ slab deck is required for the determination of Ω_3.

Equations 9.2 and 9.3 provide an upper limit to P_i, and Equations 9.4 and 9.5 establish a minimum level of prestress in the precast element.

After the in-situ slab has set, the composite cross-section resists all subsequent loading. There is a change both in the size and the properties of the cross-section and a stress discontinuity exists at the element interface. If cracking is to be avoided under the service loads, a limit F_t (say $0.25\sqrt{f_c'}$) is placed on the magnitude of the extreme fibre tensile stress at the end of load stage 5, i.e. after all prestress losses and under full service loads. This requirement places another, perhaps more severe limit on the minimum amount of prestress compared to that imposed by Equation 9.4. Alternatively, this requirement may suggest that an additional prestressing force is required on the composite member, i.e. the member may need to be further post-tensioned after the in-situ slab has developed its target strength.

The bottom fibre tensile stress immediately before the establishment of composite action may be approximated by:

$$\sigma_{b3} = -\frac{\Omega_3 P_i}{A_{pc}}\left(1 + \frac{A_{pc}e_{pc}}{Z_{b.pc}}\right) + \frac{M_3}{Z_{b.pc}} \tag{9.6}$$

If the maximum additional moment applied to the composite cross-section in load stage 4 is M_4 and the prestressing force reduces to ΩP_i with time, then the final maximum bottom fibre stress at the end of load stage 5 may be approximated by:

$$\sigma_{b5} = -\frac{\Omega P_i}{A_{pc}}\left(1 + \frac{A_{pc}e_{pc}}{Z_{b.pc}}\right) + \frac{M_3}{Z_{b.pc}} + \frac{M_4}{Z_{b.comp}} \tag{9.7}$$

where $Z_{b.comp}$ is the section modulus for the bottom fibre of the composite cross-section. If the bottom fibre stress in load stage 5 is to remain less than the stress limit F_t, then Equation 9.7 can be rearranged to give:

$$P_i \geq \frac{A_{pc}\left[(M_3/Z_{b.pc}) + (M_4/Z_{b.comp}) - F_t\right]}{\Omega(\alpha_{b.pc}e_{pc} + 1)} \tag{9.8}$$

Equation 9.8, together with Equations 9.2, 9.3 and 9.5, can be used to establish a suitable combination of P_i and e_{pc}. In some cases, the precast section may be proportioned so that the prestress and eccentricity satisfy all stress limits prior to composite action (i.e. Equations 9.2 through 9.5). However, when the additional requirement of Equation 9.8 is included, no combination of P_i and e_{pc} can be found to satisfy all the stress limits. In such cases, additional prestress may be applied to the composite member after the in-situ slab is in place.

If cracking can be tolerated in the composite member under full service loads, a cracked section analysis may be required to check for crack control and to determine the reduction of stiffness and its effect on deflection. Care must be taken in such an analysis to model stresses accurately in the various parts of the cross-section and the stress discontinuity at the slab–girder interface.

In many cases, cracking may be permitted under the full live load but not under the permanent sustained load. In such a case, M_4 in Equation 9.8 can be replaced by the sustained part of the moment applied at load stage 4 ($M_{4.sus}$) and the so modified Equation 9.8 can be used to determine the minimum level of prestress on a *partially prestressed* composite section.

9.5 METHODS OF ANALYSIS AT SERVICE LOADS

9.5.1 Introductory remarks

After the size of the concrete elements and the quantity and disposition of prestressing steel have been determined, the behaviour of the composite member at service loads should be investigated to determine the deflection (and shortening) at the various load stages (and times), and also to check for the possibility of cracking. The short-term and time-dependent analyses of uncracked composite cross-sections can be carried out conveniently using procedures similar to those described in Sections 5.6.2 and 5.7.3 for non-composite cross-sections. The approaches described here were also presented by Gilbert and Ranzi [2].

Consider a cross-section made up of a precast, pretensioned girder (element 1) and a cast in-situ reinforced concrete slab (element 2), as shown in Figure 9.3. The concrete in each element has different deformation characteristics. This particular cross-section contains four layers of non-prestressed

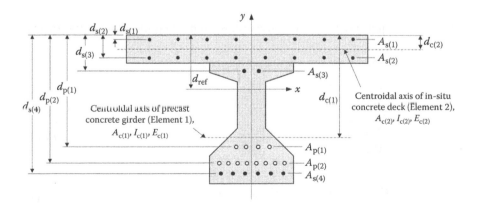

Figure 9.3 Typical prestressed concrete composite cross-section.

reinforcement and two layers of prestressing steel, although any number of steel layers can be handled without added difficulty. As was demonstrated in Tables 5.1 and 5.2, the presence of non-prestressed reinforcement may affect the time-dependent deformation of the section significantly and cause a reduction of the compressive stresses in the concrete. In the following analyses, no slip is assumed to occur between the two concrete elements or between the steel reinforcement and the concrete.

9.5.2 Short-term analysis

As outlined in Section 5.6.2 the constitutive relationships for each material for use in the instantaneous analysis are (Equations 5.21 through 5.24):

$$\sigma_{c(i),0} = E_{c(i),0}\varepsilon_0 \tag{9.9}$$

$$\sigma_{s(i),0} = E_{s(i)}\varepsilon_0 \tag{9.10}$$

$$\sigma_{p(i),0} = E_{p(i)}(\varepsilon_0 + \varepsilon_{p(i),\text{init}}) \tag{9.11}$$

Similar to Equation 5.27, the internal actions carried by the i-th concrete element (for inclusion in the equilibrium equations) can be expressed as:

$$N_{c(i),0} = \int_{A_{c(i)}} \sigma_{c(i),0}dA = \int_{A_{c(i)}} E_{c(i),0}(\varepsilon_{r,0} - y\kappa_0)dA = A_{c(i)}E_{c(i),0}\varepsilon_{r,0} - B_{c(i)}E_{c(i),0}\kappa_0 \tag{9.12}$$

$$M_{c(i),0} = \int_{A_{c(i)}} -y\sigma_{c(i),0}dA = \int_{A_{c(i)}} -E_{c(i),0}y(\varepsilon_{r,0} - y\kappa_0)dA = -B_{c(i)}E_{c(i),0}\varepsilon_{r,0} + I_{c(i)}E_{c(i),0}\kappa_0 \tag{9.13}$$

and, as in Equation 5.38, the governing system of equilibrium equations is:

$$\mathbf{r}_{\text{ext},0} = \mathbf{D}_0\varepsilon_0 + \mathbf{f}_{p,\text{init}} \tag{9.14}$$

where:

$$\mathbf{r}_{\text{ext},0} = \begin{bmatrix} N_{\text{ext},0} \\ M_{\text{ext},0} \end{bmatrix} \tag{9.15}$$

$$\mathbf{D}_0 = \begin{bmatrix} R_{A,0} & -R_{B,0} \\ -R_{B,0} & R_{I,0} \end{bmatrix} \tag{9.16}$$

$$\varepsilon_0 = \begin{bmatrix} \varepsilon_{r,0} \\ \kappa_0 \end{bmatrix} \tag{9.17}$$

$$f_{p,init} = \sum_{i=1}^{m_p} \begin{bmatrix} A_{p(i)}E_{p(i)}\varepsilon_{p(i),init} \\ y_{p(i)}A_{p(i)}E_{p(i)}\varepsilon_{p(i),init} \end{bmatrix} = \sum_{i=1}^{m_p} \begin{bmatrix} P_{init(i)} \\ y_{p(i)}P_{init(i)} \end{bmatrix} \tag{9.18}$$

Solving for the unknown strain variables gives (Equation 5.43):

$$\varepsilon_0 = D_0^{-1}(r_{ext,0} - f_{p,init}) = F_0(r_{ext,0} - f_{p,init}) \tag{9.19}$$

where:

$$F_0 = \frac{1}{R_{A,0}R_{I,0} - R_{B,0}^2} \begin{bmatrix} R_{I,0} & R_{B,0} \\ R_{B,0} & R_{A,0} \end{bmatrix} \tag{9.20}$$

The cross sectional rigidities forming the D_0 and F_0 matrices are:

$$R_{A,0} = \sum_{i=1}^{m_c} A_{c(i)}E_{c(i),0} + \sum_{i=1}^{m_s} A_{s(i)}E_{s(i)} + \sum_{i=1}^{m_p} A_{p(i)}E_{p(i)} = \sum_{i=1}^{m_c} A_{c(i)}E_{c(i),0} + R_{A,s} + R_{A,p} \tag{9.21}$$

$$R_{B,0} = \sum_{i=1}^{m_c} B_{c(i)}E_{c(i),0} + \sum_{i=1}^{m_s} y_{s(i)}A_{s(i)}E_{s(i)} + \sum_{i=1}^{m_p} y_{p(i)}A_{p(i)}E_{p(i)}$$

$$= \sum_{i=1}^{m_c} B_{c(i)}E_{c(i),0} + R_{B,s} + R_{B,p} \tag{9.22}$$

$$R_{I,0} = \sum_{i=1}^{m_c} I_{c(i)}E_{c(i),0} + \sum_{i=1}^{m_s} y_{s(i)}^2 A_{s(i)}E_{s(i)} + \sum_{i=1}^{m_p} y_{p(i)}^2 A_{p(i)}E_{p(i)} = \sum_{i=1}^{m_c} I_{c(i)}E_{c(i),0} + R_{I,s} + R_{I,p} \tag{9.23}$$

where for convenience, the following notation is introduced for the rigidities of the reinforcement and tendons:

$$R_{A,s} = \sum_{i=1}^{m_s} A_{s(i)}E_{s(i)} \tag{9.24}$$

$$R_{B,s} = \sum_{i=1}^{m_s} y_{s(i)}A_{s(i)}E_{s(i)} \tag{9.25}$$

$$R_{I,s} = \sum_{i=1}^{m_s} y_{s(i)}^2 A_{s(i)} E_{s(i)} \tag{9.26}$$

$$R_{A,p} = \sum_{i=1}^{m_p} A_{p(i)} E_{p(i)} \tag{9.27}$$

$$R_{B,p} = \sum_{i=1}^{m_p} y_{p(i)} A_{p(i)} E_{p(i)} \tag{9.28}$$

$$R_{I,p} = \sum_{i=1}^{m_p} y_{p(i)}^2 A_{p(i)} E_{p(i)} \tag{9.29}$$

The stress distribution is calculated from Equations 9.9 through 9.11:

$$\sigma_{c(i),0} = E_{c(i),0}\varepsilon_0 = E_{c(i),0}[1-y]\varepsilon_0 \tag{9.30}$$

$$\sigma_{s(i),0} = E_{s(i)}\varepsilon_0 = E_{s(i)}[1-y_{s(i)}]\varepsilon_0 \tag{9.31}$$

$$\sigma_{p(i),0} = E_{p(i)}(\varepsilon_0 + \varepsilon_{p(i),init}) = E_{p(i)}[1-y_{p(i)}]\varepsilon_0 + E_{p(i)}\varepsilon_{p(i),init} \tag{9.32}$$

where $\varepsilon_0 = \varepsilon_{r,0} - y\kappa_0 = [1-y]\varepsilon_0$.

9.5.3 Time-dependent analysis

For the analysis of stresses and deformations on a composite concrete–concrete cross-section at time τ_k after a period of sustained loading, the *age-adjusted effective modulus method* may be used, as outlined in Sections 5.7.2 and 5.7.3. The stress–strain relationships for each concrete element and for each layer of reinforcement and tendons at τ_k are as follows (Equations 5.65 through 5.67):

$$\sigma_{c(i),k} = \overline{E}_{e(i),k}(\varepsilon_k - \varepsilon_{sh(i),k}) + \overline{F}_{e(i),0}\sigma_{c(i),0} \tag{9.33}$$

$$\sigma_{s(i),k} = E_{s(i)}\varepsilon_k \tag{9.34}$$

$$\sigma_{p(i),k} = E_{p(i)}(\varepsilon_k + \varepsilon_{p(i),init} - \varepsilon_{p.rel(i),k}) \tag{9.35}$$

In this case, the contribution of the i-th concrete component to the internal axial force and moment can be determined as (similar to Equation 5.73):

$$N_{c(i),k} = \int\limits_{A_{c(i)}} \sigma_{c(i),k} dA = \int\limits_{A_{c(i)}} \left[\overline{E}_{e(i),k}\left(\varepsilon_{r,k} - y\kappa_k - \varepsilon_{cs(i),k}\right) + \overline{F}_{e(i),0}\sigma_{c(i),0} \right] dA$$

$$= A_{c(i)}\overline{E}_{e(i),k}\varepsilon_{r,k} - B_{c(i)}\overline{E}_{e(i),k}\kappa_k - A_{c(i)}\overline{E}_{e(i),k}\varepsilon_{cs(i),k} + \overline{F}_{e(i),0}N_{c(i),0} \tag{9.36}$$

$$M_{c(i),k} = \int\limits_{A_{c(i)}} -y\sigma_{c(i),k} dA = \int\limits_{A_{c(i)}} -y\left[\overline{E}_{e(i),k}\left(\varepsilon_{r,k} - y\kappa_k - \varepsilon_{cs(i),k}\right) + \overline{F}_{e(i),0}\sigma_{c(i),0} \right] dA$$

$$= -B_{c(i)}\overline{E}_{e(i),k}\varepsilon_{r,k} + I_{c(i)}\overline{E}_{e(i),k}\kappa_k - B_{c(i)}\overline{E}_{e(i),k}\varepsilon_{cs(i),k} + \overline{F}_{e(i),0}M_{c(i),0} \tag{9.37}$$

The equilibrium equations are (Equation 5.82):

$$\mathbf{r}_{ext,k} = \mathbf{D}_k\boldsymbol{\varepsilon}_k + \mathbf{f}_{cr,k} - \mathbf{f}_{cs,k} + \mathbf{f}_{p,init} - \mathbf{f}_{p.rel,k} \tag{9.38}$$

where:

$$\mathbf{r}_{ext,k} = \begin{bmatrix} N_{ext,k} \\ M_{ext,k} \end{bmatrix} \tag{9.39}$$

$$\mathbf{D}_k = \begin{bmatrix} R_{A,k} & -R_{B,k} \\ -R_{B,k} & R_{I,k} \end{bmatrix} \tag{9.40}$$

$$\boldsymbol{\varepsilon}_k = \begin{bmatrix} \varepsilon_{r,k} \\ \kappa_k \end{bmatrix} \tag{9.41}$$

In Equation 9.38, the effects of creep and shrinkage of the concrete elements are included in the vectors $\mathbf{f}_{cr,k}$ and $\mathbf{f}_{cs,k}$ (refer Equations 5.85 and 5.86):

$$\mathbf{f}_{cr,k} = \sum_{i=1}^{m_c} \overline{F}_{e(i),0} \begin{bmatrix} N_{c(i),0} \\ M_{c(i),0} \end{bmatrix} = \sum_{i=1}^{m_c} \overline{F}_{e(i),0}E_{c(i),0} \begin{bmatrix} A_{c(i)}\varepsilon_{r,0} - B_{c(i)}\kappa_0 \\ -B_{c(i)}\varepsilon_{r,0} + I_{c(i)}\kappa_0 \end{bmatrix} \tag{9.42}$$

$$f_{cs,k} = \sum_{i=1}^{m_c} \begin{bmatrix} A_{c(i)} \\ -B_{c(i)} \end{bmatrix} \bar{E}_{e(i),k}\varepsilon_{cs(i),k} \tag{9.43}$$

The initial strain in the prestressing steel and the relaxation are accounted for using (Equations 5.87 and 5.88):

$$f_{p,init} = \sum_{i=1}^{m_p} \begin{bmatrix} P_{init(i)} \\ -y_{p(i)}P_{init(i)} \end{bmatrix} \tag{9.44}$$

$$f_{p.rel,k} = \sum_{i=1}^{m_p} \begin{bmatrix} P_{i(i)}\varphi_{p(i)} \\ -y_{p(i)}P_{i(i)}\varphi_{p(i)} \end{bmatrix} \tag{9.45}$$

Solving gives the strain at time τ_k (Equation 5.89):

$$\varepsilon_k = D_k^{-1}(r_{ext,k} - f_{cr,k} + f_{cs,k} - f_{p,init} + f_{p.rel,k}) = F_k(r_{ext,k} - f_{cr,k} + f_{cs,k} - f_{p,init} + f_{p.rel,k}) \tag{9.46}$$

where:

$$\varepsilon_k = \begin{bmatrix} \varepsilon_{r,k} \\ \kappa_k \end{bmatrix} \tag{9.47}$$

$$F_k = \frac{1}{R_{A,k}R_{I,k} - R_{B,k}^2} \begin{bmatrix} R_{I,k} & R_{B,k} \\ R_{B,k} & R_{A,k} \end{bmatrix} \tag{9.48}$$

and the cross-sectional rigidities at τ_k are:

$$R_{A,k} = \sum_{i=1}^{m_c} A_{c(i)}\bar{E}_{e(i),k} + \sum_{i=1}^{m_s} A_{s(i)}E_{s(i)} + \sum_{i=1}^{m_p} A_{p(i)}E_{p(i)} = \sum_{i=1}^{m_c} A_{c(i)}\bar{E}_{e(i),k} + R_{A,s} + R_{A,p} \tag{9.49}$$

$$R_{B,k} = \sum_{i=1}^{m_c} B_{c(i)}\bar{E}_{e(i),k} + \sum_{i=1}^{m_s} y_{s(i)}A_{s(i)}E_{s(i)} + \sum_{i=1}^{m_p} y_{p(i)}A_{p(i)}E_{p(i)}$$

$$= \sum_{i=1}^{m_c} B_{c(i)}\bar{E}_{e(i),k} + R_{B,s} + R_{B,p} \tag{9.50}$$

$$R_{I,k} = \sum_{i=1}^{m_c} I_{c(i)}\bar{E}_{e(i),k} + \sum_{i=1}^{m_s} y_{s(i)}^2 A_{s(i)}E_{s(i)} + \sum_{i=1}^{m_p} y_{p(i)}^2 A_{p(i)}E_{p(i)} = \sum_{i=1}^{m_c} I_{c(i)}\bar{E}_{e(i),k} + R_{I,s} + R_{I,p}$$

(9.51)

The stress distributions at time τ_k in each concrete element and in the reinforcement and tendons are (Equations 5.91 through 5.93):

$$\sigma_{c(i),k} = \bar{E}_{e(i),k}(\varepsilon_k - \varepsilon_{cs(i),k}) + \bar{F}_{e(i),0}\sigma_{c(i),0} = \bar{E}_{e(i),k}\left\{[1-y]\varepsilon_k - \varepsilon_{cs(i),k}\right\} + \bar{F}_{e(i),0}\sigma_{c(i),0}$$

(9.52)

$$\sigma_{s(i),k} = E_{s(i)}\varepsilon_k = E_{s(i)}[1 - y_{s(i)}]\varepsilon_k$$

(9.53)

$$\sigma_{p(i),k} = E_{p(i)}(\varepsilon_k + \varepsilon_{p(i),init} - \varepsilon_{p.rel(i),k}) = E_{p(i)}[1 - y_{p(i)}]\varepsilon_k + E_{p(i)}\varepsilon_{p(i),init} - E_{p(i)}\varepsilon_{p.rel(i),k}$$

(9.54)

where $\varepsilon_k = \varepsilon_{r,k} - y\kappa_k = [1-y]\varepsilon_k$.

EXAMPLE 9.1

The cross-section of a composite footbridge consists of a precast, pretensioned trough girder and a cast in-situ slab, as shown in Figure 9.4. The precast section is cast and moist cured for 4 days prior to transfer. The cross-section is subjected to the following load history.

At t = 4 days: The total prestressing force of 2000 kN is transferred to the trough girder. The centroid of all the pretensioned strands is located 100 mm above the bottom fibre, as shown. The moment on the section caused by the self-weight of the girder $M_1 = 320$ kNm is introduced at transfer. Shrinkage of the concrete also begins to develop at this time.

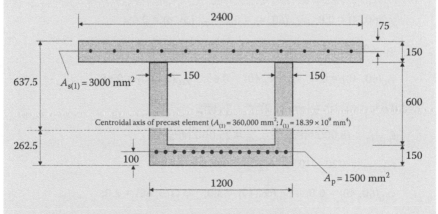

Figure 9.4 Details of composite cross-section (Example 9.1).

At t = 40 days: The in-situ slab deck is cast and cured and the moment caused by the weight of the deck is applied to the precast section, $M_3 = 300$ kNm.

At t = 60 days: A wearing surface is placed and all other superimposed dead loads are applied to the bridge, thereby introducing an additional moment $M_4 = 150$ kNm.

At t > 60 days: The moment remains constant from 60 days to time infinity.

Composite action gradually begins to develop as soon as the concrete in the deck sets. Full composite action may not be achieved for several days. However, it is assumed here that the in-situ deck and the precast section act compositely at all times after $t = 40$ days. The stress and strain distributions on the composite cross section are to be calculated:

  i. immediately after the application of the prestress at $t = 4$ days;
 ii. just before the slab deck is cast at $t = 40$ days;
iii. immediately after the slab deck is cast at $t = 40$ days;
 iv. just before the road surface is placed at $t = 60$ days;
  v. immediately after the road surface is placed at $t = 60$ days; and
 vi. at time infinity after all creep and shrinkage strains have developed.

For the precast section (element 1): $f_c' = 40$ MPa

$$E_{c(1),4} = E_{c(1)}(4) = 25{,}000 \text{ MPa}; \quad E_{c(1),40} = 31{,}500 \text{ MPa};$$
$$E_{c(1),60} = 33{,}000 \text{ MPa};$$

$$\varepsilon_{cs(1),40} = \varepsilon_{cs(1)}(40) = -150 \times 10^{-6}; \quad \varepsilon_{cs(1),60} = -200 \times 10^{-6};$$
$$\varepsilon_{cs(1),\infty} = -500 \times 10^{-6};$$

$$\varphi_{(1)}(40, 4) = 0.9; \quad \varphi_{(1)}(60, 4) = 1.2; \quad \varphi_{(1)}(\infty, 4) = 2.4;$$

$$\chi_{(1)}(40, 4) = 0.8; \quad \chi_{(1)}(60, 4) = 0.7; \quad \chi_{(1)}(\infty, 4) = 0.65;$$

$$\varphi_{(1)}(60, 40) = 0.5; \quad \varphi_{(1)}(\infty, 40) = 1.6; \quad \varphi_{(1)}(\infty, 60) = 1.2;$$

$$\chi_{(1)}(60, 40) = 0.8; \quad \chi_{(1)}(\infty, 40) = 0.65; \quad \chi_{(1)}(\infty, 60) = 0.65.$$

For the in-situ slab (element 2): $f_c' = 25$ MPa

$$E_{c(2),40} = 18{,}000 \text{ MPa}; \quad E_{c(2),60} = 25{,}000 \text{ MPa};$$

$$\varepsilon_{cs(2),60} = -120 \times 10^{-6}; \quad \varepsilon_{cs(2),\infty} = -600 \times 10^{-6};$$

$$\varphi_{(2)}(60, 40) = 0.8; \quad \varphi_{(2)}(\infty, 40) = 3.0; \quad \varphi_{(2)}(\infty, 60) = 2.0;$$

$$\chi_{(2)}(60, 40) = 0.8; \quad \chi_{(2)}(\infty, 40) = 0.65; \quad \chi_{(2)}(\infty, 60) = 0.65;$$

To account for relaxation in the prestressing tendons, we take the creep coefficient to be

$$\varphi_p(40) = 0.01; \quad \varphi_p(60) = 0.015; \quad \varphi_p(\infty) = 0.025$$

and the elastic moduli for the reinforcement and tendons are $E_s = E_p = 200,000$ MPa.

(i) At t = 4 days:
In this example, the reference x-axis is taken as the centroidal axis of the precast cross-section. The properties of the concrete part of the cross-section (with respect to the x-axis) are:

$$A_{c(1)} = A_{(1)} - A_p = 360,000 - 1,500 = 358,500 \text{ mm}^2$$

$$B_{c(1)} = A_{(1)}y_c - A_p y_p = 360,000 \times 0 - 1,500 \times (-162.5) = 243,750 \text{ mm}^3$$

$$I_{c(1)} = I_{(1)} - A_p y_p^2 = 18.39 \times 10^9 - 1,500 \times (-162.5)^2 = 18.35 \times 10^9 \text{ mm}^4$$

For this member with bonded prestressing tendons, Equations 9.21 through 9.23 give the rigidities of the cross-section at first loading (age 4 days):

$$R_{A,4} = A_{c(1)}E_{c(1),4} + A_p E_p = 358,500 \times 25,000 + 1,500 \times 200,000 = 9,263 \times 10^6 \text{ N}$$

$$R_{B,4} = B_{c(1)}E_{c(1),4} + y_p A_p E_p = 243,750 \times 25,000 + 1,500 \times (-162.5) \times 200,000$$

$$= -42,660 \times 10^6 \text{ Nmm}$$

$$R_{I,4} = I_{c(1)}E_{c(1),4} + y_p^2 A_p E_p = 18.35 \times 10^9 \times 25,000 + (-162.5)^2 \times 1,500 \times 200,000$$

$$= 466.8 \times 10^{12} \text{ Nmm}^2$$

and from Equation 9.20:

$$\mathbf{F}_1 = \frac{1}{R_{A,4}R_{I,4} - R_{B,4}^2}\begin{bmatrix} R_{I,4} & R_{B,4} \\ R_{B,4} & R_{A,4} \end{bmatrix}$$

$$= \frac{1}{9,263 \times 10^6 \times 466.8 \times 10^{12} - (-42,660 \times 10^6)^2}\begin{bmatrix} 466.8 \times 10^{12} & -42,660 \times 10^6 \\ -42,660 \times 10^6 & 9,263 \times 10^6 \end{bmatrix}$$

$$= \begin{bmatrix} 108.0 \times 10^{-12} & -9.870 \times 10^{-15} \\ -9.870 \times 10^{-15} & 2.143 \times 10^{-15} \end{bmatrix}$$

The vector of internal actions at first loading is (Equation 9.15):

$$\mathbf{r}_{ext,4} = \begin{bmatrix} N_{ext,4} \\ M_{ext,4} \end{bmatrix} = \begin{bmatrix} 0 \\ M_i \end{bmatrix} = \begin{bmatrix} 0 \\ 320 \times 10^6 \end{bmatrix}$$

and the initial strain in the prestressing steel due to the initial prestressing force is obtained as follows (Equation 5.25):

$$\varepsilon_{p,init} = \frac{P_{init}}{A_p E_p} = \frac{2,000 \times 10^3}{1,500 \times 200,000} = 0.00667$$

The vector of internal actions caused by the initial prestress $\mathbf{f}_{p,init}$ is given by Equation 9.18:

$$\mathbf{f}_{p,init} = \begin{bmatrix} P_{init} \\ -y_p P_{init} \end{bmatrix} = \begin{bmatrix} 2000 \times 10^3 \\ -(-162.5) \times 2000 \times 10^3 \end{bmatrix} = \begin{bmatrix} 2000 \times 10^3 \\ 325 \times 10^6 \end{bmatrix}$$

and the strain vector at first loading ε_4 containing the unknown strain variables is determined from Equation 9.19:

$$\varepsilon_4 = \mathbf{F}_4 (\mathbf{r}_{ext,4} - \mathbf{f}_{p,init})$$

$$= \begin{bmatrix} 108.0 \times 10^{-12} & -9.870 \times 10^{-15} \\ -9.870 \times 10^{-15} & 2.143 \times 10^{-15} \end{bmatrix} \left(\begin{bmatrix} 0 \\ 320 \times 10^6 \end{bmatrix} - \begin{bmatrix} 2000 \times 10^3 \\ 325 \times 10^6 \end{bmatrix} \right)$$

$$= \begin{bmatrix} -216.0 \times 10^{-6} \\ 0.00902 \times 10^{-6} \end{bmatrix}$$

The strain at the reference axis and the curvature at first loading are therefore:

$$\varepsilon_{r(1),4} = -216.0 \times 10^{-6} \quad \text{and} \quad \kappa_{(1),4} = +0.00902 \times 10^{-6} \text{ mm}^{-1}$$

and the strains at the top fibre of the precast section (at $y = +487.5$ mm) and at the bottom fibre (at $y = -262.5$ mm) are:

$$\varepsilon_{(1),4(top)} = \varepsilon_{r(1),4} - 487.5 \times \kappa_{(1),4} = (-216.0 - 487.5 \times 0.00902) \times 10^{-6}$$

$$= -220.4 \times 10^{-6}$$

$$\varepsilon_{(1),4(btm)} = \varepsilon_{r(1),4} - (-262.5) \times \kappa_{(1),4} = (-216.0 + 262.5 \times 0.00902) \times 10^{-6}$$

$$= -213.6 \times 10^{-6}$$

The strain in the bonded prestressing steel is:

$$\varepsilon_{p,4} = \varepsilon_{p,init} + (\varepsilon_{r(1),4} - y_p \kappa_{(1),4}) = 0.00667$$

$$+ [-216.0 - (-162.5) \times 0.00902] \times 10^{-6} = 0.00645$$

The top and bottom fibre stresses in the concrete and the stress in the pre-stressing steel are obtained from Equations 9.9 and 9.11, respectively:

$$\sigma_{c(1),4(top)} = E_{c(1),4}\varepsilon_{(1),4(top)} = 25,000 \times (-220.4 \times 10^{-6}) = -5.51\,\text{MPa}$$

$$\sigma_{c(1),4(btm)} = E_{c(1),4}\varepsilon_{(1),4(btm)} = 25,000 \times (-213.6 \times 10^{-6}) = -5.34\,\text{MPa}$$

$$\sigma_{p,4} = E_p\varepsilon_{p,4} = 200,000 \times 0.00645 = +1,290\,\text{MPa}$$

The stress and strain distributions immediately after transfer at time $t = 4$ days are shown in Figure 9.5b.

(ii) At t = 40 days (prior to casting the in-situ slab):
The age-adjusted effective modulus at this time is (Equation 5.55):

$$\bar{E}_{e(1),40} = \frac{E_{c,0}}{1 + \chi_{(1)}(40,4)\varphi_{(1)}(40,4)} = \frac{25,000}{1 + 0.8 \times 0.9} = 14,535\,\text{MPa}$$

and from Equation 5.58:

$$\bar{F}_{e(1),40} = \frac{\varphi_{(1)}(40,4)[\chi_{(1)}(40,4) - 1]}{1 + \chi_{(1)}(40,4)\varphi_{(1)}(40,4)} = \frac{0.9 \times (0.8 - 1.0)}{1.0 + 0.8 \times 0.9} - -0.1047$$

With the properties of the concrete part of the section determined in part (i) as $A_c = 358,500\,\text{mm}^2$, $B_c = 243,750\,\text{mm}^3$ and $I_c = 18.35 \times 10^9\,\text{mm}^4$, the cross-sectional rigidities are obtained from Equations 9.49 through 9.51:

$$R_{A,40} = A_{c(1)}\bar{E}_{e(1),40} + A_pE_p = 358,500 \times 14,535 + 1,500 \times 200,000$$

$$= 5,511 \times 10^6\,\text{N}$$

$$R_{B,40} = B_{c(1)}\bar{E}_{e(1),40} + y_pA_pE_p = 243,750 \times 14,535 + 1,500 \times (-162.5) \times 200,000$$

$$= -45,210 \times 10^6\,\text{Nmm}$$

$$R_{I,40} = I_{c(1)}\bar{E}_{e(1),40} + y_p^2A_pE_p = 18.35 \times 10^9 \times 14,535 + (-162.5)^2 \times 1,500 \times 200,000$$

$$= 274.7 \times 10^{12}\,\text{Nmm}^2$$

With $\varepsilon_{r(1),4} = -216.0 \times 10^{-6}$ and $\kappa_{(1),4} = +0.00902 \times 10^{-6}$ mm^{-1}, Equation 9.42 gives:

$$\mathbf{f}_{cr(1),40} = \bar{F}_{e(1),4} E_{c(1),4} \begin{bmatrix} A_c \varepsilon_{r(1),4} - B_c \kappa_{(1),4} \\ -B_c \varepsilon_{r(1),4} + I_c \kappa_{(1),4} \end{bmatrix}$$

$$= -0.1047$$

$$\times 25{,}000 \begin{bmatrix} 358{,}500 \times (-216.0 \times 10^{-6}) - 243{,}750 \times 0.00902 \times 10^{-6} \\ -243{,}750 \times (-216.0 \times 10^{-6}) + 18.35 \times 10^9 \times 0.00902 \times 10^{-6} \end{bmatrix}$$

$$= \begin{bmatrix} +202.6 \times 10^3 \text{ N} \\ -0.5711 \times 10^6 \text{ Nmm} \end{bmatrix}$$

and from Equation 9.43:

$$\mathbf{f}_{cs(1),40} = \begin{bmatrix} A_c \\ -B_c \end{bmatrix} \bar{E}_{e(1),40} \varepsilon_{cs(1),40}$$

$$= \begin{bmatrix} 358{,}500 \times 14{,}535 \times (-150 \times 10^{-6}) \\ -243{,}750 \times 14{,}535 \times (-150 \times 10^{-6}) \end{bmatrix} = \begin{bmatrix} -781.6 \times 10^3 \text{ N} \\ 0.5314 \times 10^6 \text{ Nmm} \end{bmatrix}$$

The vectors of initial prestressing actions and relaxation actions are given by Equations 9.44 and 9.45 as, respectively:

$$\mathbf{f}_{p,init} = \begin{bmatrix} 2000 \times 10^3 \text{ N} \\ 325 \times 10^6 \text{ Nmm} \end{bmatrix}$$

$$\mathbf{f}_{p.rel,40} = \begin{bmatrix} \sigma_{p4} A_p \varphi_p(40) \\ -y_p \sigma_{p4} A_p \varphi_p(40) \end{bmatrix} = \begin{bmatrix} 1290 \times 10^3 \times 1500 \times 0.01 \\ -(-162.5) \times 1290 \times 10^3 \times 1500 \times 0.01 \end{bmatrix}$$

$$= \begin{bmatrix} 19.4 \times 10^3 \text{ N} \\ 3.14 \times 10^6 \text{ Nmm} \end{bmatrix}$$

and Equation 9.48 gives:

$$\mathbf{F}_{40} = \frac{1}{R_{A,40} R_{I,40} - R_{B,40}^2} \begin{bmatrix} R_{I,40} & R_{B,40} \\ R_{B,40} & R_{A,40} \end{bmatrix}$$

$$= \frac{1}{5{,}511 \times 10^6 \times 274.7 \times 10^{12} - (-45{,}210 \times 10^6)^2}$$

$$\times \begin{bmatrix} 274.7 \times 10^{12} & -45{,}210 \times 10^6 \\ -45{,}210 \times 10^6 & 5{,}511 \times 10^6 \end{bmatrix}$$

$$= \begin{bmatrix} 181.7 \times 10^{-12} \text{ N}^{-1} & -29.90 \times 10^{-15} \text{ N}^{-1}\text{mm}^{-1} \\ -29.90 \times 10^{-15} \text{ N}^{-1}\text{mm}^{-1} & 3.645 \times 10^{-15} \text{ N}^{-1}\text{mm}^{-2} \end{bmatrix}$$

The vector of internal actions at 40 days before the casting of the slab is the same as at age 4 days:

$$\mathbf{r}_{ext,40} = \begin{bmatrix} N_{ext,40} \\ M_{ext,40} \end{bmatrix} = \begin{bmatrix} 0 \\ M_I \end{bmatrix} = \begin{bmatrix} 0 \\ 320 \times 10^6 \text{ Nmm} \end{bmatrix}$$

and the strain ε_{40} at time τ_k = 40 days is determined using Equation 9.46:

$$\varepsilon_{(1),40} = \mathbf{F}_{40}\left(\mathbf{r}_{ext,40} - \mathbf{f}_{cr(1),40} + \mathbf{f}_{cs(1),40} - \mathbf{f}_{p,init} + \mathbf{f}_{p.rel,40}\right)$$

$$= \begin{bmatrix} 181.7 \times 10^{-12} & -29.90 \times 10^{-15} \\ -29.90 \times 10^{-15} & 3.645 \times 10^{-15} \end{bmatrix}$$

$$\times \begin{bmatrix} (0 - 202.6 - 781.6 - 2000 + 19.40) \times 10^3 \\ (320 + 0.5711 + 0.5314 - 325 + 3.14) \times 10^6 \end{bmatrix}$$

$$= \begin{bmatrix} -538.7 \times 10^{-6} \\ +0.0859 \times 10^{-6} \text{ mm}^{-1} \end{bmatrix}$$

The strain at the reference axis and the curvature at time τ_k = 40 days are therefore $\varepsilon_{r(1),40} = -538.7 \times 10^{-6}$ and $\kappa_{(1),40} = +0.0859 \times 10^{-6}$ mm^{-1}, respectively, and the strains at the top fibre of the precast section (at y = +487.5 mm) and at the bottom fibre (at y = −262.5 mm) are:

$$\varepsilon_{(1)40-(top)} = \varepsilon_{r(1),40} - 487.5 \times \kappa_{(1),40} = [-538.7 - (487.5 \times 0.0859)] \times 10^{-6}$$

$$= -580.6 \times 10^{-6}$$

$$\varepsilon_{(1)40-(btm)} = \varepsilon_{r,(1)40} - (-262.5) \times \kappa_{(1),40} = (-536.7 + 262.5 \times 0.0859) \times 10^{-6}$$

$$= -516.2 \times 10^{-6}$$

The concrete stress distribution at time τ_k = 40 days is calculated using Equation 9.52:

$$\sigma_{c(1),40-(top)} = E_{e(1),40}\left(\varepsilon_{(1)40-(top)} - \varepsilon_{cs(1),40}\right) + F_{e(1),40}\sigma_{c(1),4(top)}$$

$$= 14{,}535 \times [-580.6 - (-150)] \times 10^{-6} + (-0.1047) \times (-5.51) = -5.68 \text{ MPa}$$

$$\sigma_{c(1),40-(btm)} = \overline{E}_{e(1),40}\left(\varepsilon_{(1)40-(btm)} - \varepsilon_{cs(1),40}\right) + \overline{F}_{e(1),40}\sigma_{c(1),4(btm)}$$

$$= 14{,}535 \times [-516.2 - (-150)] \times 10^{-6} + (-0.1047) \times (-5.34) = -4.76 \text{ MPa}$$

The final stress in the prestressing steel at time $\tau_k = 40$ days is obtained from Equation 9.35:

$$\sigma_{p,40} = E_p[(\varepsilon_{r,40} - y_p \kappa_{40}) + \varepsilon_{p,init} - \varepsilon_{p,rel(40)}]$$

$$= 200{,}000 \times \left[(-538.7 - (-162.5) \times 0.0859) + 0.00667 - 0.000065\right] \times 10^{-6}$$

$$= 1{,}216 \text{ MPa}$$

where the relaxation strain $\varepsilon_{p,rel(40)}$ is calculated as $\varepsilon_{p,rel(40)} = \sigma_{p,4}\, \varphi_p(40)/E_p$.

The stress and strain distributions at $t = 40$ days before the in-situ slab is cast are shown in Figure 9.5c. The increments of strain and stress that have developed in the precast girder during the sustained load period from 4 to 40 days are therefore:

$$\Delta\varepsilon_{r(1),(40-4)} = \varepsilon_{r(1),40} - \varepsilon_{r(1),4} = [-538.7 - (-216.0)] \times 10^{-6} = -322.7 \times 10^{-6}$$

$$\Delta\kappa_{(1),(40-4)} = \kappa_{(1),40} - \kappa_{(1),4} = (0.0859 - 0.00902) \times 10^{-6} \text{ mm}^{-1}$$

$$= 0.0769 \times 10^{-6} \text{ mm}^{-1}$$

$$\Delta\varepsilon_{(1),(40-4)(top)} = \varepsilon_{(1),40(top)} - \varepsilon_{(1),4(top)} = [-580.6 - (-220.4)] \times 10^{-6}$$

$$= -360.2 \times 10^{-6}$$

$$\Delta\varepsilon_{(1),(40-4)(btm)} = \varepsilon_{(1),40(btm)} - \varepsilon_{(1),4(btm)} = [-516.2 - (-213.6)] \times 10^{-6}$$

$$= -302.6 \times 10^{-6}$$

$$\Delta\varepsilon_{p,(40-4)} = \varepsilon_{p,40} - \varepsilon_{p,4} = 0.00608 - 0.00645 = -0.00037$$

$$\Delta\sigma_{c(1),(40-4)(top)} = \sigma_{c(1),40(top)} - \sigma_{c(1),4(top)} = -5.68 - (-5.51) = -0.17 \text{ MPa}$$

$$\Delta\sigma_{c(1),(40-4)(btm)} = \sigma_{c(1),40(btm)} - \sigma_{c(1),4(btm)} = -4.76 - (-5.34) = +0.58 \text{ MPa}$$

$$\Delta\sigma_{p,(40-4)} = \sigma_{p,40} - \sigma_{p,4} = 1216 - 1290 = -74 \text{ MPa}$$

(iii) At t = 40 days (after casting the in-situ slab):
The increments of stress and strain caused by $M_3 = 300$ kNm applied to the precast section at age 40 days are calculated using the same procedure as was outlined in part (i) of this example, except that the elastic modulus of the precast concrete has now increased.

The short-term rigidities of the cross-section at age 40 days are:

$$R_{A,40} = A_{c(1)}E_{c(1),40} + A_p E_p = 358{,}500 \times 31{,}500 + 1{,}500 \times 200{,}000 = 11{,}593 \times 10^6 \text{ N}$$

$$R_{B,40} = B_{c(1)}E_{c(1),4} + y_p A_p E_p = 243{,}750 \times 31{,}500 + 1{,}500 \times (-162.5) \times 200{,}000$$

$$= -41{,}072 \times 10^6 \text{ Nmm}$$

$$R_{I,40} = I_{c(I)}E_{c(I),40} + y_p^2 A_p E_p = 18.35 \times 10^9 \times 31,500 + (-162.5)^2 \times 1,500 \times 200,000$$

$$= 586.1 \times 10^{12} \text{ Nmm}^2$$

and from Equation 9.20:

$$\mathbf{F}_{40} = \frac{1}{R_{A,40}R_{I,40} - R_{B,40}^2}\begin{bmatrix} R_{I,40} & R_{B,40} \\ R_{B,40} & R_{A,40} \end{bmatrix} = \begin{bmatrix} 86.28 \times 10^{-12} & -6.047 \times 10^{-15} \\ -6.047 \times 10^{-15} & 1.707 \times 10^{-15} \end{bmatrix}$$

For the load increment applied at 40 days, the vector of internal actions is:

$$\mathbf{r}_{ext} = \begin{bmatrix} 0 \\ M_3 \end{bmatrix} = \begin{bmatrix} 0 \\ 300 \times 10^6 \end{bmatrix}$$

and the vector of instantaneous strain caused by the application of M_3 at age 40 days is:

$$\Delta\varepsilon_{40} = \mathbf{F}_{40}\mathbf{r}_{ext} = \begin{bmatrix} 86.28 \times 10^{-12} & -6.047 \times 10^{-15} \\ -6.047 \times 10^{-15} & 1.707 \times 10^{-15} \end{bmatrix}\begin{bmatrix} 0 \\ 300 \times 10^6 \end{bmatrix} = \begin{bmatrix} -1.8 \times 10^{-6} \\ 0.5120 \times 10^{-6} \end{bmatrix}$$

The increment of instantaneous strain at the reference axis and the increment of curvature caused by the application of M_3 at age 40 days are therefore:

$$\Delta\varepsilon_{r,40} = -1.8 \times 10^{-6} \quad \text{and} \quad \Delta\kappa_{40} = +0.512 \times 10^{-6} \text{ mm}^{-1}$$

and the increment of instantaneous strain at the top fibre of the precast section (at $y = +487.5$ mm) and at the bottom fibre (at $y = -262.5$ mm) are:

$$\Delta\varepsilon_{(I),40(top)} = \Delta\varepsilon_{r,40} - 487.5 \times \Delta\kappa_{40} = (-1.8 - 487.5 \times 0.512) \times 10^{-6}$$

$$= -251.4 \times 10^{-6}$$

$$\Delta\varepsilon_{(I),40(btm)} = \Delta\varepsilon_{r,40} - (-262.5) \times \Delta\kappa_{40} = (-1.8 + 262.5 \times 0.512) \times 10^{-6}$$

$$= +132.6 \times 10^{-6}$$

The increment of top and bottom fibre stresses in the concrete and the increment of stress in the prestressing steel are obtained from Equations 9.9 and 9.11:

$$\Delta\sigma_{c(I),40(top)} = E_{c(I),40}\Delta\varepsilon_{(I),40(top)} = 31,500 \times (-251.4 \times 10^{-6}) = -7.92 \text{ MPa}$$

$$\Delta\sigma_{c(I),40(btm)} = E_{c(I),40}\Delta\varepsilon_{(I),40(btm)} = 31,500 \times (+132.6 \times 10^{-6}) = +4.18 \text{ MPa}$$

$$\Delta\sigma_{p,40} = E_p\Delta\varepsilon_{p,40} = E_p(\Delta\varepsilon_{r,40} - y_p\Delta\kappa_{40})$$

$$= 200,000 \times \left[-1.8 - (-162.5) \times 0.512 \right] \times 10^{-6} = +16 \text{ MPa}$$

The extreme fibre concrete strains and stresses and the strain and stress in the tendons in the precast girder immediately after placing the in-situ slab at

$t = 40$ days are obtained by summing the respective increments calculated in parts (i), (ii) and (iii):

$$\varepsilon_{(1),40+(top)} = \varepsilon_{(1),4(top)} + \Delta\varepsilon_{(1),(40-4)(top)} + \Delta\varepsilon_{(1),40(top)}$$

$$= (-220.4 - 360.2 - 251.4) \times 10^{-6} = -832.0 \times 10^{-6}$$

$$\varepsilon_{(1),40+(btm)} = \varepsilon_{(1),4(btm)} + \Delta\varepsilon_{(1),(40-4)(btm)} + \Delta\varepsilon_{(1),40(btm)}$$

$$= (-213.6 - 302.6 + 132.6) \times 10^{-6} = -383.6 \times 10^{-6}$$

$$\sigma_{c(1),40+(top)} = \sigma_{c(1),4(top)} + \Delta\sigma_{c(1),(40-4)(top)} + \Delta\sigma_{c(1),40(top)}$$

$$= -5.51 - 0.17 - 7.92 = -13.60 \text{ MPa}$$

$$\sigma_{c(1),40+(btm)} = \sigma_{c(1),4(btm)} + \Delta\sigma_{c(1),(40-4)(btm)} + \Delta\sigma_{c(1),40(btm)}$$

$$= -5.34 + 0.58 + 4.18 = -0.58 \text{ MPa}$$

$$\sigma_{p,40+} = \sigma_{p,4} + \Delta\sigma_{p,(40-4)} + \Delta\sigma_{p,40} = 1290 - 74 + 16 = 1232 \text{ MPa}$$

The stress and strain distributions at age 40 days immediately after the in-situ slab is cast are shown in Figure 9.5d. Stress levels in the precast girder are satisfactory at all stages prior to and immediately after placing the in-situ slab. Cracking will not occur, and compressive stress in the top fibre is not excessive. However, with a sustained compressive stress of -13.60 MPa in the top fibre, a relatively large subsequent creep differential will exist between the precast and the in-situ elements.

(iv) *At t = 60 days (prior to placement of the wearing surface):*
The change of stress and strain during the time interval from $t = 40$ to 60 days is to be calculated here. During this period, the precast section and the in-situ slab are assumed to act compositely. The concrete stress increments in the precast section, calculated in aforementioned parts (i), (ii) and (iii), are applied at different times and are therefore associated with different creep coefficients.

For the stresses applied at $t = 4$ days in part (i), the creep coefficient for this time interval is $\Delta\varphi_{(1)}(60 - 40, 4) = \varphi_{(1)}(60, 4) - \varphi_{(1)}(40, 4) = 0.30$, and from Equations 5.55 and 5.58:

$$\bar{E}_{e(1),60} = \frac{E_{c(1),4}}{1 + \chi_{(1)}(60, 40)\Delta\varphi_{(1)}(60 - 40, 4)} = \frac{25,000}{1 + 0.8 \times 0.30} = 20,161 \text{ MPa}$$

$$\bar{F}_{e(1),60} = \frac{\Delta\varphi_{(1)}(60 - 40, 4)[\chi_{(1)}(60, 40) - 1]}{1 + \chi_{(1)}(60, 40)\Delta\varphi_{(1)}(60 - 40, 4)} = \frac{0.30 \times (0.8 - 1.0)}{1.0 + 0.8 \times 0.30} = -0.0484$$

The stress increment calculated in part (ii), which is in fact gradually applied between $t = 4$ and 40 days, may be accounted for by assuming that it is suddenly applied at $t = 4$ days and using the reduced creep coefficient given by $\chi_{(1)}(40, 1)[\varphi_{(1)}(60, 4) - \varphi_{(1)}(40, 4)] = 0.24$, and from Equations 5.55 and 5.58:

$$\bar{E}_{e(1),60} = \frac{25,000}{1 + 0.8 \times 0.24} = 20,973\,\text{MPa}$$

$$\bar{F}_{e(1),60} = \frac{0.24 \times (0.8 - 1.0)}{1.0 + 0.8 \times 0.24} = -0.0403$$

For the stress increment calculated in part (iii) (and caused by M_3), the appropriate creep coefficient for the precast girder is $\varphi_{(1)}(60, 40) = 0.5$, and from Equations 5.55 and 5.58:

$$\bar{E}_{e(1),60} = \frac{31,500}{1 + 0.8 \times 0.5} = 22,500\,\text{MPa}$$

$$\bar{F}_{e(1),60} = \frac{0.5 \times (0.8 - 1.0)}{1.0 + 0.8 \times 0.5} = -0.0714$$

For the in-situ slab, the creep coefficient used in this time interval is $\varphi_{(2)}(60, 40) = 0.8$, and from Equations 5.55 and 5.58:

$$\bar{E}_{e(2),60} = \frac{18,000}{1 + 0.8 \times 0.8} = 10,976\,\text{MPa}$$

$$\bar{F}_{e(2),60} = \frac{0.8 \times (0.8 - 1.0)}{1.0 + 0.8 \times 0.8} = -0.0976$$

The shrinkage strains that develop in the precast section and the in-situ slab during this time interval are, respectively, $\varepsilon_{cs(1),60} - \varepsilon_{cs(1),40} = -50 \times 10^{-6}$ and $\varepsilon_{cs(2),60} = -120 \times 10^{-6}$.

The creep coefficient associated with this time interval for the prestressing steel is $\varphi_{p,60} - \varphi_{p,40} = 0.005$.

The section properties of the concrete part of the precast girder (element 1) and the in-situ slab (element 2) with respect to the centroidal axis of the precast girder are:

$A_{c(1)} = 358,500\,\text{mm}^2,\quad B_{c(1)} = 243,750\,\text{mm}^3,$

$I_{c(1)} = 18.35 \times 10^9\,\text{mm}^4$

$A_{c(2)} = 357,000\,\text{mm}^2,\quad B_{c(2)} = 200.8 \times 10^6\,\text{mm}^3,$

$I_{c(2)} = 113.6 \times 10^9\,\text{mm}^4$

To determine the internal actions required to restrain creep, shrinkage and relaxation, the initial elastic strain distribution caused by each of the previously calculated stress increments in each concrete element must be determined.

In the in-situ slab: $\varepsilon_{i,r(2),40} = 0$ and $\kappa_{i,(2),40} = 0$ since the slab at $t = 40$ days is unloaded.

In the precast section: For the stresses applied at 4 days, calculated in part (i): $\varepsilon_{r(1),4} = -216.0 \times 10^{-6}$; $\kappa_{(1),4} = +0.00902 \times 10^{-6}$ mm^{-1}. For the stress increment calculated in part (ii) and assumed to be applied at 4 days, the increments of instantaneous elastic strain at the top and bottom of the precast section are $\Delta\varepsilon_{(1),4(top)} = \Delta\sigma_{c(1),(40-4)(top)}/E_{c(1),4} = -0.17/25,000 = -6.92 \times 10^{-6}$ and $\Delta\varepsilon_{i(1),4(btm)} = \Delta\sigma_{c(1),(40-4)(btm)}/E_{c(1),4} = +0.58/25,000 = 23.1 \times 10^{-6}$. Therefore, the increment of elastic curvature is $\Delta\kappa_{i,(1)} = [-(-6.92 \times 10^{-6}) + 23.1 \times 10^{-6}]/750 = +0.040 \times 10^{-6}$ mm^{-1} and the instantaneous strain at the reference axis is $\Delta\varepsilon_{i,r(1)} = +12.6 \times 10^{-6}$. For the stress increment applied to the precast element at 40 days [part (iii)], $\Delta\varepsilon_{r(1),40} = -1.8 \times 10^{-6}$ and $\Delta\kappa_{(1),40} = +0.512 \times 10^{-6}$ mm^{-1}.

The actions required to restrain creep due to these initial stresses are obtained from Equation 9.42:

$$f_{cr,60} = -0.0484 \times 25,000 \begin{bmatrix} 358,500 \times (-216.0 \times 10^{-6}) - 243,750 \times 0.00902 \times 10^{-6} \\ -243,750 \times (-216.0 \times 10^{-6}) + 18.35 \times 10^{9} \times 0.00902 \times 10^{-6} \end{bmatrix}$$

$$-0.0403 \times 25,000 \begin{bmatrix} 358,500 \times (+12.6 \times 10^{-6}) - 243,750 \times 0.040 \times 10^{-6} \\ -243,750 \times (+12.6 \times 10^{-6}) + 18.35 \times 10^{9} \times 0.040 \times 10^{-6} \end{bmatrix}$$

$$-0.0714 \times 31,500 \begin{bmatrix} 358,500 \times (-1.8 \times 10^{-6}) - 243,750 \times 0.512 \times 10^{-6} \\ -243,750 \times (-1.8 \times 10^{-6}) + 18.35 \times 10^{9} \times 0.512 \times 10^{-6} \end{bmatrix}$$

$$= \begin{bmatrix} 90.88 \times 10^{3} \text{ kN} \\ -22.14 \times 10^{6} \text{ kNm} \end{bmatrix}$$

and the actions required to restrain shrinkage in each concrete element and relaxation of the tendons during this time period are obtained from Equations 9.43 and 9.45, respectively:

$$f_{cs,60} = \begin{bmatrix} 358,500 \times 22,500 \times (-50 \times 10^{-6}) + 357,000 \times 10,976 \times (-120 \times 10^{-6}) \\ -243,750 \times 22,500 \times (-50 \times 10^{-6}) - 200.8 \times 10^{6} \times 10,976 \times (-120 \times 10^{-6}) \end{bmatrix}$$

$$= \begin{bmatrix} -873.5 \times 10^{3} \text{ kN} \\ 264.5 \times 10^{6} \text{ kNm} \end{bmatrix}$$

$$f_{p.rel,60} = \begin{bmatrix} 9.7 \times 10^{3} \text{ kN} \\ 1.573 \times 10^{6} \text{ kNm} \end{bmatrix}$$

The cross-sectional rigidities of the composite cross-section for the period τ_k = 40–60 days are obtained from Equations 9.49 through 9.51:

$$R_{A,60} = 358{,}500 \times 22{,}500 + 357{,}000 \times 10{,}975 + 3{,}000 \times 200{,}000 + 1{,}500 \times 200{,}000$$

$$= 12{,}885 \times 10^6 \text{ N}$$

$$R_{B,60} = 243{,}750 \times 22{,}500 + 200.8 \times 10^6 \times 10{,}975 + 562.5 \times 3{,}000 \times 200{,}000$$

$$+ (-162.5) \times 1{,}500 \times 200{,}000 = 2.498 \times 10^{12} \text{ Nmm}$$

$$R_{I,60} = 18.35 \times 10^9 \times 22{,}500 + 113.6 \times 10^9 \times 10{,}975 + 562.5^2 \times 3{,}000 \times 200{,}000$$

$$+ (-162.5)^2 \times 1{,}500 \times 200{,}000 = 1{,}857.9 \times 10^{12} \text{ Nmm}^2$$

and from Equation 9.48:

$$\mathbf{F}_{60} = \frac{1}{R_{A,60}R_{I,60} - R_{B,60}^2} \begin{bmatrix} R_{I,60} & R_{B,60} \\ R_{B,60} & R_{A,60} \end{bmatrix}$$

$$= \begin{bmatrix} 105.0 \times 10^{-12} \text{ N}^{-1} & 141.2 \times 10^{-15} \text{ N}^{-1}\text{mm}^{-1} \\ 141.2 \times 10^{-15} \text{ N}^{-1}\text{mm}^{-1} & 0.728 \times 10^{-15} \text{ N}^{-1}\text{mm}^{-2} \end{bmatrix}$$

The vector of the change in strain that occurs between 40 and 60 days on the composite section is obtained from Equation 9.46:

$$\Delta\varepsilon_{(60-40)} = \mathbf{F}_{60}(-\mathbf{f}_{cr,60} + \mathbf{f}_{cs,60} + \mathbf{f}_{p.rel,60})$$

$$= \begin{bmatrix} 105.0 \times 10^{-12} & 141.2 \times 10^{-15} \\ 141.2 \times 10^{-15} & 0.728 \times 10^{-15} \end{bmatrix} \begin{bmatrix} (-90.88 - 873.5 + 9.7) \times 10^3 \\ (+22.14 + 264.5 + 1.573) \times 10^6 \end{bmatrix}$$

$$= \begin{bmatrix} -59.5 \times 10^{-6} \\ 0.0753 \times 10^{-6} \end{bmatrix}$$

The increment of strain at the reference axis and the increment of curvature that occur between τ_k = 40 and 60 days are therefore $\Delta\varepsilon_{r,(60-40)} = -59.5 \times 10^{-6}$ and $\Delta\kappa_{(60-40)} = +0.0753 \times 10^{-6}$ mm^{-1}, respectively, and the increments of strains at the top fibre of the precast section (at y = +487.5 mm) and at the bottom fibre (at y = −262.5 mm) are:

$$\Delta\varepsilon_{(1)60-(top)} = [-59.5 - (487.5 \times 0.0753)] \times 10^{-6} = -96.2 \times 10^{-6}$$

$$\Delta\varepsilon_{(1)60-(btm)} = [-59.5 + 262.5 \times 0.0753] \times 10^{-6} = -39.7 \times 10^{-6}$$

The increments of strains at the top fibre of the in-situ slab (at $y = +637.5$ mm) and at the bottom fibre (at $y = +487.5$ mm) are:

$$\Delta\varepsilon_{(2)60-(top)} = [-59.5 - (637.5 \times 0.0753)] \times 10^{-6} = -107.5 \times 10^{-6}$$

$$\Delta\varepsilon_{(2)60-(btm)} = [-59.5 - (487.5 \times 0.0753)] \times 10^{-6} = -96.2 \times 10^{-6}$$

The increments of concrete stress that develop between $\tau_k = 40$ and 60 days are calculated using Equation 9.52. In the precast element:

$$\Delta\sigma_{c(1),60-(top)} = \bar{E}_{e(1),60}\left(\Delta\varepsilon_{(1),60-(top)} - \Delta\varepsilon_{cs(1),60-40}\right) + \bar{F}_{e(1),60}\sigma_{c(1),40+(top)}$$

$$= 22,500 \times [-96.2 - (-50)] \times 10^{-6} + (-0.0714) \times (-13.60) = -0.07 \text{ MPa}$$

$$\Delta\sigma_{c(1),60-(btm)} = \bar{E}_{e(1),60}\left(\Delta\varepsilon_{(1),60-(btm)} - \Delta\varepsilon_{cs(1),60-40}\right) + \bar{F}_{e(1),60}\sigma_{c(1),40+(btm)}$$

$$= 22,500 \times [-39.7 - (-50)] \times 10^{-6} + (-0.0714) \times (-0.58) = +0.27 \text{ MPa}$$

and in the in-situ slab:

$$\Delta\sigma_{c(2),60-(top)} = \bar{E}_{e(2),60}\left(\Delta\varepsilon_{(2),60-(top)} - \Delta\varepsilon_{cs(2),60}\right) + \bar{F}_{e(2),60}\sigma_{c(2),40+(top)}$$

$$= 10,976 \times [-107.5 - (-120)] \times 10^{-6} - 0.0976 \times 0 = +0.14 \text{ MPa}$$

$$\Delta\sigma_{c(2),60-(btm)} = \bar{E}_{e(2),60}\left(\Delta\varepsilon_{(2),60-(btm)} - \Delta\varepsilon_{cs(2),60}\right) + \bar{F}_{e(2),60}\sigma_{c(2),40+(btm)}$$

$$= 10,976 \times [-96.2 - (-120)] \times 10^{-6} - 0.0976 \times 0 = +0.26 \text{ MPa}$$

The increment of stress in the bonded prestressing steel that develops between 40 and 60 days is:

$$\Delta\sigma_{p,60-} = E_p\left(\Delta\varepsilon_{r,(60-40)} - y_p\Delta\kappa_{(60-40)} - \Delta\varepsilon_{p,rel(60-40)}\right) = -15.9 \text{ MPa}$$

and in the reinforcing steel in the in-situ slab:

$$\Delta\sigma_{s,60-} = E_s\left(\Delta\varepsilon_{r,(60-40)} - y_s\Delta\kappa_{(60-40)}\right) = -20.4 \text{ MPa}$$

The total stresses and strains at age 60 days before the application of the wearing surface are:

In the precast girder:

$$\varepsilon_{(1),60-(top)} = \varepsilon_{(1),40+(top)} + \Delta\varepsilon_{(1)60-(top)} = -832.0 \times 10^{-6} - 96.2 \times 10^{-6}$$

$$= -928.2 \times 10^{-6}$$

$$\varepsilon_{(1),60-(btm)} = \varepsilon_{(1),40+(btm)} + \Delta\varepsilon_{(1)60-(btm)} = -423.3 \times 10^{-6}$$

$$\sigma_{c(1),60-(top)} = \sigma_{c(1),40+(top)} + \Delta\sigma_{c(1),60-(top)} = -13.60 - 0.07 = -13.67 \text{ MPa}$$

$$\sigma_{c(1),60-(btm)} = \sigma_{c(1),40+(btm)} + \Delta\sigma_{c(1),60-(btm)} = -0.31 \text{ MPa}$$

$$\sigma_{p,60-} = \sigma_{p,40+} + \Delta\sigma_{p,60-} = 1216 \text{ MPa}$$

In the in-situ slab:

$$\varepsilon_{(2),60-(\text{top})} = \Delta\varepsilon_{(2)60-(\text{top})} = -107.5 \times 10^{-6}$$

$$\varepsilon_{(2),60-(\text{btm})} = \Delta\varepsilon_{(2)60-(\text{btm})} = -96.2 \times 10^{-6}$$

$$\sigma_{c(2),60-(\text{top})} = \Delta\sigma_{c(2),60-(\text{top})} = +0.14 \text{ MPa}$$

$$\sigma_{c(2),60-(\text{btm})} = \Delta\sigma_{c(2),60-(\text{btm})} = +0.26 \text{ MPa}$$

$$\sigma_{s,60-} = \Delta\sigma_{s,60-} = -20.4 \text{ MPa}$$

The stress and strain distributions at age 60 days immediately before the wearing surface is placed are shown in Figure 9.5e.

There is a complex interaction taking place between the two concrete elements. The in-situ slab is shrinking at a faster rate than the precast element and, if this were the only effect, the in-situ slab would suffer a tensile restraining force and an equal and opposite compressive force would be imposed on the precast girder. Because of the high initial compressive stresses in the top fibres of the precast section, however, the precast concrete at the element interface is creeping more than the in-situ concrete and, as a result, the in-situ slab is being compressed by the creep deformations in the precast girder, resulting in a decrease in the tension caused by restrained shrinkage. In this example, the magnitude of the tensile force on the in-situ slab as a result of restrained shrinkage is a little larger than the magnitude of the compressive force due to the creep differential. The result is that relatively small tensile stresses develop in the in-situ slab with time and the magnitude of the compressive stress in the top fibre of the precast girder increases slightly.

(v) At t = 60 days *(immediately after placement of the wearing surface):*
The instantaneous increments of stress and strain caused by $M_4 = 150$ kNm applied to the composite section at age 60 days are calculated here. The short-term rigidities of the composite cross-section at age 60 days (with respect to the reference axis at the centroid of the precast element) are:

$$R_{A,60} = A_{c(1)}E_{c(1),60} + A_{c(2)}E_{c(2),60} + A_pE_p + A_sE_s$$

$$= 358,500 \times 33,000 + 357,000 \times 25,000 + 1,500 \times 200,000 + 3,000 \times 200,000$$

$$= 21,660 \times 10^6 \text{ N}$$

$$R_{B,60} = B_{c(1)}E_{c(1),60} + B_{c(2)}E_{c(2),60} + y_pA_pE_p + y_sA_sE_s = 5.317 \times 10^{12} \text{ Nmm}$$

$$R_{I,60} = I_{c(1)}E_{c(1),60} + I_{c(2)}E_{c(2),60} + y_p^2A_pE_p + y_s^2A_sE_s = 3,644 \times 10^{12} \text{ Nmm}^2$$

and from Equation 9.20:

$$\mathbf{F}_{40} = \frac{1}{R_{A,60}R_{I,60} - R_{B,60}^2}\begin{bmatrix} R_{I,60} & R_{B,60} \\ R_{B,60} & R_{A,60} \end{bmatrix} = \begin{bmatrix} 71.96 \times 10^{-12} & 0.105 \times 10^{-12} \\ 0.105 \times 10^{-12} & 0.428 \times 10^{-15} \end{bmatrix}$$

For the load increment applied at 60 days, the vector of internal actions is:

$$\mathbf{r}_{ext} = \begin{bmatrix} 0 \\ M_4 \end{bmatrix} = \begin{bmatrix} 0 \\ 150 \times 10^6 \end{bmatrix}$$

and the vector of instantaneous strain caused by the application of M_4 at age 60 days is:

$$\Delta \varepsilon_{60} = \mathbf{F}_{60}\mathbf{r}_{ext} = \begin{bmatrix} 71.96 \times 10^{-12} & 0.105 \times 10^{-12} \\ 0.105 \times 10^{-12} & 0.428 \times 10^{-15} \end{bmatrix}\begin{bmatrix} 0 \\ 150 \times 10^6 \end{bmatrix} = \begin{bmatrix} +15.7 \times 10^{-6} \\ 0.0641 \times 10^{-6} \end{bmatrix}$$

The increment of instantaneous strain at the reference axis and the increment of curvature caused by the application of M_4 at age 60 days are therefore:

$$\Delta \varepsilon_{r,60} = +15.7 \times 10^{-6} \quad \text{and} \quad \Delta \kappa_{60} = +0.0641 \times 10^{-6} \text{ mm}^{-1}$$

and the increment of strains at the top fibre of the precast section (at $y = +487.5$ mm) and at the bottom fibre (at $y = -262.5$ mm) are:

$$\Delta \varepsilon_{(1)60+(top)} = [+15.7 - (487.5 \times 0.0641)] \times 10^{-6} = -15.5 \times 10^{-6}$$

$$\Delta \varepsilon_{(1)60+(btm)} = [+15.7 + (262.5 \times 0.0641)] \times 10^{-6} = +32.6 \times 10^{-6}$$

The increment of strains at the top fibre of the in-situ slab (at $y = +637.5$ mm) and at the bottom fibre (at $y = +487.5$ mm) are:

$$\Delta \varepsilon_{(2)60+(top)} = [+15.7 - (637.5 \times 0.0641)] \times 10^{-6} = -25.1 \times 10^{-6}$$

$$\Delta \varepsilon_{(2)60+(btm)} = [+15.7 - (487.5 \times 0.0641)] \times 10^{-6} = -15.5 \times 10^{-6}$$

The increments of concrete stress that develop at $\tau_k = 60$ days due to the wearing surface are the following:

In the precast element:

$$\Delta \sigma_{c(1),60+(top)} = E_{c(1),60}\Delta \varepsilon_{(1)60+(top)} = -0.51 \text{ MPa}$$

$$\Delta \sigma_{c(1),60+(btm)} = E_{c(1),60}\Delta \varepsilon_{(1)60+(btm)} = +1.08 \text{ MPa}$$

In the in-situ slab:

$$\Delta\sigma_{c(2),60+(top)} = E_{c(2),60}\Delta\varepsilon_{(2)60+(top)} = -0.63\,\text{MPa}$$

$$\Delta\sigma_{c(2),60+(btm)} = E_{c(2),60}\Delta\varepsilon_{(2)60+(btm)} = -0.39\,\text{MPa}$$

The increment of stress in the bonded prestressing steel at $\tau_k = 60$ days due to the wearing surface is:

$$\Delta\sigma_{p,60+} = E_p(\Delta\varepsilon_{r,60} - y_p\Delta\kappa_{60}) = +5.2\,\text{MPa}$$

and in the reinforcing steel in the in-situ slab:

$$\Delta\sigma_{s,60+} = E_s(\Delta\varepsilon_{r,60} - y_s\Delta\kappa_{60}) = -4.1\,\text{MPa}$$

The total stresses and strains at age 60 days after the application of the wearing surface are:

In the precast girder:

$$\varepsilon_{(1),60+(top)} = \varepsilon_{(1),60-(top)} + \Delta\varepsilon_{(1)60+(top)} = -943.7 \times 10^{-6}$$

$$\varepsilon_{(1),60+(btm)} = \varepsilon_{(1),60-(btm)} + \Delta\varepsilon_{(1)60+(btm)} = -390.7 \times 10^{-6}$$

$$\sigma_{c(1),60+(top)} = \sigma_{c(1),60-(top)} + \Delta\sigma_{c(1),60+(top)} = -14.18\,\text{MPa}$$

$$\sigma_{c(1),60+(btm)} = \sigma_{c(1),60-(btm)} + \Delta\sigma_{c(1),60+(btm)} = +0.77\,\text{MPa}$$

$$\sigma_{p,60+} = \sigma_{p,60-} + \Delta\sigma_{p,60+} = 1220\,\text{MPa}$$

In the in-situ slab:

$$\varepsilon_{(2),60+(top)} = \varepsilon_{(2),60-(top)} + \Delta\varepsilon_{(2)60+(top)} = -132.6 \times 10^{-6}$$

$$\varepsilon_{(2),60+(btm)} = \varepsilon_{(2),60-(btm)} + \Delta\varepsilon_{(2)60+(btm)} = -111.7 \times 10^{-6}$$

$$\sigma_{c(2),60+(top)} = \sigma_{c(2),60-(top)} + \Delta\sigma_{c(2),60-(top)} = -0.49\,\text{MPa}$$

$$\sigma_{c(2),60+(btm)} = \sigma_{c(2),60-(btm)} + \Delta\sigma_{c(2),60+(btm)} = -0.13\,\text{MPa}$$

$$\sigma_{s,60+} = \sigma_{s,60-} + \Delta\sigma_{s,60+} = -24.5\,\text{MPa}$$

and the stress and strain distributions at age 60 days immediately after the wearing surface has been placed are shown in Figure 9.5f.

(vi) At t = ∞:
The change of stress and strain on the composite cross-section during the time interval from $t = 60$ days to $t = \infty$ is to be calculated here.

The relevant creep coefficients for each of the previously calculated stress increments determined in parts (i)–(v), together with the corresponding age-adjusted effective moduli and creep factors (from Equations 5.55 and 5.58), are as follows:

For the precast girder:
Part (i): $\varphi_{(1)}(\infty - 60, 4) = \varphi_{(1)}(\infty, 4) - \varphi_{(1)}(60, 4) = 1.2$

$$E_{c(1),4} = 25{,}000 \text{ MPa}$$

$$\overline{E}_{e(1),\infty} = \frac{E_{c(1),4}}{1 + \chi_{(1)}(\infty,60)\Delta\varphi_{(1)}(\infty - 60,4)} = \frac{25{,}000}{1 + 0.65 \times 1.2}$$

$$= 14{,}045 \text{ MPa}$$

$$\overline{F}_{e(1),\infty} = \frac{1.2 \times (0.65 - 1.0)}{1.0 + 0.65 \times 1.2} = -0.236$$

Part (ii): $\chi_{(1)}(40, 4)[\varphi_{(1)}(\infty, 4) - \varphi_{(1)}(60, 4)] = 0.96$

$$E_{c(1),4} = 25{,}000 \text{ MPa}, \quad \overline{E}_{e(1),\infty} = 15{,}394 \text{ MPa} \quad \text{and} \quad \overline{F}_{e(1),\infty} = -0.2069$$

Part (iii): $\varphi_{(1)}(\infty, 40) - \varphi_{(1)}(60, 40) = 1.1$

$$E_{c(1),40} = 31{,}500 \text{ MPa}, \quad \overline{E}_{e(1),\infty} = 18{,}367 \text{ MPa} \quad \text{and} \quad \overline{F}_{e(1),\infty} = -0.2245$$

Part (iv): $\chi_{(1)}(60, 40)[\varphi_{(1)}(\infty, 40) - \varphi_{(1)}(60, 40)] = 0.88$

$$E_{c(1),40} = 31{,}500 \text{ MPa}, \quad \overline{E}_{e(1),\infty} = 20{,}038 \text{ MPa} \quad \text{and} \quad \overline{F}_{e(1),\infty} = -0.1958$$

Part (v): $\varphi_{(1)}(\infty, 60) = 1.2$

$$E_{c(1),60} = 33{,}000 \text{ MPa}, \quad \overline{E}_{e(1),\infty} = 18{,}539 \text{ MPa} \quad \text{and} \quad \overline{F}_{e(1),\infty} = -0.236$$

For the in-situ slab:

Part (iv): $\chi_{(2)}(60, 40) [\varphi_{(2)}(\infty, 40) - \varphi_{(2)}(60, 40)] = 1.76$

$$E_{c(2),40} = 18,000 \text{ MPa}, \quad \bar{E}_{e(2),\infty} = 8,396 \text{ MPa} \quad \text{and} \quad \bar{F}_{e(2),\infty} = -0.2873$$

Part (v): $\varphi_{(2)}(\infty, 60) = 2.0$

$$E_{c(2),60} = 25,000 \text{ MPa}, \quad \bar{E}_{e(2),\infty} = 10,870 \text{ MPa} \quad \text{and} \quad \bar{F}_{e(2),\infty} = -0.3043$$

Note that the stress increments calculated in parts (ii) and (iv) are accounted for by assuming that they are suddenly applied at $t = 40$ and 60 days, respectively, using the appropriate reduced creep coefficients.

The shrinkage strains that develops during the time period after $t = 60$ days are:

$$\varepsilon_{cs(1),\infty} - \varepsilon_{cs(1),40} = -300 \times 10^{-6} \quad \text{and} \quad \varepsilon_{cs(2),\infty} - \varepsilon_{cs(2),40} = -480 \times 10^{-6}$$

and the creep coefficient associated with this time interval for the prestressing steel is $\varphi_{p,\infty} - \varphi_{p,60} = 0.01$.

To determine the internal actions required to restrain creep, shrinkage and relaxation, the initial elastic strain distribution caused by each of the previously calculated stress increments in each concrete element must be determined. The elastic strains due to the stress increments applied in parts (i)–(v) are:

Part (i): $\varepsilon_{r(1),4} = -216.0 \times 10^{-6} \, \kappa_{(1),4} = +0.00902 \times 10^{-6} \text{ mm}^{-1}$

Part (ii): $\Delta\varepsilon_{i,r(1),4} = +12.6 \times 10^{-6} \, \Delta\kappa_{i(1)} = +0.0402 \times 10^{-6} \text{ mm}^{-1}$

Part (iii): $\Delta\varepsilon_{i,r(1),40} = -1.8 \times 10^{-6} \, \Delta\kappa_{i(1),40} = +0.512 \times 10^{-6} \text{ mm}^{-1}$

Part (iv): $\Delta\varepsilon_{i,r(1),40} = +4.9 \times 10^{-6} \, \Delta\kappa_{i(1),40} = +0.0145 \times 10^{-6} \text{ mm}^{-1}$

$$\Delta\varepsilon_{i,r(2),40} = +36.9 \times 10^{-6} \, \Delta\kappa_{i(2),40} = +0.0459 \times 10^{-6} \text{ mm}^{-1}$$

Part (v): $\Delta\varepsilon_{i,r(1),40} = +15.7 \times 10^{-6} \, \Delta\kappa_{i(1),40} = +0.0641 \times 10^{-6} \text{ mm}^{-1}$

$$\Delta\varepsilon_{i,r(2),40} = +15.7 \times 10^{-6} \, \Delta\kappa_{i(2),40} = +0.0641 \times 10^{-6} \text{ mm}^{-1}$$

The actions required to restrain creep due to these initial stresses during the period after $t = 60$ days are obtained from Equation 9.42:

$$\mathbf{f}_{cr,\infty} = -0.2360 \times 25,000 \begin{bmatrix} 358,500 \times (-216.0 \times 10^{-6}) - 243,750 \times 0.00902 \times 10^{-6} \\ -243,750 \times (-216.0 \times 10^{-6}) + 18.35 \times 10^{9} \times 0.00902 \times 10^{-6} \end{bmatrix}$$

$$-0.2069 \times 25,000 \begin{bmatrix} 358,500 \times (+12.6 \times 10^{-6}) - 243,750 \times 0.0402 \times 10^{-6} \\ -243,750 \times (+12.6 \times 10^{-6}) + 18.35 \times 10^{9} \times 0.0402 \times 10^{-6} \end{bmatrix}$$

$$-0.2245 \times 31,500 \begin{bmatrix} 358,500 \times (-1.8 \times 10^{-6}) - 243,750 \times 0.512 \times 10^{-6} \\ -243,750 \times (-1.8 \times 10^{-6}) + 18.35 \times 10^{9} \times 0.512 \times 10^{-6} \end{bmatrix}$$

$$-0.1958 \times 31,500 \begin{bmatrix} 358,500 \times (+4.9 \times 10^{-6}) - 243,750 \times 0.0144 \times 10^{-6} \\ -243,750 \times (+4.9 \times 10^{-6}) + 18.35 \times 10^{9} \times 0.0145 \times 10^{-6} \end{bmatrix}$$

$$-0.2873 \times 18,000 \begin{bmatrix} 357,000 \times (+36.9 \times 10^{-6}) - 200.8 \times 10^{6} \times 0.0459 \times 10^{-6} \\ -200.8 \times 10^{6} \times (+36.9 \times 10^{-6}) + 113.6 \times 10^{9} \times 0.0459 \times 10^{-6} \end{bmatrix}$$

$$-0.2360 \times 33,000 \begin{bmatrix} 358,500 \times (+15.7 \times 10^{-6}) - 243,750 \times 0.0641 \times 10^{-6} \\ -243,750 \times (+15.7 \times 10^{-6}) + 18.35 \times 10^{9} \times 0.0641 \times 10^{-6} \end{bmatrix}$$

$$-0.3043 \times 25,000 \begin{bmatrix} 357,000 \times (+15.7 \times 10^{-6}) - 200.8 \times 10^{6} \times 0.0641 \times 10^{-6} \\ -200.8 \times 10^{6} \times (+15.7 \times 10^{-6}) + 113.6 \times 10^{9} \times 0.0641 \times 10^{-6} \end{bmatrix}$$

$$= \begin{bmatrix} 419.0 \times 10^{3} \text{ kN} \\ -102.4 \times 10^{6} \text{ kNm} \end{bmatrix}$$

The actions required to restrain shrinkage in each concrete element and relaxation of the tendons during this time period are obtained from Equations 9.43 and 9.45, respectively:

$$\mathbf{f}_{cs,\infty} = \begin{bmatrix} 358,500 \times 18,539 \times (-300 \times 10^{-6}) + 357,000 \times 10,870 \times (-480 \times 10^{-6}) \\ -243,750 \times 18,539 \times (-300 \times 10^{-6}) - 200.8 \times 10^{6} \times 10,870 \times (-480 \times 10^{-6}) \end{bmatrix}$$

$$= \begin{bmatrix} -3,856.5 \times 10^{3} \text{ kN} \\ 1,049.1 \times 10^{6} \text{ kNm} \end{bmatrix}$$

$$\mathbf{f}_{p.rel,\infty} = \begin{bmatrix} 19.4 \times 10^{3} \text{ kN} \\ 3.15 \times 10^{6} \text{ kNm} \end{bmatrix}$$

The cross-sectional rigidities of the composite cross-section for the period $\tau_k = 60$ days to time infinity are obtained from Equations 9.49 through 9.51:

$$R_{A,\infty} = 358{,}500 \times 18{,}539 + 357{,}000 \times 10{,}870 + 3{,}000 \times 200{,}000 + 1{,}500 \times 200{,}000$$

$$= 11{,}427 \times 10^6 \ N$$

$$R_{B,\infty} = 243{,}750 \times 18{,}539 + 200.8 \times 10^6 \times 10{,}870 + 562.5 \times 3{,}000 \times 200{,}000$$

$$+ (-162.5) \times 1{,}500 \times 200{,}000 = 2.476 \times 10^{12} \ Nmm$$

$$R_{I,\infty} = 18.35 \times 10^9 \times 18{,}539 + 113.6 \times 10^9 \times 10{,}870 + 562.5^2 \times 3{,}000 \times 200{,}000$$

$$+ (-162.5)^2 \times 1{,}500 \times 200{,}000 = 1{,}773.2 \times 10^{12} \ Nmm^2$$

and from Equation 9.48:

$$\mathbf{F}_\infty = \frac{1}{R_{A,\infty}R_{I,\infty} - R_{B,\infty}^2} \begin{bmatrix} R_{I,\infty} & R_{B,\infty} \\ R_{B,\infty} & R_{A,\infty} \end{bmatrix}$$

$$= \begin{bmatrix} 125.5 \times 10^{-12} & 175.2 \times 10^{-15} \\ 175.2 \times 10^{-15} & 0.8086 \times 10^{-15} \end{bmatrix}$$

The vector of the change in strain that occurs after 60 days on the composite section due to creep, shrinkage and relaxation is obtained from Equation 9.46:

$$\Delta\varepsilon_{(\infty-60)} = \mathbf{F}_\infty(-\mathbf{f}_{cr,\infty} + \mathbf{f}_{cs,\infty} + \mathbf{f}_{p.rel,\infty})$$

$$= \begin{bmatrix} 125.5 \times 10^{-12} & 175.2 \times 10^{-15} \\ 175.2 \times 10^{-15} & 0.8086 \times 10^{-15} \end{bmatrix} \begin{bmatrix} (-419.0 - 3856.5 + 19.4) \times 10^3 \\ (+102.4 + 1049.1 + 3.15) \times 10^6 \end{bmatrix}$$

$$= \begin{bmatrix} -331.8 \times 10^{-6} \\ 0.1879 \times 10^{-6} \end{bmatrix}$$

The increment of strain at the reference axis and the increment of curvature that occur between $\tau_k = 60$ days and time infinity are $\Delta\varepsilon_{r,\infty-60} = -331.8 \times 10^{-6}$ and $\Delta\kappa_{\infty-60} = +0.1934 \times 10^{-6} \ mm^{-1}$, respectively, and the increments of strains at the top fibre of the precast section (at $y = +487.5$ mm) and at the bottom fibre (at $y = -262.5$ mm) are:

$$\Delta\varepsilon_{(1),\infty(top)} = [-331.8 - (487.5 \times 0.1879)] \times 10^{-6} = -423.4 \times 10^{-6}$$

$$\Delta\varepsilon_{(1),\infty(btm)} = [-331.8 + 262.5 \times 0.1879] \times 10^{-6} = -282.4 \times 10^{-6}$$

The increments of strains at the top fibre of the in-situ slab (at $y = +637.5$ mm) and at the bottom fibre (at $y = +487.5$ mm) are:

$$\Delta\varepsilon_{(2),\infty(\text{top})} = [-331.8 - (637.5 \times 0.1879)] \times 10^{-6} = -451.5 \times 10^{-6}$$

$$\Delta\varepsilon_{(2),\infty(\text{btm})} = [-331.8 - (487.5 \times 0.1879)] \times 10^{-6} = -423.4 \times 10^{-6}$$

The increments of concrete stress that develop between $\tau_k = 60$ days and time infinity are calculated using Equation 9.52. In the precast element:

$$\Delta\sigma_{c(1),\infty(\text{top})} = \overline{E}_{e(1),\infty}\left(\Delta\varepsilon_{(1),\infty(\text{top})} - \Delta\varepsilon_{cs(1),\infty-60}\right) + \overline{F}_{e(1),\infty}\sigma_{c(1),60+(\text{top})}$$

$$= 18{,}539 \times [-423.4 - (-300)] \times 10^{-6} + (-0.236) \times (-14.18) = +1.06\ \text{MPa}$$

$$\Delta\sigma_{c(1),\infty(\text{btm})} = \overline{E}_{e(1),\infty}\left(\Delta\varepsilon_{(1),\infty(\text{btm})} - \Delta\varepsilon_{cs(1),\infty-60}\right) + \overline{F}_{e(1),\infty}\sigma_{c(1),60+(\text{btm})}$$

$$= 18{,}539 \times [-282.4 - (-300)] \times 10^{-6} + (-0.236) \times (+0.77) = +0.15\ \text{MPa}$$

and in the in-situ slab:

$$\Delta\sigma_{c(2),\infty(\text{top})} = \overline{E}_{e(2),\infty}\left(\Delta\varepsilon_{(2),\infty(\text{top})} - \Delta\varepsilon_{cs(2),\infty-60}\right) + \overline{F}_{e(2),\infty}\sigma_{c(2),60+(\text{top})}$$

$$= 10{,}870 \times [-451.5 - (-480)] \times 10^{-6} - 0.3043 \times (-0.49) = +0.46\ \text{MPa}$$

$$\Delta\sigma_{c(2),\infty(\text{btm})} = \overline{E}_{e(2),\infty}\left(\Delta\varepsilon_{(2),\infty(\text{btm})} - \Delta\varepsilon_{cs(2),\infty-60}\right) + \overline{F}_{e(2),\infty}\sigma_{c(2),60+(\text{btm})}$$

$$= 10{,}870 \times [-423.4 - (-480)] \times 10^{-6} - 0.3043 \times (-0.13) = +0.65\ \text{MPa}$$

The increment of stress in the bonded prestressing steel that develops between 60 days and time infinity is:

$$\Delta\sigma_{p,\infty} = E_p\left(\Delta\varepsilon_{r,(\infty-60)} - y_p\Delta\kappa_{(\infty-60)} - \Delta\varepsilon_{p,\text{rel},(\infty-60)}\right) = -73.2\ \text{MPa}$$

and in the reinforcing steel in the in-situ slab:

$$\Delta\sigma_{s,\infty} = E_s\left(\Delta\varepsilon_{r,(\infty-60)} - y_s\Delta\kappa_{(\infty-60)}\right) = -87.5\ \text{MPa}$$

The total stresses and strains at time infinity are summarized in the following:

In the precast girder:

$$\varepsilon_{(1),\infty(\text{top})} = \varepsilon_{(1),60+(\text{top})} + \Delta\varepsilon_{(1),\infty(\text{top})} = -1367.1 \times 10^{-6}$$

$$\varepsilon_{(1),\infty(\text{btm})} = \varepsilon_{(1),60+(\text{btm})} + \Delta\varepsilon_{(1),\infty(\text{btm})} = -672.9 \times 10^{-6}$$

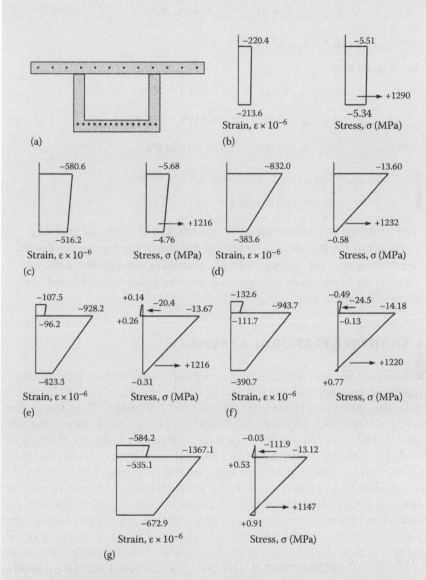

Figure 9.5 Stresses and strains on the composite cross-section of Example 9.1.
(a) Cross-section. (b) Strain and stress at t = 4 days. (c) At t = 40 days
(before casting in-situ slab). (d) At t = 40 days (after casting in-situ slab).
(e) At t = 60 days (before wearing surface). (f) At t = 60 days (after wearing surface). (g) At t = ∞.

$$\sigma_{c(1),\infty(top)} = \sigma_{c(1),60+(top)} + \Delta\sigma_{c(1),\infty(top)} = -13.12 \text{ MPa}$$

$$\sigma_{c(1),\infty(btm)} = \sigma_{c(1),60+(btm)} + \Delta\sigma_{c(1),\infty(btm)} = +0.91 \text{ MPa}$$

$$\sigma_{p,\infty} = \sigma_{p,60+} + \Delta\sigma_{p,\infty} = 1147 \text{ MPa}$$

In the in-situ slab:

$$\varepsilon_{(2),\infty(top)} = \varepsilon_{(2),60+(top)} + \Delta\varepsilon_{(2),\infty(top)} = -584.2 \times 10^{-6}$$

$$\varepsilon_{(2),\infty(btm)} = \varepsilon_{(2),60+(btm)} + \Delta\varepsilon_{(2),\infty(btm)} = -535.1 \times 10^{-6}$$

$$\sigma_{c(2),\infty(top)} = \sigma_{c(2),60+(top)} + \Delta\sigma_{c(2),\infty(top)} = -0.03 \text{ MPa}$$

$$\sigma_{c(2),\infty(btm)} = \sigma_{c(2),60+(btm)} + \Delta\sigma_{c(2),\infty(btm)} = +0.53 \text{ MPa}$$

$$\sigma_{s,\infty} = \sigma_{s,60+} + \Delta\sigma_{s,\infty} = 111.9 \text{ MPa}$$

The stress and strain distributions at time infinity are shown in Figure 9.5g.

Note that the compressive stresses at the top of the precast member at age 60 days after the wearing surface is placed are reduced with time, with much of the compression finding its way into the non-prestressed reinforcement in the in-situ slab.

9.6 ULTIMATE FLEXURAL STRENGTH

The ultimate flexural strength of a composite cross-section may be determined in accordance with the flexural strength theory outlined in Chapter 6. If adequate provision is made to transfer the horizontal shear forces that exist on the interface between the in-situ and precast components, the ultimate strength of a cross-section such as that shown in Figure 9.4 may be calculated in the same way as for an identical monolithic cross-section with the same reinforcement quantities and material properties (see Section 6.6). The calculations are based on the full effective flange width and, in general, it is not necessary to account for variations in concrete strengths between the two components. In practice, owing to the typically wide effective compressive flange, the depth to the natural axis at ultimate is relatively small, usually less than the thickness of the in-situ slab. It is therefore appropriate to consider an idealised rectangular stress block based on the properties of the in-situ concrete rather than the precast concrete. Even in situations where the depth of the compressive zone exceeds the thickness of the slab, more complicated expressions for strength based on more accurate modelling of concrete compressive stresses are not generally necessary. As seen in Chapter 6, the flexural strength of any ductile section is primarily dependent on the quantity and strength of the steel in the tensile zone and does not depend significantly on the concrete strength.

The strain discontinuity that exists at the element interface at service loads due to the construction sequence becomes less and less significant as the moment level increases, and the discontinuity may be ignored in ultimate flexural strength calculations.

9.7 HORIZONTAL SHEAR TRANSFER

9.7.1 Discussion

As has been emphasised in the previous sections, the ability of a composite member to resist load depends on its ability to carry horizontal shear at the interface between the two components. If the components are not effectively bonded together, slip occurs at the interface, as shown in Figure 9.6a, and the two components act as separate beams, each carrying its share of the external loads by bending about its own centroidal axis. To ensure full composite action, slip at the interface must be prevented, and for this, there must be an effective means for transferring horizontal shear across the interface. If slip is prevented, full composite action is assured (as shown in Figure 9.6b) and the advantages of composite construction can be realized.

In Section 9.2, various mechanisms for shear transfer were discussed. Natural adhesion and friction are usually sufficient to prevent slip in composite members with a wide interface between the components (such as the cross-sections shown in Figures 9.1b through d). The contact surface of the precast member is often roughened during manufacture to improve bond. Where the

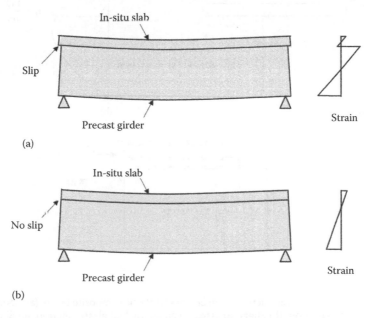

Figure 9.6 Composite behaviour. (a) Non-composite action. (b) Composite action.

contact area is smaller (as on the cross-section of Figures 9.1a and e, and 9.4), web reinforcement in the precast girder is often carried through the interface and anchored in the in-situ slab, thus providing increased frictional resistance (by clamping the contact surfaces together) and additional shear resistance through dowel action.

The theorem of complementary shear stress indicates that on the cross-section of an uncracked elastic composite member, the horizontal shear stress τ_h at the interface between the two components is equal to the vertical shear stress at that point and is given by the well-known expression:

$$\tau_h = \frac{VQ}{Ib_f} \tag{9.55}$$

where V is that part of the shear force caused by loads applied after the establishment of composite action; Q is the first moment of the area of the in-situ element about the centroidal axis of the composite cross-section; I is the moment of inertia of the gross composite cross-section; and b_f is the width of the contact surface (usually equal to the width of the top surface of the precast member).

The distribution of shear stress and the direction of the horizontal shear at the interface are shown in Figure 9.7.

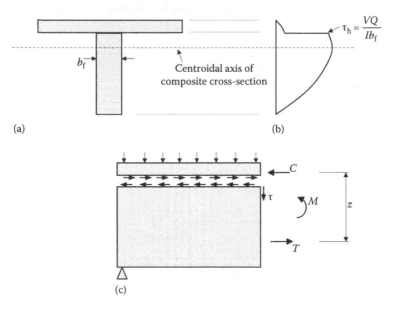

Figure 9.7 Shear stresses and actions in an uncracked elastic composite beam. (a) Composite cross-section. (b) Shear stresses on an uncracked elastic section. (c) Shear on element interface.

At overloads, concrete members crack and material behaviour becomes nonlinear and inelastic. In design, a simpler average or nominal shear stress is usually used for ultimate strength calculations and is given by:

$$\tau^* = \frac{V^*}{b_f d} \tag{9.56}$$

where V^* is the total shear force obtained using the appropriate factored load combination for the strength limit state (see Section 2.3.2). V^* is calculated from the total loads and not just the loads applied after the in-situ slab has hardened, because at ultimate loads, flexural cracking can cross the interface and horizontal shear resulting from all the applied load must be carried.

9.7.2 Provisions for horizontal shear

The design horizontal shear force V_f^* acting on the element interface depends on the position of the interface on the cross-section. AS3600-2009 [1] requires that the design shear stress (τ^*) acting on the interface be taken as:

$$\tau^* = \frac{\beta V^*}{z b_f} \tag{9.57}$$

where z is the internal moment lever arm of the section, as shown in Figure 9.7. When the interface is located in the compression zone, β is the ratio of the compressive force in the in-situ slab C_{slab} (i.e. the compressive force between the extreme compressive fibre and the interface) and the total compressive force on the cross-section C, i.e. $\beta = C_{slab}/C$. When the interface is located in the tensile zone, β is the ratio of the sum of the tensile forces in the longitudinal reinforcement and tendons in the precast member $(T_{s.pc} + T_{p.pc})$ and the total tensile force on the cross-section T, i.e. $\beta = (T_{s.pc} + T_{p.pc})/T$.

The design shear strength at the interface is due to the clamping effects produced by the shear reinforcement crossing the interface, dowel action, aggregate interlock and the effects of any transverse pressure across the interface. For adequate strength, AS3600-2009 (1) requires that τ^* must not exceed the design shear strength $\phi\tau_u$, given by:

$$\tau_u = \mu\left(\frac{A_{sf} f_{sy.f}}{s b_f} + \frac{g_p}{b_f}\right) + k_{co} f_{ct}' \leq \text{lesser of } (0.2 f_c', 10 \text{ MPa}) \tag{9.58}$$

The first term in Equation 9.58 represents the contribution of the shear reinforcement and the frictional resistance caused by external pressure. The second term represents the contribution of the concrete and depends

Table 9.1 Shear plane surface coefficients [1]

Surface condition of the shear plane	Coefficients	
	μ	k_{co}
A smooth surface, as obtained by casting against a form, or finished to a similar standard	0.6	0.1
A surface trowelled or tamped, so that the fines have been brought to the top, but where some small ridges, indentations or undulations have been left; slip-formed and vibro-beam screeded or produced by some form of extrusion technique	0.6	0.2
A surface deliberately roughened: (a) By texturing the concrete to give a pronounced profile (b) By compacting but leaving a rough surface with coarse aggregate protruding but firmly fixed in the matrix (c) By spraying when wet, to expose the coarse aggregate without disturbing it (d) By providing mechanical shear keys	0.9	0.4
Monolithic construction	0.9	0.5

on the roughness of the interface. The symbol g_p is the permanent distributed load normal to the shear interface per unit length (in N/mm), μ is a coefficient of friction given in Table 9.1, k_{co} is a cohesion coefficient also given in Table 9.1, b_f is the width of the shear plane (in mm), A_{sf} is the area of fully anchored shear reinforcement crossing the interface at a spacing s, $f_{sy.f}$ is the yield stress of the shear reinforcement in MPa (not exceeding 500 MPa) and f'_{ct} is the characteristic tensile strength of the in-situ slab (as given in Equation 4.4). The strength reduction factor for shear is $\phi = 0.7$.

The Standard (1) cautions that the values of μ and k_{co} given in Table 9.1 do not apply if the beam is subjected to high levels of differential shrinkage, temperature change, axial tension or fatigue loading.

If shear reinforcement (A_{sf} in Equation 9.58) is required for strength (i.e. the concrete component in Equation 9.58 ($k_{co}f'_{ct}$) is insufficient on its own), the shear and torsional reinforcement that is already provided, and which crosses the shear plane, may be taken into account for this purpose, provided it is anchored so that it can develop its full strength at the interface. The centre-to-centre spacing of this shear reinforcement should not exceed 3.5 times the thickness of the in-situ slab (or topping) anchored by the shear reinforcement. In addition, AS3600-2009 (1) requires that the average thickness of the structural components on either side of the interface should not be less than 50 mm, with a minimum local thickness of 30 mm.

EXAMPLE 9.2

The horizontal shear transfer requirements for the beam with cross-section shown in Figure 9.4 are to be determined. The beam is simply-supported over a span of 17.2 m and is subjected to the following loads:

Self-weight of precast trough-girder:	8.64 kN/m
Self-weight of in-situ slab:	8.10 kN/m
Superimposed dead load:	4.05 kN/m
Transient live load:	9.60 kN/m

The behaviour of the cross-section at mid-span at service loads is calculated in Example 9.1. The effective prestressing force calculated in Example 9.1 is $P_e = A_p\sigma_{p,\infty} = 1721$ kN and is assumed here to be constant along the beam. Take $f_{pb} = 1860$ MPa.

The factored load combination for the strength limit state (Equation 2.2) is:

$$w^* = 1.2 \times (8.64 + 8.10 + 4.05) + 1.5 \times 9.6 = 39.4 \text{ kN/m}$$

The maximum shear force adjacent to each support is:

$$V^* = \frac{(39.4 \times 17.2)}{2} = 339 \text{ kN}$$

At the ultimate limit state in bending, using the idealised rectangular compressive stress block (see Section 6.3.2), we determine that the depth to the neutral axis at the ultimate limit state is 68.1 mm, the tensile force in the tendons is $T_p = A_p\sigma_{pu} = 2770$ kN, the compressive force in the in-situ concrete $C_c = 2952$ kN and, with the non-prestressed reinforcement (A_s) in the in-situ slab just below the neutral axis, the tensile force in A_s is $T_s = 182$ kN. The internal lever arm between the resultant compressive and tensile forces on the cross-section is $z = 698$ mm.

The neutral axis lies in the in-situ slab, just above the non-prestressed reinforcement, and so the interface between the in-situ slab and the precast girder is in the tensile zone below the neutral axis and $\beta = T_p/(T_p + T_s) = 0.94$.

From Equation 9.57, the design shear stress acting on the interface is:

$$\tau^* = \frac{0.94 \times 339 \times 10^3}{698 \times 300} = 1.52 \text{ MPa}$$

where the width of the shear plane $b_f = 300$ mm.

The design shear strength $\phi\tau_u$ is obtained from Equation 9.58. If the top surface of the precast trough has been deliberately roughened to facilitate shear transfer, from Table 9.1, $\mu = 0.9$ and $k_{co} = 0.4$. With $f'_c = 25$ MPa for the in-situ slab and $f'_{ct} = 0.36\sqrt{f'_c} = 1.8$ MPa, and taking $f_{sy.f} = 500$ MPa, Equation 9.58 gives:

$$\phi\tau_u = \phi \times \mu \left(\frac{A_{sf}f_{sy.f}}{sb_f} + \frac{g_p}{b_f} \right) + \phi k_{co}f'_{ct}$$

$$= 0.7 \times 0.9 \times \left(\frac{A_{sf} \times 500}{s \times 300} + \frac{12.15}{300} \right) + 0.7 \times 0.4 \times 1.8 \geq 1.52 \text{ MPa}$$

Therefore:

$$\frac{A_{sf}}{s} \geq 0.943 \text{ mm}^2/\text{mm}$$

If 2–12 mm bars ($f_{sy.f} = 500$ MPa) cross the shear interface, one in each web, $A_{sf} = 220$ mm², the required spacing near each support is:

$$s \leq \frac{220}{0.943} = 233 \text{ mm}$$

The spacing can be increased further into the span, as the shear force V^* decreases. It is important to ensure that these bars are fully anchored on each side of the shear plane.

If the contact surface were not deliberately roughened, but screeded and trowelled, $\mu = 0.6$ and $k_{co} = 0.2$, and Equation 9.59 gives:

$$\phi\tau_u = 0.7 \times 0.6 \times \left(\frac{A_{sf} \times 500}{s \times 300} + \frac{12.15}{300} \right) + 0.7 \times 0.2 \times 1.8 \geq 1.52 \text{ MPa}$$

and with $A_{sf} = 220$ mm²:

$$s \leq \frac{220}{1.787} = 123 \text{ mm}$$

At the quarter-span point of the beam with the deliberately roughened interface, where $V^* = 169.5$ kN, $A_{sf}/s = 0.22$ and $s \leq 1002$ mm. This is greater than the maximum recommended spacing of 3.5 times the slab thickness or 525 mm.

9.8 ULTIMATE SHEAR STRENGTH

9.8.1 Introductory remarks

The design procedures for composite members in shear and torsion are similar to those outlined in Chapter 7 for non-composite members. An additional complication arises, however, in the estimation of the diagonal cracking load for a composite member, and hence in the estimation of the contribution of the concrete to the shear strength V_{uc} (in Equation 7.2).

Before cracking, part of the applied load is resisted exclusively by the precast element (i.e. the load applied in load stages 1–3, as defined in Section 9.3) and part by the composite section (in load stages 4 and 5). In theory, these loads need to be considered separately, using the precast section properties and the composite section properties as appropriate, in order to determine the shear force existing at the onset of diagonal cracking. As discussed in Section 7.2.4, the concrete contribution to shear strength V_{uc} is usually taken as the smaller of the shear force required to produce a flexure-shear crack and the shear force required to cause a web-shear crack.

The design approach described in Sections 7.2.4 and 7.2.5 may be used for the determination of the shear strength of a composite member, provided the stress conditions existing in the precast element are taken into account in the determination of V_{uc}.

9.8.2 Web-shear cracking

The shear force required to produce web-shear cracking at a section may be calculated from the following modification to Equation 7.11:

$$V_{uc} = V_{t.comp} + V_{pc} + P_v \qquad (9.59)$$

where V_{pc} is the shear force applied to the precast member only; and P_v is the vertical component of prestress.

$V_{t.comp}$ is the shear force applied to the composite section which, when combined with the normal stresses caused by loads applied to the composite section and normal and shear stresses caused by the prestress and the external loads applied to the precast section, produces a principal tensile stress of $0.36\sqrt{f_c'}$ at either the centroidal axis of the precast section, the centroidal axis of the composite section, the level of the prestressing duct or the intersection of the flange and the web, whichever is critical.

At a particular point on the cross-section, $V_{t.comp}$ is calculated by setting $\sigma_1 = 0.36\sqrt{f_c'}$ in Equation 7.12. The normal stress, σ in Equation 7.12, is the sum of the normal stresses on the precast element caused by the prestress and moments arising from loads applied directly to the precast member in load stages 1 and 3, and the bending stress due to moments caused by the loads producing $V_{t.comp}$ applied to the composite section in load state 4, $M_{t.comp}$. Therefore:

$$\sigma = \left(-\frac{P_e}{A} + \frac{P_e e y}{I} - \frac{M_{1,3} y}{I} \right)_{precast} - \left(\frac{M_{t.comp} y}{I} \right)_{composite} \qquad (9.60)$$

The shear stress τ in Equation 7.12 is the sum of the shear stress existing on the precast section (due to V_{pc} and P_v) and that arising from $V_{t.comp}$ on the composite section. That is:

$$\tau = \left(\frac{(V_{pc} - P_v)Q}{Ib_w} \right)_{precast} + \left(\frac{V_{t.comp}Q}{Ib_w} \right)_{composite} \tag{9.61}$$

On the right-hand sides of Equations 9.60 and 9.61, the section properties used inside each bracket are those relating to either the precast or the composite cross-sections as indicated.

9.8.3 Flexure-shear cracking

The shear force V_{uc} required to produce a flexure-shear crack is given by Equation 7.9 and is made up of the shear force that exists at decompression of the extreme tensile fibre at the section under consideration V_o, an empirical term representing the additional shear force required to produce an inclined crack, and the vertical component of prestress and is repeated here:

$$V_{uc} = \beta_1\beta_2\beta_3 b_v d_o f_{cv} \left[\frac{(A_{st} + A_{pt})}{b_v d_o} \right]^{1/3} + V_o + P_v \tag{9.62}$$

When applying Equation 9.62 to composite members, consideration must be given to the loading sequence and the stresses existing in the precast element prior to composite action. Decompression may occur with the addition of dead load to the precast section in load stage 3, that is decompression may occur on the precast section even before the in-situ slab is cast and composite action begins. Alternatively, and more commonly, decompression occurs after the composite section is formed in load stage 4 or under overloads.

If decompression occurs on the precast section, V_o must be calculated using the properties of the precast section. Some portion of the additional shear force required to produce the inclined crack (represented by the first term of Equation 9.62) will be acting on the precast section, with the remaining shear acting on the composite section. Because this term is empirical, it is not sensible to try to separate the precast and composite components. If decompression occurs in the precast section prior to composite action with the cast in-situ slab, V_{uc} should be calculated using the properties of the precast section for the determination of each term in Equation 9.62.

When decompression and cracking of the tension zone do not occur until after the section is composite, V_o is the sum of the shear force caused by the loads on the precast section (at the end of load stage 3) and the additional shear force added to the composite cross-section when the extreme tensile fibre is decompressed. In this case, the empirical term in Equation 9.62

should be calculated using the properties of the full composite section (i.e. d_u in Equation 9.62 should be the depth of the tensile reinforcement from the top surface of the in-situ slab).

EXAMPLE 9.3

In this example, the beam described in Example 9.2, with cross-section shown in Figure 9.4, is checked for shear at the cross-section 2 m from the support. In accordance with AS3600-2009 (1), the factored design load for strength was determined in Example 9.2 as $w^* = 39.4$ kN/m. At 2 m from the support:

$V^* = 260$ kN and $M^* = 599$ kNm

For this member with straight tendons: $P_v = 0$.

Web-shear cracking: The load applied to the precast member in load stages 1 and 3 is $8.64 + 8.10 = 16.74$ kN/m and the corresponding shear force and bending moment at the section 2 m from the support are:

$$V_{pc} = \frac{16.74 \times 260}{39.4} = 110.5 \text{ kN} \quad \text{and} \quad M_{l,3} = \frac{16.74 \times 599}{39.4} = 254.5 \text{ kNm}$$

The centroidal axis of the precast section is located 262.5 mm above the bottom fibre (as shown in Figure 9.4), and the centroidal axis of the composite section is 543.8 mm from the bottom fibre. The properties of both the precast and composite sections about their centroidal axes are:

Precast : $I = 18,390 \times 10^6$ mm^4

$A = 360,000$ mm^2

$Q = 35.65 \times 10^6$ mm^3 at centroid of precast section

$Q = 23.78 \times 10^6$ mm^3 at level of composite centroid

Composite : $I = 76,020 \times 10^6$ mm^4

$Q = 107.6 \times 10^6$ mm^3 at centroid of composite section

$Q = 95.8 \times 10^6$ mm^3 at level of precast centroid

The moment caused by the loads producing $V_{t.comp}$ is:

$$M_{t.comp} = \left(\frac{M^*}{V^*}\right) \times V_{t.comp} = 2300 V_{t.comp}$$

The normal stresses at the centroid of the composite section σ_{comp} and at the level of the centroid of the precast section σ_{pc} are obtained from Equation 9.60:

$$\sigma_{comp} = -\frac{1,775\times10^3}{360,000} + \frac{1,775\times10^3\times162.5\times(-281.3)}{18,390\times10^6} - \frac{254.5\times10^6\times(-281.3)}{18,390\times10^6}$$

$$= -5.45\,\text{MPa}$$

$$\sigma_{pc} = -\frac{1,775\times10^3}{360,000} - \frac{2,300\times V_{t.comp}\times10^3\times(-281.3)}{76,020\times10^6}$$

$$= -4.93 + 8.511\times10^{-3}V_{t.comp}$$

The shear stresses at the level of both centroids are found using Equation 9.61:

$$\tau_{comp} = \frac{110\times10^3\times23.78\times10^6}{18,390\times10^6\times300} + \frac{V_{t.comp}\times10^3\times107.6\times10^6}{76,020\times10^6\times300}$$

$$= 0.474 + 4.72\times10^{-3}V_{t.comp}$$

$$\tau_{pc} = \frac{110\times10^3\times35.65\times10^6}{18,390\times10^6\times300} + \frac{V_{t.comp}\times10^3\times95.8\times10^6}{76,020\times10^6\times300}$$

$$= 0.711 + 4.20\times10^{-3}V_{t.comp}$$

By substituting the aforementioned expressions into Equation 7.12 and solving, with $\sigma_1 = 0.36\sqrt{f_c'} = 2.28$ MPa, the shear forces needed to be applied to the composite section to produce web-shear cracking at each critical location are obtained:

At the centroid of the composite section: $V_{t.comp} = 789$ kN

At the centroid of the precast section: $V_{t.comp} = 472$ kN

The latter value clearly governs and the total shear force required to cause web-shear cracking is obtained from Equation 9.59:

$$V_{uc} = V_{t.comp} + V_{pc} = 472 + 110.5 = 582.5\,\text{kN}$$

Flexure-shear cracking: Decompression occurs when the moment applied to the composite section $M_{o.comp}$ just causes the bottom fibre stress to be zero. That is, when:

$$\left(-\frac{P_e}{A} + \frac{P_e e y_b}{I} - \frac{M_{1.3}y_b}{I}\right)_{precast} - \left(\frac{M_{o.comp}y_b}{I}\right)_{composite} = 0$$

and therefore:

$$M_{o.comp} = \left(-\frac{1{,}775 \times 10^3}{360{,}000} + \frac{1{,}775 \times 10^3 \times 162.5 \times (-281.3)}{18{,}390 \times 10^6} - \frac{254.5 \times 10^6 \times (-281.3)}{18{,}390 \times 10^6} \right)$$

$$\times \frac{76{,}020}{-543.8} = 798 \text{ kN m}$$

The shear force at decompression is therefore:

$$V_o = V_{pc} + \frac{V^*}{M^*} M_{o.comp} = 110.5 + \frac{260}{599} \times 798 = 457 \text{ kN}$$

The shear force required to produce flexure-shear cracking is obtained from Equation 9.62:

$$V_{uc} = 1.1 \times 1.0 \times 1.0 \times 300 \times 800 \times 40^{1/3} \left[\frac{1500}{300 \times 800} \right]^{1/3} \times 10^{-3} + 457 = 623 \text{ kN}$$

Evidently, V_{uc} is governed by web-shear cracking at this cross-section and is equal to 582.5 kN. The design strength is:

$$\phi V_{uc} = 0.7 \times 582.5 = 407.8 \text{ kN}$$

which is greater than the design action V^* and only minimum shear reinforcement is required.

REFERENCES

1. AS3600-2009. (2009). *Australian Standard for Concrete Structures*. Standards Association of Australia, Sydney, New South Wales, Australia.
2. Gilbert, R.I. and Ranzi, G. (2011). *Time-Dependent Behaviour of Concrete Structures*. London, U.K.: Spon Press.

Chapter 10

Design procedures for determinate beams

10.1 INTRODUCTION

The variables that must be established in the design of a statically determinate prestressed concrete beam are the material properties and specifications, the shape and size of the section, the amount and location of both the prestressed steel and the non-prestressed reinforcement and the magnitude of the prestressing force. The designer is constrained by the various design requirements for the strength, serviceability, stability and durability limit states.

The optimal design is the particular combination of design variables that satisfies all the design constraints at a minimum cost. The cost of a particular design depends on local conditions at the time of construction, and variations in the costs of materials, formwork, construction expertise, labour, plant hire, transportation, etc., can change the optimal design from one site to another and also from one time to another.

It is difficult, therefore, to fix hard and fast rules to achieve the optimal design. It is difficult even to determine confidently when prestressed concrete becomes more economic than reinforced concrete or when prestressed concrete that is cracked at service loads is a better solution than uncracked prestressed concrete. However, it is possible to give some broad guidelines to achieve feasible design solutions for both fully and partially-prestressed members, i.e. for members that are either uncracked or cracked at service loads. In this chapter, such guidelines are presented and illustrated by examples.

10.2 TYPES OF SECTION

Many types of cross-section are commonly used for prestressed girders. The choice depends on the nature of the applied loads, the function or usage of the member, the availability and cost of formwork, aesthetic considerations and ease of construction. Some commonly used cross-sections are shown in Figure 10.1.

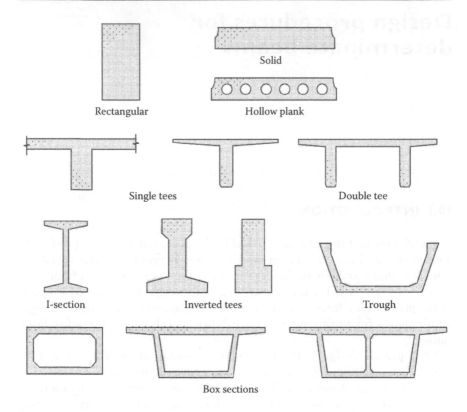

Figure 10.1 Some common prestressed concrete beam cross-sections.

Most in situ prestressed concrete beam sections are rectangular (or slab and beam T-sections with rectangular webs). Rectangular sections are not particularly efficient in bending. The self-weight of a rectangular section is larger than for an I- or T-section of equivalent stiffness, and the prestress required to resist an external moment also tends to be larger. The formwork costs for a rectangular section, however, are generally lower and steel fixing is usually easier.

For precast prestressed concrete, where re-usable formwork is available, the more efficient flanged sections are commonly used. T-sections and double T-sections are ideal for simply-supported members in situations where the self-weight of the beam is a significant part of the total load. If the moment at transfer due to self-weight (plus any other external load) is not significant, care must be taken to avoid excessive compressive stresses in the bottom fibres at transfer in T-shaped sections.

Inverted T-sections can accommodate large initial compressive forces in the lower fibres at transfer and, whilst being unsuitable by themselves for resisting positive moment, they are usually used with a cast in situ composite concrete deck. The resulting composite section is very efficient in positive bending.

For continuous members, where both positive and negative moments exist in different regions of the beam, I-sections and closed box sections are efficient. Box-shaped sections are laterally stable and have found wide application as medium and long-span bridge girders. In addition, box sections can carry efficiently the torsional moments caused by eccentric traffic loading.

10.3 INITIAL TRIAL SECTION

10.3.1 Based on serviceability requirements

A reliable initial trial cross-section is required at the beginning of a design in order to estimate self-weight accurately and to avoid too many design iterations.

For a fully prestressed member, i.e. a member in which tensile stress limits are set in order to eliminate cracking at service loads, Equation 5.5 provides an estimate of the minimum section modulus required to satisfy the selected stress limits at the critical section both at transfer and under the full service loads. If the time-dependent loss of prestress is assumed (usually conservatively) to be 25%, Equation 5.5 simplifies to:

$$Z_b \geq \frac{M_T - 0.75 M_o}{F_t - 0.75 F_{cp}} \tag{10.1}$$

Remember that the compressive stress limit at transfer F_{cp} in this expression is a negative number.

For a member containing a parabolic cable profile, a further guide to the selection of an initial trial section may be obtained by considering the deflection requirements for the member. The deflection of an uncracked prestressed beam under a uniformly distributed unbalanced load w_{ub} may be crudely approximated by:

$$v = \beta \frac{w_{ub} L^4}{E_c I} + \lambda \beta \frac{w_{ub.sus} L^4}{E_c I} \tag{10.2}$$

where $w_{ub.sus}$ is the sustained part of the unbalanced load; β is a deflection coefficient; L is the span of the beam; E_c is the elastic modulus of concrete; I is the moment of inertia of the gross cross-section about its centroidal axis; and λ is a long-term deflection multiplication factor, which should not be taken to be less than 3.0 for an uncracked prestressed member.

The deflection coefficient β is equal to 5/384 for a uniformly loaded simply-supported member. For a continuous member, β depends on the support conditions, the relative lengths of the adjacent spans and the load pattern. When the variable part of the unbalanced load is not greater than the sustained part, the deflection coefficients for a continuous beam with equal adjacent spans may be taken as $\beta = 2.75/384$ for an end span and $\beta = 1.8/384$ for an interior span.

Equation 10.2 can be re-expressed as:

$$v = \beta \frac{w_{tot}L^4}{E_c I} \tag{10.3}$$

where:

$$w_{tot} = w_{ub} + \lambda w_{ub.sus} \tag{10.4}$$

If v_{max} is the maximum permissible total deflection, then from Equation 10.3, the initial gross moment of inertia must satisfy the following:

$$I \geq \beta \frac{w_{tot}L^4}{E_c v_{max}} \tag{10.5}$$

All the terms in Equation 10.5 are generally known at the start of a design, except for an estimate of λ (in Equation 10.4), which may be taken initially to equal 3 for an uncracked member. Since self-weight is usually part of the load being balanced by prestress, it does not form part of w_{tot}.

For a cracked partially prestressed member, λ should be taken as not more than 2, for the reasons discussed in Section 5.11.4. After cracking the effective moment of inertia I_{ef} depends on the quantity of tensile steel and the level of maximum moment. If I_{ef} is taken to be $0.5I$, which is usually conservative, an initial estimate of the gross moment of inertia of the partially-prestressed section can be obtained from:

$$I \geq 2\beta \frac{w_{tot}L^4}{E_c v_{max}} \tag{10.6}$$

10.3.2 Based on strength requirements

An estimate of the section size for a prestressed member can be obtained from the flexural strength requirements of the critical section. The ultimate moment of a ductile rectangular section containing both non-prestressed and prestressed tensile steel may be found using Equation 6.22. By taking moments of the internal tensile forces in the steel about the level of the resultant compressive force in the concrete, the ultimate moment may be expressed as:

$$M_{uo} = \sigma_{pu}A_p\left(d_p - \frac{\gamma d_n}{2}\right) + f_{sy}A_{st}\left(d_o - \frac{\gamma d_n}{2}\right) \tag{10.7}$$

For preliminary design purposes, this expression can be simplified if the stress in the prestressing steel at ultimate σ_{pu} is assumed (say $\sigma_{pu} = 0.9f_{pb}$) and the internal lever arm between the resultant tension and compression forces is estimated (say $0.9d$, where d is the effective depth to the resultant tensile force at the ultimate limit state). With these simplifications, Equation 10.7 becomes:

$$M_{uo} = 0.9d(0.9f_{pb}A_p + f_{sy}A_{st})$$

Dividing both sides by $f'_c bd^2$ gives:

$$\frac{M_{uo}}{f'_c bd^2} = 0.9\left(\frac{0.9f_{pb}}{f'_c}\frac{A_p}{bd} + \frac{f_{sy}}{f'_c}\frac{A_{st}}{bd}\right)$$

and therefore:

$$bd^2 = \frac{M_{uo}}{0.9f'_c(q_p + q_s)} \tag{10.8}$$

where:

$$q_p = \frac{0.9f_{pb}}{f'_c}\frac{A_p}{bd} \tag{10.9}$$

$$q_s = \frac{f_{sy}}{f'_c}\frac{A_{st}}{bd} \tag{10.10}$$

Knowing that the design strength ϕM_{uo} must exceed the factored design moment M^*, Equation 10.8 becomes:

$$bd^2 \geq \frac{M^*}{0.9\phi f'_c(q_p + q_s)} \tag{10.11}$$

The quantity $q_p + q_s$ is the combined steel index and a value of $q_p + q_s$ of about 0.2 will usually provide a ductile section. With this assumption, Equation 10.11 may be simplified to:

$$bd^2 \geq \frac{M^*}{0.18\phi f_c'} \tag{10.12}$$

Equation 10.12 can be used to obtain preliminary dimensions for an initial trial section. The design moment M^* in Equation 10.12 must include an initial estimate of self-weight.

With the cross-sectional dimensions so determined, the initial prestress and the area of prestressing steel can then be selected based on serviceability requirements. Various criteria can be adopted. For example, the prestress required to cause decompression (i.e. zero bottom fibre stress) at the section of maximum moment under full dead load could be selected. Alternatively, load balancing could be used to calculate the prestress required to produce zero deflection under a selected portion of the external load. With the level of prestress determined and the serviceability requirements for the member satisfied, the amount of non-prestressed steel required for strength is calculated.

The size of the web of a beam is frequently determined from shear strength calculations. In arriving at a preliminary cross-section for a thin-webbed member, preliminary checks in accordance with the procedures outlined in Chapter 7 should be carried out to ensure that adequate shear strength can be provided. In addition, the arrangement of the tendon anchorages at the ends of the beam often determines the shape of the section in these regions. Consideration must be given therefore to the anchorage zone requirements (in accordance with the principles discussed in Chapter 8) even in the initial stages of design.

10.4 DESIGN PROCEDURES: FULLY PRESTRESSED BEAMS

For the design of a fully prestressed member, stress limits both at transfer and under full loads must be selected to ensure that cracking under in-service conditions does not occur at any stage. There are relatively few situations that specifically require *no cracking* as a design requirement. Depending on the span and load combinations, however, a fully prestressed design may well prove to be the most economic solution.

For long-span members, where self-weight is a major part of the design load, relatively large prestressing forces are required to produce an economic design and fully prestressed members frequently result. Fully prestressed construction is also desirable if a crack-free or water-tight structure is required or if the structure needs to possess high fatigue strength. In building structures, however, where the spans are generally small to medium, full prestressing may lead to excessive camber, and partial prestressing, where cracking may occur at service loads, is often a better solution.

When the critical sections have been proportioned so that the selected stress limits are satisfied at all stages of loading, checks must be made on the magnitude of the losses of prestress, the deflection and the flexural, shear and torsional strengths. In addition, the anchorage zone must be designed.

10.4.1 Beams with varying eccentricity

The following steps will usually lead to the satisfactory design of a statically determinate, fully prestressed beam with a draped tendon profile:

1. Determine the loads on the beam both at transfer and under the most severe load combination for the serviceability limit states. Next determine the moments at the critical section(s) both at transfer and under the full service loads (M_o and M_T, respectively). An initial estimate of self-weight is required here.
2. Make an initial selection of concrete strength and establish material properties. Using Equation 10.1, choose an initial trial cross-section.
3. Select the maximum permissible total deflection v_{max} caused by the estimated unbalanced loads w_{ub}. This is a second serviceability requirement in addition to the no cracking requirement that prompted the fully prestressed design. Next use Equation 10.5 to check that the gross moment of inertia of the section selected in Step 2 is adequate.
4. Estimate the time-dependent losses of prestress (see Section 5.10.3) and, using the procedure outlined in Section 5.4.1, determine the prestressing force and eccentricity at the critical section(s). With due consideration of the anchorage zone and other construction requirements, select the size and number of prestressing tendons.
5. Establish suitable cable profile(s) by assuming the friction losses and obtaining bounds to the cable eccentricity using Equations 5.12 through 5.15.
6. Calculate both the immediate and time-dependent losses of prestress. Ensure that the calculated losses are less than those assumed in steps 4 and 5. Repeat steps 4 and 5, if necessary.
7. Check the deflection at transfer and the final long-term deflection under maximum and minimum loads. If necessary, consider the inclusion of non-prestressed steel to reduce time-dependent deformations (top steel to reduce downward deflection, bottom steel to reduce time-dependent camber). Adjust the section size or the prestress level (or both), if the calculated deflection is excessive. Where an accurate estimate of time-dependent deflection is required, the time analysis described in Section 5.7 is recommended.

8. Check the ultimate strength in bending at each critical section (see Chapter 6). If necessary, additional non-prestressed tensile reinforcement may be used to increase strength. Add compressive reinforcement to improve ductility, as required.
9. Check the shear strength of the beam (and torsional strength if applicable) in accordance with the provisions outlined in Chapter 7. Design suitable shear reinforcement where required.
10. Design the anchorage zone using the procedures presented in Chapter 8.

Note: Durability and fire protection requirements are usually satisfied by an appropriate choice of concrete strength and cover to the tendons made very early in the design procedure (usually at about Step 2).

EXAMPLE 10.1   FULLY-PRESTRESSED POST-TENSIONED DESIGN (DRAPED TENDON)

A slab and beam floor system consists of post-tensioned, simply-supported T-beams spanning 18.5 m and spaced 4 m apart. A 140 mm thick, continuous, reinforced concrete, one-way slab spans from beam to beam. An elevation and a cross-section of a typical T-beam are shown in Figure 10.2. The beam is to be designed as a fully prestressed member. The floor supports a superimposed permanent dead load of 2 kPa and a live load of 3 kPa (of which 1 kPa is considered permanent). Material properties are $f'_c = 32$ MPa, $f'_{cp} = 25$ MPa, $f_{pb} = 1870$ MPa, $E_c = 28{,}600$ MPa, $E_{cp} = 25{,}300$ MPa and $E_p = 195{,}000$ MPa.

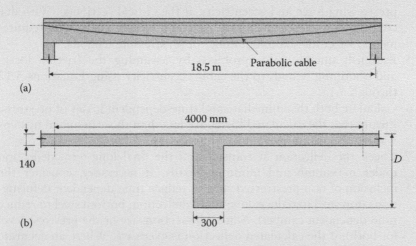

(a)

(b)

Figure 10.2 Beam details (Example 10.1). (a) Elevation. (b) Cross-section.

For this fully prestressed design, the following stress limits have been selected.

At transfer: $F_{tp} = 1.25$ MPa and $F_{cp} = -12.5$ MPa.
After all losses: $F_t = 1.5$ MPa and $F_c = -16.0$ MPa.

(1) Mid-span moments:

Due to self-weight: To estimate the self-weight of the floor w_{sw}, an initial trial depth $D = 1100$ mm is assumed (about span/17). If the concrete floor weighs 24 kN/m³:

$$w_{sw} = 24 \times [4 \times 0.14 + 0.3 \times (1.1 - 0.14)] = 20.4 \text{ kN/m}$$

and the mid-span moment due to self-weight is:

$$M_{sw} = \frac{20.4 \times 18.5^2}{8} = 871 \text{ kNm}$$

Due to 2.0 kPa superimposed dead load:

$$w_G = 2 \times 4 = 8 \text{ kN/m} \quad \text{and} \quad M_G = \frac{8 \times 18.5^2}{8} = 342 \text{ kNm}$$

Due to 3.0 kPa live load:

$$w_Q = 3 \times 4 = 12 \text{ kN/m} \quad \text{and} \quad M_Q = \frac{12 \times 18.5^2}{8} = 513 \text{ kNm}$$

At transfer:

$$M_o = M_{sw} = 871 \text{ kNm}$$

Under full loads:

$$M_T = M_{sw} + M_G + M_Q = 1726 \text{ kNm}$$

(2) Trial section size:

From Equation 10.1:

$$Z_b \geq \frac{(1726 - 0.75 \times 871) \times 10^6}{1.5 - 0.75 \times (-12.5)} = 98.6 \times 10^6 \text{ mm}^3$$

Choose the trial cross-section shown in Figure 10.3.

The revised self-weight is 20.7 kN/m and therefore the revised design moments are $M_o = 886$ kNm and $M_T = 1741$ kNm.

Note: This section just satisfies the requirement for the effective width of T-beam flanges in AS3600-2009 [1] (see Equation 6.34), namely that the flange width does not exceed the web width plus 0.2 times the span.

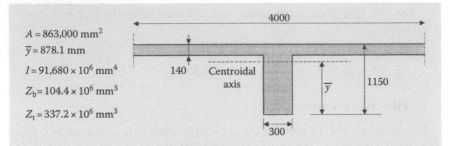

$A = 863,000 \text{ mm}^2$

$\bar{y} = 878.1 \text{ mm}$

$I = 91,680 \times 10^6 \text{ mm}^4$

$Z_b = 104.4 \times 10^6 \text{ mm}^3$

$Z_t = 337.2 \times 10^6 \text{ mm}^3$

Figure 10.3 Trial cross-section (Example 10.1).

(3) Check deflection requirements:

For this particular floor, the maximum deflection v_{max} is taken to be span/500 = 37 mm. If it is assumed that only the self-weight of the floor is balanced by prestress, the unbalanced load is:

$$w_{ub} = w_G + w_Q = 20 \text{ kN/m}$$

With one-third of the live load specified as permanent, the sustained unbalanced load is taken to be (refer to Section 2.3.4):

$$w_{ub.sus} = w_G + \psi_\ell w_Q = 8 + 0.333 \times 12 = 12 \text{ kN/m}$$

With the long-term deflection multiplier taken as $\lambda = 3$, Equation 10.4 gives:

$$w_{tot} = 20 + 3 \times 12 = 56 \text{ kN/m}$$

and from Equation 10.5:

$$I \geq \frac{5}{384} \frac{56 \times 18,500^4}{28,600 \times 37} = 80,700 \times 10^6 \text{ mm}^4$$

The trial cross-section satisfies this requirement and excessive deflection is unlikely.

(4) Determine the prestressing force and eccentricity required at mid-span:

The procedure outlined in Section 5.4.1 is used for the satisfaction of the selected stress limits. The section properties α_t and α_b are given by:

$$\alpha_t = \frac{A}{Z_t} = 0.00256 \quad \text{and} \quad \alpha_b = \frac{A}{Z_b} = 0.00827$$

and Equations 5.1 and 5.2 provide upper limits on the magnitude of prestress at transfer:

$$P_i \leq \frac{863,000 \times 1.25 + 0.00256 \times 886 \times 10^6}{0.00256e - 1} = \frac{3.347 \times 10^6}{0.00256e - 1}$$

(10.1.1)

$$P_i \leq \frac{-863,000 \times (-12.5) + 0.00827 \times 886 \times 10^6}{0.00827e + 1} = \frac{18.11 \times 10^6}{0.00827e + 1}$$

(10.1.2)

Equations 5.3 and 5.4 provide lower limits on the prestress under full service loads. If the time-dependent loss of prestress is assumed to be 20% (i.e., $\Omega = 0.80$), then:

$$P_i \geq \frac{-863,000 \times 1.5 + 0.00827 \times 1741 \times 10^6}{0.8 \times (0.00827e + 1)} = \frac{16.38 \times 10^6}{0.00827e + 1}$$

(10.1.3)

$$P_i \geq \frac{863,000 \times (-16.0) + 0.00256 \times 1741 \times 10^6}{0.8 \times (0.00256e - 1)} = \frac{-11.69 \times 10^6}{0.00256e - 1}$$

(10.1.4)

Two cables are assumed with duct diameters of 80 mm and with 40 mm minimum cover to the ducts. The position of the ducts at mid-span and the location of the resultant prestressing force P are illustrated in Figure 10.4. The resultant force in each tendon is assumed to be located at one quarter of the duct diameter below the top of the duct. The maximum eccentricity to the resultant prestressing force is therefore:

$$e_{max} = 878 - 155 = 723 \text{ mm}$$

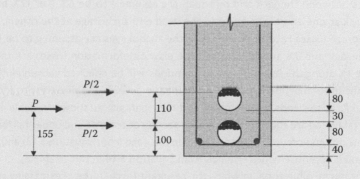

Figure 10.4 Cable locations and relevant dimensions at mid-span.

With this eccentricity at mid-span, Equations 10.1.1 through 10.1.4 give, respectively:

$$P_i \leq 3{,}935 \text{ kN} \quad P_i \leq 2{,}596 \text{ kN} \quad P_i \geq 2{,}347 \text{ kN} \quad P_i \geq -13{,}743 \text{ kN}$$

and the minimum required prestressing force at mid-span that satisfies all four equations is obtained from Equation 10.1.3:

$$P_i = 2347 \text{ kN}$$

If the immediate losses at mid-span are assumed to be 10%, then the required jacking force is:

$$P_j = \frac{P_i}{0.9} = 2608 \text{ kN}$$

From Table 4.8, the cross-sectional area of a 12.7 mm diameter 7-wire strand is 98.6 mm^2, the minimum breaking load is 184 kN and, therefore, the maximum jacking force is 0.85 × 184 = 156.4 kN. The minimum number of 7-wire strands is therefore 2608/156.4 = 16.7.

Try two cables each containing nine strands, that is, A_p = 1774.8 mm^2 (or A_p = 887.4 mm^2/cable).

(5) Establish cable profiles:

Since the member is simply-supported and uniformly loaded, and because the friction losses are only small, parabolic cable profiles with a sufficiently small resultant eccentricity at each end and an eccentricity of 723 mm at mid-span will satisfy the stress limits at every section along the beam. In order to determine the zone in which the resultant prestressing force must be located (see Figure 5.3), it is first necessary to estimate the prestress losses. The cables are to be stressed from one end only. From preliminary calculations (involving the determination of losses for a trial cable profile), the friction losses between the jack and mid-span are assumed to be 6% (i.e. 12% from the jack at one end of the beam to the dead end anchorage at the other), the anchorage losses resulting from slip at the anchorages are assumed to be 14% at the jack and 3% at mid-span and the elastic deformation losses are taken to be 1% along the beam. These assumptions will be checked subsequently.

If the time-dependent losses are assumed to be 20%, the prestressing forces P_i and P_e at the ends, quarter-span and mid-span are as shown in Table 10.1. Also tabulated are the moments at each section at transfer and under full loads, the maximum eccentricity (determined in this case from Equation 5.13) and the minimum eccentricity (determined from Equation 5.14 in this example).

The permissible zone, in which the resultant force in the prestressing steel must be located, is shown in Figure 10.5. The individual cable profiles are

Table 10.1 Bounds on the eccentricity of prestress (Example 10.1)

Distance from jack (mm)	0	4625	9250	13,875	18,500
Estimated short-term losses (%)	15	13	10	10	13
P_i (kN)	2217	2269	2347	2,347	2,269
P_e (kN)	1774	1815	1878	1,878	1,815
M_o (kNm)	0	664	886	664	0
M_T (kNm)	0	1306	1741	1,306	0
e_{max} (mm)	468	747	813	718	454
e_{min} (mm)	−209	512	723	491	−207

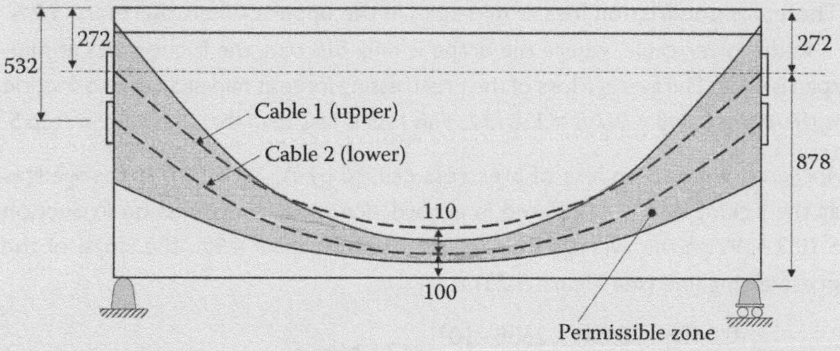

Figure 10.5 Parabolic cable profiles (Example 10.1).

also shown. The cables are separated sufficiently at the ends of the beam to easily accommodate the anchorages for the two cables.

(6) Check losses of prestress:

Immediate losses:

Elastic deformation: At mid-span, the initial prestress in each cable is $P_i = 2347/2 = 1174$ kN. The upper cable is the first to be stressed and therefore suffers elastic deformation losses when the second (lower) cable is subsequently stressed. The prestressing force in the lower cable causes an axial compressive strain at the centroidal axis of the cross-section of $\varepsilon_a = P_i/(AE_{cp})$, and with the average value of P_i in the bottom cable taken to be 1174 kN/m, $\varepsilon_a = 53.0 \times 10^{-6}$. The overall shortening of the member due to this elastic strain causes a shortening of the unbonded top cable and an elastic deformation loss of about:

$$\Delta\sigma_p A_p = \varepsilon_a E_p A_p = 53.8 \times 10^{-6} \times 195,000 \times 887.4 \times 10^{-3} = 9.3 \text{ kN}$$

This is about 0.36% of the total jacking force. The loss of force in the lower cable due to elastic shortening is zero.

Friction losses: The change in slope of the tendon between the support and mid-span is obtained using Equation 1.6. For the upper cable, the drape is 668 mm and therefore the angular change between the support and mid-span is:

$$\alpha_{tot} = \frac{4 \times 668}{18,500} = 0.144 \text{ rad}$$

With $\mu = 0.2$ and $\beta_p = 0.013$, the friction loss at mid-span is calculated using Equation 5.136:

$$\sigma_{pa} = \sigma_{pj} e^{-0.2(0.144 + 0.013 \times 9.25)} = 0.948\sigma_{pj}$$

Therefore, the friction loss at mid-span in the upper cable is therefore 5.2%.

In the lower cable, where the drape is only 518 mm, the friction loss at mid-span is 4.6%. The average loss of the prestressing force at mid-span due to friction is therefore $0.049 \times 2608 = 128$ kN. This loss is less than that assumed in step 5.

Anchorage losses: The loss of prestress caused by $\Delta_{slip} = 6$ mm at the wedges at the jacking end is calculated in accordance with the discussion in Section 5.10.2.4. With the average friction loss at midspan of 4.9%, the slope of the prestressing line (see Figure 5.25) is:

$$\alpha = \frac{0.049P_i}{L/2} = \frac{0.049 \times 2608 \times 10^3}{9250} = 13.8 \text{ N/mm}$$

The length of beam L_{di} over which the anchorage slip affects the prestress is found using Equation 5.138:

$$L_{di} = \sqrt{\frac{195,000 \times 2 \times 887.4 \times 6}{13.8}} = 12,270 \text{ mm}$$

The immediate loss of prestress at the anchorage caused by Δ_{slip} is obtained from Equation 5.139:

$$(\Delta P_{di})_{L_{di}} = 2 \times 13.8 \times 12,270 \times 10^{-3} = 338.6 \text{ kN } (=13.0\% \text{ loss})$$

and the anchorage loss at mid-span is:

$$(\Delta P_{di})_{mid} = 2\alpha\left(L_{di} - \frac{L}{2}\right) = 2 \times 13.8 \times (12,267 - 9,250) \times 10^{-3} = 83.3 \text{ kN } (=3.2\%)$$

The anchorage losses are compatible with the assumptions made in Step 5.

Jacking force: From step 4, the required prestress at mid-span immediately after transfer is $P_i = 2347$ kN. Adding the elastic shortening, friction and anchorage losses, the minimum force required at the jack is:

$$P_j = 2347 + 9.3 + 128 + 83.3 = 2568 \text{ kN} = 1284 \text{ kN/cable}$$

and this is very close to the value assumed in steps 4 and 5. The minimum tendon stress at the jack is:

$$\sigma_{pj} = \frac{P_j}{A_p} = \frac{1284 \times 10^3}{887.4} = 1446 \text{ MPa} = 0.773 f_{pb}$$

which is less than $0.85 f_{pb}$ and is therefore acceptable.

Time-dependent losses:

An accurate time analysis of the cross-section at mid-span can be carried out using the procedure outlined in Section 5.7.3 (and illustrated in Example 5.5). In this example, the more approximate procedures discussed in Section 5.10.3 are used to check time-dependent losses. First we need to estimate the shrinkage and creep characteristics of the concrete and to do this we will use the procedures specified in AS3600-2009 [1] (see Sections 4.2.5.3 and 4.2.5.4).

The hypothetical thickness of the web of this beam is defined in Section 4.2.5.3 and is taken as:

$$t_h = \frac{2A_g}{u_e} = \frac{2 \times 300 \times 1150}{2 \times (300 + 1150 - 140)} = 263 \text{ mm}$$

It would be conservative in this case to include the slender flange in the determination of the hypothetical thickness.

Assuming an air-conditioned (arid) environment in Sydney ($k_4 = 0.7$), and with $f_c' = 32$ MPa, the final autogenous shrinkage (Equation 4.27) is $\varepsilon_{cse}^* = 46 \times 10^{-6}$. From Equation 4.28, the basic drying shrinkage is $\varepsilon_{csd.b} = (1.0 - 0.008 \times 32) \times 800 \times 10^{-6} = 595 \times 10^{-6}$. From Figure 4.10, after 10,000 days of drying, $k_1 = 1.095$ and Equation 4.29 gives $\varepsilon_{csd}^* = 1.095 \times 0.7 \times 595 \times 10^{-6} = 456 \times 10^{-6}$. The final shrinkage strain is obtained from Equation 4.25:

$$\varepsilon_{cs}^* = \varepsilon_{cse}^* + \varepsilon_{csd}^* = (46 + 456) \times 10^{-6} = 502 \times 10^{-6}$$

From Table 4.4, the basic creep coefficient $\phi_{cc.b} = 3.4$. From Figure 4.8, $k_2 = 1.11$. If we assume the member is first stressed at age 7 days (i.e. the age at transfer), Equation 4.22 gives $k_3 = 1.46$. With $k_5 = 1.0$ (Equation 4.23), the final long-term creep coefficient is given by Equation 4.21:

$$\varphi_{cc}^* = 1.11 \times 1.46 \times 0.7 \times 1.0 \times 3.4 = 3.86$$

Taking the final aging coefficient to be 0.65, the age-adjusted effective modulus of concrete is obtained from Equation 5.55 as $\bar{E}_{e,k} = 8150$ MPa and the age-adjusted modular ratio is $\bar{n}_p = E_p / \bar{E}_{e,k} = 23.9$.

Shrinkage losses: The web reinforcement ratio is $p = A_p/(b_w d_p) = 0.0059$ and, from Equation 5.144, the loss of steel stress due to shrinkage may be taken as:

$$\Delta\sigma_{p.cs} = \frac{195,000 \times (-502 \times 10^{-6})}{1 + 23.9 \times 0.0059[1 + 12(723/1,150)^2]} = -54.1\,\text{MPa}$$

Creep losses: The concrete stress at the centroid of the prestressing steel at mid-span (i.e. at $e = 723$ mm) immediately after the application of the full sustained load ($w_{sus} = w_{sw} + w_G + \psi_\ell w_Q = 32.7$ kN/m) is:

$$\sigma_{c0.p} = -\frac{P_i}{A} - \frac{P_i e^2}{I} + \frac{M_{sus}e}{I} = -\frac{2,347 \times 10^3}{863,000} - \frac{2,347 \times 10^3 \times 723^2}{91,680 \times 10^6}$$

$$+ \frac{1,399 \times 10^6 \times 723}{91,680 \times 10^6}$$

$$= -5.07\,\text{MPa}$$

From Equation 5.147, an estimate of the loss of stress in the tendon at mid-span due to creep is:

$$\Delta\sigma_{p.cc} = 0.8 \times 195,000 \times 3.86 \times \left(\frac{-5.07}{28,600}\right) = -106.7\,\text{MPa}$$

Relaxation losses: The basic relaxation for low relaxation strand is obtained from Table 4.9 as $R_b = 3.0$. The final relaxation is obtained from Equation 4.30 (i.e. $R = k_4 k_5 k_6 R_b$). At $t = 10,000$ days, $k_4 = \log[5.4(t)^{1/6}] = 1.4$. With the tendon stress immediately after transfer $\sigma_{p.init} = 1322$ MPa $= 0.707 f_{pb}$, from Figure 4.16, $k_5 = 1.04$. Assuming the temperature at transfer is 20°C (i.e. $k_6 = 1.0$), the final relaxation is:

$$R = 1.4 \times 1.04 \times 1.0 \times 3.0 = 4.37\%$$

The loss of steel stress due to relaxation may be approximated using Equation 5.148:

$$\Delta\sigma_{p.rel} = -0.0437 \times \left(1 - \frac{|-54.1 - 106.7|}{1322}\right) \times 1322 = -50.7\,\text{MPa}$$

Total time-dependent losses: The total loss of stress in the tendon with time at mid-span is $\Delta\sigma_p = -54.1 - 106.7 - 50.7 = -211.5$ MPa (which is 16.0% of the prestress immediately after transfer). This is smaller than the time-dependent losses of 20% assumed in Steps 3, 4 and 5 and a slightly smaller value of P_i, and hence P_j, could be selected. However, given the approximate nature of these serviceability calculations, the original estimate of P_i is considered satisfactory. Should a more accurate estimate of the time-dependent losses of prestress be required, the analysis procedure outlined in Section 5.7.3 is recommended.

(7) Deflection check:

At transfer: The average drape for the two cables is 593 mm and the transverse force exerted on the beam by the draped tendons at transfer is obtained using Equation 1.7:

$$w_{p.i} = \frac{8 P_i e}{L^2} = \frac{8 \times 2{,}347 \times 10^3 \times 593}{18{,}500^2} = 32.5 \text{ N/mm (kN/m)} \uparrow$$

This overestimates the upward load on the member by a small amount, since the prestressing force at mid-span is taken as an average for the span.

The self-weight of the floor was calculated in Step 2 as $w_{sw} = 20.7$ kN/m $\downarrow$. Immediately after transfer, $f'_{cp} = 25$ MPa, and the elastic modulus of concrete at this time is $E_{cp} = 25{,}300$ MPa. The mid-span deflection at transfer is therefore:

$$v_i = \frac{5}{384} \frac{(32.5 - 20.7) \times 18{,}500^4}{25{,}300 \times 91{,}680 \times 10^6} = 7.8 \text{ mm} \uparrow$$

For most structures, an upward deflection of this magnitude at transfer will be satisfactory.

Under full loads: The effective prestress at mid-span after all losses is $P_e = 1971$ kN. The transverse load exerted on the beam by the tendons is therefore:

$$w_{p.e} = \frac{8 \times 1{,}971 \times 10^3 \times 593}{18{,}500^2} = 27.3 \text{ N/mm (kN/m)} \uparrow$$

The sustained gravity loads are $w_{sus} = w_{sw} + w_G + \psi_\ell w_Q = 32.7$ kN/m ($\downarrow$) and the short-term curvature and deflection at mid-span caused by all the sustained loads are:

$$\kappa_{i.sus} = \frac{(w_{sus} - w_{pe})L^2}{8 E_c I} = \frac{(32.7 - 27.3) \times 18{,}500^2}{8 \times 28{,}600 \times 91{,}680 \times 10^6} = 0.0876 \times 10^{-6} \text{ mm}^{-1}$$

$$v_{i.sus} = \frac{5}{384} \frac{(w_{sus} - w_{pe})L^4}{E_c I} = \frac{5}{384} \frac{(32.7 - 27.3) \times 18{,}500^4}{28{,}600 \times 91{,}680 \times 10^6} = 3.1 \text{ mm} \downarrow$$

Under the sustained loads, the initial curvature is small on all sections and the short-term and long-term deflections will also be small.

The creep induced curvature may be approximated using Equation 5.175. The member is uncracked, and with $p = A_p/(b_w d_o) = 0.0056$, the factor α in Equation 5.177 is taken as 1.22:

$$\kappa_{cc} = \frac{\varphi^*_{cc}}{\alpha} \kappa_{i.sus} = \frac{3.86}{1.22} \times 0.0876 \times 10^{-6} = 0.277 \times 10^{-6} \text{ mm}^{-1}$$

With the creep induced curvature at each end of the member equal to zero, the creep deflection is obtained from Equation 5.157:

$$v_{cc} = 10\kappa_{cc}\frac{L^2}{96} = 10 \times 0.277 \times 10^{-6} \times \frac{18,500^2}{96} = 9.9 \text{ mm } (\downarrow)$$

From Equation 5.181, the shrinkage coefficient $k_r = 0.40$ and an estimate of the average shrinkage induced curvature at mid-span is obtained from Equation 5.179:

$$\kappa_{cs}^* = -\frac{k_r \times \varepsilon_{cs}^*}{D} = \frac{0.40 \times 502 \times 10^{-6}}{1150} = 0.175 \times 10^{-6} \text{ mm}^{-1}$$

At each support, the prestressing steel is located at or near the centroidal axis and the shrinkage induced curvature is zero. This positive load-independent curvature at mid-span causes a downward deflection of (Equation 5.157):

$$v_{cs} = 10\kappa_{cs}^*\frac{L^2}{96} = 10 \times 0.175 \times 10^{-6} \times \frac{18,500^2}{96} = 6.2 \text{ mm } (\downarrow)$$

The final deflection due to the sustained load and shrinkage is therefore:

$$v_{i.sus} + v_{cc} + v_{cs} = 19.2 \text{ mm } (\downarrow)$$

The deflection that occurs on application of the 2.0 kPa variable live load (=8 kN/m) is:

$$v_{i.var} = \frac{5}{384}\frac{8 \times 18,500^4}{28,600 \times 91,680 \times 10^6} = 4.7 \text{ mm } (\downarrow)$$

It is evident that the beam performs satisfactorily at service loads with a maximum final deflection of $v_{max} = 19.2 + 4.7 = 23.9$ mm ($\downarrow$) = span/776. This conclusion was foreshadowed in the preliminary deflection check in Step 3.

(8) Check ultimate strength in bending at mid-span:

Using the load factors specified in Section 2.3.2 (Equation 2.2), the design load is:

$$w^* = 1.2(w_{sw} + w_G) + 1.5w_Q = 52.44 \text{ kN/m}$$

and the design moment at mid-span is:

$$M^* = \frac{52.44 \times 18.5^2}{8} = 2243 \text{ kNm}$$

The cross-section at mid-span contains a total area of prestressing steel $A_p = 1774.8$ mm^2 at an effective depth $d_p = 995$ mm. The ultimate moment is calculated using the approximate procedure outlined in Section 6.4.1. With $f'_c = 32$ MPa, Equations 6.2 and 6.3 give, respectively, $\alpha_2 = 0.85$ and $\gamma = 0.826$. For ordinary strand, $k_1 = 0.4$ and Equation 6.20 gives:

$$k_2 = \frac{1774.8 \times 1870}{4000 \times 995 \times 32} = 0.0261$$

The steel stress at ultimate is given by Equation 6.19:

$$\sigma_{pu} = 1870 \times \left(1 - \frac{0.4 \times 0.0261}{0.826}\right) = 1846 \text{ MPa}$$

and the resultant tensile force in the tendon is $T_p = 1774.8 \times 1846 = 3276$ kN. Assuming the neutral axis lies within the slab flange, the depth to the neutral axis is given by Equation 6.21:

$$d_n = \frac{1774.8 \times 1846}{0.85 \times 32 \times 0.826 \times 4000} = 36.5 \text{ mm}$$

which is in fact within the flange. For this section, the quantity of tensile steel is only small and the member will be very ductile. With $d_n = 0.037d_p$, the ductility requirement of AS3600-2009 [1] (i.e. $d_n \leq 0.36d_o$) is easily satisfied.

By taking moments of the internal forces about any point on the cross-section (e.g. the level of the resultant compressive force located $\gamma d_n/2$ below the top surface), the ultimate moment is found:

$$M_{uo} = 1846 \times 1774.8 \times \left(995 - \frac{0.826 \times 36.5}{2}\right) = 3210 \text{ kNm}$$

With the capacity reduction factor for bending $\phi = 0.8$, the design strength is:

$$\phi M_{uo} = 2568 \text{ kNm} > M^*$$

and, therefore, the cross-section at mid-span has adequate flexural strength and no non-prestressed longitudinal steel is required. At least two non-prestressed longitudinal reinforcement bars will be located in the top and bottom of the web of the beam in the corners of the transverse stirrups that are required for shear.

(9) Check shear strength:

As in step 8, $w^* = 52.44$ kN/m. Shear strength is here checked at the section 1 m from the support, where $V^* = 432.6$ kN and $M^* = 458.9$ kNm. At this section, the average depth of the prestressing steel below the centroidal axis of the cross-section is $y = e = 251$ mm and its slope is $y' = 0.114$ rad (refer to Figure 10.5). The effective prestress is $P_e = 1870$ kN and the vertical component of prestress is $P_v = P_e y' = 213$ kN.

Flexure-shear cracking: The decompression moment at this section is:

$$M_o = Z_b\left(\frac{P_e}{A} + \frac{P_e e}{Z_b}\right) = 696 \text{ kNm}$$

and the corresponding shear force is $V_o = V^*(M_o/M^*) = 656$ kN.

If two 20 mm diameter reinforcing bars are located in each bottom corner of the stirrups ($A_{st} = 620$ mm^2), then from Equation 7.9:

$$V_{uc} = 1.1 \times 1.0 \times 1.0 \times 260 \times 1100 \times 32^{1/3} \left[\frac{(620 + 1774.8)}{260 \times 1100}\right]^{1/3} \times 10^{-3} + 656 + 213$$

$$= 1072 \text{ kN}$$

in which the effective web width for shear (defined after Equation 7.6) is $b_v = 300 - (0.5 \times 80) = 260$ mm and the depth to the centroid of A_{st} is $d_o = 1100$ mm.

Web-shear cracking: At the centroidal axis, $Q = 0.5 \times 300 \times 878^2 = 116 \times 10^6$ mm^3, $\sigma = P_e/A = -2.17$ MPa and $\tau = V_t Q/(Ib) = 4.22 \times 10^{-6} V_t$. With $\sigma_1 = 0.36\sqrt{f_c'} = 2.04$ MPa, Equation 7.18 gives $V_t = 694.5$ kN and, therefore, from Equation 7.11, $V_{uc} = 694.5 + 213 = 907.5$ kN.

At this section, the concrete contribution to the design strength of the section in shear is governed by web-shear cracking and is equal to:

$$\phi V_{uc} = 0.7 \times 907.5 = 635 \text{ kN}$$

This is greater than the design shear force V^* and minimum shear reinforcement only is required. Checks at other sections along the span indicate that the minimum reinforcement requirements are sufficient throughout the length of the member. If 10 mm closed stirrups are used (two vertical legs with $A_{sv} = 157$ mm^2 and $f_{sy.f} = 250$ MPa), then the required spacing of stirrups is found using Equation 7.7:

$$s \leq \frac{157 \times 250}{0.35 \times 260} = 431 \text{mm}$$

Use 10 mm closed stirrups at 430 mm maximum centres throughout.

(10) Design anchorage zone:

The bearing plates at the end of each cable are 220 mm^2 as shown in Figure 10.6. The centroid of each plate lies on the vertical axis of symmetry, the upper plate being located on the centroidal axis of the cross-section and the lower plate centred 260 mm below the centroidal axis, as shown.

The distribution of forces on the anchorage zone after the upper cable is stressed is shown in Figure 10.7a, together with the bursting moments induced within the anchorage zone. The depth of the symmetrical prism

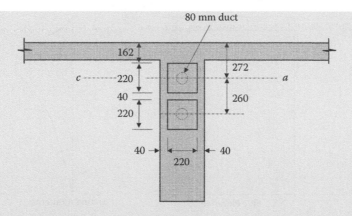

Figure 10.6 End elevation showing size and location of bearing plates.

behind the upper anchorage plate is 544 mm as shown. The transverse tension within the symmetrical prism caused by the bursting moment behind the anchorage plate (M_b = 142.6 kNm) is:

$$T_b = \frac{M_b}{D_e/2} = \frac{142.6 \times 10^6}{544/2} \times 10^{-3} = 524.1 \text{kN}$$

and the area of transverse steel required within a length of beam equal to $0.8D_e$ = 435 mm is obtained from Equation 8.12:

$$A_{sb} = \frac{524.1 \times 10^3}{150} = 3494 \text{ mm}^2$$

Using 4–16 mm diameter vertical stirrup legs at each stirrup location (800 mm²), the required spacing is (435 × 800)/3494 = 99.6 mm.

The distribution of forces on the anchorage zone when both cables are stressed is shown in Figure 10.7b. The maximum bursting moment is 284.6 kNm and the depth of the symmetrical prism behind the combined anchorage plates is 804 mm. The vertical tension and the required area of transverse steel (needed within a length of beam equal to $0.8D_e$ = 643 mm) are:

$$T_b = \frac{284.6 \times 10^6}{804/2} \times 10^{-3} = 708 \text{ kN} \quad \text{and} \quad A_{sb} = \frac{708 \times 10^3}{150} = 4720 \text{ mm}^2$$

The maximum spacing of the vertical stirrups (800 mm²/stirrup location) is (643 × 800)/4720 = 109 mm.

Use two 16 mm diameter stirrups every 100 mm from the end face of the beam to 800 mm there from.

The horizontal dispersion of prestress into the slab flange creates transverse tension in the slab, as indicated in the plan in Figure 10.8.

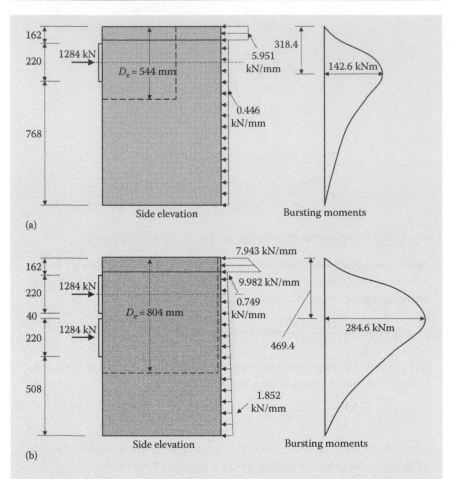

Figure 10.7 Forces and moments in anchorage zone. (a) Upper cable only stressed. (b) Both cables stressed.

From Figure 10.7b, the total force in the flange is 1254.8 kN when both cables are stressed. From the truss analogy shown, the transverse tension is 313.7 kN and the required area of steel is:

$$A_s = \frac{313.7 \times 10^3}{150} = 2091\,\text{mm}^2$$

This steel must be placed horizontally in the slab within a length of 0.8 × 4000 = 3200 mm. Use 16 mm diameter bars at (3200 × 200/2091) = 306 mm centres within 4 m of the free edge of the slab.

The reinforcement details within the anchorage zone are shown in the elevation and cross-section in Figure 10.9.

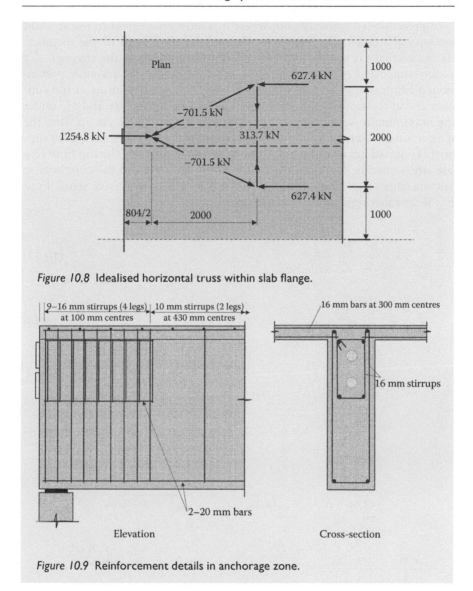

Figure 10.8 Idealised horizontal truss within slab flange.

Figure 10.9 Reinforcement details in anchorage zone.

10.4.2 Beams with constant eccentricity

The procedure described in Section 5.4.1 is a convenient technique for the satisfaction of concrete stress limits at any section at any stage of loading. However, the satisfaction of stress limits at one section does not guarantee satisfaction at other sections. If P_i and e are determined at the section of maximum moment M_o and if e is constant over the full length of the beam, the stress limits F_{cp} and F_{tp} may be exceeded in regions where the moment is less than the maximum value.

In pretensioned construction, where it is most convenient to use straight tendons at a constant eccentricity throughout the length of the member, the eccentricity is usually determined from conditions at the support of a simply-supported member where the moment is zero. In a simple pretensioned beam of constant cross-section, the stress distributions at the support and at the section of maximum moment (M_o at transfer and M_T under the maximum in-service loads) are shown in Figure 10.10. At transfer, the maximum concrete tensile and compressive stresses both occur at the support. To guard against unwanted cracking at the support, the top fibre tensile stress must be less than the tensile stress limit F_{tp} and the compressive bottom fibre stress should also be limited by the compressive stress limit F_{cp}. Remembering that F_{cp} is negative:

$$\sigma_{i.top} = -\frac{P_i}{A} + \frac{P_i e}{Z_t} \leq F_{tp} \tag{10.13}$$

$$\sigma_{i.btm} = -\frac{P_i}{A} - \frac{P_i e}{Z_b} \geq F_{cp} \tag{10.14}$$

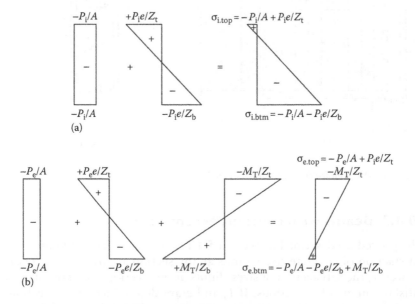

Figure 10.10 Concrete stresses in members with constant eccentricity of prestress. (a) At the support immediately after transfer. (b) At mid-span after all losses and under full loads ($P_e = \Omega P_i$).

By rearranging Equations 10.13 and 10.14 to express P_i as a linear function of e, the following design equations similar to Equations 5.1 and 5.2 (with $M_o = 0$) are obtained:

$$P_i \le \frac{AF_{tp}}{\alpha_t e - 1} \tag{10.15}$$

$$P_i \le \frac{-AF_{cp}}{\alpha_b e + 1} \tag{10.16}$$

where:

$$\alpha_t = \frac{A}{Z_t}$$

$$\alpha_b = \frac{A}{Z_b}$$

After all the time-dependent losses have taken place, the maximum tensile stress occurs in the bottom concrete fibre at mid-span (i.e. $\sigma_{e.btm}$ in Figure 10.10b) and, if cracking is to be avoided, must be limited to the tensile stress limit F_t:

$$\sigma_{e.btm} = -\frac{\Omega P_i}{A} - \frac{\Omega P_i e}{Z_b} + \frac{M_T}{Z_b} \le F_t \tag{10.17}$$

This may be rearranged to give the design equation (identical to Equation 5.3):

$$P_i \ge \frac{-AF_t + \alpha_b M_T}{\Omega(\alpha_b e + 1)} \tag{10.18}$$

By selecting a value of P_i that satisfies Equations 10.15, 10.16 and 10.18, the selected stress limits will be satisfied both at the support at transfer and at the critical section of maximum moment under the full service loads. The compressive stress limit F_c at the critical section is rarely of concern in a pretensioned member of constant cross-section.

To find the minimum sized cross-section required to satisfy the selected stress limits both at the support and at mid-span at all stages of loading, Equation 10.14 may be substituted into Equation 10.17 to give:

$$\Omega F_{cp} + \frac{M_T}{Z_b} \le F_t$$

and therefore:

$$Z_b \geq \frac{M_T}{F_t - \Omega F_{cp}} \tag{10.19}$$

Equation 10.19 can be used to select an initial trial cross-section, and then the required prestressing force and the maximum permissible eccentricity can be determined using Equations 10.15, 10.16 and 10.18.

Note the difference between Equation 10.1 (where $\Omega = 0.75$) and Equation 10.19. The minimum section modulus obtained from Equation 10.1 is controlled by the *incremental moment* $(M_T - \Omega M_o)$ since the satisfaction of stress limits are considered only at the critical section. The stress limits on all other sections are automatically satisfied by suitably varying the eccentricity along the span. If the eccentricity varies such that $P_i e$ is numerically equal to the moment at transfer M_o at all sections, then only the change in moment $(M_T - \Omega M_o)$ places demands on the flexural rigidity of the member. ΩM_o is balanced by the eccentricity of prestress. However, for a beam with constant eccentricity, e is controlled by the stress limits at the support (where M_o is zero). It is therefore the total moment at the critical section M_T that controls the minimum section modulus, as indicated in Equation 10.19.

In order to avoid excessive concrete stresses at the supports at transfer, tendons are often *debonded* near the ends of pretensioned members. In this way, a constant eccentricity greater than the limit by Equations 10.13 and 10.14 is possible.

For a simply-supported member containing straight tendons at a constant eccentricity, the following design steps are appropriate:

1. Determine the loads on the beam both at transfer and under the most severe load combination for the serviceability limit states. Hence determine the moments M_o and M_T at the critical section (an initial estimate of self-weight is required here).
2. Make an initial selection of concrete strength and establish material properties. Using Equation 10.19, choose an initial trial cross-section.
3. Estimate the time-dependent losses and use Equations 10.15, 10.16 and 10.18 to determine the prestressing force and eccentricity at the critical section.
4. Calculate both the immediate and time-dependent losses. Ensure that the calculated losses are less than those assumed in step 3. Repeat Step 3, if necessary.
5. Check the deflection at transfer and the final long-term deflection under maximum and minimum loads. Consider the inclusion of non-prestressed steel to reduce the long-term deformation, if necessary. Adjust section size and/or prestress level, if necessary.

6. Check the ultimate flexural strength at the critical sections. Calculate the quantities of non-prestressed reinforcement required for strength and ductility.
7. Check shear strength of beam (and torsional strength if applicable) in accordance with the provisions outlined in Chapter 7. Design suitable stirrups where required.
8. Design the anchorage zone using the procedures presented in Chapter 8.

EXAMPLE 10.2   FULLY-PRESTRESSED PRETENSIONED DESIGN (STRAIGHT TENDONS)

Simply-supported fully prestressed planks, with a typical cross-section shown in Figure 10.11, are to be designed to span 6.5 m. The planks are to be placed side by side to form a precast floor and are to be pretensioned with straight tendons at a constant eccentricity. The planks are assumed to be long enough for the full prestress to develop at each support (although this is frequently not the case in practice). The floor is to be subjected to a superimposed dead load of 1.2 kPa and a live load of 3.0 kPa (of which 0.7 kPa may be considered to be permanent and the remainder transitory). As in Example 10.1, material properties are $f'_c = 32$ MPa, $f'_{cp} = 25$ MPa, $f_{pb} = 1870$ MPa, $E_c = 28,600$ MPa, $E_{cp} = 25,300$ MPa and $E_p = 195,000$ MPa and the selected stress limits are $F_{tp} = 1.25$ MPa, $F_{cp} = -12.5$ MPa, $F_t = 1.5$ MPa and $F_c = -16.0$ MPa.

(1) Mid-span moments:
Due to self-weight: If the initial depth of the plank is assumed to be $D = $ span/40 = 160 mm, and the plank is assumed to weigh 24 kN/m³, then $w_{sw} = 24 \times 0.16 \times 1.05 = 4.03$ kN/m and at mid-span:

$$M_{sw} = \frac{4.03 \times 6.5^2}{8} = 21.28 \text{ kNm}$$

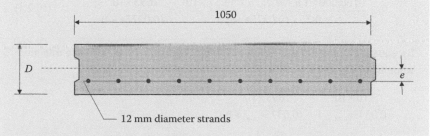

Figure 10.11 Cross-section of pretension plank (Example 10.2).

Due to superimposed dead and live load:

$$w_G = 1.2 \times 1.05 = 1.26 \text{ kN/m} \quad \text{and} \quad M_G = \frac{1.26 \times 6.5^2}{8} = 6.65 \text{ kNm}$$

$$w_Q = 3 \times 1.05 = 3.15 \text{ kN/m} \quad \text{and} \quad M_Q = \frac{3.15 \times 6.5^2}{8} = 16.64 \text{ kNm}$$

At transfer: $M_o = M_{sw} = 21.28$ kNm

Under full loads: $M_T = M_{sw} + M_G + M_Q = 44.57$ kNm

(2) Trial section size:

From Equation 10.19:

$$Z_b \geq \frac{44.57 \times 10^6}{1.5 - 0.75 \times (-12.5)} = 4.098 \times 10^6 \text{ mm}^3$$

and therefore:

$$D \geq \sqrt{\frac{Z_b \times 6}{b}} = \sqrt{\frac{4.098 \times 10^6 \times 6}{1050}} = 153 \text{ mm}$$

Try $D = 160$ mm as originally assumed.

(3) Determine the prestressing force and the eccentricity:

With $D = 160$ mm, the section properties are $A = 168 \times 10^3$ mm^2, $I = 358.4 \times 10^6$ mm^4, $Z_t = Z_b = 4.48 \times 10^6$ mm^3 and $\alpha_t = \alpha_b = 0.0375$. Substituting into Equations 10.15 and 10.16 gives the upper limits to P_i:

$$P_i \leq \frac{168,000 \times 1.25}{0.0375e - 1} = \frac{210 \times 10^3}{0.0375e - 1} \quad \text{and} \quad P_i \leq \frac{-168,000 \times (-12.5)}{0.0375e + 1} = \frac{2,100 \times 10^3}{0.0375e + 1}$$

and substituting into Equation 10.18 gives the lower limit to P_i:

$$P_i \geq \frac{-168,000 \times 1.5 + 0.0375 \times 44.57 \times 10^6}{0.75 \times (0.0375e + 1)} = \frac{1,893 \times 10^3}{(0.0375e + 1)} \tag{10.2.1}$$

By equating the maximum value of P_i from Equation 10.15 with the minimum value from Equation 10.18, we obtain the maximum eccentricity:

$$\frac{210 \times 10^3}{0.0375e_{max} - 1} = \frac{1893 \times 10^3}{(0.0375e_{max} + 1)}$$

and solving gives $e_{max} = 33.4$ mm.

Taking $e = 33$ mm, the corresponding minimum prestress P_i is obtained from Equation 10.2.1:

$$P_i \geq \frac{1893 \times 10^3}{(0.0375 \times 33 + 1)} = 846 \text{ kN}$$

Assuming 5% immediate losses at mid-span, the minimum jacking force is $P_j = P_i/0.95 = 891$ kN. Using 12.7 mm diameter 7-wire strands each with breaking load = 184 kN (see Table 4.8), the maximum jacking force is $0.85 \times 184 = 156.4$ kN/strand. The minimum number of strands is $891/(156.4) = 5.7$.

Use six 12.7 mm diameter strands at $e = 33$ mm (i.e. $A_p = 6 \times 98.6 = 591.6$ mm², $d_p = 113$ mm), with an initial jacking stress of $\sigma_{pj} = P_j/A_p = 891 \times 10^3/591.6 = 1506$ MPa $(=0.805 f_{pb})$.

(4) Calculate losses of prestress:

Immediate losses: For this pretensioned member with straight tendons, the immediate loss of prestress is due to elastic shortening. The concrete stress at the steel level at mid-span immediately after transfer is:

$$\sigma_{cp.0} = -\frac{846 \times 10^3}{168 \times 10^3} - \frac{846 \times 10^3 \times 33^2}{358.4 \times 10^6} + \frac{21.28 \times 10^6 \times 33}{358.4 \times 10^6} = -5.65 \text{ MPa}$$

and from Equation 5.134:

$$\Delta\sigma_{p.0} = \frac{195,000}{25,300} \times (-5.65) = -43.5 \text{ MPa}$$

The loss of prestress at mid-span due to elastic shortening is therefore:

$$\Delta\sigma_{p.0}A_p = -43.5 \times 591.6 \times 10^{-3} = -25.7 \text{ kN}$$

(At the supports, where the moment caused by external loads is zero, $\sigma_{cp.0} = -7.61$ MPa, $\Delta\sigma_{p.0} = -58.7$ MPa and $\Delta\sigma_{p.0}A_p = -34.7$ kN.)

Jacking force: From step 4, the required minimum prestress at mid-span immediately after transfer is $P_i = 846$ kN and $\sigma_{pi} = P_i/A_p = 1430$ MPa. Adding the elastic shortening losses, the minimum force required at the jack is:

$$P_j = 846 + 25.7 = 872 \text{ kN} = 145.3 \text{ kN/strand}$$

With this jacking force, the stress in the strand at the jack is:

$$\sigma_{pj} = \frac{P_j}{A_p} = \frac{872 \times 10^3}{591.6} = 1474 \text{ MPa} = 0.788 f_{pb}$$

which is less than $0.85f_{pb}$ and is therefore acceptable.
In summary:

$$P_j = 872 \text{ kN}$$

$$P_i = 846 \text{ kN at mid-span (and } P_i = 837 \text{ kN at the supports)}$$

Time-dependent losses: The hypothetical thickness of an isolated plank is $t_h = 138.8$ mm. Assuming an air-conditioned (arid) environment in Sydney, using the procedure outlined in Section 4.2.5.4, the final shrinkage strain is $\varepsilon_{cs}^* = 629 \times 10^{-6}$.

If we assume the member is first stressed at age 7 days (i.e. the age at transfer), from Table 4.4 and Equation 4.21, the final long-term creep coefficient is $\varphi_{cc}^* = 4.72$. Taking the final aging coefficient to be 0.65, the age-adjusted effective modulus of concrete is obtained from Equation 5.55 as $\bar{E}_{e,k} = 7030$ MPa and the age-adjusted modular ratio is $\bar{n}_p = E_p/\bar{E}_{e,k} = 27.7$.

Shrinkage losses: The reinforcement ratio is $p = A_p/(bd_p) = 0.00499$ and, from Equation 5.144, the loss of stress in the strands due to shrinkage is:

$$\Delta\sigma_{p.cs} = \frac{195,000 \times (-629 \times 10^{-6})}{1 + 27.7 \times 0.00499(1 + 12(33/160)^2)} = -101.5 \text{ MPa}$$

Creep losses: The sustained load is $w_{sus} = 6.03$ kN/m and the sustained moment at mid-span is $M_{sus} = 31.8$ kNm. The concrete stress at the centroid of the prestressing steel at mid-span (at $e = 33$ mm) immediately after the application of the full sustained load is:

$$\sigma_{c0.p} = -\frac{P_i}{A} - \frac{P_i e^2}{I} + \frac{M_{sus}e}{I} = -4.68 \text{ MPa}$$

From Equation 5.147, an estimate of the loss of stress in the tendon at mid-span due to creep is:

$$\Delta\sigma_{p.cc} = 0.8 \times 195,000 \times 4.72 \times \left(\frac{-4.68}{28,600}\right) = -120.5 \text{ MPa}$$

At the supports, where the moment caused by external loads is zero, $\sigma_{c0.p} = -7.53$ MPa and $\Delta\sigma_{p.cc} = -193.7$ MPa.

Relaxation losses: With the tendon stress at mid-span immediately after transfer $\sigma_{pi} = 1474 - 43.5 = 1430$ MPa $= 0.765f_{pb}$, from Figure 4.16, $k_5 = 1.325$. The basic relaxation for low relaxation strand is $R_b = 3.0$ and the final relaxation is at t = 10,000 days is calculated (using Equation 4.30) as R = 1.4 × 1.325 × 1.0 × 3.0 = 5.57%. The loss of steel stress due to relaxation may be approximated using Equation 5.148:

$$\Delta\sigma_{p.rel} = -0.0557 \times \left(1 - \frac{|-101.5 - 120.5|}{1430}\right) \times 1430 = -67.3\,\text{MPa}$$

At the supports, σ_{pi} = 1415 MPa = 0.757f_{pb}, R = 5.39% and $\Delta\sigma_{p.rel}$ = −60.4 MPa.

Total time-dependent losses: The total loss of stress in the tendon with time at mid-span is $\Delta\sigma_p$ = −101.5 −120.5 −67.3 = −289.3 MPa (which is 20.2% of the prestress immediately after transfer). This is smaller than the time-dependent losses of 25% assumed in Step 3 and is therefore acceptable. The total time-dependent loss at the supports is −355.6 MPa.

The prestressing force after all losses P_e is:

At mid-span: $P_e = P_i + A_p\Delta\sigma_p$ = 846 − 591.6 × 289.3 × 10⁻³ = 675 kN

At the supports: $P_e = P_i + A_p\Delta\sigma_p$ = 837 − 591.6 × 355.6 × 10⁻³ = 627 kN

(5) Deflection check

At transfer: The curvature immediately after transfer at each support is:

$$\kappa_{i.s} = \frac{-P_i e}{E_{cp}I} = \frac{-837 \times 10^3 \times 33}{25,300 \times 358.4 \times 10^6} = -3.05 \times 10^{-6}\,\text{mm}^{-1}$$

and at mid-span:

$$\kappa_{i.m} = \frac{M_o - P_i e}{E_{cp}I} = \frac{21.28 \times 10^6 - 846 \times 10^3 \times 33}{25,300 \times 358.4 \times 10^6} = -0.732 \times 10^{-6}\,\text{mm}^{-1}$$

The corresponding deflection at mid-span may be calculated using Equation 5.157:

$$v_{i.m} = \frac{6500^2}{96}[-3.05 + 10 \times (-0.732) - 3.05] \times 10^{-6} = -5.91\,\text{mm}\,(\uparrow)$$

which is likely to be satisfactory in most practical situations.

Under full loads: The instantaneous curvature caused by the effective pre-stress at the supports is:

$$\kappa_{i.sus.s} = \frac{-P_e e}{E_c I} = \frac{-627 \times 10^3 \times 33}{28,600 \times 358.4 \times 10^6} = -2.02 \times 10^{-6}\,\text{mm}^{-1}$$

With $p = 0.00499$, Equation 5.177 gives $\alpha = 1.20$ and the final creep induced curvature at the supports is estimated using Equation 5.175:

$$\kappa_{cc.s} = \kappa_{i.sus.s}\frac{\varphi_{cc}^*}{\alpha} = -2.02 \times 10^{-6} \times \frac{4.72}{1.20} = -7.95 \times 10^{-6} \text{ mm}^{-1}$$

The final load-dependent curvature at the supports is:

$$\kappa_{sus.s} = \kappa_{i.sus.s} + \kappa_{cc.s} = -9.97 \times 10^{-6} \text{ mm}^{-1}$$

The moment at mid-span caused by the sustained loads is $M_{sus} = M_{sw} + M_G + (0.7/3.0)M_Q = 31.8$ kNm, and the instantaneous curvature caused by the effective prestress and the sustained moment is:

$$\kappa_{i.sus.m} = \frac{M_{sus} - P_e\,e}{E_c I} = \frac{31.8 \times 10^6 - 675 \times 10^3 \times 33}{28,600 \times 358.4 \times 10^6} = 0.929 \times 10^{-6} \text{ mm}^{-1}$$

With $\alpha = 1.20$, the final creep induced curvature at mid-span is:

$$\kappa_{cc.m} = 0.929 \times 10^{-6} \times \frac{4.72}{1.20} = 3.65 \times 10^{-6} \text{ mm}^{-1}$$

The final load-dependent curvature at mid-span is:

$$\kappa_{sus.m} = \kappa_{i.sus.m} + \kappa_{cc.m} = 4.58 \times 10^{-6} \text{ mm}^{-1}$$

The moment at mid-span due to the variable part of the live load is $(2.3/3.0)M_Q = 12.76$ kNm and the corresponding curvature at mid-span is:

$$\kappa_{var.m} = \frac{M_{var}}{E_{cp} I} = \frac{12.76 \times 10^6}{28,600 \times 358.4 \times 10^6} = 1.24 \times 10^{-6} \text{ mm}^{-1}$$

The shrinkage-induced curvature is constant along the span (since the bonded steel is at a constant eccentricity). From Equation 5.181, the shrinkage coefficient $k_r = 0.18$ and an estimate of the average shrinkage induced curvature at mid-span is obtained from Equation 5.179:

$$\kappa_{cs}^* = \frac{0.18 \times 629 \times 10^{-6}}{160} = 0.708 \times 10^{-6} \text{ mm}^{-1}$$

The final curvatures at each end and at mid-span are the sum of the load-dependent and shrinkage curvatures:

$$\kappa_s = \kappa_{sus.s} + \kappa_{cs}^* = -9.26 \times 10^{-6} \text{ mm}^{-1}$$

$$\kappa_m = \kappa_{sus.m} + \kappa_{var.m} + \kappa_{cs}^* = +6.53 \times 10^{-6} \text{ mm}^{-1}$$

From Equation 5.157, the final maximum mid-span deflection is:

$$v_{max.m} = \frac{6500^2}{96}(-9.26 + 10 \times 6.53 - 9.26) \times 10^{-6} = +20.6 \text{ mm } (\downarrow) = \frac{span}{316}$$

A deflection of this magnitude is probably satisfactory, provided that the floor does not support any brittle partitions or finishes. A small quantity of non-prestressed steel in the top of the plank (say 4–12 mm bars [A_{sc} = 440 mm²] at a depth of 30 mm) would reduce the time-dependent part of the total deflection significantly.

(6) Check ultimate strength in bending at mid-span:

Using the same procedure as outlined in Step 8 of Example 8.1, the design strength in bending of the cross-section containing A_p = 591.6 mm² at d_p = 113 mm and A_{sc} = 440 mm² at d_{sc} = 30 mm is:

$$\phi M_{uo} = 75.8 \text{ kNm}$$

with d_n = 39.55 mm, σ_{pu} = 1685 MPa and σ_{sc} = −145 MPa. ϕM_{uo} is greater than the design moment M^* = 58.5 kNm, and hence the flexural strength of the plank is adequate. With d_n = 0.35d_p, the ductility requirements of AS3600-2009 [1] are also satisfied (but only just).

(7) Check shear strength:
For this wide shallow plank, the design shear force V^* is much less than the design strength ϕV_u on each cross-section and no transverse steel is required.

10.5 DESIGN PROCEDURES: PARTIALLY-PRESTRESSED BEAMS

10.5.1 Discussion

In the design of a partially-prestressed member, concrete stresses at transfer should be checked to ensure undesirable cracking or excessive compressive stresses do not occur during and immediately after the stressing operation. However, under full service loads, cracking may occur and so a smaller level of prestress than that required for a fully-prestressed structure may be adopted. It is often convenient to approach the design from an ulti-mate strength point of view in much the same way as for a conventionally reinforced member. Equations 10.6 and 10.12 can both be used to select an initial section size in which tensile reinforcement (both prestressed and non-prestressed) may be added to provide adequate strength and ductility. The various serviceability requirements can then be used to determine the level of prestress. The designer may choose to limit tension under the sustained load or some portion of it. Alternatively, the designer may select a part of the total load to be balanced by the prestress. Under this balanced load, the curvature induced on a cross-section by the eccentric prestress is equal and opposite to the curvature caused by the load. Losses are calculated and the area of prestressing steel is determined.

It should be remembered that the cross-section obtained using Equation 10.12 is only a trial section. Serviceability requirements may indicate that a larger section is needed or that a smaller section would be satisfactory. If the latter is the case, the strength and ductility requirements can usually still be met by the inclusion of non-prestressed reinforcement, either compressive or tensile, or both.

After establishing the dimensions of the cross-section, and after the magnitude of the prestressing force and the size and location of the prestressed steel have been determined, the non-prestressed steel required to provide the necessary additional strength and ductility is calculated. Checks are made with regard to deflection and crack control, and finally, the shear reinforcement and anchorage zones are designed.

The following steps usually lead to a satisfactory design:

1. Determine the loads on the beam including an initial estimate of self-weight. Hence determine the in-service moments at the critical section, both at transfer M_o and under the full loads M_T. Also calculate the design ultimate moment M^* at the critical section.

2. Make an initial selection of concrete strength and establish material properties. Using Equation 10.12, determine suitable section dimensions. Care should be taken when using Equation 10.12. If the neutral axis at ultimate is outside the flange in a T-beam or I-beam, the approximation of a rectangular compression zone may not be acceptable. For long-span, lightly loaded members, deflection and not strength will usually control the size of the section.

3. By selecting a suitable load to be balanced, the unbalanced load can be calculated and Equation 10.6 can be used to check the initial trial section selected in Step 2. Adjust section dimensions, if necessary.

4. Determine the prestressing force, the area of prestressing steel and the cable profile to suit the serviceability requirements. For example, no tension may be required under a portion of the service load, such as the dead load. Alternatively, the load at which deflection is zero may be the design criterion.

5. Calculate the immediate and time-dependent losses of prestress and ensure that the serviceability requirements adopted in Step 4 and the stress limits at transfer are satisfied.

6. To supplement the prestressing steel determined in Step 4, calculate the non-prestressed reinforcement (if any) required to provide adequate flexural strength and ductility.

7. Check crack control and deflections both at transfer and under full loads. A cracked section analysis is usually required to determine I_{ef} and to check the increment of steel stress after cracking.
8. Design for shear (and torsion) at the critical sections in accordance with the design provisions in Chapter 7.
9. Design the anchorage zone using the procedures outlined in Chapter 8.

EXAMPLE 10.3 PARTIALLY-PRESTRESSED BEAM (DRAPED TENDON)

The fully-prestressed T-beam designed in Example 10.1 is re-designed here as a partially-prestressed beam. A section and an elevation of the beam are shown in Figure 10.2, and the material properties and floor loadings are as described in Example 10.1. At transfer, the stress limits are $F_{tp} = 1.25$ MPa and $F_{cp} = -12.5$ MPa.

(1) Mid-span moments:

As in Step 1 of Example 10.1, $w_G = 8$ kN/m, $w_Q = 12$ kN/m, $M_G = 342$ kNm and $M_Q = 513$ kNm.

Since the deflection of the fully-prestressed beam designed in Example 10.1 is only small, a section of similar size may be acceptable even after cracking. The same section will be assumed here in the estimate of self-weight. Therefore, $w_{sw} = 20.7$ kN/m, $M_{sw} = 886$ kNm and the moments at mid-span at transfer and under full loads are as calculated previously:

$$M_o = 886 \text{ kNm} \quad \text{and} \quad M_T = 1741 \text{ kNm}$$

The design ultimate moment at mid-span is calculated as in Step 8 of Example 10.1, i.e. $M^* = 2243$ kNm.

(2) Trial section size based on strength considerations:

From Equation 10.12:

$$bd^2 \geq \frac{2243 \times 10^6}{0.18 \times 0.8 \times 32} = 486.8 \times 10^6 \text{ mm}^3$$

For $b = 4000$ mm, the required effective depth is $d > 348.8$ mm.

Clearly, strength and ductility are easily satisfied (as is evident in Step 8 of Example 10.1). Deflection requirements will control the beam depth.

(3) Trial section size based on acceptable deflection:

In Example 10.1, the balanced load was $w_{p.e} = 27.3$ kN/m (see step 7). If we are permitting cracking, then less prestress will be provided than in Example 10.1. For this cracked, partially prestressed member, we will adopt enough prestress to balance a load of $w_{p.e} = 20$ kN/m. Therefore, the maximum unbalanced load is $w_{ub} = w_{sw} + w_G + w_Q - w_{p.e} = 20.7$ kN/m and the sustained unbalanced load is $w_{ub.sus} = w_{sw} + w_G + \psi_\ell w_Q - 20 = 12.7$ kN/m. The choice of balanced load is somewhat arbitrary, with $w_{p.e} = 28.6$ kN/m representing the balanced load for a fully-prestressed design and $w_{p.e} = 0$ representing a reinforced concrete design (with zero prestress).

Anticipating the inclusion of some compression steel to limit long-term deflection and, hence taking $\lambda = 1.5$, Equation 10.4 gives:

$$w_{tot} = 20.7 + 1.5 \times 12.7 = 39.8 \text{ kN/m}$$

If the maximum total deflection v_{max} is to be limited to 50 mm (= span/370), then from Equation 10.6, the initial gross moment of inertia must satisfy the following:

$$I \geq 2 \times \frac{5}{384} \times \frac{39.8 \times 18,500^4}{28,600 \times 50} = 84,900 \times 10^6 \text{ mm}^4$$

Choose the same trial section as was used for the fully-prestressed design (as shown in Figure 10.3).

(4) Determine prestressing force and cable profile:

A single prestressing cable is to be used, with sufficient prestress to balance a load of 20 kN/m. The cable is to have a parabolic profile with zero eccentricity at each support and $e = 778$ mm at mid-span (i.e. at mid-span $d_p = 1050$ mm and the duct has the same cover at mid-span as the lower cable shown in Figure 10.5). The duct diameter is therefore taken to be 80 mm with 40 mm concrete cover to the duct.

The effective prestress required at mid-span to balance $w_{p.e} = 20$ kN/m is calculated using Equation 1.7:

$$P_e = \frac{w_{p.e}L^2}{8e} = \frac{20.0 \times 18.5^2}{8 \times 0.778} = 1100 \text{ kN}$$

Since the initial stress in the concrete at the steel level is lower than that in Example 10.1, due to the reduced prestressing force, the creep losses will be lower. The time-dependent losses are here assumed to be 15%. If the immediate losses at mid-span (friction plus anchorage draw-in) are assumed to be 10%, the prestressing force at mid-span immediately after transfer P_i and the required jacking force P_j are:

$$P_i = \frac{P_e}{0.85} = 1294 \text{ kN} \quad \text{and} \quad P_j = \frac{P_i}{0.9} = 1438 \text{ kN}$$

and the number of 12.7 mm diameter strands n is:

$$n = \frac{1438 \times 10^3}{0.85 \times 184} = 9.2$$

Try ten 12.7 mm diameter ordinary 7-wire strands ($A_p = 986$ mm²).

(5) Calculate losses of prestress:

Immediate losses: With only one prestressing cable, elastic deformation losses are zero. Using the same procedures as demonstrated in Example 10.1, the friction loss between the jack and mid-span is 5.6% and the anchorage (draw-in) loss at mid-span is 2.8%. The total immediate loss is therefore 8.4%.

Time-dependent losses: The concrete has the same shrinkage and creep characteristics as in Example 10.1. The web reinforcement ratio is $p = A_p/(b_w d_p) = 0.00313$ and, from Equation 5.144, the loss of steel stress due to shrinkage may be taken as:

$$\Delta\sigma_{p.cs} = \frac{195,000 \times (-502 \times 10^{-6})}{1 + 23.9 \times 0.00313 \times [1 + 12(778/1,150)^2]} = -65.9 \text{ MPa}$$

The concrete stress at the centroid of the prestressing steel at mid-span (i.e. at $e = 778$ mm) immediately after the application of the full sustained load ($w_{sus} = w_{sw} + w_G + \psi_\ell w_Q = 32.7$ kN/m) is $\sigma_{c0.p} = +1.83$ MPa (tensile) and, hence tensile creep will cause a small gain in stress in the steel. Assuming the creep coefficient in tension is the same as in compression, from Equation 5.147, an estimate of the gain of stress in the tendon at mid-span due to creep is:

$$\Delta\sigma_{p.cc} = 0.8 \times 195,000 \times 3.86 \times \left(\frac{+1.83}{28,600}\right) = +38.5 \text{ MPa}$$

With the tendon stress immediately after transfer $\sigma_{p.init} = 1312$ MPa $= 0.702 f_{pb}$, from Figure 4.16, $k_5 = 1.01$. Assuming the temperature at transfer is 20°C, the final relaxation is $R = 1.4 \times 1.01 \times 1.0 \times 3.0 = 4.24\%$. The loss of steel stress due to relaxation may be approximated using Equation 5.148:

$$\Delta\sigma_{p.rel} = -0.0424 \times \left(1 - \frac{|-65.9 + 38.5|}{1312}\right) \times 1312 = -54.5 \text{ MPa}$$

Total time-dependent losses: The total loss of stress in the tendon with time at mid-span is $\Delta\sigma_p = -65.9 + 38.5 - 54.5 = -81.9$ MPa (which is 6.2% of the prestress immediately after transfer). This is smaller than the time-dependent losses assumed in Step 4 and a slightly smaller value of P_i, and hence P_j, may be appropriate.

With $P_e = 1100$ kN as calculated in Step 4, the revised estimates of P_i and P_j are:

$$P_i = \frac{P_e}{0.938} = 1173 \text{ kN} \quad \text{and} \quad P_j = \frac{P_i}{0.916} = 1280 \text{ kN}$$

and the required minimum number of strands is $n = 8.2$.

Use nine 12.7 mm diameter ordinary 7-wire strands (A_p = 887.4 mm^2).

By comparison with the beam in Example 10.1 (that has almost double the jacking force), the concrete stress limits at transfer are clearly satisfied.

(6) Design for flexural strength:

As stated in Step 1, the design moment at mid-span is M^* = 2243 kNm and the minimum required ultimate strength is M_{uo} = M^*/ϕ = 2805 kNm, with ϕ = 0.8 as specified in AS3600-2009 [1]. Using the approximate procedure described in Section 6.4.1, the ultimate strength of the cross-section containing A_p = 887.4 mm^2 at d_p = 1050 mm is M_{uo1} = 1720 kN (with d_n = 18.4 mm). Clearly, additional non-prestressed tensile steel is required to ensure adequate strength. If the depth of the non-prestressed tensile reinforcement is d_o = 1080 mm, then the required steel area may be obtained from Equations 6.26 and 6.27:

$$A_{st} = \frac{(2805 - 1720) \times 10^6}{500 \times 0.9 \times (1080 - 0.826 \times 18.4)} = 2265 \text{ mm}^2$$

Try four 28 mm diameter bottom reinforcing bars (A_{st} = 2480 mm^2, f_{sy} = 500 MPa) in two layers, as shown in Figure 10.12.

Checking the strength of this proposed cross-section gives T_p = 1642 kN, T_s = 1240 kN, d_n = 32.1 mm and M_{uo} = 3025 kNm, and therefore ϕM_{uo} = 2420 kNm > M^*. The proposed section at mid-span has adequate strength and ductility.

(7) Check deflection and crack control:

The maximum moment at mid-span due to the full service load is M_T = 1741 kNm and the moment at mid-span caused by the sustained load

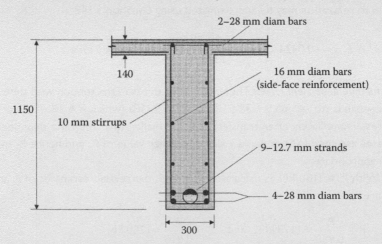

Figure 10.12 Proposed steel layout at mid-span (Example 10.3).

is M_{sus} = 1398 kNm. With P_e = 1100 kN, the tensile strength of concrete taken to be $0.6\sqrt{f_c'}$ = 3.39 MPa and with the web reinforcement ratios, p_w = $(A_{st} + A_p)/b_wd$ = (2480 + 887.4)/(300 × 1063) = 0.0106 and $p_{cw} = A_{sc}/b_wd$ = 1240/(300 × 1063) = 0.0039, the tensile stress that develops in the bottom fibre due to shrinkage may be approximated by Equation 5. 171:

$$\sigma_{cs} = \left(\frac{2.5 \times 0.0106 - 0.8 \times 0.0039}{1 + 50 \times 0.0106} \times 200,000 \times 502 \times 10^{-6} \right) = 1.53\,\text{MPa}$$

The cracking moment may be approximated using Equation 5.170:

$$M_{cr} = \left[104.4 \times 10^6 \left[3.39 - 1.53 + \frac{1,100 \times 10^3}{863,000} \right] + 1,100 \times 10^3 \times 778 \right] \times 10^{-6}$$

$$= 1,183\,\text{kNm}$$

Cracking occurs at mid-span since the cracking moment is less than the sustained moment.

Using the cracked section analysis described in Section 5.8.2, the response of the cracked section at mid-span to the full service moment (M_T = 1741 kNm) is as follows:

The top fibre stress and strain: $\sigma_{c,0(top)}$ = −4.94 MPa

$\varepsilon_{0(top)}$ = −173 × 10^{-6}

The depth to the neutral axis: d_n = 182.4 mm

The stress in the bottom layer of
non-prestressed steel: $\sigma_{s(1),0}$ = 170.1 MPa

The stress in the prestressed steel: $\sigma_{p,0}$ = 1482 MPa

The average moment of inertia: I_{av} = 30,570 × 10^6 mm^4

The effective moment of inertia (using Equation 5.169 to account for tension stiffening):

I_{ef} = 54,490 × 10^6 mm^4

Since the maximum stress in the non-prestressed steel is less than 200 MPa, flexural crack control should not be a problem. Side-face reinforcement as shown in Figure 10.12 should be included to control flexural cracking in the web of the beam above the bottom steel.

The upward transverse force exerted by the prestress on the member is $w_{p.e}$ = 20 kN/m and the maximum gravity load is 40.7 kN/m. An estimate of the maximum short-term deflection v_i caused by the full service load is:

$$v_i = \frac{5}{384} \times \frac{(40.7 - 20.0) \times 18,500^4}{28,600 \times 54,490 \times 10^6} = 20.3\,\text{mm}\ (\downarrow)$$

Under the sustained loads, the loss of stiffness due to cracking will not be as great. The cracks will partially close, and the depth of the compression zone will increase as the variable live load is removed. For the calculation of the short-term deflection due to the sustained loads (32.7 kN/m), the magnitude of I_{ef} is higher than that used above. However, using $I_{ef} = 54,490 \times 10^6$ mm^4 will result in a conservative overestimate of deflection:

$$v_{i.sus} = \frac{5}{384} \times \frac{(32.7 - 20.0) \times 18,500^4}{28,600 \times 54,490 \times 10^6} = 12.4 \text{ mm} (\downarrow)$$

With two 28 mm diameter bars included in the compression zone ($A_{sc} = 1240$ mm^2), Equation 5.177 gives $\alpha = 2.68$ and the creep induced deflection can be approximated by:

$$v_{cc} = \frac{\varphi_{cc}^*}{\alpha} v_{i.sus} = \frac{3.86}{2.68} \times 12.4 = 17.9 \text{ mm} (\downarrow)$$

At the supports, where the cross-section is uncracked and the prestressing steel is located at the centroidal axis, Equation 5.181 gives $k_r = 0.33$, and an estimate of the average shrinkage induced curvature is:

$$\kappa_{cs}^* = -\frac{k_r \varepsilon_{cs}^*}{D} = \frac{0.33 \times 502 \times 10^{-6}}{1150} = 0.144 \times 10^{-6} \text{ mm}^{-1}$$

At mid-span, where the cross-section has cracked and the prestressing cable is near the bottom of the section, Equation 5.183 gives $k_r = 0.47$, and an estimate of the average shrinkage induced curvature is:

$$\kappa_{cs}^* = -\frac{k_r \varepsilon_{cs}^*}{D} = \frac{0.47 \times 502 \times 10^{-6}}{1150} = 0.205 \times 10^{-6} \text{ mm}^{-1}$$

The shrinkage induced deflection is obtained from Equation 5.157:

$$v_{cs} = \frac{18,500^2}{96} (0.144 + 10 \times 0.205 + 0.144) \times 10^{-6} = 8.3 \text{ mm} (\downarrow)$$

The maximum final deflection is therefore:

$$v_i + v_{cc} + v_{cs} = 46.5 \text{ mm} (\downarrow) = \frac{\text{span}}{398}$$

Deflections of this order may be acceptable for most floor types and occupancies.

The design for shear strength and the design of the anchorage zone for this beam are similar to the procedures illustrated in steps 9 and 10 of Example 10.1.

It should be noted that the same cross-sectional dimensions are required for both the fully-prestressed solution in Example 10.1 and the partially-prestressed solution in Example 10.3, provided the deflection calculated earlier is acceptable. Both satisfy strength and serviceability requirements. With significantly less prestress, the partially-prestressed beam is most likely to be the more economical solution.

REFERENCE

1. AS3600-2009 (2009). *Australian Standard for Concrete Structures*. Standards Association of Australia, Sydney, New South Wales, Australia.

Chapter 11

Statically indeterminate members

11.1 INTRODUCTION

The previous chapters have been concerned with the behaviour of individual cross-sections and the analysis and design of statically determinate members. In such members, the deformation of individual cross-sections can take place without restraint being introduced at the supports, and reactions and internal actions can be determined using only the principles of statics. For any set of loads on a statically determinate structure, there is one set of reactions and internal actions that satisfies equilibrium, i.e. there is a single load path.

In this chapter, attention is turned towards the analysis and design of statically indeterminate or continuous members, where the number of unknown reactions is greater than the number of equilibrium equations available from statics. The internal actions in a continuous member depend on the relative stiffnesses of the individual regions and, in structural analysis, consideration must be given to the material properties, the geometry of the structure and geometric compatibility, as well as equilibrium. For any set of loads applied to a statically indeterminate structure, there are an infinite number of sets of reactions and internal actions that satisfy equilibrium, but only one set that also satisfies geometric compatibility and the stress–strain relationships for the constituent materials at each point in the structure. Imposed deformations cause internal actions in statically indeterminate members and methods for determining the internal actions caused by both imposed loads and imposed deformations are required for structural design.

By comparison with simply-supported members, continuous members enjoy certain structural and aesthetic advantages. Maximum bending moments are significantly smaller and deflections are substantially reduced for a given span and load. The reduced demand on strength and the increase in overall stiffness permit a shallower cross-section for an indeterminate member for any given serviceability requirement, and this leads to greater flexibility in sizing members for aesthetic considerations.

In reinforced concrete structures, these advantages are often achieved without an additional cost, since continuity is an easily achieved consequence of in-situ construction. Prestressed concrete, on the other hand, is very often not cast in-situ, but is precast, and continuity is not a naturally achieved consequence. In precast construction, continuity is obtained with extra expense and care in construction. When prestressed concrete is cast in-situ, or when continuity can be achieved by stressing precast units together over several supports, continuity can result in significant cost savings. By using single cables for several spans, the number of anchorages can be reduced significantly, as can the labour costs involved in the stressing operation.

Continuity provides increased resistance to transient loads and also to progressive collapse resulting from wind, explosion or earthquake. In continuous structures, failure of one member or cross-section does not necessarily jeopardise the entire structure, and a redistribution of internal actions may occur. When overload of the structure or member in one area occurs, a redistribution of forces may take place, provided that the structure is sufficiently ductile and an alternative load path is available.

In addition to the obvious advantages of continuous construction, there are several notable disadvantages. Some of the disadvantages are common to all continuous structures, and others are specific to the characteristics of prestressed concrete. Among the disadvantages common to all continuous beams and frames are: (a) the occurrence of a region of both high shear and high moment adjacent to each internal support; (b) high localised moment peaks over the internal supports; and (c) the possibility of high moments and shears resulting from imposed deformations caused by foundation or support settlement, temperature changes and restrained shrinkage.

In continuous beams of prestressed concrete, the quantity of prestressed reinforcement can often be determined from conditions at mid-span, with additional non-prestressed reinforcement included at each interior support to provide the additional strength required in these regions. The length of beam associated with the high local moment at each interior support is relatively small, so that only short lengths of non-prestressed reinforcement are usually required. In this way, economical partially-prestressed concrete continuous structures can be proportioned.

When cables are stressed over several spans in a continuous member, the loss of prestress caused by friction along the duct may be large. The tendon profile usually follows the moment diagram and the tendon suffers relatively large angular changes as the sign of the moment changes along the member from span to span and the distance from the jacking end of the tendon increases. In the design of long continuous members, the loss of prestress that occurs during the stressing operation must therefore be carefully checked. Attention must also be given to the accommodation of the axial deformation that takes place as the member is stressed. Prestressed concrete members shorten as a result of the longitudinal prestress, and this can require special structural details at the supports of continuous members to allow for this movement.

There are other disadvantages or potential problems that may arise as a result of continuous construction. Often beams are built into columns or walls in order to obtain continuity, thereby introducing large additional lateral forces and moments in these supporting elements.

Perhaps the most significant difference between the behaviour of statically indeterminate and statically determinate prestressed concrete structures is the restraining actions that develop in continuous structures as a result of imposed deformations. As a statically indeterminate structure is prestressed, the supports provide restraint to the deformations caused by prestress (both axial shortening and curvature) and reactions may be introduced at the supports. The supports also provide restraint to volume changes of the concrete caused by temperature variations and shrinkage. The reactions induced at the supports during the prestressing operation are self-equilibrating, and they introduce additional moments and shears in a continuous member, called *secondary moments and shears*. These secondary actions may or may not be significant in design. Methods for determining the magnitudes of the secondary effects and their implications in the design for both strength and serviceability are discussed in this chapter.

11.2 TENDON PROFILES

The tendon profile used in a continuous structure is selected primarily to maximise the beneficial effects of prestress and to minimise the disadvantages discussed in Section 11.1. The shape of the profile may be influenced by the techniques adopted for construction. Construction techniques for prestressed concrete structures have changed considerably over the past half-century with many outstanding and innovative developments. Continuity can be achieved in many ways. Some of the more common construction techniques and the associated tendon profiles are briefly discussed here. The methods presented later in the chapter for the analysis of continuous structures are not dependent, however, on the method of construction.

Figure 11.1a represents the most basic tendon configuration for continuous members and is used extensively in slabs and relatively short, lightly loaded beams. Because of the straight soffit, simplicity of formwork is the main advantage of this type of construction. The main disadvantage is the high immediate loss of prestress caused by friction between the tendon and the duct. With the tendon profile following the shape of the moment diagram, the tendon undergoes large angular changes over the length of the member. Tensioning from both ends can be used to reduce the maximum friction loss in long continuous members.

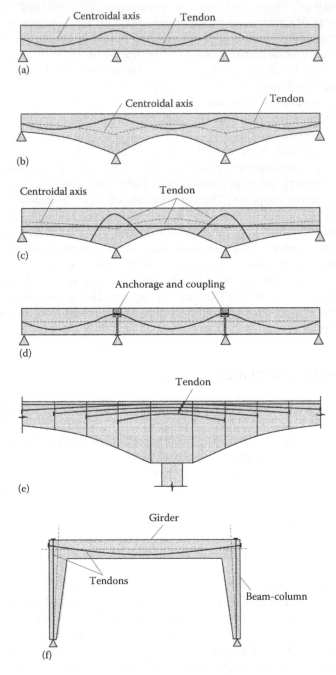

Figure 11.1 Representative tendon profiles. (a) Prismatic beam. (b) Haunched beam. (c) Haunched beam with overlapping tendons. (d) Segmental beam construction. (e) Cantilever construction. (f) Portal frame.

Figure 11.1b indicates an arrangement that has considerable use in longer span structures subjected to heavy applied loads. By haunching the beam as shown, large eccentricities of prestress can be obtained in the regions of high negative moment. This arrangement permits the use of shallower cross sections in the mid-span region and the reduced cable drape can lead to smaller friction losses.

Techniques for overlapping tendons or providing cap cables are numerous. Figure 11.1c shows a tendon layout where the regions of high negative moment are provided with extra prestressing. Continuity of the structure is maintained even though there may be considerable variation of prestress along the member. This general technique can eliminate some of the disadvantages associated with the profiles shown in Figure 11.1a and b where the prestressing force is gradually decreasing along the member. However, any structural benefits that arise by tendon layouts of the type shown in Figure 11.1c are gained at the expense of extra prestressing and additional anchorages.

Many types of segmental construction are available, and a typical case is represented in Figure 11.1d. Precast or cast in situ segments are stressed together using prestress couplers to achieve continuity. The couplers and hydraulic jacks are accommodated during the stressing operation within cavities located in the end surface of the individual segments. The cavities are later filled with concrete, cement grout or other suitable compounds, as necessary.

In large-span structures, such as bridges spanning highways, rivers and valleys, construction techniques are required where falsework is restricted to a minimum. The *cantilever construction method* permits the erection of prestressed concrete segments without the need for major falsework systems. Figure 11.1e illustrates diagrammatically the tendon profiles for a method of construction where precast elements are positioned alternatively on either side of the pier and stressed against the previously placed elements, as shown. The structure is designed initially to sustain the erection forces and construction loads as simple balanced cantilevers on each side of the pier. When the structure is completed and the cantilevers from adjacent piers are joined, the design service loads are resisted by the resulting continuous haunched girders. Construction and erection techniques, such as *balanced cantilevered construction*, are continually evolving and considerable ingenuity is evident in the development of these applications.

Figure 11.1f shows a typical tendon profile for a prestressed concrete portal frame. Prestressed concrete portal frames have generally not had widespread use. With the sudden change of direction of the member axis at each corner of the frame, it is difficult to prestress the columns and beams in a continuous fashion. The horizontal beam and vertical columns are therefore usually stressed separately, with the beam and column tendons crossing at the frame corners and the anchorages positioned on the end and top outside faces of the frame, as shown.

11.3 CONTINUOUS BEAMS

11.3.1 Effects of prestress

As mentioned in Section 11.1, the deformation caused by prestress in a statically determinate member is free to take place without any restraint from the supports. In statically indeterminate members, however, this is not necessarily the case. The redundant supports impose additional geometric constraints, such as zero deflection at intermediate supports (or some prescribed non-zero settlement) or zero slope at a built-in end. During the stressing operation, the geometric constraints may cause additional reactions to develop at the supports, which in turn change the distribution and magnitude of the moments and shears in the member. The magnitudes of these additional reactions (usually called *hyperstatic reactions*) depend on the magnitude of the prestressing force, the support configuration and the tendon profile. For a particular structure, a prestressing tendon with a profile that does not cause hyperstatic reactions is called a *concordant tendon*. Concordant tendons are discussed further in Section 11.3.2.

The moment induced by prestress on a particular cross-section in a statically indeterminate structure may be considered to be made up of two components:

1. The first component is the product of the prestressing force P and its eccentricity from the centroidal axis e. This is the moment that acts on the concrete part of the cross-section when the geometric constraints imposed by the redundant supports are removed. The moment Pe is known as the *primary moment*.
2. The second component is the moment caused by the hyperstatic reactions, i.e. the additional moment required to achieve deformations that are compatible with the support conditions of the indeterminate structure. The moments caused by the hyperstatic reactions are the *secondary moments*.

In a similar way, the shear force caused by prestress on a cross-section in a statically indeterminate member can be divided into primary and secondary components. The primary shear force in the concrete is equal to the prestressing force P times the slope θ of the tendon at the cross-section under consideration. For a member containing only horizontal tendons ($\theta = 0$), the primary shear force on each cross-section is zero. The secondary shear force at a cross-section is caused by the hyperstatic reactions.

The resultant internal actions caused by prestress at any cross-section are the algebraic sums of the primary and secondary effects.

Since the secondary effects are caused by hyperstatic reactions at the supports, it follows that the secondary moments always vary linearly between the supports in a continuous prestressed concrete member and the secondary shear forces are constant in each span.

11.3.2 Determination of secondary effects using virtual work

In the design and analysis of continuous prestressed concrete members, it is usual to make the following simplifying assumptions (none of which introduce significant errors for normal applications):

1. the concrete behaves in a linear-elastic manner within the range of stresses considered;
2. plane sections remain plane throughout the full range of loading;
3. the effects of external loading and prestress on the member can be calculated separately and added to obtain the final conditions, i.e. the principle of superposition is valid; and
4. the magnitude of the eccentricity of prestress is small in comparison with the member length, and hence, the horizontal component of the prestressing force is assumed to be equal to the prestressing force at every cross-section.

Consider the two-span beam shown in Figure 11.2a with straight prestressing tendons at a constant eccentricity e below the centroidal axis. Prior to prestressing, the beam rests on the three supports at A, B and C. On each

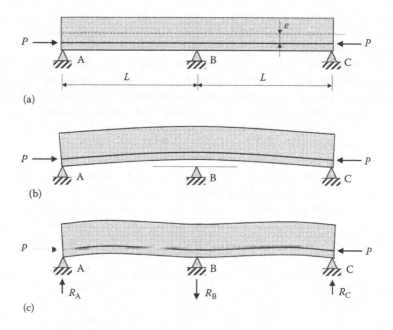

Figure 11.2 Two-span prestressed beam with straight tendon (constant e). (a) Beam elevation. (b) Unrestrained deflection due to primary moment (support B removed). (c) Restrained deflection.

cross-section, prestress causes an axial force P on the concrete and a negative primary moment Pe. If the support at B was removed, the hogging curvature associated with the primary moment would cause the beam to deflect upward at B, as shown in Figure 11.2b.

In the real beam, the deflection at B is zero, as indicated in Figure 11.2c. To satisfy this geometric constraint, a downward reaction is induced at support B, together with equilibrating upward reactions at supports A and C.

To determine the magnitude of these hyperstatic reactions, one of a number of different methods of structural analysis can be used. For one- or two-fold indeterminate structures, the force method (or flexibility method) is a convenient approach. For multiply redundant structures, a displacement method (e.g. such as moment distribution) is more appropriate.

Moment-area methods can be used for estimating the deflection of beams from known curvatures. The principle of virtual work can also be used and is often more convenient. The principle is briefly outlined in the following. For a more comprehensive discussion of virtual work, the reader is referred to textbooks on structural analysis, such as Reference [1].

The principle of virtual work states that if a structure is subjected to an equilibrium force field (i.e. a force field in which the external forces are in equilibrium with the internal actions) and a geometrically consistent displacement field (i.e. a displacement field in which the external displacements are compatible with the internal deformations and the boundary conditions), then the external work product W of the two fields is equal to the internal work product of the two fields U. The force field may be entirely independent of the compatible displacement field.

In the applications discussed here, the compatible displacement field is the actual strain and curvature on each cross-section caused by the external loads and prestress, together with the corresponding external displacements. The equilibrium force field consists of a unit external force (or couple) applied to the structure at the point and in the direction of the displacement being determined, together with any convenient set of internal actions that are in equilibrium with this unit force (or couple). The unit force is called a *virtual force* and is introduced at a particular point in the structure to enable the rapid determination of the real displacement at that point. The bending moments caused by the virtual force are designated $\bar{M}$.

To illustrate the principle of virtual work, consider again the two-span beam of Figure 11.2. In order to determine the hyperstatic reaction at B, it is first necessary to determine the upward deflection v_B caused by the primary moment when the support at B is removed (as illustrated in Figure 11.3a). If the prestress is assumed to be constant throughout the length of the beam, the curvature caused by the primary moment is as shown in Figure 11.3b. A unit virtual force is introduced at B in the direction of v_B, as indicated in Figure 11.3c, and the corresponding virtual moments are illustrated in Figure 11.3d.

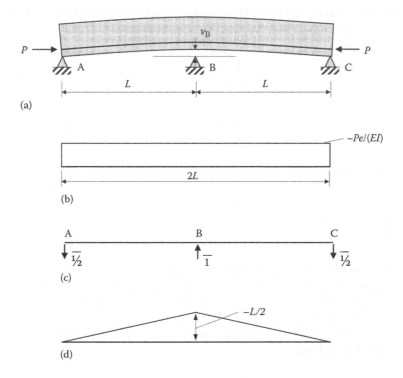

Figure 11.3 Virtual forces on a two-span beam. (a) Unrestrained deflection due to primary moment (support B removed). (b) Curvature caused by primary moment. (c) Virtual force at position of redundant support. (d) Virtual moment diagram, $\bar{M}$.

The external work is the product of the virtual forces and their corresponding displacements:

$$W = 1 \times v_B = v_B \tag{11.1}$$

In this example, the internal work is the integral over the length of the beam of the product of the virtual moments $\bar{M}$ and the real deformations $-Pe/(EI)$:

$$U = \int_0^{2L} \bar{M}\left(\frac{-Pe}{EI}\right) dx \tag{11.2}$$

If the virtual force applied to a structure produces virtual axial forces $\bar{N}$, in addition to virtual bending moments, then internal work is also done by the virtual axial forces and the real axial deformation. For any length of beam, ΔL, a more general expression for internal work is:

$$U = \int_0^{\Delta L} \bar{M}\left(\frac{M}{EI}\right) dx + \int_0^{\Delta L} \bar{N}\left(\frac{N}{EA}\right) dx \qquad (11.3)$$

where M/EI and N/EA are the real curvature and axial strain, respectively; and $\bar{M}$ and $\bar{N}$ are the virtual internal actions.

An integral of the form:

$$\int_0^{\Delta L} \bar{F}(x)F(x)dx \qquad (11.4)$$

may be considered as the volume of a solid of length ΔL whose plan is the function $F(x)$ and whose elevation is the function $\bar{F}(x)$. Consider the two functions $F(x)$ and $\bar{F}(x)$ illustrated in Figure 11.4 and the notation also shown. The volume integral (Equation 11.4) can be evaluated exactly using Simpson's rule if the shape of the function $F(x)$ is linear or parabolic and the shape $\bar{F}(x)$ is linear. Thus:

$$\int_0^{\Delta L} \bar{F}(x)F(x)dx = \frac{\Delta L}{6}\left(\bar{F}_L F_L + 4\bar{F}_M F_M + \bar{F}_R F_R\right) \qquad (11.5)$$

In the example considered here, the function $F(x)$ is constant and equal to $-Pe/EI$ (i.e. $F_L = F_M = F_R = -Pe/EI$) and the function $\bar{F}(x)$ is the virtual moment diagram $\bar{M}$, which is also negative and varies linearly from A to B and from B to C, as shown in Figure 11.3d. Evaluating the internal work in the spans AB and BC, Equation 11.2 gives:

$$U = U_{AB} + U_{BC} = 2 \times \int_0^L \bar{M}\left(\frac{-Pe}{EI}\right) dx \qquad (11.6)$$

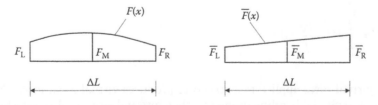

Figure 11.4 Notation for volume integration.

With $\bar{F}_L = 0$, $\bar{F}_M = -L/4$ and $\bar{F}_R = -L/2$, Equation 11.5 gives:

$$U = 2 \times \frac{L}{6}\left[\left(0 \times \frac{-Pe}{EI}\right) + \left(4 \times \frac{-L}{4} \times \frac{-Pe}{EI}\right) + \left(\frac{-L}{2} \times \frac{-Pe}{EI}\right)\right] = \frac{PeL^2}{2EI} \qquad (11.7)$$

The principle of virtual work states that:

$$W = U \qquad (11.8)$$

and substituting Equations 11.1 and 11.7 into Equation 11.8 gives:

$$v_B = \frac{PeL^2}{2EI} \qquad (11.9)$$

It is next necessary to calculate the magnitude of the redundant reaction R_B required to restore compatibility at B, i.e. the value of R_B required to produce a downward deflection at B equal in magnitude to the upward deflection given in Equation 11.9. It is convenient to calculate the *flexibility coefficient* f_B associated with the released structure. The flexibility coefficient f_B is the deflection at B caused by a unit value of the redundant reaction at B. The curvature diagram caused by a unit vertical force at B has the same shape as the moment diagram shown in Figure 11.3d. That is the curvature diagram caused by a unit force at B (M/EI) and the virtual moment diagram $\bar{M}$ have the same shape and the same sign. Using the principle of virtual work and Equation 11.5 to evaluate the volume integral, we get:

$$f_B = \int_0^{2L} \bar{M}\left(\frac{M}{EI}\right) dx = 2 \times \frac{L}{6EI}\left[\left(4 \times \frac{-L}{4} \times \frac{-L}{4}\right) + \left(\frac{-L}{2} \times \frac{-L}{2}\right)\right] = \frac{L^3}{6EI} \qquad (11.10)$$

Compatibility requires that the deflection of the real beam at B is zero:

$$v_B + f_B R_B = 0 \qquad (11.11)$$

and therefore:

$$R_B = -\frac{v_B}{f_B} = -\frac{PeL^2}{2EI} \times \frac{6EI}{L^3} = -\frac{3Pe}{L} \qquad (11.12)$$

The negative sign indicates that the hyperstatic reaction is downward (or opposite in direction to the unit virtual force at B). With the hyperstatic reactions thus calculated, the secondary moments and shears are determined readily. The effects of prestress on the two-span beam under consideration are shown in Figure 11.5.

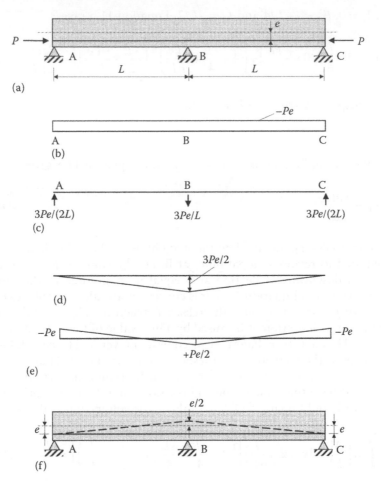

Figure 11.5 Effects of prestress. (a) Elevation of two-span beam with constant eccentricity. (b) Primary moment diagram. (c) Hyperstatic reactions. (d) Secondary moment diagram. (e) Total moment diagram caused by prestress (primary + secondary). (f) The pressure line.

In a statically determinate beam under the action of prestress only, the resultant force on the concrete at a particular cross-section is a compressive force C equal in magnitude to the prestressing force and located at the position of the tendon. The distance of the force C from the centroidal axis is therefore equal to the primary moment divided by the prestressing force $Pe/P = e$. In a statically indeterminate member, if secondary moments exist at a section, the location of C does not coincide with the position of the tendon. The distance of C from the centroidal axis is the total moment due to prestress (primary plus secondary) divided by the prestressing force.

For the beam shown in Figure 11.5a, the total moment due to prestress is illustrated in Figure 11.5e. The position of the stress resultant C varies as the total moment varies along the beam. At the two exterior supports (ends A and C), the stress resultant C is located at the tendon level (i.e. a distance e below the centroidal axis), since the secondary moment at each end is zero. At the interior support B, the secondary moment is $3Pe/2$ and the force C is located at $e/2$ above the centroidal axis (or $3e/2$ above the tendon level). In general, at any section of a continuous beam, the distance of the resultant C from the level of the tendon is equal to the secondary moment divided by the prestressing force. If the position of C at each section is plotted along the beam, a line known as the pressure line is obtained. The pressure line for the beam of Figure 11.5a is shown as the dashed line in Figure 11.5f.

If the prestressing force produces hyperstatic reactions, and hence secondary moments, the pressure line does not coincide with the tendon profile. If, however, the pressure line and the tendon profile do coincide at every section along a beam, there are no secondary moments and the tendon profile is said to be *concordant*. In a statically determinate member, of course, the pressure line and the tendon profile always coincide.

11.3.3 Linear transformation of a tendon profile

The two-span beam shown in Figure 11.6 is similar to the beam in Figure 11.2a (and Figure 11.5a), except that the eccentricity of the tendon is not

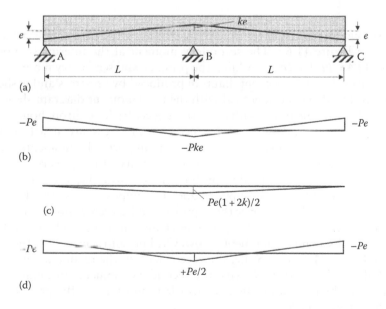

Figure 11.6 Moments induced by prestress in a two-span beam with linearly varying tendon profile. (a) Linear variation of tendon eccentricity. (b) Primary moment diagram. (c) Secondary moment diagram. (d) Total moment diagram caused by prestress.

constant but varies linearly in each span. At the exterior supports, the eccentricity is e (as in the previous examples) and at the interior support, the eccentricity is ke, where k is arbitrary. If the tendon is above the centroidal axis at B, as shown, k is negative.

The primary moment at a section is the product of the prestressing force and the tendon eccentricity and is shown in Figure 11.6b. If the support at B is removed, the deflection at B (v_B) caused by the primary moment may be calculated using the principle of virtual work. The virtual moment diagram $\overline{M}$ is shown in Figure 11.3d.

Using Equation 11.5 to perform the required volume integration, we get:

$$v_B = 2 \times \frac{L}{6EI} \left[\left(4 \times \frac{-L}{4} \times \left(\frac{-Pe - Pke}{2} \right) \right) + \left(\frac{-L}{2} \times (-Pke) \right) \right] = \frac{PeL^2}{EI} \left(\frac{1 + 2k}{6} \right)$$

$$(11.13)$$

The flexibility coefficient associated with a release at support B is given by Equation 11.10, and the compatibility condition of zero deflection at the interior support is expressed by Equation 11.11. Substituting Equations 11.10 and 11.13 into Equation 11.11 gives the hyperstatic reaction at B:

$$R_B = -\frac{Pe}{L}(1 + 2k) \qquad\qquad (11.14)$$

The secondary moments produced by this downward reaction at B are shown in Figure 11.6c. The secondary moment at the interior support is $(R_B \times 2L)/4 = Pe(1 + 2k)/2$. Adding the primary and secondary moment diagrams gives the total moment diagram produced by prestress and is shown in Figure 11.6d. This is identical with the total moment diagram shown in Figure 11.5e for the beam with a constant eccentricity e throughout.

Evidently, the total moments induced by prestress are unaffected by variations in the eccentricity at the interior support. The moments due to prestress are produced entirely by the eccentricity of the prestress at each end of the beam. If the tendon profile remains straight, variation of the eccentricity at the interior support does not impose transverse loads on the beam (except directly over the supports) and therefore does not change the moments caused by prestress. It does change the magnitudes of both the primary and secondary moments, however, but not their sum. If the value of k in Figure 11.6 is -0.5 (i.e. the eccentricity of the tendon at support B is $e/2$ above the centroidal axis), the secondary moments in Figure 11.6c disappear. The tendon profile is concordant and follows the pressure line shown in Figure 11.5f.

A change in the tendon profile in any beam that does not involve a change in the eccentricities at the free ends and does not change the tendon curvature at any point along a span will not affect the total moments due to prestress. Such a change in the tendon profile is known as *linear transformation*, since it involves a change in the tendon eccentricity at each cross-section by an amount that is linearly proportional to the distance of the cross-section from the end of each span.

Linear transformation can be used in any beam to reduce or eliminate secondary moments. For any statically indeterminate beam, the tendon profile in each span can be made concordant by linearly transforming the profile so that the total moment diagram and the primary moment diagram are the same. The tendon profile and the pressure line for the beam will then coincide.

Stresses at any section in an uncracked structure due to the prestressing force can be calculated as follows:

$$\sigma = -\frac{P}{A} \pm \frac{Pe^*y}{I} \tag{11.15}$$

The term e^* is the eccentricity of the pressure line from the centroidal axis of the member, and not the actual eccentricity of the tendon (unless the tendon is concordant and the pressure line and tendon profile coincide). The significance of the pressure line is now apparent. It is the location of the concrete stress resultant caused by the axial prestress, the moment caused by the tendon eccentricity and the moment caused by the hyperstatic support reactions.

11.3.4 Analysis using equivalent loads

In the previous section, the force method was used to determine the hyperstatic reaction in a one-fold indeterminate structure. This method is useful for simple structures, but is not practical for manual solution when the number of redundants is greater than two or three.

A procedure more suited in determining the effects of prestress in highly indeterminate structures is the *equivalent load method*. In this method, the forces imposed on the concrete by the prestressing tendons are considered as externally applied loads. The structure is then analysed under the action of these *equivalent loads* using moment distribution or an equivalent method of structural analysis. The equivalent loads include the loads imposed on the concrete at the tendon anchorage (which may include the axial prestress, the shear force resulting from a sloping tendon and the moment due to an eccentrically placed anchorage) and the transverse forces exerted on the member wherever the tendon changes direction. Commonly occurring tendon profiles and their equivalent loads are illustrated in Figure 11.7.

The total moment caused by prestress at any cross-section is obtained by analysing the structure under the action of the equivalent loads in each span. The moment due to prestress is caused only by moments applied at

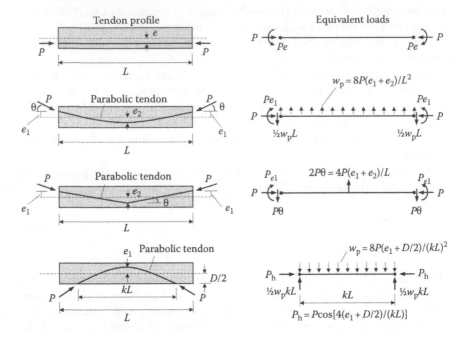

Figure 11.7 Tendon profiles and equivalent loads.

each end of a member (due to an eccentrically located tendon anchorage) and by transverse loads resulting from changes in the direction of the tendon anywhere between the supports. Changes in tendon direction at a support (such as at support B in Figure 11.6a) do not affect the moment caused by prestress, since the transverse load passes directly into the support. This is why the total moments caused by prestress in the beams of Figures 11.5 and 11.6 are identical.

The primary moment at any section is the product of the prestress and its eccentricity Pe. The secondary moment may therefore be calculated by subtracting the primary moment from the total moment caused by the equivalent loads.

11.3.4.1 Moment distribution

Moment distribution is a relaxation procedure developed by Cross [2] for the analysis of statically indeterminate beams and frames. It is a displacement method of analysis that is ideally suited to manual calculation. Although the method has been replaced in many applications by other analysis procedures implemented in computer programs, it remains a valuable tool for practising engineers, because it is simple, easy to use and provides an insight into the physical behaviour of the structure.

Initially, the rotational stiffness of each member framing into each joint in the structure is calculated. Joints in the structure are then locked against rotation by the introduction of imaginary restraints. With the joints locked, fixed-end moments (FEMs) develop at the ends of each loaded member. At a locked joint, the imaginary restraint exerts a moment on the structure equal to the unbalanced moment, which is the resultant of all the FEMs at the joint. The joints are then released, one at a time, by applying a moment to the joint equal and opposite to the unbalanced moment. This balancing moment is distributed to the members framing into the joint in proportion to their rotational stiffnesses. After the unbalanced moment at a joint has been balanced, the joint is relocked. The moment distributed to each member at a released joint induces a carry-over moment at the far end of the member. These carry-over moments are the source of new unbalanced moments at adjacent locked joints. Each joint is unlocked, balanced and then relocked, in turn, and the process is repeated until the unbalanced moments at every joint are negligible. The final moment in a particular member at a joint is obtained by summing the initial FEM and all the increments of distributed and carry-over moments. With the moment at each end of a member thus calculated, the moments and shears at any point along the member can be obtained from statics.

Consider the member AB shown in Figure 11.8a. When the couple M_{AB} is applied to the rotationally released end at A, the member deforms as shown and a moment M_{BA} is induced at the fixed support B at the far end of the member. The relationships between the applied couple M_{AB} and the rotation at A (θ_A) and between the couples at A and B may be expressed as:

$$M_{AB} = k_{AB}\theta_A \tag{11.16}$$

$$M_{BA} = CM_{AB} \tag{11.17}$$

where k_{AB} is the stiffness coefficient for the member AB; and the term C is the carry-over factor.

For a prismatic member, it is a simple matter (using virtual work) to show that for the beam in Figure 11.8a:

$$k = \frac{4EI}{L} \tag{11.18}$$

$$C = 0.5 \tag{11.19}$$

Expressions for the stiffness coefficient and carry-over factor for members with other support conditions are shown in Figure 11.8b through d. FEMs for members carrying distributed and concentrated loads are shown in Figure 11.8e through i.

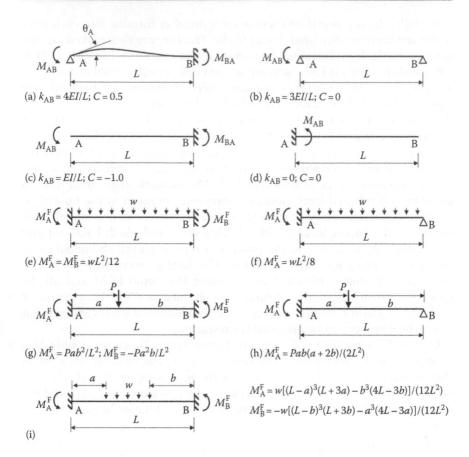

(a) $k_{AB} = 4EI/L$; $C = 0.5$

(b) $k_{AB} = 3EI/L$; $C = 0$

(c) $k_{AB} = EI/L$; $C = -1.0$

(d) $k_{AB} = 0$; $C = 0$

(e) $M_A^F = M_B^F = wL^2/12$

(f) $M_A^F = wL^2/8$

(g) $M_A^F = Pab^2/L^2$; $M_B^F = -Pa^2b/L^2$

(h) $M_A^F = Pab(a + 2b)/(2L^2)$

(i)

$M_A^F = w[(L - a)^3(L + 3a) - b^3(4L - 3b)]/(12L^2)$
$M_B^F = -w[(L - b)^3(L + 3b) - a^3(4L - 3a)]/(12L^2)$

Figure 11.8 Stiffness coefficients, carry-over factors and FEMs for prismatic members. (a) Propped cantilever with moment applied at simple support. (b) Simply-supported beam with moment at one end. (c) Cantilever with moment at free end. (d) Cantilever with moment at fixed end. (e) Uniformly loaded fixed-ended beam. (f) Uniformly loaded propped cantilever. (f) Fixed-ended beam with concentrated load. (h) Propped cantilever with concentrated load. (i) Fixed-ended beam with part uniformly loaded.

The stiffness coefficient for each member framing into a joint in a continuous beam or frame is calculated and summed to obtain the total rotational stiffness of the joint Σk. The *distribution factor* for a member at the joint is the fraction of the total balancing moment distributed to that particular member each time the joint is released. Since each member meeting at a joint rotates by the same amount, the distribution factor for member AB is $k_{AB}/\Sigma k$. The sum of the distribution factors for each member at a joint is therefore unity.

An example of moment distribution applied to a continuous beam is given in the following example. For a more detailed description of moment distribution, the reader is referred to textbooks on structural analysis such as Ref. [1].

EXAMPLE 11.1   CONTINUOUS BEAM

The continuous beam shown in Figure 11.9a has a rectangular cross-section 400 mm wide and 900 mm deep. If the prestressing force is assumed to be constant along the length of the beam and equal to 1800 kN, calculate the bending moment and shear force diagrams induced by prestress. The tendon profile

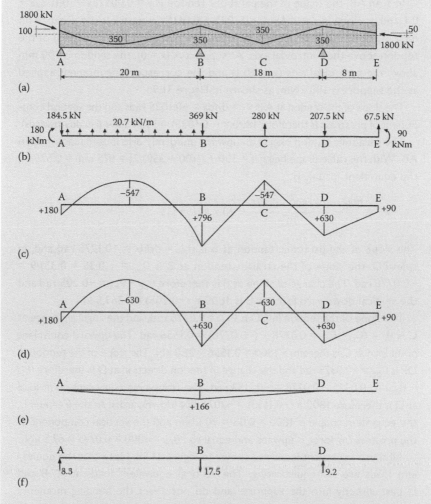

(a)

(b)

(c)

(d)

(e)

(f)

Figure 11.9 Equivalent loads and actions induced by prestress (Example 11.1). (a) Beam elevation and idealised tendon profile. (b) Equivalent loads exerted on concrete beam tendon. (c) Total moment caused by prestress M_p (kNm). (d) Primary moment, Pe (kNm). (e) Secondary moment, $M_p - Pe$ (kNm). (f) Hyperstatic reactions caused by prestress (kN).

shown in Figure 11.9a is adopted for illustrative purposes only. In practice, a post-tensioned tendon profile with sharp kinks or sudden changes in direction would not be used. Relatively short lengths of more gradually curved tendons would be used instead of the kinks shown at B, C and D. The results of an analysis using the idealised tendon profile do, however, provide a reasonable approximation of the behaviour of a more practical beam with continuous curved profiles at B, C and D.

In span AB, the shape of the parabolic tendon is $y = 0.00575x^2 - 0.1025x + 0.1$ and its slope is $y' = dy/dx = -0.1025 + 0.0115x$, where x is the distance (in metres) along the beam from support A and y is the depth (in metres) of the tendon above the centroidal axis. At support A ($x = 0$), the tendon is 100 mm above the centroidal axis ($y = +0.1$) and the corresponding moment applied at the support is 180 kNm, as shown in Figure 11.9b.

The slope of the tendon at A is $\theta_A = dy/dx = -0.1025$ rads and the vertical component of prestress is therefore $1800 \times (-0.1025) = -184.5$ kN (i.e. downwards).

The parabolic tendon exerts an upward uniformly distributed load on span AB. With the cable drape being $h = 350 + [(100 + 350)/2] = 575$ mm $= 0.575$ m, the equivalent load w_p is:

$$w_p = \frac{8Ph}{L^2} = \frac{8 \times 1800 \times 0.575}{20^2} = 20.7 \text{ kN/m } (\uparrow)$$

The slope of the parabolic tendon at B is $\theta_{BA} = dy/dx = +0.1275$ rad and, in span BD, the slope of the straight tendon at B is $\theta_{BC} = -(0.35 + 0.35)/9 = -0.0778$ rad. The change of slope at B is therefore $\theta_{BC} - \theta_{BA} = -0.205$ rad and the vertical downward force at B is $1800 \times (-0.205) = -369.5$ kN.

The slope of the tendon in CD is $\theta_{CD} = 0.0778$ rad, and the angular change at C is $\theta_C = \theta_{CD} - \theta_{BC} = 0.0778 - (-0.0778) = 0.1556$ rad. The upward equivalent point load at C is therefore $1800 \times 0.1556 = 280$ kN. The slope of the tendon in DE is $\theta_{DE} = -0.0375$ rad and the change in tendon direction at D is therefore $\theta_{DE} - \theta_{CD} = -0.0375 - 0.0778 = -0.1153$ rad. The transverse equivalent point load at D is therefore $1800 \times (-0.1153) = -207.5$ kN (downwards). At the free end E, the equivalent couple is $1800 \times 0.05 = 90$ kNm and the vertical component of the prestressing force is upward and equal to $P\theta_{DE} = 1800 \times 0.0375 = 67.5$ kN.

All these equivalent loads are shown in Figure 11.9b. Note that the equivalent loads are self-equilibrating. The vertical equivalent loads at A, B and D pass directly into the supports and do not affect the bending moments induced in the member by prestress.

The continuous beam is analysed under the action of the equivalent loads using moment distribution as outlined in Table 11.1.

The total moment diagram caused by prestress (as calculated in Table 11.1) and the primary moments are illustrated in Figures 11.9c and d, respectively.

Table 11.1 Moment distribution table (Example 11.1)

	AB		BA	BD		DB	DE		ED
Stiffness coefficients			$\dfrac{3EI}{20}$	$\dfrac{4EI}{18}$		$\dfrac{3EI}{18}$	0		
Carry-over factor	0.5		0	0.5		0.5	0		
Distribution factor	0		0.403	0.597		1.0	0		0
FEM (kNm)	−180		+1035	−630		+630	−630		+90
			−90						
			−127	−188					
						−94			
						+94			
				+47					
			−19	−28					
						−14			
						+14			
				+7					
			−3	−4					
Final moments (kNm)	−180		+796	−796		+630	−630		+90

The secondary moment diagram in Figure 11.9e is obtained by subtracting the primary moments from the total moments, and the hyperstatic reactions shown in Figure 11.9f are deduced from the secondary moment diagram.

The secondary shear force diagram corresponding to the hyperstatic reactions is illustrated in Figure 11.10a. The total shear force diagram is obtained

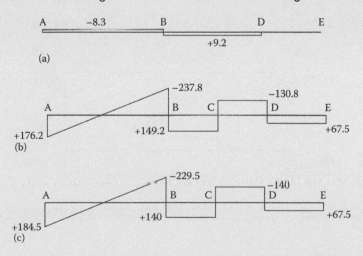

Figure 11.10 Shear force components caused by prestress (Example 11.1). (a) Secondary shear force diagram (kN). (b) Total shear force diagram (kN). (c) Primary shear force diagram *Pθ* (kN).

from statics using the total moments calculated by moment distribution and is given in Figure 11.10b. By subtracting the secondary shear force from the total shear force at each section, the primary shear force diagram shown in Figure 11.10c is obtained. Note that the primary shear force at any section is the vertical component of prestress $P\theta$.

EXAMPLE 11.2   FIXED-ENDED BEAM

The beams shown in Figures 11.11a, 11.12 and 11.13a are rotationally restrained at each end but are not restrained axially. Determine the moments induced by prestress in each member. Assume that the prestressing force is constant throughout and the member has a constant EI.

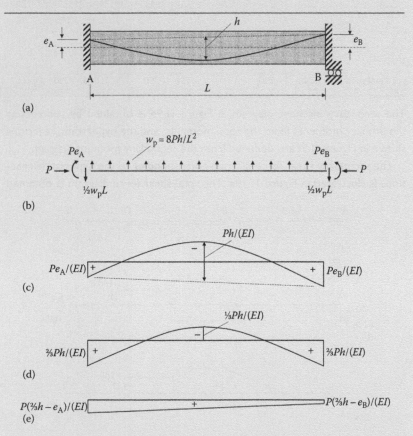

Figure 11.11 Moments caused by a parabolic tendon in a fixed-ended beam. (a) Elevation. (b) Equivalent loads. (c) Primary curvature diagram. (d) Total curvature diagram. (e) Secondary curvature diagram.

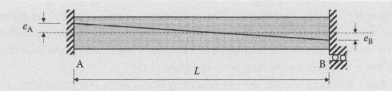

Figure 11.12 Fixed-ended beam with straight tendon.

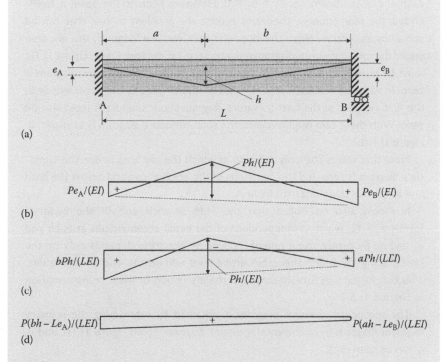

(a)

(b)

(c)

(d)

Figure 11.13 Curvature induced by a harped tendon in a fixed-end beam.
(a) Elevation. (b) Primary curvature diagram. (c) Total curvature diagram. (d) Secondary curvature diagram.

Case (a): The beam shown in Figure 11.11a is prestressed with a parabolic tendon profile with unequal end eccentricities. The equivalent loads on the structure are illustrated in Figure 11.11b, with end moments of Pe_A and Pe_B, as shown, and an equivalent uniformly distributed upward load of $w_p = 8Ph/L^2$.

If the rotational restraints at each end of the beam are released, the curvature is due entirely to the primary moment and is directly proportional to the tendon eccentricity, as shown in Figure 11.11c. The final curvature diagram is obtained by adding the curvature caused by the primary moments to the curvature caused by the restraining secondary moments at each end of the beam M_A^s and M_B^s, respectively. The secondary curvature caused by these secondary moments varies linearly over the length of the beam (provided EI is uniform), so that the final curvature involves a linear shift in the base line of the primary curvature diagram (see Figure 11.11c).

From structural analysis, for example using virtual work, it is possible to calculate the restraining moments M_A^s and M_B^s required to produce zero slope at each end of the beam, i.e. $\theta_A = \theta_B = 0$. However, because the beam is fixed-ended, the moment-area theorems reduce the problem to one that can be solved by inspection. Since the slopes at each end are identical, the net area under the total curvature diagram must be zero, i.e. the base line in Figure 11.11c must be translated and rotated until the area under the curvature diagram is zero. In addition, because support A lies on the tangent to the beam axis at B, the first moment of the final curvature diagram about support A must also be zero. With these two requirements, the total curvature diagram is as shown in Figure 11.11d.

Note that this is the only solution in which the net area under the curvature diagram is zero and the centroids of the areas above and below the base line are at the same distance from A.

It should also be noted that the FEM at each end of the beam is $\frac{2}{3}Ph = w_pL^2/12$, which is independent of the initial eccentricities at each end e_A and e_B. Evidently, the moment induced by prestress depends only on the prestressing force and the cable drape, and not on the end eccentricities. This conclusion was foreshadowed in the discussion of linear transformation in Section 11.3.3.

The secondary moment diagram is obtained by subtracting the primary moment diagram from the total moment diagram. From Figure 11.11c and d, it can be seen that:

$$M_A^s = P\left(\frac{2}{3}h - e_A\right) \quad \text{and} \quad M_B^s = P\left(\frac{2}{3}h - e_B\right) \tag{11.20}$$

The secondary curvature diagram caused by the linearly varying secondary moments is shown in Figure 11.11e.

Case (b): The beam in Figure 11.12 is prestressed with a single straight tendon with arbitrary end eccentricities. This beam is essentially the same as that in the previous example, except that the tendon drape is zero, i.e. the tendon is straight. To satisfy the moment-area theorems in this case, the base line of the total curvature diagram coincides with the primary curvature diagram, i.e. the total moment induced by prestress is everywhere zero, and the primary and secondary moments at each cross-section are equal in magnitude and opposite in sign. By substituting $h = 0$ in Equation 11.20, the secondary moments at each end of the beam of Figure 11.12 are:

$$M_A^s = -M_A^p = -Pe_A \quad \text{and} \quad M_B^s = -M_B^p - Pe_B \qquad (11.21)$$

Case (c): The beam in Figure 11.13a is prestressed with the harped tendon shown. The primary curvature diagram is shown in Figure 11.13b and the total curvature diagram, established for example by satisfaction of the moment-area theorems, is illustrated in Figure 11.13c. As for the previous case, the total curvature (moment) induced by prestress is independent of the end eccentricities e_A and e_B. The curvature induced by the secondary moments is given in Figure 11.13d, and the secondary moments at each support are:

$$M_A^s = P\left(\frac{b}{L}h - e_A\right) \quad \text{and} \quad M_B^s = P\left(\frac{a}{L}h - e_B\right)$$

11.3.5 Practical tendon profiles

In a span of a continuous beam, it is rarely possible to use a tendon profile that consists of a single parabola, as shown in Figure 11.11a. A more realistic tendon profile consists of a series of segments each with a different shape. Frequently, the tendon profile is a series of parabolic segments, concave in the spans and convex over the interior supports, as illustrated in Figure 11.1a. The convex segments are required to avoid sharp kinks in the tendon at the supports.

Consider the span shown in Figure 11.14, with a tendon profile consisting of three parabolic segments. Adjacent segments are said to be compatible at their point of intersection if the slope of each segment is the same. Compatible segments are desirable to avoid kinks in the tendon profile. In Figure 11.14, B is the point of maximum eccentricity e_1 and is located at a distance of $\alpha_1 L$ from the interior support.

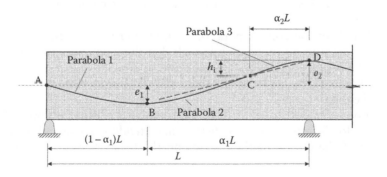

Figure 11.14 Tendon profile with parabolic segments.

Both parabolas 1 and 2 have zero slope at B. The point of inflection at C between the concave parabola 2 and the convex parabola 3 is located at a distance $\alpha_2 L$ from the interior support. Parabolas 2 and 3 have the same slope at C. Over the internal support at D, the eccentricity is e_2 and the slope of parabola 3 is zero. By equating the slopes of parabolas 2 and 3 at C, it can be shown that:

$$h_i = \frac{\alpha_2}{\alpha_1}(e_1 + e_2) \tag{11.22}$$

and that the point C lies on the straight line joining the points of maximum eccentricity, B and D. The slope of parabolas 2 and 3 at C is:

$$\theta_C = \frac{2(e_1 + e_2)}{\alpha_1 L} \tag{11.23}$$

The curvatures of each of the three parabolic segments (κ_{p1}, κ_{p2} and κ_{p3}) are given by:

$$\kappa_{p1} = \frac{1}{R_1} = \frac{2e_1}{L^2(1 - \alpha_1)^2} \quad \text{(concave)} \tag{11.24}$$

$$\kappa_{p2} = \frac{1}{R_2} = \frac{2(e_1 + e_2 - h_i)}{L^2(\alpha_1 - \alpha_2)^2} \quad \text{(concave)} \tag{11.25}$$

$$\kappa_{p3} = \frac{1}{R_3} = \frac{2(e_1 + e_2)}{\alpha_1 \alpha_2 L^2} \quad \text{(convex)} \tag{11.26}$$

where R_1, R_2 and R_3 are the radius of curvature of parabolas 1, 2 and 3, respectively.

The length of the convex parabola $\alpha_2 L$ should be selected so that the radius of curvature of the tendon is such that the duct containing the tendon can be bent to the desired profile without damage. For a multi-strand system, R_3 should be greater than about $75d_d$, where d_d is the inside diameter of the duct.

Equations 11.22 through 11.26 are useful for the calculation of the equivalent loads imposed by a realistic draped tendon profile and the determination of the effects of these loads on the behaviour of a continuous structure.

EXAMPLE 11.3

Determine the total and secondary moments caused by prestress in the fixed-end beam shown in Figure 11.15a. The tendon profile ACDEB consists of three parabolic segments, and the prestressing force is 2500 kN throughout the 16 m span. The convex segments of the tendon at each end of the beam are identical, with zero slope at A and B and a radius of curvature $R_3 = 8$ m. The tendon eccentricity at mid-span and at each support is 300 mm, that is $e_1 = e_2 = 0.3$ m, and α_1 (as defined in Figure 11.14) equals 0.5.

From Equation 11.26:

$$\alpha_2 = \frac{2R_3(e_1 + e_2)}{\alpha_1 L^2} = \frac{2 \times 8 \times (0.3 + 0.3)}{0.5 \times 16^2} = 0.075$$

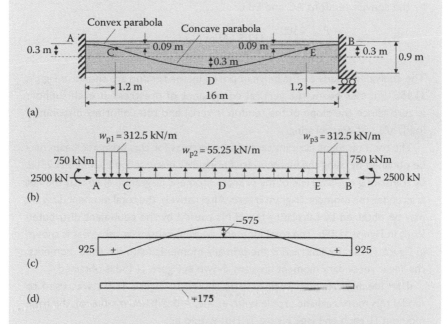

(a)

(b)

(c)

(d)

Figure 11.15 Fixed-end beam with curvilinear tendon profile (Example 11.3). (a) Elevation. (b) Equivalent loads. (c) Total moment diagram caused by prestress (kNm). (d) Secondary moment diagram (kNm).

The convex parabolic segments therefore extend for a distance $\alpha_2 L = 1.2$ m at each end of the span, as shown. The depth of the points of inflection (points C and E) below the tendon level at each support is obtained using Equation 11.22:

$$h_i = \frac{0.075}{0.5}(0.3+0.3) = 0.09 \text{ m}$$

The curvature of the concave parabolic segment CDE extending over the middle $16 - (2 \times 1.2) = 13.6$ m of the span is given by Equation 11.25:

$$\kappa_{p2} = \frac{2 \times (0.3 + 0.3 - 0.09)}{16^2 \times (0.5 - 0.075)^2} = 0.0221 \text{ m}^{-1}$$

and the equivalent uniformly distributed upward load exerted by the concrete tendon is:

$$w_{p2} = P\kappa_{p2} = 2500 \times 0.0221 = 55.25 \text{ kN/m} (\uparrow)$$

The equivalent load w_{p2} acts over the middle 13.6 m of the span. The equivalent downward uniformly distributed load imposed at each end of the beam by the convex tendons AC and EB is:

$$w_{p3} = P\kappa_{p3} = \frac{P}{R_3} = \frac{2500}{8} = 312.5 \text{ kN/m} (\downarrow)$$

The equivalent loads on the beam imposed by the tendon are shown in Figure 11.15b. For this beam, the vertical component of prestress at each support is zero (since the slope of the tendon is zero) and the uniformly distributed loads are self-equilibrating.

The total moment diagram caused by prestress for this prismatic beam may be obtained by using the moment-area principles discussed in Example 11.2, i.e. by translating the base line of the primary moment diagram (Pe) so that the net area under the moment diagram is zero. Alternatively, the total moment diagram may be obtained by calculating the FEMs caused by the equivalent distributed loads in Figure 11.15b. The total moment diagram caused by prestress is shown in Figure 11.15c. By subtracting the primary moments from the total moments, the linear secondary moment diagram shown in Figure 11.15d is obtained.

If an idealised tendon such as that shown in Figure 11.11a was used to model this more realistic profile (with $e_A = e_B = 0.3$ m and $h = 0.6$ m), the total moment at each end (see Figure 11.11d) would be:

$$\frac{2Ph}{3} = \frac{2 \times 2500 \times 0.6}{3} = 1000 \text{ kN}$$

which is about 8% higher than the value shown in Figure 11.15c.

11.3.6 Members with varying cross-sectional properties

The techniques presented for the analysis of continuous structures hold equally well for members with non-uniform section properties. Section properties may vary owing to haunching or changes in member depth (as illustrated in Figure 11.1b, c and e), or from varying web and flange thicknesses or simply from cracking in regions of high moment.

Increasing the member depth by haunching is frequently used to increase the tendon eccentricity in the peak moment regions at the interior supports. In such members, the position of the centroidal axis varies along the member. If the tendon profile is a smooth curve and the centroidal axis suffers sharp changes in direction or abrupt steps (where the member depth changes suddenly), the total moment diagram caused by prestress also exhibits corresponding kinks or steps, as shown in Figure 11.16.

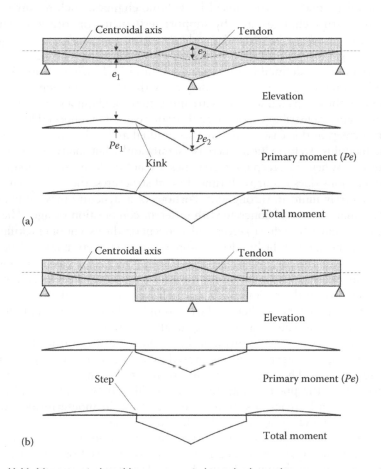

Figure 11.16 Moments induced by prestress in haunched members.

To determine the FEMs and carry-over factors for members with varying section properties and to calculate the member displacements, the principle of virtual work may be used. The internal work is readily calculated using Equation 11.3 by expressing the section properties (EI and EA) as functions of position x.

By dividing the structure into small segments, Equation 11.5 can be used in many practical problems to provide a close approximation of the volume integral for internal work in a non-prismatic member.

11.3.7 Effects of creep

When a statically indeterminate member is subjected to an imposed deformation, the resulting internal actions are proportional to the member stiffness. Since creep gradually reduces stiffness, the internal actions caused by an imposed deformation in a concrete structure decrease with time. Imposed deformations are caused by volume changes, such as shrinkage and temperature changes, and by support settlements or rotations. Under these deformations, the time-dependent restraining actions can be estimated using a reduced or effective modulus for concrete. The age-adjusted effective modulus defined in Equations 4.18 and 5.55 (see Sections 4.2.4.3 and 5.7.2) may be used to model adequately the effects of creep.

Provided the creep characteristics are uniform throughout a structure, creep does not cause redistribution of internal actions caused by imposed loads. The effect of creep in this case is similar to a gradual and uniform change in the elastic modulus. Deformations increase significantly, but internal actions are unaffected. When the creep characteristics are not uniform, redistribution of internal actions does occur with time. In real structures, the creep characteristics are rarely uniform throughout. Portions of a structure may be made of different materials or of concrete with different composition or age. The rate of change of curvature due to creep is dependent on the extent of cracking and the size and position of the bonded reinforcement. The creep characteristics are therefore not uniform if part of the structure has cracked or when the bonded reinforcement layout varies along the member. In general, internal actions are redistributed from the regions with the higher creep rate to the regions with the lower creep rate. Nevertheless, the creep-induced redistribution of internal actions in indeterminate structures is generally relatively small.

Since prestress imposes equivalent loads on structures rather than fixed deformations, the internal actions caused by prestress are not significantly affected by creep. The internal actions are affected in so far as creep causes a reduction of the prestressing force by anything between 0% and about 12%. Hyperstatic reactions induced by prestress in indeterminate structures are not therefore significantly relieved by creep.

If the structural system changes after the application of some of the prestress, creep may cause a change in the hyperstatic reactions. For example the two-span beam shown in Figure 11.17 is fabricated as two precast units

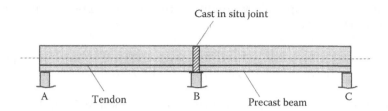

Cast in situ joint

A   Tendon   B   Precast beam   C

Figure 11.17 Providing continuity at an interior support.

of length L and joined together at the interior support by a cast in-situ joint. Creep causes the gradual development of hyperstatic reactions with time and the resulting secondary moments and shears. After the in-situ joint is constructed, the structure is essentially the same as that shown in Figure 11.5a.

Before the joint in Figure 11.17 is cast, the two precast units are simply-supported, with zero deflection and some non-zero slope at the interior support. Immediately after the joint is made and continuity is established, the primary moment in the structure is the same as that shown in Figure 11.5b, but the secondary moment at B (and elsewhere) is zero. With time, creep causes a gradual change in the curvature on each cross-section. If the support at B was released, the member would gradually deflect upward due to the creep-induced hogging curvature associated with the primary moment Pe. If it is assumed that the creep characteristics are uniform and that the prestressing force is constant throughout, the time-dependent upward deflection caused by prestress is obtained by multiplying the deflection given in Equation 11.9 by the creep coefficient (adjusted to include the restraint offered to creep by the bonded reinforcement):

$$v_B(t) = \frac{PeL^2}{2E_c I}\left[\frac{\varphi(t,\tau)}{\alpha}\right] \tag{11.27}$$

where the parameter α was introduced in Equation 5.175, and defined in Section 5.11.4.1, and is typically in the range 1.1–1.5 for an uncracked member containing bonded reinforcement.

The short-term deflection at B caused by a unit value of the redundant force applied at the release B is given in Equation 11.10. Owing to creep, however, the redundant at B is gradually applied to the structure. It is therefore appropriate to use the age-adjusted effective modulus ($\bar{E}_e$ given in Equations 4.18 and 5.55) to determine the corresponding time-dependent deformations (elastic plus creep). Substituting $\bar{E}_e$ for E in Equation 11.10 gives:

$$f_B(t) = \frac{L^3[1 + \chi\varphi(t,\tau)/\alpha]}{6E_c I} \tag{11.28}$$

To enforce the compatibility condition that the deflection at B is zero, Equation 11.11 gives:

$$R_B(t) = -\frac{v_B(t)}{f_B(t)} = -\frac{3Pe}{L}\frac{\varphi(t,\tau)/\alpha}{1+\chi\varphi(t,\tau)/\alpha} = R_B\frac{\varphi(t,\tau)/\alpha}{1+\chi\varphi(t,\tau)/\alpha} \qquad (11.29)$$

where $R_B(t)$ is the creep-induced hyperstatic reaction at B; and R_B is the hyperstatic reaction that would have developed at B if the structure was initially continuous and then prestressed with a straight tendon.

The reaction R_B is shown in Figure 11.5 and given in Equation 11.12. For a prestressed element with typical long-term values of the creep, aging and restraint coefficients (say $\varphi(t, 7) = 2.5$, $\chi = 0.65$ and $\alpha = 1.2$), Equation 11.29 gives:

$$R_B(t) = R_B\frac{2.5/1.2}{1+0.65\times2.5/1.2} = 0.885R_B$$

In general, if R is any hyperstatic reaction or the restrained internal action that would occur at a point due to prestress in a continuous member, and $R(t)$ is the corresponding creep-induced value if the member is made continuous only after the application of the prestress, then:

$$R(t) = \frac{\varphi(t,\tau)/\alpha}{1+\chi\varphi(t,\tau)/\alpha}\,R \qquad (11.30)$$

If the creep characteristics are uniform throughout the structure, then Equation 11.30 may be applied to systems with any number of redundants.

When continuity is provided at the interior supports of a series of simple precast beams not only is the time-dependent deformation caused by prestress restrained, but the deformation due to the external loads is also restrained. For all external loads applied after continuity has been established, the effects can be calculated by moment distribution or an equivalent method of analysis. Due to the loads applied prior to casting the joints, when the precast units are simply-supported (such as self-weight), the moment at each interior support is initially zero. However, after the joint has been cast, the creep-induced deformation resulting from the self-weight moments in the spans is restrained and moments develop at the supports. For the beam shown in Figure 11.5a, the moment at B due to self-weight is $M_B = w_{sw}L^2/8$. For the segmental beam shown in Figure 11.17, it can be shown that the restraining moment that develops at support B due to creep and self-weight is:

$$M_B(t) = \frac{\varphi(t,\tau)/\alpha}{1+\chi\varphi(t,\tau)/\alpha}\,M_B$$

EXAMPLE 11.4

Consider two simply-supported pretensioned concrete planks, erected over two adjacent spans as shown in Figure 11.17. An in-situ reinforced concrete joint is cast at the interior support to provide continuity. Each plank is 1000 mm wide by 150 mm thick and pretensioned with straight strands at a constant depth of 110 mm below the top fibre, with A_p = 400 mm and $P_{p,init}$ = 500 kN. A typical cross-section is shown in Figure 11.18. The span of each plank is L = 6 m. If it is assumed that the continuity is provided immediately after the transfer of prestress, the reactions that develop with time due to creep are to be determined. For convenience, and to better illustrate the effects of creep in this situation, shrinkage is not considered and the self-weight of the planks is also ignored. The material properties are: E_c = 30,000 MPa; f_{ct} = 3.0 MPa; $\varphi(\tau_k, \tau_0)$ = 2.0; $\chi(\tau_k, \tau_0)$ = 0.65; ε_{sh} = 0; and $\bar{E}_e$ = 13,040 MPa. From Equation 5.177, α = 1.15.

Immediately after transfer, and before any creep has taken place, the curvature on each cross-section caused by the eccentric prestress may be calculated using the procedure outlined in Section 5.6.2 giving κ_0 = −2.02 × 10^{-6} mm^{-1}. The top and bottom fibre concrete stresses are $\sigma_{top,0}$ = 1.30 MPa and $\sigma_{top,0}$ = −7.81 MPa, so that cracking has not occurred.

In the absence of any restraint at the supports, the curvature on each cross-section would change with time due to creep from κ_0 to κ_k = −5.81 × 10^{-6} mm^{-1} (calculated using the procedure outlined in Section 5.7.3). If the support at B in Figure 11.17 was removed, the deflection at B ($\Delta v_B(\tau_k)$) that would occur with time due to the change in curvature due to creep (($\kappa_k)_{cr}$ = $\kappa_k − \kappa_0$ = −3.79 × 10^{-6} mm^{-1}) can be calculated from Equation 5.157 for the planks that are now spanning 12 m:

$$\Delta v_B(\tau_k) = \frac{12,000^2}{96}\left[-3.79 + 10 \times (-3.79) - 3.79\right] \times 10^{-6}$$

$$= -68.2 \text{ mm} \quad \text{(i.e. upward)}$$

The deflection at B caused by a unit value of the vertical redundant reaction force gradually applied at the release at B is calculated using the procedures

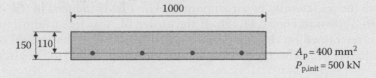

Figure 11.18 Cross section of planks (Example 11.4).

of Section 11.3.7 and for this example is given by Equation 11.28. With the second moment of area of the age-adjusted transformed cross-section $\bar{I} = 288.3 \times 10^6$ mm^4, Equation 11.28 gives:

$$\bar{f}_1 = \frac{L^3}{6\bar{E}_e\bar{I}} = \frac{6,000^3}{6 \times 13,040 \times 288.3 \times 10^6} = 9.58 \times 10^{-3} \text{mm/N}$$

and the redundant force at B that gradually develops with time $R_B(\tau_k)$ is therefore:

$$R_B(\tau_k) = -\frac{\Delta v_B(\tau_k)}{\bar{f}_1} = -\frac{-68.2}{0.00958} = 7117 \text{ N}$$

The reaction that would have developed at B due to prestress immediately after transfer, if the member had initially been continuous, is $R_B = 8621$ N (Equation 11.14) and the approximation of Equation 11.29 gives:

$$R_B(\tau_k) \approx \frac{2.0/1.15}{1 + 0.65 \times 2.0/1.15} \times 8621 = 7038 \text{ N}$$

In this example, Equation 11.29 gives a value for $R_B(\tau_k)$ that is within 2% of the value determined earlier and provides a quick and reasonable estimate of the effects of creep. The secondary moment at B that develops with time due to prestress is 21.4 kNm, and this is 83% of the secondary moment that would have developed if the two planks had been continuous at transfer.

11.4 STATICALLY INDETERMINATE FRAMES

The equivalent load method is a convenient approach for the determination of primary and secondary moments in framed structures. In the treatment of continuous beams in the previous section, it was assumed that all members were free to undergo axial shortening. This is often not the case in real structures. When the horizontal member of a portal frame, for example is prestressed, significant restraint to axial shortening may be provided by the flexural stiffness of the vertical columns. Moment distribution can be used to determine the internal actions that develop in the structure as a result of the axial restraint.

Consider the single-bay portal frame shown in Figure 11.19a. Owing to the axial shortening of the girder BC during prestressing, the top of each column moves laterally by an amount Δ. The FEMs induced in the structure are shown in Figure 11.19b. If the girder BC was free to shorten (i.e. was unrestrained by the columns), the displacement Δ that would occur immediately after the application of a prestressing force P to the girder is:

$$\Delta = \frac{P}{E_c A_b} \frac{L_b}{2} \tag{11.31}$$

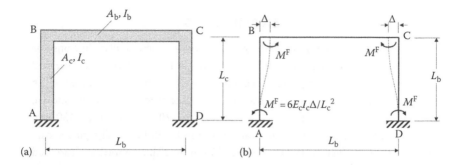

Figure 11.19 FEMs in a fixed-base frame due to axial shortening of the girder. (a) Single-bay portal frame. (b) Equivalent loads.

This value of Δ is usually used as a starting point in the analysis. The FEMs in the supporting columns due to a relative lateral end displacement of Δ are given by:

$$M^F = \frac{6E_cI_c}{L_c^2}\,\Delta \tag{11.32}$$

and a moment distribution is performed to calculate the restraining actions produced by the FEMs. If the base of the frame at A was pinned rather than fixed, the fixed end moment at B due to the rotation at A would be $(3E_cI_c/L_c^2)\Delta$. In addition to bending in the beam and in the columns, an outward horizontal reaction is induced at the base of each column and the girder BC is therefore subjected to tension. The tension in BC will reduce the assumed axial shortening, usually by a small amount when the columns are relatively slender. If the reduction in Δ is significant, a second iteration could be performed using the reduced value for Δ to obtain a revised estimate of the FEMs and, hence, a more accurate estimate of the axial restraint.

The magnitude of the axial restraining actions depends on the relative stiffness of the columns and girder. The stiffer the columns, the greater is the restraint to axial shortening of the girder, and hence the larger is the reduction in prestress in the girder. On the other hand, slender columns offer less resistance to deformation and less restraint to the girder.

Axial shortening of the girder BC can also occur due to creep and shrinkage. A time analysis to include these effects can be made by using the age-adjusted effective modulus for concrete, instead of the elastic modulus, to model the gradually applied restraining actions caused by creep and shrinkage.

The internal actions that arise in a prestressed structure as a result of the restraint to axial deformation are sometimes called *tertiary effects*. These effects are added to the primary and secondary effects (calculated using the equivalent load method) to obtain the total effect of prestress in a framed structure.

EXAMPLE 11.5

Determine the primary, secondary and tertiary moment distributions for the single bay fixed-base portal frame shown in Figure 11.20a. The vertical columns AB and ED are prestressed with a straight tendon profile, while the horizontal girder BD is post-tensioned with a parabolic profile, as shown. The girder BD has a rectangular cross-section 1200 mm deep by 450 mm wide, and the column dimensions are 900 mm by 450 mm. The girder carries a uniformly distributed live load of 10 kN/m, a superimposed dead load

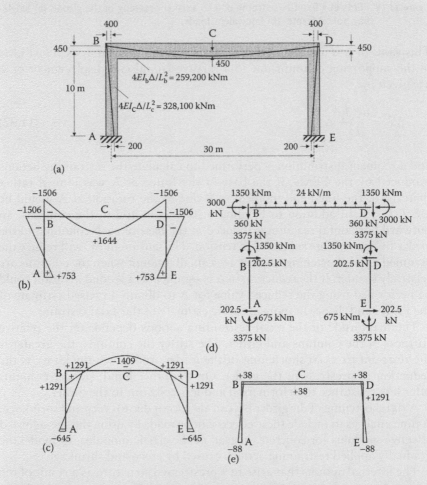

Figure 11.20 Actions in fixed-base portal frame (Example 11.5). (a) Elevation. (b) Moments due to gravity loads (kNm). (c) Primary + secondary moments due to prestress (kNm). (d) Equivalent loads due to prestress. (e) Moments caused by axial restraint (kNm).

of 5 kN/m and the self-weight of the girder is 13 kN/m. If E_c = 30,000 MPa, the moments caused by the total uniformly distributed load on the girder (live load + dead load + self-weight = 28 kN/m) are calculated using moment distribution and are shown in Figure 11.20b.

By satisfying the serviceability requirements (as discussed in Chapter 5), an estimate of the prestressing force and the tendon profile can be made for both the girder BD and the columns. For the girder, the tendon profile shown in Figure 11.20a is selected and the effective prestress $P_{e.BD}$ required to balance the self-weight plus dead load is determined:

$$P_{e.BD} = \frac{18 \times 30^2}{8 \times 0.9} = 2250 \text{ kN}$$

If the time-dependent losses are taken as 25%, the average prestressing force in the girder immediately after transfer is $P_{i.BD}$ = 3000 kN.

To determine the effective prestress in the columns, the primary moments in the girder and in the columns at the corner connections B and D are taken to be the same. If the eccentricity in the column at B (to the centroidal axis of the column) is 400 mm, as shown in Figure 11.20a, then $0.4P_{e.AB} = 0.45P_{e.BD}$. Therefore:

$$P_{e.AB} = \frac{2250 \times 0.45}{0.4} = 2531 \text{ kN}$$

The time-dependent losses in the columns are also taken as 25%, and the prestressing force immediately after transfer is therefore $P_{i.AB}$ = 3375 kN.

The equivalent load method and moment distribution are used here to calculate the primary and secondary moments caused by prestress. The equivalent loads imposed by the tendon on the concrete members immediately after prestressing are shown in Figure 11.20c. The FEM caused by prestress at each end of span BD is obtained using the results of the fixed-ended beam analysed in Example 11.2 – case (a) (and illustrated in Figure 11.11) and is given by:

$$M_{BD}^F = \frac{2P_{i.BD}h_{BD}}{3} = \frac{2 \times 3000 \times 0.9}{3} = 1800 \text{ kNm}$$

The FEMs in the vertical columns due to the straight tendon profile are zero, as was determined for the fixed-end beam analysed in Example 11.2 – case (b). From a moment distribution, the primary and secondary moments caused by prestress are calculated and are illustrated in Figure 11.20d.

To calculate the tertiary effect of axial restraint, the axial shortening of BD immediately after prestressing is estimated using Equation 11.31:

$$\Delta = \frac{3000 \times 10^3}{30,000 \times 1200 \times 450} \frac{30,000}{2} = 2.78 \text{ mm}$$

The FEM in the columns is obtained from Equation 11.32 and is given by:

$$M^F = \frac{6 \times 30,000 \times 900^3 \times 450}{10,000^2 \times 12} \times 2.78 = 136.8 \text{ kNm}$$

Moment distribution produces the tertiary moments shown in Figure 11.20e. The restraining tensile axial force induced in the girder BD is only 12.6 kN and, compared with the initial prestress, is insignificant in this case.

11.5 DESIGN OF CONTINUOUS BEAMS

11.5.1 General

The design procedures outlined in Chapter 10 for statically determinate beams can be extended readily to cover the design of indeterminate beams. The selection of tendon profile and magnitude of prestress in a continuous beam is based on serviceability considerations, as is the case for determinate beams. Load balancing is a commonly used technique for making an initial estimate of the level of prestress required to control deflections. The design of individual cross-sections for bending and shear strength, the estimation of losses of prestress and the design of the anchorage zones are the same for all types of beams, irrespective of the number of redundants.

In continuous beams, the satisfaction of concrete stress limits for crack control must involve consideration of both the primary and secondary moments caused by prestress. Concrete stresses resulting from prestress should be calculated using the pressure line, rather than the tendon profile, as the position of the resultant prestress in the concrete.

Because of the relatively large number of dependent and related variables, the design of continuous beams tends to be more iterative than the design of simple beams, and more dependent on the experience and engineering judgement of the designer. A thorough understanding of the behaviour of continuous prestressed beams and a knowledge of the implications of each design decision is of great benefit.

11.5.2 Service load range – Before cracking

Prior to cracking, the behaviour of a continuous beam is essentially linear and the principle of superposition can be used in the analysis. This means that the internal actions and deformations caused by prestress and those caused by the external loads can be calculated separately using linear analyses and the combined effects obtained by simple summation.

Just as for simple beams, a designer must ensure that a continuous beam is serviceable at the two critical loading stages: immediately after transfer (when the prestress is at its maximum and the applied service loads are usually small) and under the full loads after all losses have taken place (when the prestress is at a minimum and the applied loads are at a maximum).

In order to obtain a good estimate of the in-service behaviour, the prestressing force must be accurately known at each cross-section. This involves a reliable estimate of losses, both short-term and long-term. It is also important to know the load at which flexural cracking is likely to occur. In Section 5.7.3, it was observed that creep and shrinkage gradually relieve the concrete of prestress and transfer the resultant compression from the concrete to the bonded reinforcement. Therefore, a reliable estimate of the cracking moment at a particular cross-section must involve consideration of the time-dependent effects of creep and shrinkage.

Prior to cracking, load balancing can be used in design to establish a suitable effective prestressing force and tendon profile. The concept of load balancing was introduced in Section 1.4.3 and involves balancing a preselected portion of the applied load (and self-weight) with the transverse equivalent load imposed on the beam by the draped tendons. Under the balanced load, w_b, the curvature on each cross-section is zero, the beam does not therefore deflect and each cross-section is subjected only to the longitudinal axial prestress applied at the anchorages.

By selecting a parabolic tendon profile with the drape h as large as cover requirements permit, the minimum prestressing force required to balance w_b is calculated by rearranging Equation 1.7 to give:

$$P = \frac{w_b L^2}{8h}$$ (11.33)

In order to control the final deflection of a continuous beam, the balanced load w_b is often taken to be the sustained or permanent load (or some significant percentage of it).

Because of its simplicity, load balancing is probably the most popular approach for determining the prestressing force in a continuous member. Control of deflection is an obvious attraction. However, load balancing does not guard against cracking caused by the unbalanced loads and it does not ensure that individual cross-sections possess adequate strength. If the balanced load is small, and hence the prestressing force and prestressing steel quantities are also small, significant quantities of non-prestressed steel may be required to increase the strength of the critical cross-sections and to limit crack widths under the full service loads.

At service loads prior to cracking, the concrete stresses on any cross-section of a continuous beam can be calculated easily by considering only the unbalanced load and the longitudinal prestress. The transverse loads imposed on the beam by the draped tendons have been effectively cancelled by w_b. The total moment diagram due to prestress (primary + secondary moments) is equal and opposite to the moment diagram caused by w_b. The primary and secondary moments induced by prestress need not, therefore, enter into the calculations, and there is no need to calculate the hyperstatic reactions at this stage (at least for the determination of concrete stresses). In Example 11.6, the load balancing approach is applied to a two-span continuous member.

In the discussion to this point, the prestressing force has been assumed to be constant throughout the member. In long members, friction losses may be significant and the assumption of constant prestress may lead to serious errors. To account for variations in the prestressing force with distance from the anchorage, a continuous member may be divided into segments. Within each segment, the prestressing force may be assumed constant and equal to its value at the mid-point of the segment. In many cases, it may be acceptable to adopt each individual span as a segment of constant prestress. In other cases, it may be necessary to choose smaller segments to model the effects of variations in prestress more accurately.

It is possible, although rarely necessary, to calculate the equivalent loads due to a continuously varying prestressing force. With the shape of the tendon profile throughout the member and the variation of prestress due to friction and draw-in determined previously, the transverse equivalent load at any point is equal to the curvature of the tendon (obtained by differentiating the equation for the tendon shape twice) times the prestressing force at that point. The effect of prestress due to these non-uniform equivalent transverse loads can then be determined using the same procedures as for uniform loads.

EXAMPLE 11.6   LOAD BALANCING

The idealised parabolic tendons in the two-span beam shown in Figure 11.21 are required to balance a uniformly distributed gravity load of 20 kN/m. The beam cross-section is rectangular: 800 mm deep and 300 mm wide. Determine the concrete stress distribution on the cross-section at B over the interior support when the total uniformly distributed gravity load is 25 kN/m. Assume that the prestressing force is constant throughout.

In span AB, the tendon sag is $h_{AB} = 325 + (0.5 \times 325) = 487.5$ mm and the required prestressing force is obtained from Equation 11.33:

$$P = \frac{20 \times 16^2}{8 \times 0.4875} = 1313 \text{ kN}$$

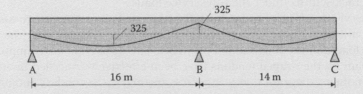

Figure 11.21 Two-span beam (Example 11.6).

If P is constant throughout, the required sag in BC may also be obtained from Equation 11.33:

$$h_{BC} = \frac{20 \times 14^2}{8 \times 1313} = 0.373 \text{ m} = 373 \text{ mm}$$

and the eccentricity of the tendon at the mid-point of the span BC is equal to:

$$373 - (0.5 \times 325) = 210.5 \text{ mm (below the centroidal axis)}$$

Under the balanced load of 20 kN/m, the beam is subjected only to the axial prestress applied at each anchorage. The concrete stress on every cross-section is uniform and equal to:

$$\sigma = -\frac{P}{A} = -\frac{1313 \times 10^3}{800 \times 300} = -5.47 \text{ MPa}$$

The bending moment at B due to the uniformly distributed unbalanced load of kN/m is −142.5 kNm (obtained by moment distribution or an equivalent method of analysis) and the extreme fibre concrete stresses at B caused by the unbalanced moment are:

$$\sigma = \pm\frac{M}{Z} = \pm\frac{142.5 \times 10^6 \times 6}{800^2 \times 300} = \pm4.45 \text{ MPa}$$

The resultant top and bottom fibre stresses at B caused by prestress and the applied load of 25 kN/m are therefore:

$$\sigma_t = -5.47 + 4.45 = -1.02 \text{ MPa}$$

$$\sigma_b = -5.47 - 4.45 = -9.92 \text{ MPa}$$

The same result could have been obtained by adding the total stresses caused by the equivalent loads (longitudinal plus transverse forces imposed by prestress) to the stresses caused by a uniformly distributed gravity load of 25 kN/m.

11.5.3 Service load range – After cracking

When the balanced load is relatively small, the unbalanced load may cause cracking in the peak moment regions over the interior supports and at mid-span. When cracking occurs, the stiffness of the member is reduced in the vicinity of the cracks. The change in relative stiffness between the positive and negative moment regions causes a redistribution of bending moments. In prestressed members, the reduction of stiffness caused by cracking in a particular region is not as great as in an equivalent reinforced concrete member and the redistribution of bending moments under short-term service loads can usually be ignored. It is therefore usual to calculate beam moments using a linear analysis both before and after cracking.

The effect of cracking should not be ignored, however, when calculating the deflection of the member. A cracked section analysis (see Section 5.8.3) can be used to determine the effective moment of inertia of the cracked section (see Section 5.11.3) and the corresponding initial curvature. After calculating the initial curvature at each end and at the mid-point of a span, the short-term deflection may be determined using Equation 5.157.

Under the sustained loads, the extent of cracking is usually not great. In many partially-prestressed members, the cracks over the interior supports (caused by the peak loads) are completely closed for most of the life of the member. The time-dependent change in curvature caused by creep, shrinkage and relaxation at each support and at mid-span can be calculated using the time analysis of Section 5.7.3 (or Section 5.9.2 if the cracks remain open under the permanent loads). With the final curvature determined at the critical sections, the long-term deflection can also be calculated using Equation 5.157.

Alternatively, long-term deflections may be estimated from the short-term deflections using the approximate expressions outlined in Section 5.11.4.

The control of flexural cracking in a cracked prestressed beam is easily achieved by suitably detailing the bonded reinforcement in the cracked region. According to AS3600-2009 [3], crack widths may be considered to be satisfactory for interior exposure conditions, provided the change in stress in the bonded tensile steel is less than that given in Table 5.3 (in Section 5.12.1) as the load is increased from its value when the extreme concrete tensile fibre is at zero stress to the short-term service load value. The change in tensile steel stress may be calculated in a cracked section analysis. In addition, the centre-to-centre spacing of the bonded steel should be less than 300 mm.

11.5.4 Overload range and ultimate strength in bending

11.5.4.1 Behaviour

The behaviour of a continuous beam in the overload range depends on the ductility of the cross-sections in the regions of maximum moment. If the cross-sections are ductile, their moment–curvature relationships are similar to that shown in Figure 11.22.

Consider the propped cantilever shown in Figure 11.23a. Each cross-section is assumed to possess a ductile moment–curvature relationship. At service loads, bending moments in the beam, even in the post-cracking range, may be approximated reasonably using elastic analysis. The magnitude of the negative elastic moment at A caused by the uniformly distributed load w is $wL^2/8$. When the load w causes yielding of the reinforcement

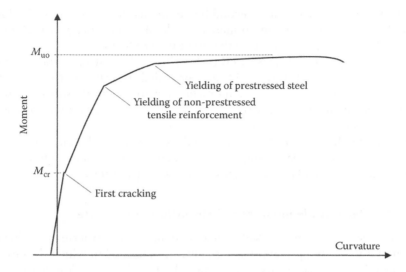

Figure 11.22 Moment–curvature relationship for a ductile prestressed cross-section.

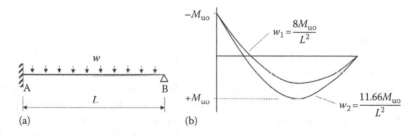

Figure 11.23 Moment redistribution in a propped cantilever. (a) Beam layout. (b) Bending moment diagrams.

on the cross-section at A, a sudden loss of stiffness occurs (as illustrated by the kinks in the moment–curvature relationship in Figure 11.22). Any further increase in load will cause large increases in curvature at support A, but only small increases in moment. A constant-moment hinge (or *plastic hinge*) develops at A as the moment capacity is all but exhausted and the curvature becomes large. In reality, the moment at the hinge is not constant, but the rate of increase in moment with curvature in the post-yield range is very small. As loading increases and the moment at the support A remains constant or nearly so, the moment at mid-span increases until it too reaches its ultimate value M_{uo} and a second plastic hinge develops. The formation of two constant-moment hinges reduces a one-fold indeterminate structure to a mechanism and collapse occurs. If an elastic-perfectly plastic moment–curvature relationship is assumed with the same moment capacity M_{uo} at both hinge locations, the moment diagrams associated with the formation of the first and second hinges are as shown in Figure 11.23b. The ductility at A results in an increase in load-carrying capacity of 46% above the load required to cause the first hinge to form.

Plastic analysis techniques can therefore be used to estimate the collapse load of a continuous prestressed beam, provided the critical cross-sections are ductile, that is provided the moment–curvature relationships can reasonably be assumed to be elastic–plastic and the critical cross-sections possess the necessary rotational capacity.

By subdividing a member into small segments and calculating the moment–curvature relationship for each segment, an incremental analysis may be used to calculate the collapse load more accurately.

11.5.4.2 Permissible moment redistribution at ultimate

For the design of prestressed concrete continuous structures, a lower bound ultimate strength approach is generally specified in which the design moment M^* on every cross-section must be less than the design strength ϕM_{uo}. Design moments are usually calculated using elastic analysis and gross member stiffnesses (and are therefore very approximate). To account for the beneficial effects of moment redistribution, AS3600-2009 [3] generally permits the peak elastic moments at the interior supports of a continuous beam to be reduced, provided the cross-sections at these supports are ductile. A reduction in the magnitudes of the negative moments at the ends of a span must be associated with an increase in the positive span moment in order to maintain equilibrium. The elastically determined bending moments at any interior support may be reduced (or increased) by redistribution, provided the designer can show that there is adequate rotational capacity in the peak moment regions. This requirement is deemed to be satisfied, provided:

1. all of the longitudinal tensile steel is either tendons or ductility class N reinforcing bars;
2. the bending moment distribution before redistribution shall be obtained from an elastic analysis;
3. the maximum decrease or increase in the moment at an interior support (λ_m as a percentage of the bending moment before redistribution) depends on the neutral axis parameter at the ultimate limit state at the interior support ($k_u = d_n/d$) and is given by:
   a. $\lambda_m = 30$ when $k_u \leq 0.2$ in all peak moment regions
   b. $\lambda_m = 75(0.4 - k_u)$ when $0.2 < k_u \leq 0.4$ in one or more peak moment regions
   c. $\lambda_m = 0$ when $k_u > 0.4$ in any peak moment region
4. the positive bending moment in all spans shall be adjusted to maintain equilibrium; and
5. the static equilibrium of the structure after redistribution is used to evaluate all action effects for strength design, including shear checks.

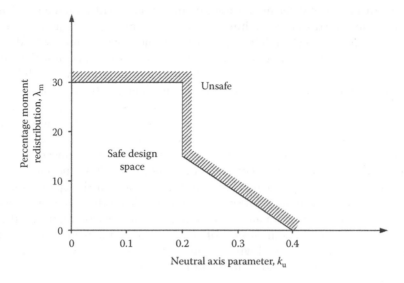

Figure 11.24 Permissible moment redistribution [3].

The values of k_u are calculated for cross-sections that have been designed based on the redistributed bending moment diagram.

Figure 11.24 shows the permissible design space according to AS3600-2009 [3]. Designers can safely choose any desired moment redistribution between 0% and 30%, provided k_u does not exceed 0.2. The reduction in allowable moment redistribution as k_u values increase above 0.2 reflects the decreasing section ductility with decreasing values of ultimate curvature $(0.003/d_n)$, i.e. with increasing values of $d_n = k_u d$.

11.5.4.3 Secondary effects at ultimate

AS3600-2009 [3] requires that the design moment M^* be calculated as the sum of the moments caused by the factored design load (dead, live, etc., as outlined in Section 2.3.2, Equations 2.1 through 2.7) and the moments resulting from the hyperstatic reactions caused by prestress (with load factor of 1.0).

Earlier in this chapter, the hyperstatic reactions and the resulting secondary moments were calculated using linear-elastic analysis. Primary moments, secondary moments and the moments caused by the applied loads were calculated separately and summed to obtain the combined effect. Superposition is only applicable, however, when the member behaviour is linear. At overloads, behaviour is highly nonlinear and it is not possible to distinguish between the moments caused by the applied loads and those caused by the hyperstatic reactions. Consider the ductile propped cantilever in Figure 11.23. After the formation of the first plastic hinge at A, the beam becomes determinate for all subsequent load increments. With no rotational restraint at the hinge at A, the magnitude of the secondary moment is not at all clear. The total moment and shears can only be determined using a refined analysis that accurately takes into account the various sources of material non-linearity. It is meaningless to try to subdivide the total moments into individual components. The treatment of secondary moments at ultimate has been studied extensively, see References [4–7].

Provided that the structure is ductile and moment redistribution occurs as the collapse load is approached, secondary moments can be ignored in ultimate strength calculations. After all, the inclusion of an uncertain estimate of the secondary moment generally amounts to nothing more than an increase in the support moments and a decrease in the span moment or vice versa, i.e. a redistribution of moments. Since the moments due to the factored loads at ultimate are calculated using elastic analysis, there is no guarantee that the inclusion of the secondary moments (also calculated using gross stiffnesses) will provide better agreement with the actual moments in the structure after moment redistribution.

On the other hand, if the critical section at an interior support is non-ductile, its design needs to be carefully considered. It is usually possible to avoid non-ductile sections by the inclusion of sufficient quantities of compressive reinforcement. If non-ductile sections cannot be avoided, it is recommended that secondary moments (calculated using linear elastic analysis and gross stiffnesses) are considered at ultimate. Where the secondary moment at an interior support has the same sign as the moment caused by the applied loads, it is conservative to include the secondary moment (with a load factor of 1.0) in the calculation of the design moment M^*. Where the secondary moment is of opposite sign to the moment caused by the applied loads, it is conservative to ignore its effect.

11.5.5 Steps in design

A suitable design sequence for a continuous prestressed concrete member is as follows:

1. Determine the loads on the beam both at transfer and under the most severe load combination for the serviceability limit states. Make an initial selection of concrete strength and establish material properties.

    Using approximate analysis techniques, estimate the maximum design moments at the critical sections in order to make an initial estimate of the cross-section size and self-weight. The moment and deflection coefficients given in Figure 11.25 may prove useful.

    Determine appropriate cross-section sizes at the critical sections. The discussion in Section 10.3 is relevant here.

    Equation 10.11 may be used to obtain cross-sectional dimensions that are suitable from the point of view of flexural strength and ductility. By

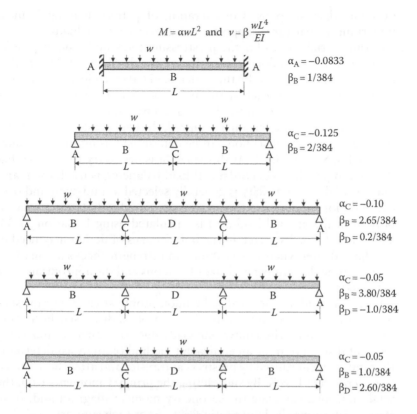

$$M = \alpha w L^2 \quad \text{and} \quad v = \beta \frac{w L^4}{EI}$$

Figure 11.25 Moment and deflection coefficients for equal span elastic beams.

estimating the maximum unbalanced load, the sustained part of the unbalanced load and by specifying a maximum deflection limit for the structure, a minimum moment of inertia may be selected from Equation 10.5 (if the member is to be crack-free) or 10.6 (if cracking does occur). If a fully prestressed (uncracked) beam is required, Equation 10.1 can be used to determine the minimum section modulus at each critical section.

For continuous beams in building structures, the span-to-depth ratio is usually in the range 24–30, but this depends on the load level and the type of cross-section.

2. Determine the bending moment and shear force envelopes both at transfer and under the full service loads. These envelopes should include the effects of self-weight, superimposed permanent dead and live loads and the maximum and minimum values caused by transient loads. Where they are significant, pattern loadings such as those shown in Figure 11.25 should be considered. For example, the minimum moment at the mid-point of a particular span may not be due to dead load only, but may result when the transient live load occurs

only on adjacent spans. Consideration of pattern loading is most important in structures supporting large transient live loads.

3. Determine trial values for the prestressing force and tendon profile. Assume idealised tendon profiles that follow the shape of the bending moment diagram caused by the anticipated balanced loads (or as near to it as practical). In each span, make the tendon drape as large as possible in order to minimise the required prestress.

   If a fully-prestressed beam is required, the trial prestress and eccentricity at each critical section can be determined using the procedure outlined in Section 5.4.1. At this stage, it is necessary to assume that the tendon profile is concordant. If load balancing is used, the maximum available eccentricity is generally selected at mid-span and over each interior support and the prestress required to balance a selected portion of the applied load (w_b) is calculated using Equation 11.33. The balanced load selected in the initial stages of design may need to be adjusted later when serviceability and strength checks are made.

   Determine the number and size of tendons and the appropriate duct diameter(s).

4. Replace the kink in the idealised tendon profile at each interior support with a short convex parabolic segment as discussed in Section 11.3.5. Determine the equivalent loads due to prestress, and using moment distribution (or an equivalent method of analysis) determine the total moment caused by prestress at transfer and after the assumed time-dependent losses. By subtracting the primary moments from the total moments, calculate the secondary moment diagram and, from statics, determine the hyperstatic reactions at each support.

5. Concrete stresses at any cross-section caused by prestress (including both primary and secondary effects) and the applied loads may now be checked at transfer and after all losses. If the beam is fully-prestressed, the trial estimate of prestress made in Step 3 was based on the assumption of a concordant tendon profile and secondary moments were ignored. If secondary moments are significant, stresses calculated here may not be within acceptable limits and a variation of either the prestressing force or the eccentricity may be required.

6. Calculate the losses of prestress and check the assumptions made earlier.

7. Check the ultimate strength in bending at each critical section. If necessary, additional non-prestressed tensile reinforcement may be used to increase strength. Add compressive reinforcement to improve ductility, if required. Some moment redistribution at ultimate may be permissible to reduce peak negative moments at interior supports, provided that cross-sections at the supports have adequate ductility.

8. Check the deflection at transfer and the final long-term deflection. For partially prestressed designs, check crack control in regions of peak moment. Consider the inclusion of non-prestressed steel to reduce

time-dependent deformations, if necessary. Adjust the section size or the prestress level (or both), if the calculated deflection is excessive.
9. Check shear strength of beam (and torsional strength if applicable) in accordance with the provisions outlined in Chapter 7. Design suitable shear reinforcement where required.
10. Design the anchorage zone using the procedures presented in Chapter 8.

Note: Durability and fire protection requirements are usually satisfied by an appropriate choice of concrete strength and cover to the tendons in Steps 1 and 3.

EXAMPLE 11.7

Design the four-span beam shown in Figure 11.26. The beam has a uniform I-shaped cross-section and carries a uniformly distributed dead load of 25 kN/m (not including self-weight) and a transient live load of 20 kN/m. Controlled cracking is to be permitted at peak loads. The beam is prestressed by jacking simultaneously from each end, thereby maintaining symmetry of the prestressing force about the central support C and avoiding excessive friction losses. Take $f_c' = 40$ MPa, $f_{cp}' = 30$ MPa and $f_{pb} = 1870$ MPa.

---

1 (and 2). The bending moments caused by the applied loads must first be determined. Because the beam is symmetrical about the central support at C, the bending moment envelopes can be constructed from the moment diagrams shown in Figure 11.27 caused by the distributed load patterns shown. These moment diagrams were calculated for a unit distributed load (1 kN/m) using moment distribution.

If the self-weight is estimated at 15 kN/m and the total dead load is 40 kN/m, then factored design loads are:

$$w_G^* = 1.2 \times 40 = 48 \text{ kN/m} \quad \text{and} \quad w_Q^* = 1.5 \times 20 = 30 \text{ kN/m}$$

The maximum design moment M^* occurs over the support C, when the transient live load is on only the adjacent spans BC and CB'. Therefore, using the moment coefficients in Figure 11.27:

$$M^* = 80.9 \times 48 + (-46.3 - 46.3) \times 30 = 6661 \text{ kNm}$$

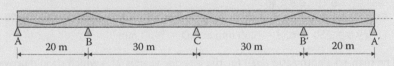

Figure 11.26 Elevation of beam (Example 11.7).

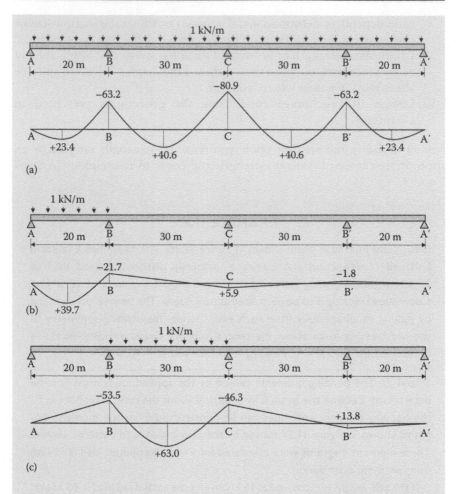

Figure 11.27 Bending moment diagrams (kNm) due to unit distributed loads (Example 11.7). (a) Load case 1: 1 kN/m throughout. (b) Load case 2: 1 kN/m on span AB only. (c) Load case 3: 1 kN/m on span BC only.

The overall dimensions of the cross-section are estimated using Equation 10.11 (which is valid, provided the compressive stress block at ultimate is within the flange of the I-section):

$$bd^2 \geq \frac{6661 \times 10^6}{0.18 \times 0.8 \times 40} = 1156 \times 10^6 \text{ mm}^3$$

Try b = 750 mm, d = 1250 mm and D = 1400 mm

The span-to-depth ratio (L/D) for the interior span is 21.4 which should prove acceptable from a serviceability point of view.

To obtain a trial flange thickness, find the depth of the compressive stress block at ultimate. The volume of the stress block is $C_c = \alpha_2 f'_c \gamma d_n b$ (see Equation 6.4), with $\alpha_2 = 0.85$ and $\gamma = 0.77$ for 40 MPa concrete (Table 6.1). With the lever arm between C_c and the resultant tensile force taken to be $0.85d$, then:

$$\gamma d_n = \frac{M^*}{\phi \times 0.85 d \times 0.85 f'_c \; b} = \frac{6661 \times 10^6}{0.8 \times 0.85 \times 1250 \times 0.85 \times 40 \times 750} = 307 \text{ mm}$$

Adopt a tapering flange 250 mm thick at the tip and 350 mm thick at the web.

To ensure that the web width is adequate for shear, it is necessary to ensure that web crushing does not occur. If the vertical component of the prestressing force P_v is ignored, then Equation 7.8 gives:

$$V^* \leq \phi V_{u.max} = \phi \times 0.2 f'_c \, b_v d_o \text{ and therefore } b_v \geq \frac{7.14 \, V^*}{f'_c d_o}$$

The maximum shear force V^* also occurs adjacent to support C when live load is applied to span BC and CB' and is equal to 1251 kN. Therefore:

$$b_v \geq \frac{7.14 \times 1251 \times 10^3}{40 \times 1340} = 167 \text{ mm}$$

It is advisable to select a web width significantly greater than this minimum value in order to avoid unnecessarily large quantities of transverse steel and the resulting steel congestion. Duct diameters of about 100 mm are anticipated, with only one duct in any horizontal plane through the web. With these considerations, the web width is taken to be:

$$b_w = b_v + \frac{1}{2} d_d = 300 \text{ mm}$$

The trial cross-section and section properties are shown in Figure 11.28. The self-weight is $24 \times 0.69 = 16.6$ kN/m, which is 10% higher than originally assumed. The revised value of M^* is 6985 kNm.

3. If 100 mm ducts are assumed (side by side in the flanges), the cover to the reinforcement is 40 and 12 mm stirrups are used, the maximum eccentricity over an interior support and at mid-span is:

$$e_{max} = 700 - 40 - 12 - \left(\frac{3}{4} \times 100 \right) = 573 \text{ mm}$$

The maximum drape in the spans BC and CB' is therefore:

$$(h_{BC})_{max} = 2 \times 573 = 1146 \text{ mm}$$

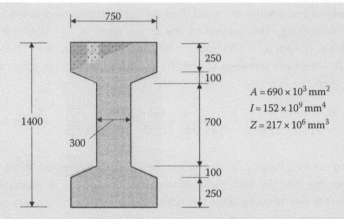

Figure 11.28 Trial section dimensions and properties.

The balanced load is taken to be 32 kN/m (which is equal to self-weight plus about 60% of the additional dead load). From Equation 11.33, the required average effective prestress in span BC is:

$$(P_e)_{BC} = \frac{32 \times 30^2}{8 \times 1.146} = 3141 \text{ kN}$$

If the friction loss between the mid-point of span BC and the mid-point of AB is 15% (to be subsequently checked), then:

$$(P_e)_{AB} = \frac{3141}{0.85} = 3694 \text{ kN}$$

and the required drape in span AB is:

$$h_{AB} = \frac{32 \times 20^2}{8 \times 3694} = 0.433 \text{ m} = 433 \text{ mm}$$

The idealised tendon profiles for spans AB and BC are shown in Figure 11.29, together with the corresponding tendon slopes and friction losses (calculated from Equation 5.136 with $\mu = 0.2$ and $\beta_p = 0.01$). The friction losses at the mid-span of BC are 17.3%, and if the time-dependent losses in BC are assumed to be 20%, then the required jacking force is:

$$P_j = \frac{3141}{0.827 \times 0.8} = 4748 \text{ kN}$$

The maximum jacking force for a 12.7 mm diameter strand is 0.85 × 184 = 156.4 kN (see Table 4.8). The minimum number of strands is therefore 4748/156.4 = 30.

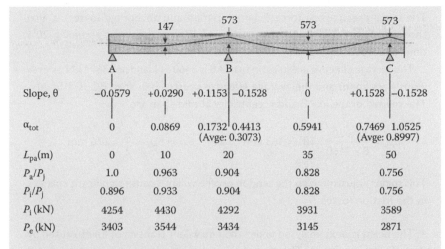

Slope, θ	−0.0579	+0.0290	+0.1153	−0.1528	0	+0.1528	−0.1528
α_{tot}	0	0.0869	0.1732	0.4413	0.5941	0.7469	1.0525
			(Avge: 0.3073)			(Avge: 0.8997)	
L_{pa}(m)	0	10	20		35	50	
P_a/P_j	1.0	0.963	0.904		0.828	0.756	
P_i/P_j	0.896	0.933	0.904		0.828	0.756	
P_i (kN)	4254	4430	4292		3931	3589	
P_e (kN)	3403	3544	3434		3145	2871	

Figure 11.29 Friction losses and tendon forces.

Try two cables each containing 15 strands (A_p = 1479 mm² per strand).

The two cables are to be positioned so that they are located side by side in the top flange over the interior supports and in the bottom flange at mid-span of BC, but are located one above the other in the web. The position of the resultant tension in the tendons should follow the desired tendon profile.

The loss of prestress due to a 6 mm draw-in at the anchorage is calculated as outlined in Section 5.10.2.4. The slope of the prestress line adjacent to the anchorage at A is:

$$\alpha = \frac{0.037 \times P_i}{L_{AB}/2} = 17.57 \text{ N/mm}$$

and, from Equation 5.138, the length of beam associated with the draw-in losses is:

$$L_{di} = \sqrt{\frac{195,000 \times 2 \times 1,479 \times 6}{17.57}} = 14,035 \text{ mm}$$

The loss of force at the jack due to slip at the anchorage is given by Equation 5.139:

$$(\Delta P_{di})_{L_{di}} = 2\alpha L_{di} = 2 \times 17.57 \times 14,035 \times 10^{-3} = 493 \text{ kN } (-0.104 P_j)$$

and at mid-span:

$$(\Delta P_{di})_{AB} = 2\alpha \left(L_{di} - \frac{L_{AB}}{2} \right) = 2 \times 17.57 \times (14,035 - 10,000) \times 10^{-3}$$

$$= 142 \text{ kN } (=0.030 P_j)$$

The initial prestressing force P_i (after friction and anchorage losses) is also shown in Figure 11.29, together with the effective prestress assuming 20% time-dependent losses.

The average effective prestress in span AB is 3460 kN (and not 3694 kN as previously assumed) and the average effective prestress in span BC is 3150 kN. The revised drape in AB and eccentricity at mid-span are:

$$h_{AB} = \frac{32 \times 20^2}{8 \times 3460} \times 10^3 = 463 \text{ mm} \quad \text{and} \quad e_{AB} = h_{AB} - \frac{e_B}{2} = 176 \text{ mm}$$

This minor adjustment to the tendon profile will not cause significant changes in the friction losses.

4. The beam is next analysed under the equivalent transverse loads caused by the effective prestress. The sharp kinks in the tendons over the supports B and C are replaced by short lengths with a convex parabolic shape, as illustrated and analysed in Example 11.3. In this example, it is assumed that the idealised tendons provide a close enough estimate of moments due to prestress.

The equivalent uniformly distributed transverse load due to the effective prestress is approximately 32 kN/m (upward). Using the moment diagram in Figure 11.27a, the total moments due to prestress at B and C are:

$$(M_{pt})_B = +63.2 \times 32 = 2022 \text{ kNm} \quad \text{and} \quad (M_{pt})_C = +80.9 \times 32 = 2589 \text{ kNm}$$

The secondary moments at B and C are obtained by subtracting the primary moments corresponding to the average prestress in each span (as was used for the calculation of total moments):

$$(M_{ps})_B = (M_{pt})_B - (P_e e)_B = 2022 - 0.5 \times (3150 + 3460) \times 0.573 = 128 \text{ kNm}$$

$$(M_{ps})_C = (M_{pt})_C - (P_e e)_C = 2589 - 3150 \times 0.573 = 784 \text{ kNm}$$

The total and secondary moment diagrams are shown in Figure 11.30, together with the corresponding hyperstatic reactions. It should be noted that, in the real beam, the equivalent transverse load varies along the beam as the prestressing force varies and the moment diagrams shown in Figure 11.30 are only approximate. A more accurate estimate of the moments due to prestress and the hyperstatic reactions can be made by dividing each span into smaller segments (say four per span) and assuming constant prestress in each of these segments.

5. It is prudent to check the concrete stresses at transfer. The equivalent transverse load at transfer is 32/0.8 = 40 kN/m (↑) and the self-weight is

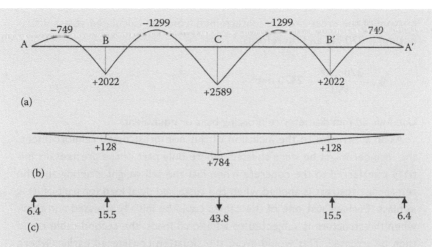

(a)

(b)

(c)

Figure 11.30 Moments and reactions caused by the average effective prestress.
(a) Total moment caused by prestress (kNm). (b) Secondary moments
(kNm). (c) Hyperstatic reactions (kN).

16.6 kN/m ($\downarrow$). Therefore, the unbalanced load is 23.4 kN/m ($\uparrow$). At support
C, the moment caused by the uniformly distributed unbalanced load is (see
Figure 11.27a):

$$(M_{ub})_C = +80.9 \times 23.4 = 1893 \text{ kNm}$$

and the initial prestressing force at C is 3589 kN. The extreme fibre concrete
stresses immediately after transfer at C are:

$$\sigma_t = -\frac{3589 \times 10^3}{690 \times 10^3} - \frac{1893 \times 10^6}{217 \times 10^6} = -5.20 - 8.72 = -13.92 \text{ MPa}$$

$$\sigma_b = -\frac{3589 \times 10^3}{690 \times 10^3} + \frac{1893 \times 10^6}{217 \times 10^6} = -5.20 + 8.72 = +3.52 \text{ MPa}$$

The flexural tensile strength at transfer is $0.6\sqrt{f'_{cp}} = 3.29$ MPa and, therefore, cracking at support C is likely to occur at transfer. Bonded reinforcement should therefore be provided in the bottom of the member over
support C to control cracking at transfer. For this level of tension, it is reasonable to calculate the resultant tensile force on the concrete (assuming
no cracking) and supply enough non-prestressed steel to carry this tension
with a steel stress of 150 MPa. In this case, the resultant tension near the

bottom of the cross-section (determined from the calculated stress distribution) is 370 kN and therefore:

$$A_{st} = \frac{370 \times 10^3}{150} = 2470 \text{ mm}^2$$

Use four 28 mm diameter reinforcing bars or equivalent.

As an alternative to the inclusion of this non-prestressed reinforcement, the member might be *stage stressed*, where only part of the prestress is initially transferred to the concrete when just the self-weight is acting and the remaining prestress is applied when the sustained dead load (or part of it) is in place. Perhaps just one of the cables could be initially stressed and then, when the structure is subjected to additional loads, the second cable could then be stressed. This would avoid the situation considered earlier where the maximum prestress is applied when the minimum external load is acting.

Similar calculations are required to check for cracking at other sections at transfer. At support B, $(M_{ubt})_C = +63.2 \times 23.4 = 1479$ kNm, $P_i = 4292$ kN and $\sigma_b = 0.60$ MPa. Cracking will not occur at B at transfer. In fact, support C is the only location where cracking is likely to occur at transfer.

For this partially-prestressed beam, before conditions under full loads can be checked (using cracked section analyses in the cracked regions), it is necessary to determine the amount of non-prestressed steel required for strength.

6. In this example, the time-dependent losses estimated earlier are assumed to be satisfactory. In practice, of course, losses should be calculated using the procedures of Section 5.10.3 and illustrated in Example 10.1.

7. The strength of each cross-section is now checked. For the purpose of this example, calculations are provided only for the critical section at support C. From Step 3, $M^* = -6985$ kNm (due to the factored dead plus live loads). The secondary moment can be included with a load factor of 1.0. Therefore:

$$M^* = -6985 + 784 = -6201 \text{ kNm}$$

The inclusion of the secondary moment here is equivalent to a redistribution of moment at C of 11.2%. The secondary moment will cause a corresponding increase in the positive moments in the adjacent spans. If the cross-section at C is ductile, a further redistribution of moment may be permissible (as outlined in Section 11.5.4). No additional redistribution of moment is considered here.

The minimum required ultimate strength at C is $M_{uo} = M^*/\phi = 7751$ kNm. Using the procedure outlined in Section 6.4.1, the strength of a cross-section with flange width b = 750 mm and containing A_p = 2958 mm^2 at d_p = 1273 mm is:

M_{uol} = 6130 kNm (with d_n = 267 mm)

Additional non-prestressed tensile reinforcement A_{st} is required in the top of the cross-section at C. If the distance from the reinforcement to the compressive face is d_o = 1330 mm, then A_{st} can be calculated using Equation 6.26:

$$A_{st} = \frac{M_{uo} - M_{uol}}{f_{sy}\ell} = \frac{(7751 - 6130) \times 10^6}{500 \times 0.9 \times (1330 - 0.77 \times 267)} = 3204 \text{ mm}^2$$

Use six 28 mm diameter bars in the top over support C (3720 mm^2). This is in addition to the four 28 mm bars required in the bottom of the section for crack control at transfer. These bottom bars in the compressive zone at the ultimate limit state will improve ductility. From an ultimate strength analysis of the proposed cross-section, with reinforcement details shown in Figure 11.31, the section satisfies strength requirements (M_{uo} = 8350 kNm > M^*/ϕ) and with d_n = 297 mm = 0.223 d_o the section is ductile enough to satisfy the requirements of AS3600-2009 [3] with regard to moment redistribution at C.

Similar calculations show that four 28 mm diameter bars are required in the negative moment region over the first interior support at B and B', but at the mid-span region in all spans, the prestressing steel alone provides adequate moment capacity.

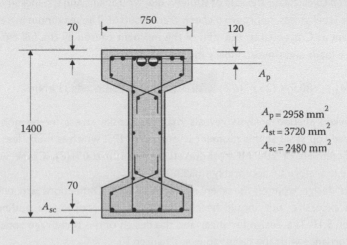

Figure 11.31 Reinforcement details for cross-section at support C (Example 11.7).

8. It is necessary to check crack control under full service loads. Results are provided for the cross-section at support C. With the effective prestress balancing 32 kN/m, the unbalanced sustained load is $w_{ub.sus} = 25 + 16.6 - 32 = 9.6$ kN/m and the unbalanced transient load is 20 kN/m. Using the moment coefficients in Figure 11.27 and the transient live load only on spans BC and CB′, the maximum unbalanced moment at support C is:

$$(M_{ub})_C = 9.6 \times (-80.9) + 20 \times (-92.6) = -2629 \text{ kNm}$$

With the effective prestress at C, $P_e = 2871$ kN, the extreme top fibre stress under the unbalanced loads is:

$$\sigma_t = -\frac{2871 \times 10^3}{690 \times 10^3} + \frac{2629 \times 10^6}{217 \times 10^6} = -4.16 + 12.12 = +7.96 \text{ MPa}$$

With the tensile strength taken as $0.6\sqrt{f_c'} = 3.8$ MPa, cracking will occur under the full unbalanced moment. The error associated with estimates of the cracking moment based on elastic stress calculation may be significantly large. As was discussed in Section 5.7.3 and illustrated in Example 5.5, creep and shrinkage may cause a large redistribution of stress on the cross-section with time, particularly when the cross-section contains significant quantities of non-prestressed reinforcement (as is the case here). If a more accurate estimate of stresses is required, a time analysis is recommended (see Section 5.7.3).

A cracked section analysis, similar to that outlined in Section 5.8.3, is required to calculate the loss of stiffness due to cracking and the increment of tensile steel stress, in order to check crack control. The maximum in-service moment at C is equal to the sum of the moment caused by the full external service loads and the secondary moment:

$$M_C = -80.9 \times (25 + 16.6) + 20 \times (-92.6) + 784 = -4433 \text{ kNm}$$

A cracked section analysis reveals that the tensile stress in the non-prestressed top steel at this moment is only 109 MPa, which is much less than the increment of 200 MPa permitted by AS3600-2009 [3]. Crack widths should therefore be acceptably small.

This design example is taken no further here. Deflections are unlikely to be excessive, but should be checked using the procedures outlined in Section 5.11. The design for shear and the design of the anchorage zones are in accordance with the discussions in Chapter 10.

REFERENCES

1. Ranzi, G. and Gilbert, R. I. (2015). *Structural Analysis: Principles, Methods and Modelling.* Boca Raton, FL: CRC Press (Taylor & Francis Group).
2. Cross, H. (1930). Analysis of continuous frames by distributing fixed-end moments. *Transactions of the American Society of Civil Engineers* Paper No. 1793, 10 pp.
3. AS3600-2009 (2009). *Australian Standard for Concrete Structures.* Standards Association of Australia, Sydney, New South Wales, Australia.
4. Lin, T. Y. and Thornton, K. (1972). Secondary moments and moment redistribution in continuous prestressed concrete beams. *Journal of the Prestressed Concrete Institute*, 17(1), 1–20.
5. Mattock, A. H. (1972). Secondary moments and moment redistribution in continuous prestressed concrete beams. Discussion of Lin and Thornton 1972. *Journal of the Prestressed Concrete Institute*, 17(4), 86–88.
6. Nilson, A. H. (1978). *Design of Prestressed Concrete.* New York: Wiley.
7. Warner, R. F. and Faulkes, K. A. (1983). Overload behaviour and design of continuous prestressed concrete beams. Presented at the *International Symposium on Non-Linearity and Continuity in Prestressed Concrete*, University of Waterloo, Waterloo, Ontario, Canada.

Chapter 12

Two-way slabs

Behaviour and design

12.1 INTRODUCTION

Post-tensioned concrete floors form a large proportion of all prestressed concrete construction and are economically competitive with reinforced concrete slabs in most practical medium- to long-span situations.

Prestressing overcomes many of the disadvantages associated with reinforced concrete slabs. Deflection, which is almost always the governing design consideration, is better controlled in post-tensioned slabs. A designer is better able to reduce or even eliminate deflection by a careful choice of prestress. More slender slab systems are therefore possible, and this may result in increased head room or reduced floor to floor heights. Prestress also inhibits cracking and may be used to produce crack-free and watertight floors. Prestressed slabs generally have simple uncluttered steel layouts. Steel fixing and concrete placing are therefore quicker and easier. In addition, prestress improves punching shear (see Chapter 7) and reduces formwork stripping times and formwork costs. On the other hand, prestressing often produces significant axial shortening of slabs and careful attention to the detailing of movement joints is frequently necessary.

In this chapter, the analysis and design of the following common types of prestressed concrete slab systems are discussed (each type is illustrated in Figure 12.1):

1. One-way slabs.
2. *Edge-supported two-way slabs* are rectangular slab panels supported on all four edges by either walls or beams. Each panel edge may be either continuous or discontinuous.
3. *Flat plate slabs* are continuous slabs of constant thickness supported by a rectangular grid of columns.

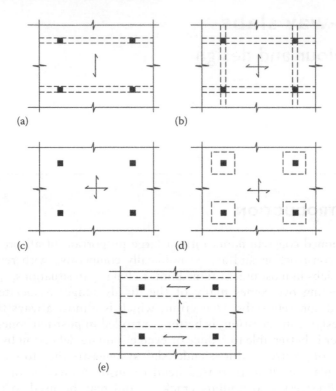

Figure 12.1 Plan views of different slab systems. (a) One-way slab. (b) Edge-supported two-way slab. (c) Flat plate. (d) Flat plate with drop panels. (e) Band beam and slab.

4. *Flat slabs with drop panels* are similar to flat plate slabs but with local increases in the slab thickness (drop panel) over each supporting column.
5. *Band-beam and slab systems* comprise wide shallow continuous prestressed beams in one direction (generally the longer span) with one-way prestressed or reinforced slabs in the transverse direction (generally the shorter span).

Almost all prestressed slabs are post-tensioned using draped tendons. In Australia and elsewhere, use is made of flat-ducted tendons, consisting of five or less strands in a flat sheath with fan-shaped anchorages, as shown in Figure 12.2. Individual strands are usually stressed one at a time using light hand-held hydraulic jacks. The flat ducts are structurally efficient and allow maximum tendon eccentricity and drape. In Australia, these ducts are almost always grouted after stressing to provide bond between the steel and the concrete.

In North America and elsewhere, unbonded construction is often used for slabs. Single plastic-coated greased tendons are generally used, resulting in

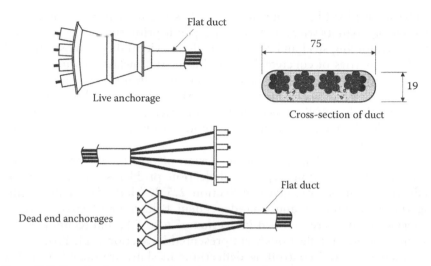

Live anchorage

Flat duct

75

19

Cross-section of duct

Flat duct

Dead end anchorages

Figure 12.2 Details of typical flat-ducted tendons and anchorages.

slightly lower costs, small increases in available tendon drape, the elimination of the grouting operation (therefore reducing cycle times) and reduced friction losses. However, unbonded construction also leads to reduced flexural strength, reduced crack control (additional bonded reinforcement is often required), possible safety problems if a tendon is lost or damaged (by corrosion, fire or accident) and increased demolition problems. Single strands are also more difficult to fix to profile.

Prestressed concrete slabs are typically thin in relation to their spans. If a slab is too thin, it may suffer excessively large deflections when fully loaded or exhibit excessive camber after transfer. The initial selection of the thickness of a slab is usually governed by the serviceability requirements for the member. The selection is often based on personal experience or on recommended maximum span-to-depth ratios. Whilst providing a useful starting point in design, such a selection of slab thickness does not necessarily ensure that serviceability requirements are satisfied. Deflections at all critical stages in the slab's history must be calculated and limited to acceptable design values. Failure to predict deflections adequately has frequently resulted in serviceability problems.

In slab design, *excessive deflection* is a relatively common type of failure. This is particularly true for slabs supporting relatively large transitory live loads or for slabs not subjected to their full service loads until some considerable time after transfer. AS3600-2009 [1] requires that the camber, deflection and vibration frequency and amplitude of slabs must be within acceptable limits at service loads. In general, however, little guidance is given as to how this is to be done and methods for calculating long-term camber or deflection for prestressed slabs are not prescribed.

The service load behaviour of a concrete structure can be predicted far less reliably than its strength. Strength depends primarily on the properties of the reinforcing steel and tendons, whilst serviceability is most affected by the properties of concrete. The nonlinear and inelastic nature of concrete complicates the calculation of deflection, even for line members such as beams. For two-way slab systems, the three-dimensional nature of the structure, the less well-defined influence of cracking and tension stiffening and the development of biaxial creep and shrinkage strains create additional difficulties.

A more general discussion of the design of prestressed structures for serviceability, including types of deflection problems and criteria for deflection control, was given in Section 2.5.2. Methods for determining the instantaneous and time-dependent behaviour of cross-sections at service loads were outlined in Sections 5.6 to 5.9, and techniques for calculating beam deflections were presented in Section 5.11. Procedures for calculating and controlling deflections in slabs are included in this chapter.

12.2 EFFECTS OF PRESTRESS

As discussed previously, the prestressing operation results in the imposition of both longitudinal and transverse forces on post-tensioned members. The concentrated longitudinal prestress P produces a complex stress distribution immediately behind the anchorage, and the design of the anchorage zone requires careful attention (see Chapter 8). At sections further away from the anchorage, the longitudinal prestress applied at the anchorage causes a linearly varying compressive stress over the depth of the slab. If the longitudinal prestress is applied at the centroidal axis (which is generally at the mid-depth of the slab), this compressive stress is uniform over the slab thickness and equal to P/A.

We have already seen that wherever a change in direction of the tendon occurs, a transverse force is imposed on the member. For a parabolic tendon profile, such as that shown in Figure 12.3a, the curvature is constant along the tendon and hence, the transverse force imposed on the member is uniform along its length (if P is assumed to be constant). From Equation 1.7, the uniformly distributed transverse force is:

$$w_p = \frac{8Ph}{L^2} \tag{12.1}$$

where h is the sag of the parabolic tendon and L is the span.

If the cable spacing is uniform across the width of a slab and P is the prestressing force per unit width of slab, then w_p is the uniform upward load per unit area.

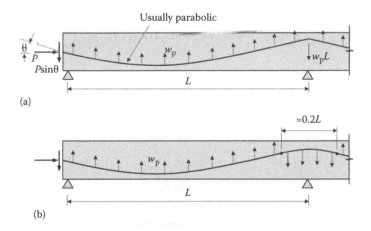

(a)

(b)

Figure 12.3 Idealised and actual tendon profiles in a continuous slab. (a) Idealised tendon profile. (b) Practical tendon profile.

The cable profile shown in Figure 12.3a, with the sharp kink located over the internal support, is an approximation of the more realistic and practical profile shown in Figure 12.3b. The difference between the effects of the idealised and practical profiles was discussed in Section 11.3.5 for continuous beams. The idealised profile is more convenient for the analysis and design of continuous slabs, and the error introduced by the idealisation is usually not great.

The transverse load w_p causes moments and shears that usually tend to be opposite in sign to those produced by the external loads. In Figure 12.4, the elevation of a prestressing tendon in a continuous slab is shown, together with the transverse loads imposed on the slab by the tendon in each span. If the slab is a two-way slab, with prestressing tendons placed in two orthogonal directions, the total transverse load caused by the prestress is the sum of w_p for the tendons in each direction.

The longitudinal prestress applied at the anchorage may also induce moments and shears in a slab. At changes of slab thickness, such as occur in a flat slab with drop panels, the anchorage force P is eccentric with

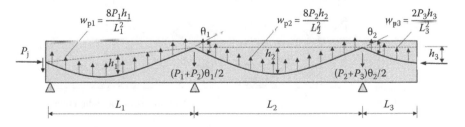

Figure 12.4 Transverse loads imposed by tendons in one direction.

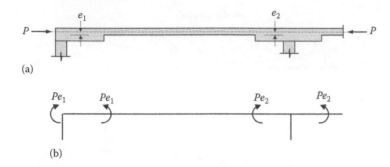

(a)

(b)

Figure 12.5 Effect of changes in slab thickness. (a) Elevation of slab. (b) Imposed moments.

respect to the centroidal axis of the section, as shown in Figure 12.5a. The moments caused by this eccentricity are indicated in Figure 12.5b and should also be considered in the analysis of the slab. However, the moments produced by relatively small changes in slab thickness tend to be small compared with those caused by cable curvature and, if the thickening is below the slab, it is conservative to ignore them.

At some distance from the slab edge, the concentrated anchorage forces have dispersed and the slab is uniformly stressed. The so-called angle of dispersion θ (shown in Figure 12.6) determines the extent of slab where the prestress is not effective.

Specifications for θ vary considerably. It is claimed in some trade literature (VSL 1988) that tests have shown θ to be 120°. In AS3600-2009 [1], θ is specified conservatively as 60°. A value of θ = 90° is usually satisfactory for design purposes.

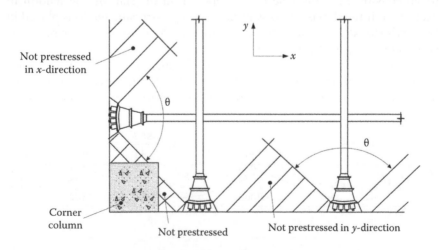

Figure 12.6 Areas of ineffective prestressing at slab edge.

Care must be taken in the design of the hatched areas of slab shown in Figure 12.6, where the prestress in one or both directions is not effective. It is a good practice to include a small quantity of bonded non-prestressed reinforcement in the bottom of the slab perpendicular to the free edge in all exterior spans. An area of non-prestressed steel of about $0.0015bD$ is usually sufficient, where b and D are the width and depth of the slab. In addition, when checking the punching shear strength at the corner column in Figure 12.6, the beneficial effect of prestress is not available. At sections remote from the slab edge, the average P/A stresses are uniform across the entire slab width and do not depend on the value of θ and variations of cable spacing from one region of the slab to another.

12.3 BALANCED LOAD STAGE

Under transverse loads, two-way panels deform into dish-shaped surfaces, as shown in Figure 12.7. The slab is curved in both principal directions and therefore, bending moments exist in both directions. In addition, part of the applied load is resisted by twisting moments which develop in the slab at all locations except the lines of symmetry.

The prestressing tendons are usually placed in two directions parallel to the panel edges, each tendon providing resistance for a share of the applied load. The transverse load on the slab produced by the tendons in one direction adds to (or subtracts from) the transverse load imparted by the tendons in the perpendicular direction. For edge-supported slabs, the portion of the load to be carried by tendons in each direction is more or less arbitrary; the only strict requirement is the satisfaction of statics. For flat slabs, the total load must be carried by tendons in each direction from column line to column line.

The concept of utilising the transverse forces, resulting from the curvature of the draped tendons, to balance a selected portion of the applied load is useful from the point of view of controlling deflections. In addition to providing the basis for establishing a suitable tendon profile, load balancing allows the determination of the prestressing force required to produce zero deflection in a slab panel under the selected balanced load.

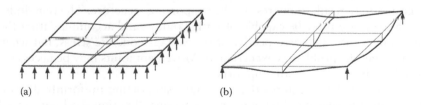

(a) (b)

Figure 12.7 Deformation of interior two-way slab panels. (a) Edge-supported two-way slab. (b) Flat slab.

At the balanced load, the slab is essentially flat (no curvature) and is subjected only to the effects of the prestressing forces applied at the anchorages. A slab of uniform thickness is subjected only to uniform compression (P/A) in the directions of the orthogonal tendons. With the state of the slab under the balanced load confidently known, the deflection due to the unbalanced portion of the load may be calculated by one of the techniques discussed later in this chapter. The calculation of the deflection of a prestressed slab is usually more reliable than for a conventionally reinforced slab, because only a portion of the total service load needs to be considered (the unbalanced portion) and, unlike reinforced concrete slabs, prestressed slabs are often uncracked at service loads.

To minimise deflection problems, the external load to be balanced is usually a significant portion of the sustained or permanent service load. If all the permanent loads are balanced, the sustained concrete stress (P/A) is uniform over the slab depth. A uniform compressive stress distribution produces uniform creep strain and, hence, little long-term load-dependent curvature or deflection. Bonded reinforcement does, of course, provide restraint to both creep and shrinkage and causes a change of curvature with time if the steel is eccentric to the slab centroid. However, the quantity of bonded steel in prestressed slabs is generally small and the time-dependent curvature caused by this restraint does not usually cause significant deflection.

Problems can arise if a relatively heavy dead load is to be applied at some time after stressing. Excessive camber after transfer, which continues to increase with time owing to creep, may cause problems prior to the application of the full balanced load. In such a case, the designer may consider *stage stressing* as a viable solution.

The magnitude of the average concrete compressive stress after all losses can indicate potential serviceability problems. If P/A is too low, the prestress may not be sufficient to prevent or control cracking due to shrinkage, temperature changes and the unbalanced loads. Some codes of practice specify minimum limits on the average concrete compressive stress after all losses. No such limit is specified in AS3600-2009 [1], but using flat-ducted tendons containing four or more strands, prestressing levels are typically in the range P/A = 1.2–2.6 MPa in each direction of a two-way slab.

If the average prestress is high, axial deformation of the slab may be large and may result in distress in the supporting structure. The remainder of the structure must be capable of withstanding and accommodating the shortening of the slab, irrespective of the average concrete stress, but when P/A is large, the problem is exacerbated. Movement joints may be necessary to isolate the slab from stiff supports.

ACI-318-11M [3] requires that, for slabs supporting uniformly distributed loads, the spacing of tendons in at least one direction shall not exceed the smaller of eight times the slab thickness and 1.5 m. AS3600-2009 [1] places no such limit on tendon spacing, but the ACI limit is recommended here, particularly when the slab contains less than minimum quantities of conventional tensile reinforcement (i.e. less than about 0.15% of the cross-sectional area of the slab).

12.4 INITIAL SIZING OF SLABS

12.4.1 Existing guidelines

At the beginning of the design of a post-tensioned floor, the designer must select an appropriate floor thickness. The floor must be stiff enough to avoid excessive deflection or camber, and it must have adequate fire resistance and durability.

In its recommendations for the design of post-tensioned slabs, the Post-Tensioning Institute [4] suggested typical span-to-depth ratios that had proved acceptable, in terms of both performance and economy, for a variety of slab types. These recommendations are summarised in Table 12.1. Note that for flat plates and flat slabs with drop panels, the longer of the two orthogonal spans is used in the determination of the span-to-depth ratio, whilst for edge-supported slabs, the shorter span is used. The minimum slab thickness that will prove acceptable in any situation depends on a variety of factors, not the least of which is the level of the superimposed load and the occupancy. The effect of load level on the limiting span-to-depth ratio of flat slabs in building structures is illustrated in Figure 12.8. The minimum thickness obtained from Table 12.1 may not therefore be appropriate in some situations. On the other hand, thinner slabs may be acceptable, if the calculated deflections, camber and vibration frequency and amplitude are acceptable. In addition to the satisfaction of serviceability requirements, strength requirements, such as punching shear at supporting columns, must also be satisfied. In addition, fire resistance and durability requirements must also be considered.

A slab exposed to fire must retain its structural adequacy and integrity for a particular *fire resistance period* (FRP). It must also be sufficiently

Table 12.1 Limiting span-to-depth ratios [3]

Floor system	Span-to-depth ratio (L/D)
Flat plate	45
Flat slab with drop panels	50
One-way slab	48
Edge-supported slab	55
Waffle slab	35
Band beams ($b \approx 3D$)	30

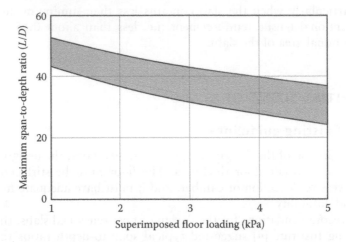

Figure 12.8 Effect of superimposed load on maximum *L/D* for flat slabs.

thick to limit the temperature on one side, when exposed to fire on the other side, that is it must provide a suitable FRP for insulation. The FRP required for a particular structure is generally specified by the local building authority and depends on the type of structure and its occupancy. The minimum effective thickness specified in AS3600-2009 [1] of a prestressed concrete flat slab (with or without drop panels) required to provide a particular FRP for structural adequacy (and insulation) is given in Table 2.11, together with the minimum axis distance to the bottom layer of reinforcement or tendons.

12.4.2 Serviceability approach for the calculation of slab thickness

For uniformly loaded slabs, a better initial estimate of slab thickness which should ensure adequate stiffness and satisfactory service load behaviour can be made using a procedure originally developed for reinforced concrete slabs [5] and extended to cover post-tensioned floor systems [6]. By rearranging the expression for the deflection of a span, a simple equation is developed for the span-to-depth ratio that is required to satisfy any specified deflection limit. The method forms the basis of the deemed to comply span-to-depth ratios for reinforced concrete slabs in AS3600-2009 [1].

If it is assumed that a prestressed concrete slab is essentially uncracked at service loads, which is most often the case, the procedure for estimating the overall depth of the slab is relatively simple. Figure 12.9 shows typical interior panels of a one-way slab, a flat-slab and an edge-supported slab. *Equivalent one-way slab strips* are also defined and illustrated for

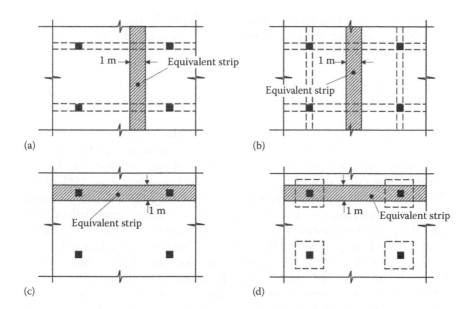

Figure 12.9 Slab types and equivalent slab strips. (a) One-way slab. (b) Edge-supported two-way slab. (c) Flat plate. (d) Flat slab with drop panels.

each slab type. For a one-way slab, the mid-span deflection is found by analysing a strip of unit width as shown in Figure 12.9a. For an edge-supported slab, the deflection at the centre of the panel may be calculated from an equivalent slab strip through the centre of the panel in the short direction, as shown in Figure 12.9b. For the flat plate and flat slab panel shown in Figure 12.9c and d, the deflection at the mid-point of the long span on the column line is found by analysing a unit wide strip located on the column line.

The stiffness of these equivalent slab strips must be adjusted for each slab type, so that the deflection of the one-way strip at mid-span is the same as the deflection of the two-way slab at that point. For an edge-supported slab, for example, the stiffness of the equivalent slab strip is increased significantly to realistically model the actual slab deflection at the mid-panel. For a flat slab, the stiffness of the slab strip must be reduced, if the maximum deflection at the centre of the panel is to be controlled rather than the deflection on the column line. The stiffness adjustment is made using a *slab system factor K* that was originally calibrated using a non-linear, finite element model [7,8].

By rearranging the equation for the mid-span deflection of a one-way slab, an expression can be obtained for the minimum slab thickness required to satisfy any specified deflection limit. The maximum deflection caused by the unbalanced uniformly distributed service loads on an uncracked one-way prestressed slab strip of width b, depth D and span L_{ef} may be estimated using Equation 10.2, which is reproduced and renumbered here for ease of reference:

$$v = \beta \frac{w_{ub}L_{ef}^4}{E_c I} + \lambda \beta \frac{w_{ub.sus}L_{ef}^4}{E_c I} \tag{12.2}$$

where E_c is the elastic modulus of concrete; I is the gross moment of inertia of the cross-section; w_{ub} is the unbalanced service load per unit length; and $w_{ub.sus}$ is the sustained portion of the unbalanced load per unit length.

In the design of a slab, $w_{ub.sus}$ should not be taken less than 25% of the self-weight of the member. This is to ensure that at least a small long-term deflection is predicted by Equation 12.2. A small long-term deflection is inevitable, even for the case when an attempt is made to balance the entire sustained load by prestress. The term β in Equation 12.2 is a deflection coefficient that depends on the support conditions and the type of load. The effective span of the slab strip L_{ef} may be taken to be the centre-to-centre distance between supports or the clear span plus the depth of the member, whichever is the smaller [1]. As discussed in Section 10.3.1, the long-term deflection multiplier λ for an uncracked prestressed member is significantly higher than for a cracked reinforced concrete member and should not be taken less than 3.

By substituting $bD^3/12$ for I and rearranging Equation 12.2, we get:

$$\frac{L_{ef}}{D} = \left[\frac{(v/L_e)bE_c}{12\beta(w_{ub} + \lambda w_{ub.sus})} \right]^{1/3} \tag{12.3}$$

If v is the deflection limit selected in design, the maximum span-to-depth ratio for the slab strip is obtained from Equation 12.3. To avoid dynamic problems, a maximum limit should be placed on the span-to-depth ratio. Upper limits of the span-to-depth ratio for slabs to avoid excessive vertical acceleration due to pedestrian traffic that may cause discomfort to occupants were recommended by Mickleborough and Gilbert [9]. This work forms the basis of the upper limits on L/D specified below in Equation 12.4.

For prestressed concrete slabs, an estimate of the minimum slab thickness may be obtained by applying Equation 12.3 to the slab strips in Figure 12.9. Equation 12.3 can be re-expressed as follows:

$$\frac{L_{ef}}{D} \leq K \left[\frac{(v/L_e)1000 E_c}{w_{ub} + \lambda w_{ub.sus}} \right]^{1/3} \tag{12.4}$$

< 50 for one-way slabs and flat slabs
< 55 for two-way edge-supported slabs

The width of the equivalent slab strip b has been taken as 1000 mm and, in the absence of any other information, the long-term deflection multiplier λ should be taken not less than 3. The loads w_{ub} and $w_{ub.sus}$ are in kPa (i.e. kN/m^2) and E_c is in MPa. The term K is the *slab system factor* that accounts for the support conditions of the slab panel, the aspect ratio of the panel, the load dispersion and the torsional stiffness of the slab. For each slab type, values for K are presented and discussed in the following.

12.4.2.1 The slab system factor, K

One-way slabs: For a one-way slab, K depends only on the support conditions and the most critical pattern of unbalanced load. From Equation 12.3:

$$K = \left(\frac{1}{12\beta}\right)^{1/3} \tag{12.5}$$

For a continuous slab, β should be determined for the distribution of unbalanced load which causes the largest deflection in each span. For most slabs, a large percentage of the sustained load (including self-weight) is balanced by the prestress and much of the unbalanced load is transitory. Pattern loading must therefore be considered in the determination of β. For a simply-supported span $\beta = 5/384$, and from Equation 12.5, $K = 1.86$. For a fully loaded *end span* of a one-way slab that is continuous over three or more equal spans and with the adjacent interior span, unloaded $\beta = 3.5/384$ (determined from an elastic analysis) and, therefore, $K = 2.09$. For a fully loaded *interior span* of a continuous one-way slab with adjacent spans, unloaded $\beta = 2.6/384$ and $K = 2.31$.

Flat slabs: For flat slabs, the values of K given earlier must be modified to account for the variation of curvature across the panel width. The moments, and hence curvatures, in the uncracked slab are greater, close to the column line than near the mid-panel of the slab in the middle strip region. For this reason, the deflection of the slab on the column line will be greater than the deflection of a one-way slab of similar span and continuity. If the deflections of the equivalent slab strips in Figure 12.9c and d are to represent accurately the deflection of the real slab on the column-line, a greater than average share of the total load on the slab must be assigned to the column strip (of which the equivalent strip forms a part). A reasonable assumption is that 65% of the total load on the slab is carried by the column strips, and so the value for K for a flat slab becomes:

$$K = \left(\frac{1}{15.6\beta}\right)^{1/3} \tag{12.6}$$

For an end span, with $\beta = 3.5/384$, Equation 12.6 gives K = 1.92. For an interior span, with $\beta = 2.6/384$, the slab system factor $K = 2.12$.

For a slab containing drop panels that extend at least $L/6$ in each direction on each side of the support centre-line and that have an overall depth not less than 1.3 times the slab thickness beyond the drops, the aforementioned values for K may be increased by 10%. If the maximum deflection at the centre of the panel is to be limited (rather than the deflection on the long-span column line), the values of K for an end span and for an interior span should be reduced to 1.75 and 1.90, respectively.

Edge-supported two-way slabs: For an edge-supported slab, values for K must be modified to account for the fact that only a portion of the total load is carried by the slab in the short span direction. In addition, torsional stiffness, and even compressive membrane action, increases the overall slab stiffness. In an earlier study of span-to-depth limits for reinforced concrete slabs, a nonlinear finite element model was used to quantify these effects [5]. Values of K depend on the aspect ratio of the rectangular edge-supported panel and the support conditions of all edges, and are given in Table 12.2.

12.4.3 Discussion

Equation 12.4 forms the basis of a useful approach for the preliminary design of prestressed slabs. When the load to be balanced and the deflection limit have been selected, an estimate of slab depth can readily be made. All parameters required for input into Equation 12.4 are usually known at the beginning of the design. An iterative approach may be required if an estimate of self-weight is needed, i.e. if the load to be balanced is less than self-weight.

Deflections at various stages in the slab history may still have to be calculated, particularly if the unbalanced load causes significant cracking or if

Table 12.2 Values of K for an uncracked two-way edge-supported slab [6]

Support conditions for slab panel	Slab system factor, K			
	Ratio of long span to short span			
	1.0	1.25	1.5	2.0
4 edges continuous	3.0	2.6	2.4	2.3
1 short edge discontinuous	2.8	2.5	2.4	2.3
1 long edge discontinuous	2.8	2.4	2.3	2.2
2 short edges discontinuous	2.6	2.4	2.3	2.3
2 long edges discontinuous	2.6	2.2	2.0	1.9
2 adjacent edges discontinuous	2.5	2.3	2.2	2.1
2 short +1 long edge discontinuous	2.4	2.3	2.2	2.1
2 long +1 short edge discontinuous	2.4	2.2	2.1	1.9
4 edges discontinuous	2.3	2.1	2.0	1.9

an unusual load history is expected. Serviceability problems, however, can be minimised by a careful choice of slab depth using Equation 12.4. This involves an understanding of the derivation of the equation and its limitations. If, for example, a designer decides to minimise deflection by balancing the entire sustained load, it would be unwise to set the sustained part of the unbalanced load $w_{ub.sus}$ to zero. In the real slab, of course, the magnitude of the sustained unbalanced load varies as the prestressing force varies with time and does not remain zero. Restraint to creep and shrinkage caused by the eccentric bonded steel will inevitably cause some time-dependent deflection (or camber). In such cases, selection of a slab depth greater than that indicated by Equation 12.4 would be prudent. It is suggested that $w_{ub.sus}$ should not be taken as less than 0.25 times the self-weight of the slab. As with the rest of the design process, sound engineering judgement is required.

EXAMPLE 12.1

Determine preliminary estimates of the thickness of a post-tensioned flat slab floor for an office building. The supporting columns are 400 mm × 400 mm in section and are regularly spaced at 9.8 m centres in one direction and 7.8 m centres in the orthogonal direction. Drop panels extending span/6 in each direction are located over each interior column. The slab supports a dead load of 1 kPa (in addition to self-weight) and a service live load of 2.5 kPa (of which 0.75 kPa is sustained or permanent). The self-weight of the slab only is to be balanced by prestress. Therefore, the unbalanced loads are:

$$w_{ub} = 3.5 \text{ kPa} \quad \text{and} \quad w_{ub.sus} = 1.75 \text{ kPa}$$

In this example, the longer effective span is calculated as clear span + D. If D is initially assumed to be about 200 mm, then $L_{ef} = 9800 - 400 + 200 = 9600$ mm. The elastic modulus for concrete is taken as $E_c = 28,000$ MPa.

In Case (a), the maximum deflection on the column line in the long-span direction is first limited to span/250, and then in Case (b), it is limited to span/500.

Case (a): The deflection in an exterior or edge panel of the slab will control the thickness. From Equation 12.6, $K = 1.92$ for an end span and this may be increased by 10% to account for the stiffening effect of the drop panels, that is $K = 2.11$. With $\lambda = 3$, Equation 12.4 gives:

$$\frac{9,600}{D} \leq 2.11 \times \left[\frac{(1/250) \times 1,000 \times 28,000}{3.5 + 3 \times 1.75} \right]^{1/3} = 49.3$$

$\therefore D \geq 195$ mm plus drop panels.

Case (b): If the slab supports brittle partitions and the deflection limit is taken to be span/500, a thicker slab than that required for (a) will be needed. Assuming $D = 250$ mm, the revised effective span is $L_{ef} = 9650$ mm and, with $\lambda = 3$, Equation 12.4 gives:

$$\frac{9,650}{D} \leq 2.11 \times \left[\frac{(1/500) \times 1,000 \times 28,000}{3.5 + 3 \times 1.75} \right]^{1/3} = 39.2$$

$\therefore D \geq 246$ mm plus drop panels.

EXAMPLE 12.2

Evaluate the slab thickness required for an edge-panel of a two-way slab with short and long effective spans of 8.5 m and 11 m, respectively. The slab is continuously supported on all four edges by stiff beams and is discontinuous on one long edge only. The slab must carry a dead load of 1.25 kPa (plus self-weight) and a service live load of 3 kPa (of which 1 kPa is sustained). As in the previous example, only the self-weight is to be balanced by prestress, and therefore:

$$w_{ub} = 4.25 \text{ kPa} \quad \text{and} \quad w_{ub.sus} = 2.25 \text{ kPa}$$

The maximum mid-panel deflection is limited to $v = 25$ mm, and the elastic modulus for concrete is $E_c = 28,000$ MPa.

With an aspect ratio of 11.0/8.5 = 1.29, the slab system factor is obtained by interpolation from Table 12.2, i.e. $K = 2.4$. With $\lambda = 3$, Equation 12.4 gives:

$$\frac{8,500}{D} \leq 2.4 \times \left[\frac{(25/8,500) \times 1,000 \times 28,000}{4.25 + 3 \times 2.25} \right]^{1/3} = 47.0$$

$\therefore D \geq 181$ mm.

12.5 OTHER SERVICEABILITY CONSIDERATIONS

12.5.1 Cracking and crack control in prestressed slabs

The effect of cracking in slabs is to reduce the flexural stiffness of the highly stressed regions and thus to increase the deflection. Prior to cracking, deflection calculations are usually based on the moment of inertia of the gross concrete section I_g, neglecting the contributions of the reinforcement. After cracking, the effective moment of inertia I_{ef}, which is less than I_g, is used. In Section 5.8, the analysis of a cracked prestressed section was presented and procedures for calculating the cracked moment of inertia and for including the tension stiffening effect were discussed in Section 5.11.3. Using these procedures, the effective moment of inertia of a cracked region of the slab can be calculated.

For prestressed concrete flat slabs, flexural cracking at service loads is usually confined to the negative moment column strip region above the supports. A mat of non-prestressed reinforcement is often placed in the top of the slab over the column supports for crack control and to increase both the stiffness and the strength of this highly stressed region. This mat of crack control reinforcement should consist of bars in each direction (or welded wire mesh), with each bar or wire being continuous across the column line and extending at least 0.25 × the clear span in each direction from the face of the column. An area of crack control reinforcement of about 0.0015 times the gross area of the cross-section is usually sufficient.

The mechanism of flexural cracking in a statically indeterminate two-way slab is complex. The direction of flexural cracking is affected to some extent by the spacing and type of bonded reinforcement, the level of prestress in each direction, the support conditions and the level and distribution of the applied loads. However, for slabs containing conventionally tied bonded reinforcement at practical spacings in both directions, flexural cracks occur in the direction perpendicular to the direction of principal tension.

If the level of prestress in a slab is sufficiently high to ensure that the tensile stresses in a slab in bending are always less than the tensile strength of concrete, flexural cracking will not occur. If the level of prestress is not sufficient, cracking occurs and bonded reinforcement at reasonable centres is necessary to control the cracks adequately. Because slabs tend to be very lightly reinforced, the maximum moments at service loads are rarely very much larger than the cracking moment. However, when cracking occurs, the stress in the bonded reinforcement increases and crack widths may become excessive if too little bonded steel is present or the steel spacing is too wide.

The requirements for flexural crack control in AS3600-2009 [1] were discussed in Section 5.12.1 and the maximum increment of stress in the steel near the tensile face (as the load is increased from its value when the extreme concrete tensile fibre is at zero stress to the short-term service load value) was given in Table 5.3. In addition, the requirements for the control of *direct tension cracking* in slabs due to restrained shrinkage and temperature changes were outlined in Section 5.12.2.

12.5.2 Long-term deflections

As discussed in Chapter 5, long-term deflections due to creep and shrinkage are influenced by many variables, including load intensity, mix proportions, slab thickness, age of slab at first loading, curing conditions, quantity of compressive steel, relative humidity and temperature.

In most prestressed slabs, the majority of the sustained load is most often balanced by the transverse force exerted by the tendons on the slab. Under this balanced load, the time-dependent deflection will not be zero because of the restraint to creep and shrinkage offered by eccentrically located bonded reinforcement. The use of a simple deflection multiplier to calculate long-term deflection is not, therefore, always satisfactory.

In Sections 4.2.4.3 and 4.2.4.4, the procedures specified in AS3600-2009 [1] for calculating the final creep coefficient of concrete φ_{cc}^* and the final shrinkage strain ε_{cs}^* were presented and a procedure for the determination of the long-term behaviour of uncracked and cracked prestressed cross-sections was presented in Sections 5.7 and 5.9. Alternative and more approximate expressions for estimating the creep and shrinkage components of the long-term deflection of beams were given in Section 5.11.4. Similar equations for determining deflections of slab strips are presented in the following text.

For uncracked prestressed concrete slabs, which usually have low quantities of steel, the increase in curvature due to creep is nearly proportional to the increase in strain due to creep. This is in contrast with the behaviour of a cracked, reinforced section. If we set α in Equation 5.175 to unity on every cross-section, the final creep induced deflection v_{cc} may be approximated by:

$$v_{cc} = \varphi_{cc}^* v_{i.sus} \tag{12.7}$$

where $v_{i.sus}$ is the short-term deflection produced by the sustained portion of the unbalanced load. Typical values for the final creep coefficient for concrete in post-tensioned slabs are φ_{cc}^* and are in the range 2.5–3.0.

The average deflection due to shrinkage of an equivalent slab strip (in the case of edge-supported slabs) or the wide beam (as discussed subsequently in Section 12.9.6 for the case of flat slabs) may be obtained from the following equation:

$$v_{cs} = \beta \kappa_{cs}^* L_{ef}^2 \tag{12.8}$$

where κ_{cs}^* is the average shrinkage-induced curvature; L_{ef} is the effective span of the slab strip under consideration; and β depends on the support conditions and equals 0.125 for a simply-supported span, 0.090 for an end span of a continuous member and 0.065 for an interior span of a continuous member.

The shrinkage curvature κ_{cs}^* is non-zero wherever the eccentricity of the bonded steel area is non-zero and varies along the span as the eccentricity of the draped tendons varies. A simple and very approximate estimate of the average shrinkage curvature for a fully-prestressed slab, which will usually produce reasonable results, is:

$$\kappa_{cs}^* = \frac{0.3\varepsilon_{cs}^*}{D} \tag{12.9}$$

For a cracked partially prestressed slab, with significant quantities of non-prestressed conventional reinforcement, the value of κ_{cs}^* is usually at least 100% higher than that indicated above.

12.6 DESIGN APPROACH – GENERAL

After making an initial selection of the slab thickness, the second step in slab design is to determine the amount and distribution of prestress. Load balancing is generally used to this end. A portion of the load on a slab is balanced by the transverse forces imposed by the draped tendons in each direction. Under the balanced load, the slab remains plane (without curvature) and is subjected only to the resultant longitudinal compressive P/A stresses. It is the remaining unbalanced load that enters into the calculation of service-load behaviour, particularly for the estimation of load-dependent deflections and for checking the extent of cracking and crack control.

At ultimate conditions, when the slab behaviour is nonlinear and superposition is no longer valid, the full factored design load must be considered. The factored design moments and shears at each critical section must be calculated and compared with the design strength of the section, as discussed in Chapters 6 (for flexure) and 7 (for shear). Slabs are usually very ductile and redistribution of moment occurs as the collapse load of the slab is approached. Under these conditions, secondary moments can usually be ignored.

In the following sections, procedures for the calculation of design moments and shears at the critical sections in the various slab types are presented. In addition, techniques and recommendations are also presented for the determination of the magnitude of the prestressing force required in each direction to balance the desired load.

12.7 ONE-WAY SLABS

A one-way slab is generally designed as a beam with cables running in the direction of the span at uniform centres. A slab strip of unit width is analysed using simple beam theory. In any span, the maximum cable sag h depends on the concrete cover requirements and the tendon dimensions. When h is determined, the prestressing force required to balance an external load w_b is calculated from Equation 11.33, which is restated and renumbered here for ease of reference:

$$P = \frac{w_b L^2}{8h}$$ (12.10)

In the transverse direction, conventional reinforcement may be used to control shrinkage and temperature cracking (see Section 5.12.2) and to distribute local load concentrations. Not infrequently, the slab is prestressed in the transverse direction to eliminate the possibility of shrinkage cracking parallel to the span and to ensure a watertight and crack-free slab.

12.8 TWO-WAY EDGE-SUPPORTED SLABS

12.8.1 Load balancing

Consider the interior panel of the two-way edge-supported slab shown in Figure 12.10. The panel is supported on all sides by walls or beams and contains parabolic tendons in both the x and y directions. If the cables in each direction are uniformly spaced, then from Equation 12.1, the upward forces per unit area exerted by the tendons are:

$$w_{px} = \frac{8P_x h_x}{L_x^2}$$ (12.11)

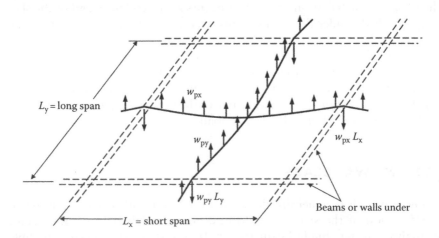

Figure 12.10 Interior edge-supported slab panel.

and

$$w_{py} = \frac{8P_y h_y}{L_y^2} \qquad (12.12)$$

where P_x and P_y are the prestressing forces in each direction per unit width; and h_x and h_y are the cable drapes in each direction.

The uniformly distributed downward load to be balanced per unit area w_b is calculated as:

$$w_b = w_{px} + w_{py} \qquad (12.13)$$

In practice, perfect load balancing is not possible, since external loads are rarely perfectly uniformly distributed. However, for practical purposes, adequate load balancing can be achieved. Any combination of w_{px} and w_{py} that satisfies Equation 12.13 can be used to make up the balanced load. The smallest quantity of prestressing steel will result if all the loads are balanced by cables in the short-span direction, i.e. $w_b = w_{px}$. However, under unbalanced loads, serviceability problems in the form of unsightly cracking may result. It is often preferable to distribute the prestress in much the same way as the load is distributed to the supports in an elastic slab, i.e. more prestress in the short-span direction than in the long-span direction. The balanced load resisted by tendons in the short direction may be estimated by:

$$w_{px} = \frac{L_y^4}{\delta L_x^4 + L_y^4} w_b \qquad (12.14)$$

where δ depends on the support conditions and is given by:

δ = 1.0 for 4 edges continuous or discontinuous
  = 1.0 for 2 adjacent edges discontinuous
  = 2.0 for 1 long edge discontinuous
  = 0.5 for 1 short edge discontinuous
  = 2.5 for 2 long +1 short edge discontinuous
  = 0.4 for 2 short +1 long edge discontinuous
  = 5.0 for 2 long edges discontinuous
  = 0.2 for 2 short edges discontinuous

Equation 12.14 is the expression obtained for that portion of any external load which is carried in the short-span direction if twisting moments are ignored and the mid-span deflections of the two orthogonal unit wide strips through the slab centre are equated.

With w_{px} and w_{py} selected, the prestressing force per unit width in each direction is calculated using Equations 12.11 and 12.12 as:

$$P_x = \frac{w_{px}L_x^2}{8h_x} \tag{12.15}$$

and

$$P_y = \frac{w_{py}L_y^2}{8h_y} \tag{12.16}$$

Equilibrium dictates that the downward forces per unit length exerted over each edge support by the reversal of cable curvature (as shown in Figure 12.10) are:

$w_{py}L_y$ (kN/m)   carried by the short span supporting beams or walls

$w_{px}L_x$ (kN/m)   carried by the long span supporting beams or walls

The total force imposed by the slab tendons that must be carried by the edge beams is therefore:

$$w_{px}L_xL_y + w_{py}L_yL_x = w_bL_xL_y$$

and this is equal to the total upward force exerted by the slab cables. Therefore, for this two-way slab system, to carry the balanced load to the supporting columns (or footings), resistance must be provided for twice the total load to be balanced, i.e. the slab tendons must resist $w_bL_xL_y$ and the supporting beams must resist $w_bL_xL_y$. This requirement is true for all two-way slab systems, irrespective of the construction type or material.

At the balanced load condition, when the transverse forces imposed by the cables exactly balance the applied external loads, the slab is subjected only to the compressive stresses imposed by the longitudinal prestress in each direction, that is $\sigma_x = P_x/D$ and $\sigma_y = P_y/D$, where D is the slab thickness.

12.8.2 Methods of analysis

For any service load above (or below) the balanced load, moments are induced in the slab and, if large enough, these moments may lead to cracking or excessive deflection. A reliable technique for estimating slab moments is therefore required to check in-service behaviour under the unbalanced loads. In addition, reliable estimates of the maximum moments and shears caused by the full factored dead and live loads must be made in order to check the flexural and shear strength of a slab.

In AS3600-2009 [1], a simplified method is proposed for the analysis of reinforced two-way edge-supported rectangular slabs subjected to uniformly distributed design ultimate loads. Moment coefficients are specified in the code that may be used to calculate the design moments. In the absence of more accurate methods of analysis, the moment coefficients may also be used to determine the design moments in prestressed concrete edge-supported slabs.

The positive design moments per unit width at the mid-span of the slab in each direction are:

$$M_x^* = \beta_x w^* L_x^2 \tag{12.17}$$

and

$$M_y^* = \beta_y w^* L_x^2 \tag{12.18}$$

where w^* is the factored uniformly distributed design load per unit area; L_x is the short span; and β_x and β_y are moment coefficients that depend on the support conditions and the aspect ratio of the panel (i.e. L_y/L_x).

Values for β_x and β_y are given in Table 12.3 or may be obtained from:

$$\beta_y = \frac{2\left[\sqrt{3 + (\gamma_x/\gamma_y)^2} - \gamma_x/\gamma_y\right]^2}{9\gamma_y^2} \tag{12.19}$$

and

$$\beta_x = \frac{\beta_y L_x}{L_y} + \frac{2\,[1 - (L_x/L_y)]}{3\gamma_y^2} \tag{12.20}$$

where:

γ_x = 2.0 if both short edges are discontinuous
γ_x = 2.5 if one short edge is discontinuous
γ_x = 3.0 if both short edges are continuous
γ_y = 2.0 if both long edges are discontinuous
γ_y = 2.5 if one long edge is discontinuous
γ_y = 3.0 if both long edges are continuous

The negative design moments at a continuous edge and at a discontinuous edge are taken to be 1.33 times the mid-span value and 0.5 times the mid-span value, respectively.

Table 12.3 Design moment coefficients for rectangular edge-supported slabs [1]

| Edge conditions of the rectangular slab panel | Short-span coefficient, β_x | | | | | | | | Long-span coefficient β_y for all values of L_y/L_x |
| | Aspect ratio, L_y/L_x | | | | | | | | |
	1.0	1.1	1.2	1.3	1.4	1.5	1.75	≥2.0	
1. Four edges continuous	0.024	0.028	0.032	0.035	0.037	0.040	0.044	0.048	0.024
2. One short edge discontinuous	0.028	0.032	0.036	0.038	0.041	0.043	0.047	0.050	0.028
3. One long edge discontinuous	0.028	0.035	0.041	0.046	0.050	0.054	0.061	0.066	0.028
4. Two short edges discontinuous	0.034	0.038	0.040	0.043	0.045	0.047	0.050	0.053	0.034
5. Two long edges discontinuous	0.034	0.046	0.056	0.065	0.072	0.078	0.091	0.100	0.034
6. Two adjacent edges discontinuous	0.035	0.041	0.046	0.051	0.055	0.058	0.065	0.070	0.035
7. Three edges discontinuous (one long edge continuous)	0.043	0.049	0.053	0.057	0.061	0.064	0.069	0.074	0.043
8. Three edges discontinuous (one short edge continuous)	0.043	0.054	0.064	0.072	0.078	0.084	0.096	0.105	0.043
9. Four edges discontinuous	0.056	0.066	0.074	0.081	0.087	0.093	0.103	0.111	0.056

The moment coefficients of Table 12.3 may also be used for serviceability calculations, with the moments caused by the unbalanced load taken as:

$$M_{x.ub} = \beta_x w_{ub} L_x^2 \qquad (12.21)$$

and

$$M_{y.ub} = \beta_y w_{ub} L_x^2 \qquad (12.22)$$

These values of unbalanced moments may be used to check for cracking at service loads. For edge-supported prestressed concrete slabs, cracking is unlikely at service loads. Even reinforced concrete slabs continuously supported on all edges are often uncracked at service loads. However, if cracking is detected, then an average effective moment of inertia I_{ef} that

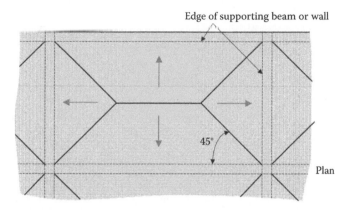

Figure 12.11 Distribution of shear forces in an edge-supported slab [1].

reflects the loss in stiffness due to cracking should be used in deflection calculations (see Section 5.11.3). For the determination of deflection of the slab strip in the shorter-span direction (shown in Figure 12.9b), a weighted average value of I_{ef} should be used and may be taken as 0.7 times the value of I_{ef} at mid-span plus 0.3 times the average of the values of I_{ef} at each end of the span. For an exterior span, a reasonable weighted average is 0.85 times the mid-span value plus 0.15 times the value at the continuous end. If a particular region is uncracked, I_{ef} for this region should be taken to be equal to I_g.

For the purposes of calculating the shear forces in a slab or the forces applied to the supporting walls or beams, AS3600-2009 [1] suggests that the uniformly distributed load on the slab is allocated to the supports as shown in Figure 12.11.

EXAMPLE 12.3

Design an exterior panel of a 180 mm thick two-way floor slab for a retail store. The rectangular panel is supported on four edges by stiff beams and is discontinuous on one long edge as shown in Figure 12.12a. The slab is post-tensioned in both directions using the draped parabolic cable profiles shown in Figures 12.12c and d. The slab supports a dead load of 1.5 kPa in addition to its own self-weight and the live load is 5.0 kPa. The level of prestress required to balance a uniformly distributed load of 5.0 kPa is required. Relevant material properties are: $f_c' = 40\,\text{MPa}$, $f_{ct}' = 3.5\,\text{MPa}$, $E_c = 30{,}000$ MPa, $f_{pb} = 1870$ MPa and $E_p = 195{,}000$ MPa.

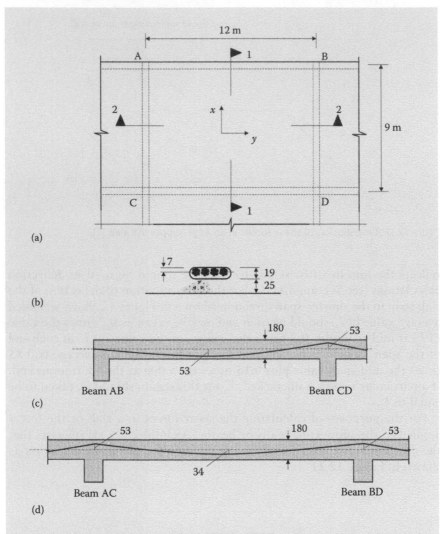

(a)

(b)

(c)

Beam AB   Beam CD

(d)

Beam AC   Beam BD

Figure 12.12 Details of edge-supported slab (Example 12.3). (a) Plan. (b) Section through a typical duct at mid-span (dimensions in mm). (c) Tendon profile in x-direction (Section 1-1). (d) Tendon profile in y-direction (Section 2-2).

Load balancing: Flat-ducted tendons containing four 12.7 mm strands are to be used with duct size 75 mm × 19 mm, as shown in Figure 12.12b. With 25 mm concrete cover to the duct, the maximum depth to the centre of gravity of the short-span tendons is:

$$d_x = 180 - 25 - (19 - 7) = 143 \text{ mm} \quad \text{(refer to Figure 12.12b)}$$

The cable drape in the short-span direction is therefore:

$$h_x = \frac{53+0}{2} + 53 = 79.5 \text{ mm}$$

The depth d_y of the long-span tendons at mid-span is less than d_x by the thickness of the duct running in the short-direction, i.e. $d_y = 143 - 19 = 124$ mm. The cable drape in the long-span direction is shown in Figure 12.12d and is given by:

$$h_y = \frac{53+53}{2} + 34 = 87.0 \text{ mm}$$

The self-weight of the slab is $24 \times 0.18 = 4.3$ kPa and if 40% of the live load is assumed to be sustained, then the total sustained load is:

$$w_{sus} = 4.3 + 1.5 + (0.4 \times 5.0) = 7.8 \text{ kPa}$$

In this example, the effective prestress in the tendons in both directions balances an external load of $w_b = 5.0$ kPa. The transverse load exerted by the tendons in the short-span direction is determined using Equation 12.14:

$$w_{px} = \frac{12^4}{2 \times 9^4 + 12^4} \times 5.0 = 3.06 \text{ kPa}$$

and the transverse load imposed by the tendons in the long-span direction is calculated using Equation 12.13:

$$w_{py} = w_b - w_{px} = 5.0 - 3.06 = 1.94 \text{ kPa}$$

The effective prestress in each direction is obtained from Equations 12.15 and 12.16:

$$P_x = \frac{3.06 \times 9^2}{8 \times 0.0795} = 390 \text{ kN/m}$$

and

$$P_y = \frac{1.94 \times 12^2}{8 \times 0.087} = 401 \text{ kN/m}$$

To determine the jacking forces and cable spacing in each direction, both the time-dependent losses and friction losses must be calculated. For the purpose of this example, it is assumed that the time-dependent losses in each direction are 15% and the immediate losses (friction, anchorage, etc.) in the x-direction are 8% and in the y-direction are 12%. Immediately after transfer, before the time-dependent losses have taken place, the prestressing forces at mid-span in each direction are:

$$P_{x.i} = \frac{390}{0.85} = 459 \text{ kN/m}$$

and

$$P_{y.i} = \frac{401}{0.85} = 472 \text{ kN/m}$$

and at the jack are:

$$P_{x.j} = \frac{459}{0.92} = 499 \text{ kN/m}$$

and

$$P_{y.j} = \frac{472}{0.88} = 536 \text{ kN/m}$$

Using four 12.7 mm strands/tendon, A_p = 394.4 mm²/tendon and the breaking load per tendon is 4 × 184 = 736 kN (see Table 4.8). If a limit of $0.85f_{pb}A_p$ is placed on the maximum force to be applied to the post-tensioned tendon during the stressing operation, the maximum jacking force/tendon is 0.85 × 736 = 626 kN and the required tendon spacing in each direction (rounded down to the nearest 10 mm) is therefore:

$$s_x = \frac{1000 \times 626}{499} = 1250 \text{ mm}$$

and

$$s_y = \frac{1000 \times 626}{536} = 1160 \text{ mm}$$

We will select a tendon spacing of 1200 mm in each direction.

This simply means that the tendons in the x-direction will balance slightly more load than previously assumed and the tendons in the y-direction slightly less.

With each tendon stressed to 626 kN, the revised prestressing forces at the jack per metre width are $P_x = P_y = 626/1.2 = 522$ kN/m, and at mid-span, after all losses are:

$$P_{x.e} = 0.85 \times 0.92 \times 522 = 408 \text{ kN/m}$$

$$P_{y.e} = 0.85 \times 0.88 \times 522 = 390 \text{ kN/m}$$

The load to be balanced is revised using Equations 12.11 and 12.12:

$$w_{px} = \frac{8 \times 408 \times 0.0795}{9^2} = 3.20 \text{ kPa}$$

$$w_{py} = \frac{8 \times 390 \times 0.087}{12^2} = 1.89 \text{ kPa}$$

and therefore $w_b = 3.20 + 1.89 = 5.09$ kPa.

Estimate maximum moment due to unbalanced load: The maximum unbalanced transverse load to be considered for short-term serviceability calculations is:

$$w_{ub} = w_{sw} + w_G + \psi_s w_Q - w_b = 4.30 + 1.5 + (0.7 \times 5.0) - 5.09 = 4.21 \text{ kPa}$$

Under this unbalanced load, the maximum moment occurs over the beam support CD. Using the moment coefficients for edge-supported slabs in Table 12.3, the maximum moment is approximated by:

$$M_{CD} = -1.33 \times 0.047 \times 4.21 \times 9^2 = -21.3 \text{ kNm/m}$$

Check for cracking: In the x-direction over support CD, the concrete stresses in the top and bottom fibres caused by the maximum moment M_{CD} after all losses are:

$$\sigma_t = -\frac{P_{x.e}}{A} + \frac{M_{CD}}{Z} = -2.27 + 3.94 = +1.67 \text{ MPa (tension)}$$

$$\sigma_b = -\frac{P_{x.e}}{A} - \frac{M_{CD}}{Z} = -2.27 - 3.94 = -6.21 \text{ MPa (compression)}$$

where A is the area of the gross cross-section per metre width (=180×10^3 mm^2/m) and Z is the section modulus per metre width (=5.4×10^6 mm^3/m).

Both tensile and compressive stresses are relatively low. Even though the moment used in these calculations is an average and not a peak moment, if cracking does occur, it will be localised and the resulting loss of stiffness will be small. Deflection calculations may be based on the properties of the uncracked cross-section.

Estimate maximum total deflection: The deflection at the mid-panel of the slab can be estimated using the so-called *crossing beam analogy*, in which the deflections of a pair of orthogonal beams (slab strips) through the centre of the panel are equated. The fraction of the unbalanced load carried by the strip in the short-span direction is given by an equation similar to Equation 12.14. With:

$$w_{ub.x} = \frac{12^4}{2.0 \times 9^4 + 12^4} \times 4.21 = 0.61 \times 4.21 = 2.57 \text{ kN/m}$$

and, with the deflection coefficient β taken as 2.6/384 (in accordance with the discussion in Section 12.4.2), the corresponding short-term deflection at mid-span of this 1 m wide slab strip in the short-span direction through the mid-panel (assuming the variable live load is removed from the adjacent slab panel) is approximated by:

$$v_i = \frac{2.6}{384} \frac{w_{ub.x} L_x^4}{E_c I} = \frac{2.6}{384} \frac{2.57 \times 9,000^4}{30,000 \times 486 \times 10^6} = 7.83 \text{ mm}$$

The sustained portion of the unbalanced load on the slab strip is:

$$\frac{L_y^4}{\delta L_x^4 + L_y^4} \times (w_{sw} + w_G + \psi_l w_Q - w_b) = 0.61 \times [4.30 + 1.5 + (0.4 \times 5.0) - 5.09]$$

$$= 1.65 \text{ kPa}$$

and the corresponding short-term deflection is:

$$v_{i.sus} = \frac{1.65}{2.57} \times v_i = 5.03 \text{ mm}$$

Assuming a creep coefficient $\varphi_{cc}^* = 2.0$ and conservatively ignoring the restraint provided by any bonded reinforcement, the creep-induced deflection given by Equation 12.7 may be estimated using the following equation:

$$v_{cc} = 2.0 \times 5.03 = 10.06 \text{ mm}$$

If the final shrinkage strain is assumed to be $\varepsilon^*_{cs} = 0.0005$. The shrinkage curvature κ^{III}_{cs} is non-zero wherever the eccentricity of the steel area is non-zero and varies along the span as the eccentricity of the draped tendons varies. A simple and very approximate estimate of the average final shrinkage curvature is made using Equation 12.9:

$$\kappa^*_{cs} = \frac{0.3\varepsilon^*_{cs}}{D} = \frac{0.3 \times 0.0005}{180} = 0.83 \times 10^{-6} \text{ mm}^{-1}$$

The average deflection of the slab strip due to shrinkage is given by Equation 12.8:

$$v_{cs} = 0.090 \times 0.83 \times 10^{-6} \times 9000^2 = 6.07 \text{ mm}$$

The maximum total deflection of the slab strip is therefore:

$$v_{tot} = v_i + v_{cc} + v_{cs} = 7.83 + 10.06 + 6.07 = 24.0 \text{ mm} = \frac{\text{span}}{375}$$

and the long-term to short-term deflection ratio is $\lambda = 2.06$. This deflection is likely to be satisfactory for a retail floor.

It is of value to examine the slab thickness predicted by Equation 12.4, if the limiting deflection is taken to be 24.0 mm. For this edge-supported slab panel, the slab system factor is obtained from Table 12.2 as $K = 2.37$, the unbalanced load $w_{ub} = 4.21$ kPa, and the sustained part of the unbalanced load is $w_{ub.sus} = 2.70$ kPa. With $\lambda = 2.06$, the minimum slab thickness required to limit the total deflection to 24 mm is obtained from Equation 12.4 as:

$$\frac{9,000}{D} \leq 2.37 \times \left[\frac{(24.0/9,000) \times 1,000 \times 30,000}{4.21 + 2.06 \times 2.70} \right]^{1/3} = 47.7$$

$$\therefore D \geq 189 \text{ mm}$$

In this example, Equation 12.4 is a little conservative.

Check flexural strength: It is necessary to check the ultimate strength of the slab. As previously calculated, the dead load is $1.5 + 4.3 = 5.8$ kPa and the live load is 5.0 kPa. The factored design load (using the load factors specified in Equation 2.2) is:

$$w^* = 1.2 \times 5.8 + 1.5 \times 5.0 = 14.46 \text{ kPa}$$

The design moments at mid-span in each direction are obtained from Equations 12.17 and 12.18, with $\beta_x = 0.047$ and $\beta_y = 0.028$ taken from Table 12.3:

$$M_x^* = 0.047 \times 14.46 \times 9^2 = 55.1 \text{ kNm/m}$$

$$M_y^* = 0.028 \times 14.46 \times 9^2 = 32.8 \text{ kNm/m}$$

The maximum design moment occurs over the beam support CD (the long continuous edge) and is:

$$(M_x^*)_{CD} = -1.33 \times 55.1 = -73.3 \text{ kNm/m}$$

A safe, lower bound solution to the problem of adequate ultimate strength is obtained if the design strength of the slab at this section exceeds the design moment.

The ultimate strength per metre width of the 180 mm thick slab containing tendons at 1200 mm centres (i.e. $A_p = 394.4/1.2 = 329 \text{ mm}^2/\text{m}$) at an effective depth of 143 mm is obtained using the procedures discussed in Chapter 6. Such an analysis indicates that the cross-section is ductile, with the depth to the neutral axis at ultimate $d_{nl} = 22.2$ mm (or $0.155d$), which is much less than the maximum limiting value of $0.4d$. The tensile force in the steel is 581 kN/m ($\sigma_{pu} = 1765$ MPa) and the design strength is:

$$\phi M_{uol} = 0.8 \times 329 \times 1765 \times \left(143 - \frac{0.77 \times 22.2}{2}\right) \times 10^{-6} = 62.5 \text{ kNm/m}$$

With $\phi M_{uol} < (M_x^*)_{CD}$, conventional reinforcement is required to supplement the prestressing steel over the beam support CD. From Equation 6.26, with the internal lever arm ℓ taken to be $0.9(d_p - \gamma d_{nl}) = 113.3$ mm, the required area of additional non-prestressed steel is approximated by:

$$A_{st} = \frac{(M_x^*)_{CD} - \phi M_{uol}}{\phi f_{sy} \ell} = \frac{(73.3 - 62.5) \times 10^6}{0.8 \times 500 \times 113.3} = 238 \text{ mm}^2/\text{m}$$

Use 12 mm diameter bars ($f_{sy} = 500$ MPa) at 300 mm centres ($A_{st} = 367 \text{ mm}^2/\text{m}$) as additional steel in the top of the slab over beam support CD.

Checking flexural strength at other critical sections indicates that:

1. At mid-span in the x-direction:

$M_x^* = 55.1$ kNm/m; $\phi M_{uo1} = 62.5$ kNm/m ($d = 143$ mm)

∴ No additional bottom reinforcement is required at mid-span in the x-direction.

2. At mid-span in the y-direction:

$M_y^* = 32.8$ kNm/m; $\phi M_{uo1} = 53.2$ kNm/m ($d = 124$ mm)

∴ No additional bottom reinforcement is required at mid-span in the y-direction.

3. At the short continuous support in the y-direction:

$(M_y^*)_{AC} = 1.33 \times 32.8 = 43.6$ kNm/m; $\phi M_{uo1} = 62.5$ kNm/m ($d = 143$ mm)

∴ No additional top reinforcement is required in the x-direction over AC and BD.

Summary of reinforcement requirements: Tendons consisting of four 12.7 mm strands at 1200 mm centres in each direction are used with the profiles shown in Figure 12.12c and d. In addition, 12 mm diameter non-prestressed reinforcing bars in the x-direction at 300 mm centres are also placed in the top of the slab over the long support CD (extending on each side of the beam to the point 0.3 times the clear span in the x-direction from the face of the support).

Check shear strength: In accordance with Figure 12.11, the maximum shear in the slab occurs at the face of the long support near its mid-length, where:

$V^* = w^* L_x / 2 = 14.46 \times 9/2 = 65.1$ kN/m

The contribution of the concrete to the shear strength in the region of low moment at the face of the discontinuous support is given by Equation 7.11 as:

$V_{uc} = V_t + P_v$

where V_t is the shear force required to cause web-shear cracking. From Equations 7.13 and 7.14:

$$\sigma = -\frac{408 \times 10^3}{180 \times 10^3} = -2.27 \text{ MPa} \quad \text{and} \quad \tau = \frac{V_t Q}{Ib} = (8.33 \times 10^{-6})V_t$$

and solving using Equation 7.12 gives $V_t = 539$ kN/m.

Clearly, V^* is much less than ϕV_{uc} and the shear strength is ample here. Shear strengths at all other sections are also satisfactory. Shear is rarely a problem in edge-supported slabs.

12.9 FLAT PLATE SLABS

12.9.1 Load balancing

Flat plates behave in a similar manner to edge-supported slabs except that the *edge beams* are strips of slab located on the column lines, as shown in Figure 12.13. The edge beams have the same depth as the remainder of the slab panel and, therefore, the system tends to be less stiff and more prone to serviceability problems. The load paths for both the flat plate and the edge-supported slab are, however, essentially the same (compare Figures 12.10 and 12.13).

In the flat plate panel of Figure 12.13, the total load to be balanced is $w_b L_x L_y$. The upward forces per unit area exerted by the slab tendons in each direction are given by Equations 12.11 and 12.12, and the slab tendons impose a total upward force of:

$$w_{px} L_x L_y + w_{py} L_y L_x = w_b L_x L_y$$

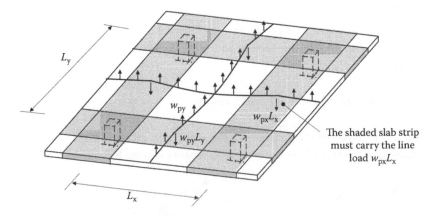

Figure 12.13 Interior flat plate panel.

Just as for edge-supported slabs, the slab tendons may be distributed arbitrarily between the x- and y-directions, provided that adequate additional tendons are placed in the slab strips to balance the line loads $w_{py}L_y$ and $w_{px}L_x$ shown on the column lines in Figure 12.13. These additional *column line* tendons correspond to the *beam* tendons in an edge-supported slab system. For perfect load balancing, the column line tendons would have to be placed within the width of slab in which the slab tendons exert downward load due to reverse curvature. However, this is not a strict requirement and considerable variation in tendon spacing can occur without noticeably affecting slab behaviour. Column line tendons are frequently spread out over a width of slab as large as one-half the shorter span, as indicated in Figure 12.14c.

The total upward force that must be provided in the slab along the column lines is:

$$w_{px}L_xL_y + w_{py}L_yL_x = w_bL_xL_y$$

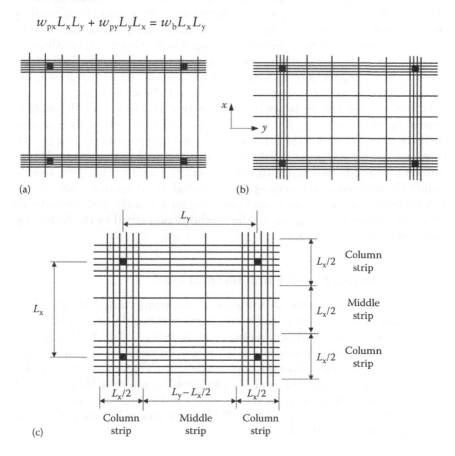

Figure 12.14 Alternative tendon layouts. (a) One-way slab arrangement. (b) Two-way slab arrangement with column line tendons in narrow band. (c) Two-way slab arrangement with column line tendons distributed over column strip.

Therefore, prestressing tendons (slab tendons plus column line tendons) must be provided in each panel to give a total upward force of $2w_b L_x L_y$. The slab tendons and column line tendons *in each direction* must, between them, provide an upward force equal to the total load to be balanced $w_b L_x L_y$. For example in the slab system shown in Figure 12.14a, the entire load to be balanced is carried by slab tendons in the x-direction, i.e. $w_{px} = w_b$ and $w_{py} = 0$. This entire load is deposited as a line load on the column lines in the y-direction and must be balanced by column line tendons in this vicinity. This slab has in effect been designed as a one-way slab spanning in the x-direction and supported by shallow heavily stressed slab strips on the y-direction column lines.

The two-way system shown in Figure 12.14b is more likely to perform better under unbalanced loads, particularly when the orthogonal spans L_x and L_y are similar and the panel is roughly square. In practice, however, steel congestion over the supporting columns and minimum spacing requirements (often determined by the size of the anchorages) make the concentration of tendons on the column lines impossible. Figure 12.14c shows a more practical and generally acceptable layout. Approximately 75% of the tendons in each direction are located in the column strips, as shown, the remainder being uniformly spread across the middle strip regions.

If the tendon layout is such that the upward force on the slab is approximately uniform, then at the balanced load, the slab has zero deflection and is subjected only to uniform compression caused by the longitudinal prestress in each direction applied at the anchorages. Under unbalanced loads, moments and shears are induced in the slab. To calculate the moments and stresses due to unbalanced service loads and to calculate the factored design moments and shears in the slab (in order to check ultimate strength), one of the methods described in the following sections may be adopted.

12.9.2 Behaviour under unbalanced load

Figure 12.15 illustrates the distribution of moments caused by an unbalanced uniformly distributed load w on an internal panel of a flat plate. The moment diagram in the direction of span L_y is shown in Figure 12.15b. The slab in this direction is considered as a wide, shallow beam of width L_x and span L_y, and carrying a load wL_x per unit length. The relative magnitudes of the negative moments M_{1-2} and M_{3-4} and positive moment M_{5-6} are found by elastic frame analysis (see Section 12.9.3) or by more approximate recommendations (see Section 12.9.4). Whichever method is used, the *total static moment* M_o is fixed by statics and is given by:

$$M_o = \frac{wL_x L_y^2}{8}$$

(12.23)

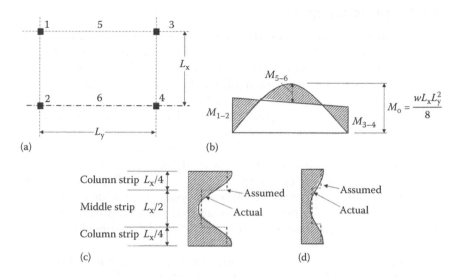

Figure 12.15 Distribution of moments in a flat plate. (a) Plan. (b) Moments in long direction. (c) Distribution of M_{1-2} across panel. (d) Distribution of M_{5-6} across panel.

In Figure 12.15c and d, variations in elastic moments across the panel at the column lines and at mid-span, respectively, are shown. At the column lines, where curvature is a maximum, the moment is also a maximum. On panel centre-line where curvature is a minimum, so too is the moment. In design, it is convenient to divide the panel into column and middle strips and to assume that the moment is constant in each strip as shown. The column strips in the L_y direction are defined as strips of width $0.25L_x$, but not greater than $0.25L_y$, on each side of the column centre-line. The middle strips are the slab strips between the column strips.

It may appear from the moment diagrams that at ultimate loads, the best distribution of tendons (and hence strength) is one in which tendons are widely spaced in the middle strips and closer together in the column strips, as shown in Figure 12.14c. However, at ultimate loads, provided that the slab is ductile, redistribution of moments takes place as the ultimate condition is approached and the final distribution of moments depends very much on the layout of the bonded steel.

After the slab cracks and throughout the overload range, superposition is no longer applicable and the concepts of balanced and unbalanced loads are not meaningful. As discussed in Section 11.5.4, at ultimate conditions, when the load factors are applied to the dead and live load moments, codes of practice often insist that secondary moments are considered with a load factor of 1.0. However, provided that the slab is ductile, and slabs are usually very ductile, secondary moments may be ignored in ultimate strength calculations. The difficulty in accurately estimating slab moments, particularly in the overload range, is rendered relatively unimportant by the ductile nature of slabs.

12.9.3 Frame analysis

A commonly used technique for the analysis of flat plates in building structures is the *equivalent frame method* (or *idealised frame method*). The
structure is idealised into a set of parallel two-dimensional frames running
in two orthogonal directions through the building. Each frame consists
of a series of vertical columns spanned by horizontal *beams*. These idealised beams consist of the strip of slab of width on each side of the column
line equal to half the distance to the adjacent parallel row of columns and
include any floor beams forming part of the floor system. The member stiffnesses are determined and the frames are analysed under any desired gravity loading using a linear-elastic frame analysis. For a flat plate building
in which shear walls or some other bracing system is provided to resist all
lateral loads, it is usually permissible to analyse each floor of the building
separately with the columns above and below the slab assumed to be fixed
at their remote ends.

The equivalent frame method provides a relatively crude model of structural behaviour, with inaccuracies being associated with each of the following assumptions: (a) a two-way plate is idealised by orthogonal one-way
strips; (b) the stiffness of a cracked slab is usually based on gross section
properties; and (c) a linear-elastic analysis is applied to a structure that is
nonlinear and inelastic both at service loads and at overloads. A simple
estimate of member stiffness, based, for example, on gross section properties, will lead to an estimate of frame moments that satisfies equilibrium
and usually provides an acceptable solution. AS3600-2009 [1] suggests that
the stiffness of the frame members should be chosen to represent conditions at the limit state under consideration. All such assumptions should be
applied consistently throughout the analysis. When such a frame analysis is
used to check bending strength, an equilibrium load path is established that
will prove to be a satisfactory basis for design, provided that the slab is ductile and the moment distribution in the real slab can redistribute towards
that established in the analysis.

For the determination of the design moments at each critical section of
the frame, variations of the load intensities on individual spans should be
considered. This includes pattern loading where the transient load is applied
to some spans and not to others. AS3600-2009 [1] requires that the loading
patterns to be considered should include at least the following:

1. Where the loading pattern is known, the frame should be analysed
   under that known loading. This includes the factored permanent dead
   load G.
2. With regard to the live loads Q, where the pattern of loaded and
   unloaded spans is variable, the factored live load should be applied:
   a. on alternate spans (this will permit the determination of the maxi-
      mum factored positive moment near the middle of the loaded spans);

b. on two-adjacent spans (this will permit the determination of the maximum factored negative moment at the interior support between the loaded spans); and
c. on all spans.

Where the live load Q is less than three-quarters of the permanent dead load, a single analysis with the full factored live load on all spans is acceptable [1].

The frame moments calculated at the critical sections of the idealised horizontal members are distributed across the floor slab into the column and middle strips (as defined in Section 12.9.2). Studies have shown that the performance of reinforced concrete flat slabs both at service loads and at overloads is little affected by variations in the fraction of the total frame moment that is assigned to the column strip [10], provided that the slab is ductile and capable of the necessary moment redistribution.

AS3600-2009 [1] specifies that the column strip shall be designed to resist the total negative or positive frame bending moment at each section multiplied by a *column strip moment factor* taken from within the ranges given in Table 12.4. An idealised frame analysis may be used to examine the serviceability of a floor slab. With the in-service moments caused by the unbalanced loads determined at all critical regions, checks for cracking and crack control and calculations of deflection may be undertaken.

When the ultimate strengths of the column and middle strips are being checked, it is advisable to ensure that the depth to the neutral axis at ultimate at any section does not exceed $0.25d$. This will ensure sufficient ductility for the slab to redistribute bending moments towards the bending moment diagram predicted by the idealised frame analysis and will also allow the designer safely to ignore the secondary moments. There are obvious advantages in allocating a large fraction of the negative moment at the supports to the column strip. The increased steel quantities that result stiffen and strengthen this critical region of the slab, thereby improving punching shear and crack control. In prestressed flat slabs, only the column strip regions over the interior columns are likely to experience significant cracking. In Australia, it is not uncommon, in the design of reinforced

Table 12.4 Fraction of frame moments distributed to the column strip [1]

Bending moment under consideration	Column strip moment factor	
	Strength limit state	Serviceability limit state
Negative moment at interior support	0.6–1.00	0.75
Negative moment at exterior support with spandrel beam	0.75–1.00	0.75
Negative moment at exterior support without spandrel beam	0.75–1.00	1.00
Positive moment in all spans	0.5–0.7	0.6

concrete flat slabs that are not exposed to the weather, to specify uniform steel in the bottom of the slab (i.e. 50% of the positive frame moments are assigned to the column strip), with all the top steel confined to the column strip (i.e. 100% of the negative frame moments in the column strip). The in-service performance of such slabs is at least as good as that of the more traditionally reinforced slabs, and significant cost savings usually result. Steel fixing is greatly simplified and, with large portions of the slab free from top steel, concrete placing is much easier.

12.9.4 Direct design method

A simple semi-empirical approach for the analysis of flat plates is the *direct design method*. The method is outlined specifically for reinforced concrete slabs in AS3600-2009 [1], but within certain limitations, the direct design method can be applied equally well to prestressed slabs and the results obtained are just as reliable as those obtained from a frame analysis.

Limitations are usually imposed on the use of the direct design method, such as the following requirements imposed by AS3600-2009 [1]:

1. There are at least two continuous spans in each direction.
2. The support grid is rectangular, or nearly so (individual supports may be offset up to a maximum of 10% of the span in the direction of the offset).
3. The ratio of the longer to shorter span measured centre-to-centre of supports within any panel is not greater than 2.0.
4. In each direction, successive span lengths do not differ by more than one-third of the longer span and in no case is an end span longer than the adjacent interior span.
5. Gravity loads are essentially uniformly distributed. Lateral loads are resisted by shear walls or braced vertical elements and do not enter into the analysis.
6. The live load q does not exceed twice the dead load g.
7. Ductility class L reinforcement is not used as the flexural reinforcement.

The slab is divided into design strips in each direction, and each strip is designed one span at a time. The total static moment, M_o, in each span of the design strip is calculated from the following equation:

$$M_o = \frac{w^* L_t L_{ef}^2}{8} \tag{12.24}$$

where w^* is the design load per unit area (factored for strength); L_{ef} is the effective span (i.e. the lesser of the centre-to-centre distance between supports and $L_n + D$); L_n is the clear span; D is the overall slab thickness;

Table 12.5 Design moment factors for an end span [1]

	Negative moment factor at an exterior support	Positive moment factor	Negative moment factor at an interior support
Flat slabs with exterior edge unrestrained	0.0	0.60	0.80
Flat slabs with exterior edge restrained by columns only	0.25	0.50	0.75
Flat slabs with exterior edge restrained by spandrel beams and columns	0.30	0.50	0.70
Flat slabs with exterior edge fully restrained	0.65	0.35	0.65
Beam-and-slab construction	0.15	0.55	0.75

Table 12.6 Design moment factors for an interior span [1]

Type of slab system	Negative moment factor	Positive moment factor
All types	0.65	0.35

and L_t is the width of the design strip measured transverse to the direction of bending.

For an interior design strip, L_t is equal to the average of the centre-to-centre distance between the supports of the adjacent transverse spans. For an edge design strip, L_t is measured from the slab edge to the point halfway to the centre-line of the next interior and parallel row of supports.

The static moment M_0 is shared between the supports (negative moments) and the mid-span (positive moment). At any critical section, the design moment is determined by multiplying M_0 by the relevant factor given in Tables 12.5 or 12.6, as appropriate. According to AS3600-2009 [1], it is permissible to modify these design moments by up to 10%, provided that the total static design moment M_0 for the span is not reduced. At any interior support, the floor slab should be designed to resist the larger of the two negative design moments determined for the two adjacent spans unless the unbalanced moment is distributed to the adjoining members in accordance with their relative stiffnesses.

The positive and negative design moments are next distributed to the column and middle strips using the column strip moment factor from Table 12.4.

12.9.5 Shear strength

Punching shear strength requirements often control the thickness of a flat slab at the supporting columns and must always be checked. The shear strength of the slabs was discussed in Chapter 7 (Section 7.4) and methods for designing the slab–column intersection were presented.

If frame analyses are performed to check the flexural strength of a slab, the design moment M_v^* transferred from the slab to a column and the design shear V^* are obtained from the relevant analyses. M_v^* is that part of the unbalanced slab bending moments transferred into the column at the support. If the direct design method is used for the slab design, M_v^* and V^* must be calculated separately. The shear force crossing the critical shear perimeter around a column support may be taken as the product of the factored design load w^* and the plan area of slab supported by the column and located outside the critical shear perimeter. AS3600-2009 [1] specifies that, at an interior support, M_v^* shall not taken less than the value given by:

$$M_v^* = 0.06\,[(1.2g + 0.75q)L_t(L_o)^2 - 1.2gL_t(L_o')^2]$$ (12.25)

where g and q are the uniformly distributed dead and live loads on the slab (per unit area), L_t is the transverse width of the slab as defined in the text following Equation 12.24, $L_o = L - 0.7a_{sup}$, $L_o' = L' - 0.7a_{sup}'$, $L(L')$ is the centre-to-centre distance between supports of the longer (shorter) adjacent span and $a_{sup}(a_{sup}')$ is 50% of the average of the column dimensions in the span direction at each end of the longer (shorter) adjacent span.

For an edge column, M_v^* is equal to the design moment at the exterior edge of the slab and may be taken as $0.25M_o$ (where M_o is the static moment for the end span of the slab calculated using Equation 12.24).

When detailing the slab–column connection, it is advisable to have at least two prestressing tendons crossing the critical shear perimeter in each direction. Additional well-anchored non-prestressed reinforcement crossing the critical perimeter will also prove beneficial (both in terms of crack control and ductility) in the event of unexpected overloads.

12.9.6 Deflection calculations

The deflection of a uniformly loaded flat slab may be estimated using the *wide beam method* which was formalised by Nilson and Walters [11]. Originally developed for reinforced concrete slabs, the method is particularly appropriate for prestressed flat slabs which are usually uncracked at service loads [12]. The basis of the method is illustrated in Figure 12.16. Deflections of the two-way slab are calculated by considering separately the slab deformations in each direction. The contributions in each direction are then added to obtain the total deflection.

In Figure 12.16a, the slab is considered to act as a wide shallow beam of width equal to the smaller panel dimension L_x and span equal to the longer panel dimension L_y. This wide beam is assumed to rest on unyielding supports. Because of variations in the moments caused by the unbalanced loads and the flexural rigidity across the width of the slab, all unit strips in the x-direction will not deform identically.

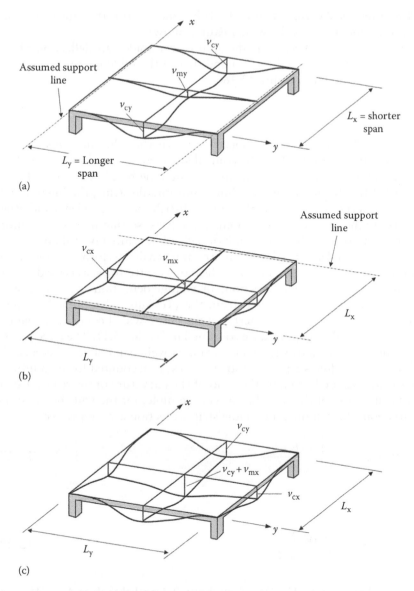

Figure 12.16 Basis of the wide beam method. (a) Bending in y-direction. (h) Rending in x-direction. (c) Combined bending [1].

Unbalanced moments and hence curvatures in the regions near the column lines (the column strip) are greater than in the middle strips. This is particularly so for uncracked prestressed concrete slabs or in prestressed slabs that are cracked only in the column strips. The deflection on the column line is therefore greater than that at the panel centre. The slab is next considered to act as a wide shallow beam spanning in the

y-direction, as shown in Figure 12.16b. Once again, the effect of varia-tion of moment across the wide beam is shown.

The mid-panel deflection is the sum of the mid-span deflection of the column strip in the long direction and that of the middle strip in the short direction, as shown in Figure 12.16c:

$$v_{\text{mid}} = v_{\text{cy}} + v_{\text{mx}} \tag{12.26}$$

The method can be used irrespective of whether the moments in each direction are determined by the equivalent frame method, frame analysis based on gross stiffnesses or the direct design method (see Sections 12.9.3 and 12.9.4). The definition of column and middle strips, the longitudinal moments in the slab, the lateral moment distribution coefficients and other details are the same as for the moment analysis, so that most of the infor-mation required for the calculation of deflection is already available.

The actual deflection calculations are more easily performed for strips of floor in either direction bounded by the panel centre-lines, as is used for the moment analysis. In each direction, an average deflection v_{avge} at mid-span of the wide beam is calculated from the previously determined moment diagram and the moment of inertia of the entire wide beam I_{beam} using the deflection calculation procedures outlined in Section 5.11. The effect of the moment variation across the wide beam, as well as possible differences in column and middle strip sizes and rigidities, is accounted for by multiply-ing the average deflection by the ratio of the curvature of the relevant strip to the curvature of the wide beam. For example, for the wide beam in the y-direction, the column and middle strip deflection are, respectively.

$$v_{\text{cy}} = v_{\text{avge.y}} \frac{M_{\text{col}}}{M_{\text{beam}}} \frac{E_c I_{\text{beam}}}{E_c I_{\text{col}}} \tag{12.27}$$

and

$$v_{\text{my}} = v_{\text{avge.y}} \frac{M_{\text{mid}}}{M_{\text{beam}}} \frac{E_c I_{\text{beam}}}{E_c I_{\text{mid}}} \tag{12.28}$$

It is usual to assume $M_{\text{col}}/M_{\text{beam}}$ is about 0.7 and therefore $M_{\text{mid}}/M_{\text{beam}}$ is about 0.3. If cracking is detected in the column strip, the effective moment of inertia of the cracked cross-section can be calculated using the analysis described in Section 5.11.3. The effective moment of inertia of the column strip I_{col} is calculated as the average of I_{ef} at the negative moment region at each end of the strip, which may include the loss of stiffness due to crack-ing and/or the stiffening effect of a drop panel, and the positive moment region, which is usually uncracked. I_{col} is then added to the moment of iner-tia of the middle strip I_{mid} (which is also usually uncracked and therefore based on gross section properties) to form the effective moment of inertia of the wide beam I_{beam}. These quantities are then used in the calculation of the short-term column and middle strip deflections in each direction using Equations 12.27 and 12.28. A reasonable estimate of the weighted average effective moment of inertia of an interior span of the wide beam is obtained by taking 0.7 times the value at mid-span plus 0.3 times the average of the values at each end of the span. For an exterior span, a reasonable weighted average is 0.85 times the mid-span value plus 0.15 times the value at the continuous end. This recommendation may also be used for the calculation of I_{col} for a cracked column strip.

The moment of inertia of the wide beam is, of course, always the sum of I_{col} and I_{mid}. Long-term deflections due to sustained unbalanced loads can also be calculated in each direction using the procedures outlined in Section 5.11.4.

Nilson and Walters [11] originally proposed to analyse a fixed-ended beam and then calculate the deflection produced by rotation at the supports. This does not significantly improve the accuracy of the model and the additional complication is not warranted.

EXAMPLE 12.4

Determine the tendons required in the 220 mm thick flat plate shown in Figure 12.17. The live load on the slab is 3.0 kPa and the dead load is 1.0 kPa plus the slab self-weight. All columns are 600 mm × 600 mm and are 4 m long above and below the slab. At the top of each column, a 300 mm column capital is used to increase the supported area, as shown. In this example, the dead load g is to be effectively balanced by prestress and is given by:

$$g = 1 \text{ kPa} + \text{self-weight} = 1 + (24 \times 0.22) = 6.3 \text{ kPa}$$

1. *Checking punching shear:*
Before proceeding too far into the design, it is prudent to make a preliminary check of punching shear at typical interior and exterior columns. Consider the interior column B in Figure 12.17. The area of slab supported by the column is $10 \times (8.5 + 10)/2 = 92.5$ m². Using the strength load factors specified in AS3600-2009 [1] (see Equation 2.2), the factored design load is:

$$w^* = 1.2g + 1.5q = (1.2 \times 6.3) + (1.5 \times 3.0) = 12.1 \text{ kN/m}^2$$

and therefore the shear force crossing the critical section may be approximated by:

$$V^* \approx 12.1 \times 92.5 = 1120 \text{ kN}$$

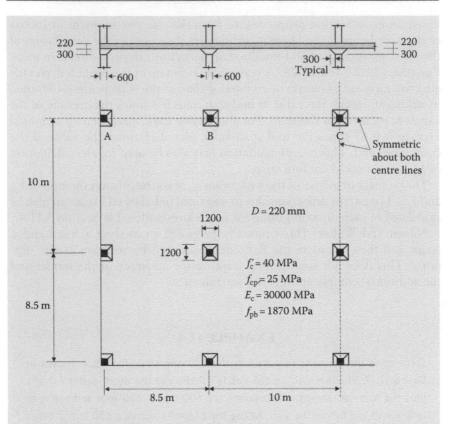

Figure 12.17 Plan and section of flat plate (Example 12.4).

From Equation 12.25, the design moment transferred to the column may be taken as:

$$M_v^* = 0.06 \left[(1.2 \times 6.3 + 0.75 \times 3.0) \times 10 \times 9.16^2 - 1.2 \times 6.3 \times 10 \times 7.77^2 \right]$$

$$= 220 \text{ kNm}$$

The average effective depth around the shear perimeter is taken to be $d_{om} = 220 - 50 = 170$ mm and the critical shear perimeter is therefore:

$$u = 4 \times (1200 + 170) = 5480 \text{ mm}$$

The average prestress in the concrete is assumed to be $\sigma_{cp} = 2.75$ MPa (this will need to be checked subsequently) and, from Equation 7.37, the limiting shear stress in the concrete is:

$$f_{cv} = 0.34\sqrt{40} = 2.15 \text{ MPa}$$

From Equation 7.36:

$$V_{uo} = 5480 \times 170 \times (2.15 + 0.3 \times 2.75) \times 10^{-3} = 2772 \text{ kN}$$

The critical section possesses adequate shear strength if the design shear V^* is less than ϕV_u, where V_u is given by Equation 7.40:

$$\phi V_u \leq \frac{0.7 \times 2772}{\left[1 + ((5480 \times 220 \times 10^6)/(8 \times 1120 \times 10^3 \times 1200 \times 170))\right]} = 1169 \text{ kN}$$

which is just greater than V^* and is considered to be acceptable at this preliminary stage. Punching shear at edge and corner columns should similarly be checked.

2. *Establish cable profiles:*
Using four 12.7 mm strands in a flat duct, with 25 mm concrete cover to the duct (the same as in Figure 12.12b), the maximum depth to the centre of gravity of the strand is:

$$d_p = 220 - (25 + 19 - 7) = 183 \text{ mm}$$

and the corresponding eccentricity is $e = 73$ mm. The maximum cable drapes in an exterior span and in an interior span are, respectively:

$$(h_{max})_{ext} = \frac{73}{2} + 73 = 109.5 \text{ mm}$$

$$(h_{max})_{int} = \frac{73 + 73}{2} + 73 = 146 \text{ mm}$$

Consider the trial cable profile shown in Figure 12.18. For the purposes of this example, it is assumed that jacking occurs simultaneously from both ends of a tendon, so that the prestressing force in a tendon is symmetrical with respect to the centre-line of the structure (shown in Figure 12.17). The friction losses have been calculated using Equation 5.136 with $\mu = 0.2$ and $\beta_p = 0.016$ for flat ducts, and the losses due to a 6 mm draw-in at the anchorage is calculated as outlined in Section 5.10.2.4 using Equation 5.139. The immediate losses (friction + draw-in) are also shown in Figure 12.18.

3. *Determine tendon layout*
It is assumed here that the average time-dependent loss of prestress in each low relaxation tendon is 15%. Of course, this assumption should be checked.

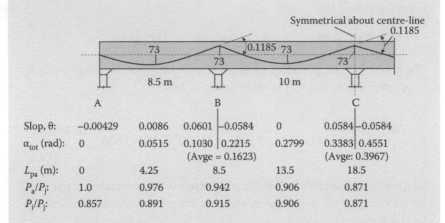

Slop, θ:	−0.00429	0.0086	0.0601	−0.0584	0	0.0584	−0.0584
α_{tot} (rad):	0	0.0515	0.1030	0.2215	0.2799	0.3383	0.4551
			(Avge = 0.1623)			(Avge: 0.3967)	
L_{pa} (m):	0	4.25	8.5		13.5	18.5	
P_a/P_j:	1.0	0.976	0.942		0.906	0.871	
P_l/P_j:	0.857	0.891	0.915		0.906	0.871	

Figure 12.18 Cable profile and immediate loss details (Example 12.4).

The effective prestressing forces per metre width required to balance 6.3 kPa using the full available drape in the exterior span (AB) and in the interior span (BC) are found using Equation 12.10:

$$(P_e)_{AB} = \frac{6.3 \times 8.5^2}{8 \times 0.1095} = 520 \text{ kN/m} \quad \text{and} \quad (P_e)_{BC} = \frac{6.3 \times 10^2}{8 \times 0.146} = 540 \text{ kN/m}$$

and the corresponding forces required at the jack prior to the instantaneous and time-dependent losses are:

$$P_j = \frac{520}{0.891 \times 0.85} = 687 \text{ kN/m} \text{ (required to balance 6.3 kPa in the exterior span)}$$

$$P_j = \frac{540}{0.906 \times 0.85} = 701 \text{ kN/m} \text{ (required to balance 6.3 kPa in the interior span)}$$

The jacking force is therefore governed by the requirements for the interior span.

For the 10 m wide panel, the total jacking force required is 701 × 10 = 7010 kN. If the maximum stress in the tendon is $0.85f_{pb}$, the total area of prestressing steel is therefore:

$$A_p = \frac{7010 \times 10^3}{0.85 \times 1870} = 4410 \text{ mm}^2$$

At least twelve flat-ducted cables are required in the 10 m wide panel (with the area of prestressing steel in each cable $A_p = 394.4$ mm²/cable) with an initial jacking force of 7010/12 = 584 kN/cable (i.e. $\sigma_{pj} = 0.79\ f_{pb}$).

The required jacking force in the 8.5 m wide panel is 701 × 8.5 = 5960 kN and therefore $A_p = 3750$ mm². At least 10 flat-ducted cables are needed in the 8.5 m wide panels ($A_p = 3944$ mm²) with an initial jacking force of 5960/10 = 596 kN/cable (i.e. $\sigma_{pj} = 0.81\ f_{pb}$).

In the interests of uniformity, all tendons will be initially stressed with a jacking force of 596 kN/cable (i.e. $\sigma_{pj} = 0.81\ f_{pb}$). This means that a slightly higher load than 6.3 kPa will be balanced in the 10 m wide panels. The average prestress at the jack in each metre width of slab is (22 × 596)/(8.5 + 10) = 709 kN/m, and the revised drape in the exterior span is:

$$h_{AB} = \frac{6.3 \times 8.5^2}{8 \times 709 \times 0.981 \times 0.85} = 0.106\ \text{m}$$

The final cable profile and effective prestress per panel after all losses are shown in Figure 12.19.

The maximum average stress in the concrete due to the longitudinal anchorage force after the deferred losses is:

$$\frac{P}{A} = \frac{5,514 \times 10^3}{10,000 \times 220} = 2.51\ \text{MPa}$$

which is within the range mentioned in Section 12.3 as being typical for flat slabs.

The cable layout for the slab is shown on the plan in Figure 12.20. For effective load balancing, about 75% of the cables are located in the column strips. The minimum spacing of tendons is usually governed by the size of

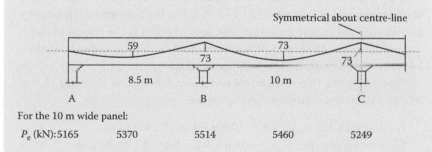

For the 10 m wide panel:

P_e (kN): 5165   5370   5514   5460   5249

Figure 12.19 Cable profile and effective prestress (Example 12.4).

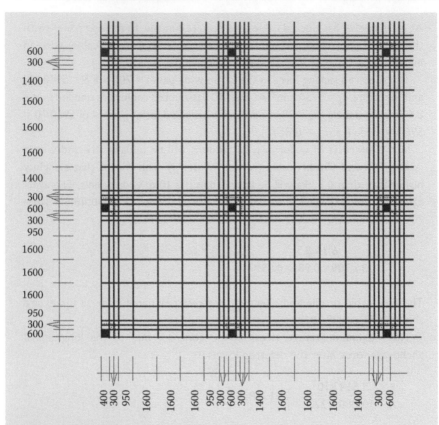

Figure 12.20 Tendon layout (Example 12.4).

the anchorage and is taken here as 300 mm, whilst a maximum spacing of 1600 mm has also been adopted.

4. *Serviceability considerations*

In practice, the time-dependent losses in the post-tensioned tendons should now be checked using the procedures outlined in Section 5.10.3 and illustrated previously in Examples 10.1 and 10.3. For the purposes of this example, we will assume that the losses have been checked at the critical sections and are as assumed in Step 3. We will now analyse the slab under the unbalanced loads and check for the likelihood of cracking.

Considering the 10 m wide frame on column-line ABC in Figure 12.17, the effective spans (clear spans + slab depth) are:

For end span AB: $(L_{ef})_{AB} = 8.5 - 0.6 - 0.6 + 0.22 = 7.52$ m
For interior span BC: $(L_{ef})_{BC} = 10.0 - 0.6 - 0.6 + 0.22 = 9.02$ m

With the full dead load balanced by the effective prestress, the maximum unbalanced load is 3 kPa (of which 1.0 kPa is assumed to be sustained). The total static moments in the end span AB and in the interior span BC caused by the maximum unbalanced load are obtained using Equation 12.24:

$$(M_o)_{AB} = \frac{3 \times 10 \times 7.52^2}{8} = 212.1 \text{ kNm}$$

$$(M_o)_{BC} = \frac{3 \times 10 \times 9.02^2}{8} = 305.1 \text{ kNm}$$

The moment diagrams for each span caused by the unbalanced load obtained using the *direct design method* are shown in Figure 12.21.

Check for cracking: For the 5 m wide and 220 mm deep column strip, the properties of the gross cross-section are $A = 1100 \times 10^3 \text{ mm}^2$, $I = 4437 \times 10^6 \text{ mm}^4$ and $Z = 40.33 \times 10^6 \text{ mm}^3$. Taking 75% of the negative frame moment at support C to be carried in the 5 m wide column strip, the column strip moment at C is $(M_C)_{col} = 0.75 \times (-198.3) = -148.7 \text{ kNm}$ and the effective prestress in the column strip at C is $P_e = 0.85 \times 0.871 \times P_i = 2625 \text{ kN}$. The tensile stress in the top concrete fibre is:

$$\sigma_{top} = -\frac{P_e}{A} + \frac{(M_C)_{col}}{Z} = -2.39 + 3.69 = +1.30 \text{ MPa (tension)}$$

This is less than the flexural tensile strength of the concrete and less than the limiting value of $0.25\sqrt{40} = 1.58$ MPa in AS3600-2009 [1] (see the discussion on flexural crack control in Section 12.5.1). The tensile stresses at the top of the slab in the columns strip over support B and in the bottom of the slab in

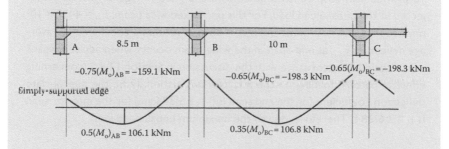

Figure 12.21 Bending moment diagram for flat slab frame ABC (Example 12.4).

the columns strip at the mid-spans of AB and BC will be less than that calculated above, because in each case, the magnitude of the effective prestress is larger than that at support C and the magnitude of the unbalance moment is smaller.

Although the tensile stresses in the column strips are less than the tensile strength of concrete, some local cracking over the interior column support is likely, since peak moments are much higher than average values. A mat of conventional non-prestressed reinforcement (rectangular in plan) is here provided over the interior supporting columns to ensure crack control. The steel in each direction will be continuous across the column support line and extend to 25% of the clear span in each direction (see discussion in Section 12.5.1). The area of non-prestressed steel area in each direction for crack control is at least:

$$A_{s.x} = A_{s.y} = 0.0015 \times 220 \times 1000 = 330 \text{ mm}^2/\text{m}$$

Unless a larger quantity of non-prestressed steel is required for strength, we will use a mat of 12 mm diameter bars at 300 mm centre spacing in each direction in the top of the slab over each interior column.

In addition, non-prestressed steel of area $0.0015bD = 330 \text{ mm}^2/\text{m}$ is required for crack control in the top of the slab perpendicular to the free edge in all exterior panels (where prestress may not be effective in accordance with the discussion in the last paragraph of Section 12.2).

Check deflection: Although some localised cracking may occur in the column strip over the interior columns of this slab, it will not be sufficient to significantly affect the slab stiffness and deflection may be calculated using the properties of the gross cross-section.

We will check deflection in the end span and in the interior spans of the 10 m wide slab strip centred on column line ABC in Figure 12.17 and subjected to the unbalanced load. For this uncracked *wide beam*: $I_{col} = 4437 \times 10^6$ mm^4, $I_{mid} = 4437 \times 10^6$ mm^4 and $I_{beam} = 8873 \times 10^6$ mm^4. The maximum average deflection v_{avge} at mid-span of the wide beam occurs when adjacent spans are unloaded. In accordance with the discussion in Section 12.4.2 (concerning effect of pattern loading on the β factor in Equation 12.6), the appropriate deflection coefficient for the end span is $\beta = 3.5/384$ and for the interior span is $\beta = 2.6/384$. Therefore, due to the maximum unbalanced load:

$$(v_{avge})_{AB} = \frac{3.5}{384} \frac{w_{ub}L_t(L_{ef})_{AB}^4}{E_c I_{beam}} = \frac{3.5}{384} \frac{3.0 \times 10 \times 7,520^4}{30,000 \times 8,873 \times 10^6} = 3.28 \text{ mm}$$

$$(v_{avge})_{BC} = \frac{2.6}{384} \frac{w_{ub}L_t(L_{ef})^4_{BC}}{E_c I_{beam}} = \frac{2.6}{384} \frac{3.0 \times 10 \times 9,020^4}{30,000 \times 8,873 \times 10^6} = 5.05 \text{ mm}$$

Taking 70% of moment in the column strip, the deflections of the column strip and the middle strip are obtained from Equations 12.27 and 12.28:

$$(v_c)_{AB} = (v_{avge})_{AB} \times 0.7 \times \frac{l_{beam}}{l_{col}} = 3.28 \times 0.7 \times 2 = 4.59 \text{ mm}$$

$$(v_m)_{AB} = (v_{avge})_{AB} \times 0.3 \times \frac{l_{beam}}{l_{mid}} = 3.28 \times 0.3 \times 2 = 1.97 \text{ mm}$$

$$(v_c)_{BC} = 5.05 \times 0.7 \times 2 = 7.07 \text{ mm}$$

$$(v_m)_{BC} = 5.05 \times 0.3 \times 2 = 3.05 \text{ mm}$$

Due to symmetry, the column and middle strip deflections in the orthogonal direction are the same for this uncracked slab.

The maximum short-term deflection at the mid-point of the panels due to the unbalanced load is obtained by adding the column strip deflection to the y-direction with the middle strip deflection in the x-direction (Equation 12.26). For the edge panel, adjacent to column line AB (with l_y = 10 m and L_x = 8.5 m), the maximum short-term deflection is:

$$(v_{i.mid})_{AB} = 7.07 + 1.97 = 9.04 \text{ mm}$$

whilst for the internal panel adjacent to column line BC (with L_y = 10 m and L_x = 10 m), the maximum short-term deflection is:

$$(v_{i.mid})_{BC} = 7.07 + 3.05 = 10.12 \text{ mm}$$

The sustained portion of the unbalanced load $w_{ub.sus}$ = 1.0 kPa and the short-term mid-panel deflections produced by the sustained unbalanced load are therefore one-third of the above values. With a creep factor of φ^*_{cc} = 2.5, the creep-induced deflection at the mid-point of each panel is estimated from Equation 12.7:

$$(v_{cc.mid})_{AB} = 2.5 \times 0.333 \times 9.04 = 7.53 \text{ mm}$$

$$(v_{cc.mid})_{BC} = 2.5 \times 0.333 \times 10.12 = 8.43 \text{ mm}$$

Assuming the final shrinkage strain in the concrete is $\varepsilon_{cs}^* = 0.0005$, the average shrinkage curvature κ_{cs}^* in each direction is estimated using Equation 12.9:

$$\kappa_{cs}^* = \frac{0.3 \times 0.0005}{220} = 0.68 \times 10^{-6} \, \text{mm}^{-1}$$

The average deflections due to shrinkage on the column centre-lines are obtained from Equation 12.8:

$$(v_{cs})_{AB} = 0.090 \times 0.68 \times 10^{-6} \times 7520^2 = 3.46 \text{ mm}$$

$$(v_{cs})_{BC} = 0.065 \times 0.68 \times 10^{-6} \times 9020^2 = 3.60 \text{ mm}$$

and the shrinkage deflections at the mid-points of the panels adjacent to column line AB and BC are the sum of the shrinkage deflection in each direction:

$$(v_{cs.mid})_{AB} = (v_{cs})_{AB} + (v_{cs})_{BC} = 7.06 \text{ mm}$$

$$(v_{cs.mid})_{BC} = (v_{cs})_{BC} + (v_{cs})_{BC} = 7.20 \text{ mm}$$

Therefore, the maximum total deflections at the mid-points of the panels adjacent to column line AB and BC are:

$$(v_{mid})_{AB} = (v_{i.mid})_{AB} + (v_{cc.mid})_{AB} + (v_{cs.mid})_{AB} = 9.04 + 7.53 + 7.06 = 23.6 \text{ mm}$$

$$(v_{mid})_{BC} = (v_{i.mid})_{BC} + (v_{cc.mid})_{BC} + (v_{cs.mid})_{BC} = 10.12 + 8.43 + 7.20 = 25.8 \text{ mm}$$

As the longer effective span in each of these two panels is $L_{ef} = 9.02$ m, the maximum deflection is $L_{ef}/350$ and this should be satisfactory for most occupancies.

5. Check shear and flexural strength

With the level of prestress determined, punching shear should also be checked at both exterior and interior columns in accordance with the procedure outlined in Sections 7.4.1 and 7.4.2. The dimensions of the column capitals may need to be modified, and shear reinforcement may be required in the spandrel strips along each free edge.

The ultimate flexural strength of the slab must also be checked. For the purposes of this example, the flexural strength of the interior panel will be compared with the design moments determined from the direct design method. As calculated in step 1, $w^* = 12.1$ kPa, the panel width is $L_t = 10$ m and the effective span of an interior panel is $L_{ef} = 9.02$ m. From Equation 12.23, the total static moment is:

$$M_o = \frac{12.1 \times 10 \times 9.02^2}{8} = 1231 \text{ kNm}$$

From Table 12.6, the negative support moment is:

$0.65M_o = 800$ kNm

Because both the positive and negative moment capacities are similar (each having the same quantity of prestressed steel at the same effective depth), it is appropriate to take advantage of the 10% permissible redistribution (reduction) in the negative support moment (as discussed in Section 12.9.4). The negative support moment is therefore taken as $0.9 \times 800 = 720$ kNm, and therefore, the positive design moment at mid-span is $1231 - 720 = 511$ kNm. From Table 12.4, the design negative moment in the column strip at the support is taken as:

$$M^* = 0.75 \times 720 = 540 \text{ kNm}$$

The 5 m wide column strip contains eight cables ($A_p = 3155$ mm^2) at an effective depth of 183 mm. The following results are obtained for the column strip at the column support in accordance with the ultimate strength procedures outlined in Chapter 6:

$$\sigma_{pu} = 1713 \text{ MPa}; \quad T_p = 5406 \text{ kN}; \quad d_n = 41.3 \text{ mm} = 0.226d_p$$

$$\phi M_{uo} = 0.8 \times 5406 \times \left(183 - \frac{0.77 \times 41.3}{2}\right) \times 10^{-3} = 723 \text{ kNm}$$

which is substantially greater than M^* and therefore, the slab possesses adequate strength at this location. The strength is also adequate at all other regions in the slab. With the maximum value of $d_n/d = 0.226$, ductility is also acceptable.

12.9.7 Yield line analysis of flat plates

Yield line analysis is a convenient tool for calculating the collapse load required to cause flexural failure in reinforced concrete slabs. The procedure was described in detail by Johansen [13,14] and is a plastic method for the analysis of two-way slabs, with yield lines (or plastic hinge lines) developing in the slab and reducing the slab to a mechanism.

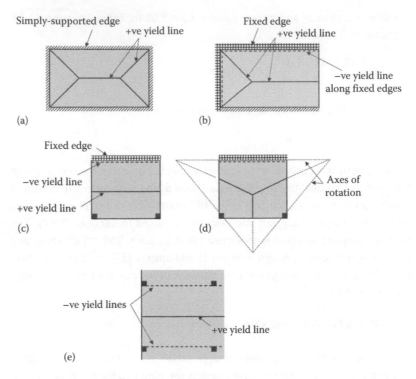

Figure 12.22 Plan views of slabs showing typical yield line patterns. (a) Panel with four simply-supported edges. (b) Panel with two fixed, one simply-supported and one free edge. (c) One fixed edge and two corner supports – Pattern 1. (d) One fixed and two corner supports – Pattern 2. (e) Edge panel of a flat plate.

Typical yield line patterns for a variety of slab types subjected to uniformly distributed loads are shown in Figure 12.22. The yield lines divide the slab into rigid segments. At collapse, each segment rotates about an axis of rotation that is either a fully supported edge or a straight line through one or more point supports, as shown. All deformation is assumed to take place on the yield lines between the rigid segments or on the axes of rotation. The yield line pattern, or the collapse mechanism, for a particular slab must be compatible with the support conditions.

The principle of virtual work is used to determine the collapse load corresponding to any possible yield line pattern. For a particular layout of yield lines, a compatible virtual displacement system is postulated. Symmetry in the slab and yield line pattern should be reflected in the virtual displacement system. The external work W done by all the external forces as the slab undergoes its virtual displacement is equal to the internal work U. The internal work associated with a particular yield line is the product of the total bending moment on the yield line and the angular rotation that takes place at the line. Since all internal deformation takes place on the yield lines, the internal work U is the sum of the work done on all yield lines.

In reinforced concrete slabs with isotropic reinforcement, the ultimate moment of resistance or plastic moment m_u (per unit length) is constant along any yield line and the internal work associated with any of the collapse mechanisms shown in Figure 12.22 is easily calculated. In prestressed concrete slabs, the depth of the orthogonal prestressing tendons may vary from point to point along a particular yield line and the calculation of U is more difficult.

For flat plate structures, however, with the yield line patterns shown in Figures 12.22e and 12.23, the prestressing tendons crossing a particular yield line do so at the same effective depth, the plastic moment per unit length of the yield line is constant and the collapse load is readily calculated.

Consider the interior span of Figure 12.23a. If it is assumed conservatively that the columns are point supports and that the negative yield lines pass through the support centre-lines and if the slab strip shown is given a unit vertical displacement at the position of the positive yield line, the external work done by the collapse loads w_u (in kN/m^2) acting on the slab strip is the total load on the strip times its average virtual displacement (which in this case is 0.5):

$$W = \frac{w_u L_t L}{2} \tag{12.29}$$

The internal work done at the negative yield line at each end of the span is the total moment $m_u' L_t$ times the angular change at the yield line θ ($=1/(L/2) = 2/L$). At the positive yield line, the angular change is 2θ ($=4/L$)

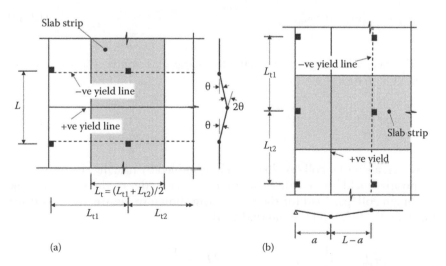

Figure 12.23 Yield line analysis of a flat plate. (a) Interior span. (b) Exterior span.

and the internal work is $m_u L_t \times 4/L$. The total internal work on all yield lines is therefore:

$$U = m_u L_t \frac{4}{L} + 2m_u' L_t \frac{2}{L} = \frac{4L_t(m_u + m_u')}{L} \tag{12.30}$$

The principle of virtual forces states that $W = U$ and therefore:

$$w_u = \frac{8}{L^2}(m_u + m_u') \tag{12.31}$$

where m_u and m_u' are the ultimate moments of resistance per unit length along the positive and negative yield lines, respectively.

When calculating m_u and m_u', it is reasonable to assume that the total quantity of prestressed and non-prestressed steel crossing the yield line is uniformly distributed across the slab strip, even though this is unlikely to be the case.

The amount of non-prestressed steel and the depth of the prestressed tendons may be different at each end of an interior span, and hence, the value of m_u' at each negative yield line may be different. When this is the case, the positive yield line will not be located at mid-span. The correct position is the one that corresponds to the smallest collapse load w_u.

Consider the exterior span in Figure 12.23b. If the positive yield line is assumed to occur at mid-span, the collapse load is given by an expression similar to Equation 12.31, except that only one negative yield line contributes to the internal work and therefore:

$$w_u = \frac{8}{L^2}(m_u + 0.5m_u') \tag{12.32}$$

For the case when m_u and m_u' have the same magnitude, the value of w_u given by Equation 12.32 is:

$$w_u = \frac{12m_u}{L^2} \tag{12.33}$$

However, a smaller collapse load can be obtained by moving the position of the positive yield line a little closer to the exterior edge of the slab strip. The minimum collapse load for the mechanism shown in Figure 12.23b occurs when $a = 0.414L$, and the internal work is:

$$U_i = m_u L_t \left(\frac{1}{0.414L} + \frac{1}{0.586L} \right) + m_u' L_t \frac{1}{0.586L} = \frac{5.83 L_t m_u}{L}$$

The external work is still given by Equation 12.29. Equating the internal and external work gives:

$$w_u = \frac{11.66 m_u}{L^2} \tag{12.34}$$

The collapse loads predicted by both Equations 12.33 and 12.34 are close enough to suggest that, for practical purposes, the positive yield line in this mechanism may be assumed to be at mid-span.

Yield line analysis is therefore an *upper bound approach* and predicts a collapse load that is equal to or greater than the theoretically correct value. It is important to check that another yield line pattern corresponding to a lower collapse load does not exist. In flat plates, a fan-shaped yield line pattern may occur locally in the slab around a column (or in the vicinity of any concentrated load), as shown in Figure 12.24.

The concentrated load P_u at which the fan mode shown in Figure 12.24c occurs is:

$$P_u = 2\pi(m_u + m'_u) \tag{12.35}$$

The loads required to cause the fan mechanisms around the columns in Figure 12.24a and b increase as the column dimensions increase. Fan mechanisms may be critical in cases where the column dimensions are both less than about 6% of the span in each direction [15].

Although yield analysis theoretically provides an upper bound to the collapse load, slabs tested to failure frequently (almost invariably) carry very much more load than that predicted. When slab deflections become large, in-plane forces develop in most slabs and the applied load is resisted by

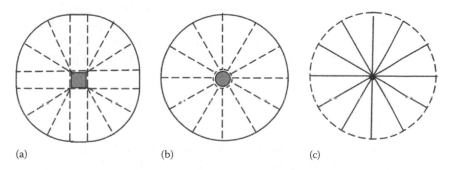

(a)   (b)   (c)

Figure 12.24 Fan mechanisms at columns or under concentrated loads. (a) Rectangular column (under slab). (b) Circular column. (c) Concentrated load (on top of slab).

membrane action in addition to bending. The collapse load predicted by yield line analysis is therefore usually rendered conservative by membrane action.

Although yield line analysis provides a useful measure of flexural strength, it does not provide any information regarding serviceability. Service-load behaviour must be examined separately.

12.10 FLAT SLABS WITH DROP PANELS

Flat slabs with drop panels behave and are analysed similarly to flat plates. The addition of drop panels improves the structural behaviour both at service loads and at overloads. Drop panels stiffen the slab, thereby reducing deflection. Drop panels also increase the flexural and shear strength of the slab by providing additional depth at the slab–column intersection. The extent of cracking in the negative moment region over the column is also reduced. The slab thickness outside the drop panel may be significantly reduced from that required for a flat plate. Drop panels, however, interrupt ceiling lines and are often undesirable from an architectural point of view.

Drop panels increase the slab stiffness in the regions over the columns and therefore affect the distribution of slab moments caused by unbalanced loads. The negative or hogging moments over the columns tend to be larger and the span moments tend to be smaller than the corresponding moments in a flat plate.

Building codes often place minimum limits on the dimensions of drop panels. For example to include the effect of drop panels when sizing a slab by limiting the span-to-depth ratios (similar to Equation 12.4 in Section 12.4.2), AS3600-2009 [1] specifies that, on each side of the column centreline, drop panels should extend a distance equal to at least one-sixth of the span in that direction (measured centre to centre of supports) and the projection of the drop below the slab should be at least 30% of the slab thickness beyond the drop.

In Figure 12.25, the moments introduced into a slab by the change in eccentricity of the horizontal prestressing force at the drop panels are illustrated. These may be readily included in the slab analysis. The fixed end moment at each support of the span shown in Figure 12.25a is given by:

$$M_{FEM} = \frac{2P_e}{(I_1/I_2) + 2} \tag{12.36}$$

and the resultant bending moment diagram is shown in Figure 12.25b. The moments of inertia of the various slab regions I_1 and I_2 are defined in Figure 12.25a. The moments in the drop panel due to this effect are positive and those in the span are negative, as shown, and, although usually relatively small, tend to reduce the moments caused by the unbalanced loads.

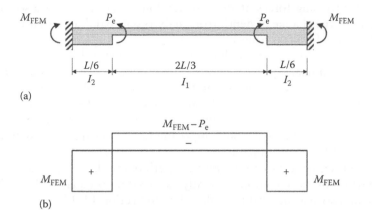

(a)

(b)

Figure 12.25 Bending moments due to eccentricity of longitudinal prestress. (a) Equivalent loads. (b) Bending moment diagram.

12.11 BAND-BEAM AND SLAB SYSTEMS

Band-beam floors are a popular form of prestressed concrete construction. A one-way prestressed or reinforced concrete slab is supported by wide shallow beams (slab-bands or band-beams) spanning in the transverse direction. The system is particularly appropriate when the spans in one direction are significantly larger than those in the other direction.

The slab-bands, which usually span in the long direction, have a depth commonly about two to three times the slab thickness and a width that may be as wide as the drop panels in a flat slab. A section through a typical band-beam floor is shown in Figure 12.26. The one-way slab is normally considered to have an effective span equal to the clear span (from band edge to band edge) plus the slab depth. If the slab is prestressed, the tendons are usually designed using a load balancing approach and have a constant eccentricity over the slab bands with a parabolic drape through the effective span as shown in Figure 12.26. The depth and width of the band beams should be carefully checked to ensure that the reaction from the slab, deposited near the edge of the band, can be safely carried back to the column line.

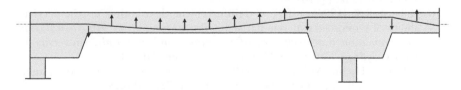

Figure 12.26 Band-beam and slab floor system

The prestressing forces at the slab tendon anchorages will also induce moments at the change of depth from slab to slab-band in the same way as was discussed for drop panels.

The slab-band is normally designed to carry the full load in the transverse direction (usually the long-span direction). The prestressing tendons in this direction are concentrated in the slab-bands, and these are also usually designed by load balancing. Because the prestress disperses out into the slab over the full panel width, the prestress anchorage should be located at the centroid of the T-section comprising the slab-band and a slab flange equal in width to the full panel.

When checking serviceability and strength of the slab-band, the effective flange width of the T-section is usually assumed to be equal to the width of the column strip as defined for a flat plate in Section 12.9.2.

REFERENCES

1. AS3600-2009. (2009). *Australian Standard for Concrete Structures*. Standards Association of Australia, Sydney, New South Wales, Australia.
2. VSL Prestressing (Aust.) Pty. Ltd. (1988). *Slab Systems*, 2nd edn. Sydney, New South Wales, Australia: V.S.L.
3. ACI 318-11M. (2011). *Building Code Requirements for Reinforced Concrete*. Detroit: American Concrete Institute.
4. Post-Tensioning Institute. (1977). *Design of Post-Tensioned Slabs*. Glenview, IL: Post-Tensioning Institute.
5. Gilbert, R.I. (1985). Deflection control of slabs using allowable span to depth ratios. *ACI Journal Proceedings*, 82, 67–72.
6. Gilbert, R.I. (1989). Determination of slab thickness in suspended post-tensioned floor systems. *ACI Journal Proceedings*, 86, 602–607.
7. Gilbert, R.I. (1979). Time-dependent behaviour of structural concrete slabs. PhD Thesis, Sydney, New South Wales, Australia: School of Civil Engineering, University of New South Wales.
8. Gilbert, R.I. (1979). Time-dependent analysis of reinforced and prestressed concrete slabs. *Proceedings of the Third International Conference in Australia on Finite Elements Methods*, University of New South Wales, Unisearch Ltd., Sydney, New South Wales, Australia, pp. 215–230.
9. Mickleborough, N.C. and Gilbert, R.I. (1986). Control of concrete floor slab vibration by L/D limits. *Proceedings of the 10th Australian Conference on the Mechanics of Structures and Materials*. University of Adelaide, Adelaide, South Australia, pp. 51–56.
10. Gilbert, R.I. (1984). Effect of reinforcement distribution on the serviceability of reinforced concrete flat slabs. *Proceedings of the 9th Australasian Conference on the Mechanics of Structures and Materials*, University of Sydney, Sydney, New South Wales, Australia, pp. 210–214.
11. Nilson, A.H. and Walters, D.B. (1975). Deflection of two-way floor systems by the equivalent frame method. *ACI Journal*, 72, 210–218.

12. Nawy, E.G. and Chakrabarti, P. (1976). Deflection of prestressed concrete flat plates. *Journal of the PCI*, 21, 86–102.
13 Johansen, K.W. (1962). *Yield-Line Theory*. London, U.K.: Cement and Concrete Association.
14. Johansen, K.W. (1972). *Yield-Line Formulae for Slabs*. London, U.K.: Cement and Concrete Association.
15. Ritz, P., Matt, P., Tellenbach, Ch., Schlub, P., and Aeberhard, H.U. (1981). *Post-Tensioned Concrete in Building Construction—Post-Tensioned Slabs*. Berne, Switzerland: Losinger.

Chapter 13

Compression and tension members

13.1 TYPES OF COMPRESSION MEMBERS

Many structural members are subjected to longitudinal compression, including columns and walls in buildings, bridge piers, foundation piles, poles, towers, shafts and web and chord members in trusses. The idea of applying prestress to a compression member may at first seem unnecessary or even unwise. In addition to axial compression, however, these members are often subjected to significant bending moments. Bending in compression members can result from a variety of load types. Moments are induced in the columns in framed structures by the gravity loads on the floor systems. Lateral loads on buildings and bridges cause bending in columns and piers and lateral earth pressures bend foundation piles. Even members that are intended to be axially loaded may be subjected to unintentional bending caused by eccentric external loading or by initial crookedness of the member itself. Most codes of practice specify a minimum eccentricity for use in design. All compression members must therefore be designed for combined bending and compression.

Prestress can be used to overcome the tension caused by bending and therefore reduce or eliminate cracking at service loads. By eliminating cracking, prestress can be used to reduce the lateral deflection of columns and piles and greatly improve the durability of these elements. Prestress also improves the handling of slender precast members and is used to overcome the tension due to rebound in driven piles. The strength of compression members is dependent on the strength of the concrete and considerable advantage can be gained by using concrete with high mechanical properties. Prestressed columns and piles are therefore commonly precast, in an environment where quality control and supervision are of a high standard.

If a structural member is subjected primarily to axial compression, with little or no bending, prestress causes a small reduction in the load-carrying capacity. For most prestressed concrete columns, the level of prestress is usually between 1.5 and 5 MPa, which is low enough not to cause significant reductions in strength. When the eccentricity of the applied load is large and bending is significant, however, prestress results in an increase in the moment capacity, in addition to improved behaviour at service loads.

13.2 CLASSIFICATION AND BEHAVIOUR OF COMPRESSION MEMBERS

Consider the pin-ended column shown in Figure 13.1. The column is subjected to an external compressive force P applied at an initial eccentricity e_o. When P is first applied, the column shortens and deflects laterally by an amount δ_i. The bending moment at each end of the column is Pe_o, but at the column mid-length the moment is $P(e_o + \delta_i)$. The moment at any section away from the column ends depends on the lateral deflection of the column, which in turn depends on the length of the column and its flexural stiffness. The initial moment Pe_o is called the primary moment and the moment caused by the lateral displacement of the column $P\delta_i$ is the secondary moment. As the applied load P increases, so too does the lateral displacement δ_i. The rate of increase of the secondary moment $P\delta_i$ is therefore faster than the rate of increase of P. This nonlinear increase in the internal actions is brought about by the change in geometry of the column and is referred to as *geometric nonlinearity*.

For a reinforced or prestressed concrete column under sustained loads, the member suffers additional lateral deflection due to creep. This time-dependent deformation leads to additional bending in the member, and this in turn causes the column to deflect further. During a period of sustained loading, an additional deflection $\Delta\delta$ develops and the resulting gradual increase in secondary moment with time $P(\delta_i + \Delta\delta)$ reduces the factor of safety.

Columns are usually classified into two categories according to their length or slenderness. Short (or stocky) columns are compression members in which the secondary moments are insignificant, that is columns that are

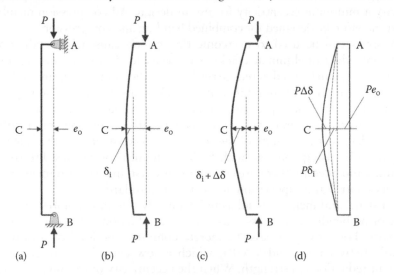

Figure 13.1 Deformation and moments in slender pin-ended column. (a) Elevation. (b) At $t = 0$. (c) At time t. (d) Moments.

geometrically linear. Long (or slender) columns are geometrically nonlinear and the secondary moment is significant, that is the lateral deflection of the column is enough to cause a significant increase in the bending moment at the critical section and, hence, a reduction in strength. For given cross-section and material properties, the magnitude of the secondary moment depends on the length of the column and its support conditions. The secondary moment in a long column may be as great as or greater than the primary moment, and the load-carrying capacity is much less than that of a short column with the same cross-section.

The strength of a stocky column is equal to the strength of its cross-section when a compressive load is applied at an eccentricity e_o. Strength depends only on the cross-sectional dimensions, the quantity and distribution of the steel reinforcement (both prestressed and non-prestressed) and the compressive strengths of both concrete and steel. Many practical concrete columns in buildings are, in fact, stocky columns. Ultimate strength analysis of a prestressed concrete column cross-section is presented in Section 13.3.

The strength of a slender column is also determined from the strength of the critical cross-section subjected to an applied compressive load at an $(e_o + \delta)$. The calculation of secondary moments $(P\delta)$ at the ultimate limit state and the treatment of slenderness effects in design are discussed in Section 13.4. Many precast, prestressed compression members, as well as some in-situ columns and piers, fall into the category of slender columns.

For very long columns, instability or buckling failure may take place before the strength of any cross-section is reached. The strength of a very slender member is not dependent on the cross-sectional strength and must be determined from a nonlinear stability analysis (which is outside the scope of this book). A very slender member may buckle under a relatively small applied load, either when the load is first applied or after a period of sustained loading. The latter type of instability is caused by excessive lateral deformation due to creep and is known as creep buckling. Upper limits on the slenderness of columns are usually specified by codes of practice in order to avoid buckling failures.

13.3 CROSS-SECTIONAL ANALYSIS: COMPRESSION AND BENDING

13.3.1 The strength interaction diagram

The ultimate strength of a prestressed concrete column cross-section in combined bending and uniaxial compression is calculated as for a conventionally reinforced concrete cross-section. Strength is conveniently represented by a plot of the axial load capacity N_u versus the moment on the section at ultimate M_u. This plot is called the *strength interaction curve*.

A typical strength interaction curve is shown in Figure 13.2 and represents the failure line or strength line. Any combination of axial force and bending moment applied to the column cross-section that falls inside the interaction curve is safe and can be carried by the cross-section. Any point outside the curve represents a combination of axial force and moment that exceeds the strength of the cross-section. Depending on the properties of the cross-section and the relative magnitudes of the axial force and bending moment, the type of failure can range from compressive, when the moment is small, to tensile or flexural, when the axial force is small and bending predominates.

Several critical points are identified on the strength interaction curve in Figure 13.2. Point 0, on the vertical axis, is the point of axial compression (zero bending), and the strength is N_{uo} (often called the *squash load*). The cross-section is subjected to a uniform compressive strain, as shown. Point 1 represents the zero tension point. The combination of axial force N_{u1} and moment $N_{u1}e_1$ at point 1 (when combined with the pre-strain caused by

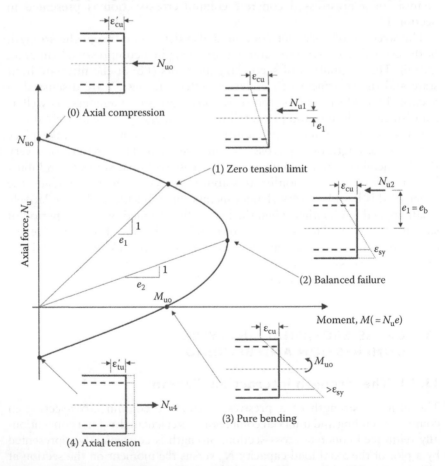

Figure 13.2 A typical strength interaction diagram.

prestress) produces zero strain in the extreme concrete fibre. The extreme fibre compressive strain at failure is ε_{cu}. Between points 0 and 1 on the curve, the entire cross-section is in compression.

When the eccentricity of the applied load is greater than e_1, bending causes tension over part of the cross-section. Point 2 is known as the balanced failure point. The strain in the extreme compressive fibre is ε_{cu} and the strain in the tensile steel is the yield strain ε_{sy} (=0.2% offset). The eccentricity of the applied load at the balanced failure point is e_2 (=e_b). When a cross-section contains both non-prestressed and prestressed tensile steel with different yield strains and located at different positions on the cross-section, the balanced failure point is not well defined. Point 2 is usually taken as the point corresponding to a strain of ε_{sy} in the steel closest to the tensile face of the cross-section and is usually at or near the point of maximum moment capacity. At any point on the interaction curve between points 0 and 2, the tensile steel has not yielded at ultimate and failure is essentially compressive. Failures that occur between points 0 and 2 (when the eccentricity is less than e_b) are sensibly known as *primary compressive failures*.

Point 3 is the pure bending point, where the axial force is zero, and point 4 is the point corresponding to direct axial tension. At any point on the interaction curve between points 2 and 4, the capacity of the tensile steel (or part of the tensile steel) is exhausted, with strains exceeding the yield strain, and the section suffers a *primary tensile failure*.

Any straight line through the origin represents a line of constant eccentricity called a loading line. Two such lines, corresponding to points 1 and 2, are drawn on Figure 13.2. The slope of each loading line is $1/e$. When a monotonically increasing compressive force N is applied to the cross-section at a particular eccentricity e_i, the plot of N versus M (=Ne_i) follows the loading line of slope $1/e_i$ until the strength of the cross-section is reached at the point where the loading line and the interaction curve intersect. If the eccentricity of the applied load is increased, the loading line becomes flatter, and the strength of the cross-section at ultimate N_u is reduced.

The general shape of the interaction curve shown in Figure 13.2 is typical for any cross-section that is under-reinforced in pure bending (i.e. where the tensile steel strain at point 3 exceeds the yield strain). A small increase in axial compression increases the internal compressive stress resultant on the section but does not appreciably reduce the internal tension, thus increasing the moment capacity, as is indicated by the part of the interaction curve between points 3 and 2. AS3600-2009 [1] states that for short columns where the axial force N^* is less than $0.1f_c'A_g$, the cross-section may be designed for bending only.

13.3.2 Ultimate strength analysis

Individual points on the strength interaction curve can be calculated using ultimate strength theory, similar to that outlined for pure bending in Section 6.3. The analysis described below is based on the assumptions listed in Section 6.3.1 and the idealised rectangular stress block specified in AS3600-2009 [1] and presented in Section 6.3.2. At any point on the interaction curve between points 1 and 3, the extreme fibre concrete compressive strain at failure is taken to be $\varepsilon_{cu} = 0.003$. For axial compression at point 0, AS3600-2009 [1] suggests that the maximum strain in the reinforcement is 0.0025. This corresponds to the yield strain of 500 Grade steel reinforcement and is close to the strain at failure of plain concrete subjected to monotonically increasing compressive load.

Calculation of the ultimate moment M_{uo} in pure bending (point 3 on the interaction curve) was discussed in Chapter 6. Other points on the strength interaction curve (between points 3 and 1) may be obtained by successively increasing the depth to the neutral axis and analysing the cross-section. With the extreme fibre strain equal to 0.003, each neutral axis position defines a particular strain distribution that corresponds to a point on the strength interaction diagram. The strain diagrams associated with points 1, 2 and 3 are also shown in Figure 13.2.

To define the interaction curve accurately, relatively few points are needed. In fact, if only points 0, 1, 2 and 3 are determined, a close approximation can be made by passing a smooth curve through each point, or even by linking successive points together by straight lines. Such an approximation is often all that is required in design.

Consider the rectangular cross-section shown in Figure 13.3a, with overall dimensions D and b. The section contains two layers of non-prestressed reinforcement of areas A_{s1} and A_{s2}, and two layers of bonded prestressing steel A_{p1} and A_{p2}, as shown. A typical ultimate strain diagram and the corresponding idealised stresses and stress resultants are illustrated in Figures 13.3b and c, respectively. These strains and stresses correspond to a resultant axial force N_u at an eccentricity e measured from the plastic centroid of the cross-section (as shown in Figure 13.3d). Longitudinal equilibrium requires that:

$$N_u = C_c + C_{s1} - T_{p1} - T_{p2} - T_{s2} \tag{13.1}$$

and moment equilibrium gives:

$$M_u = N_u e = C_c \left(d_{pc} - \frac{\gamma d_n}{2} \right) + C_{s1} \left(d_{pc} - d_{s1} \right) - T_{p1} \left(d_{pc} - d_{p1} \right)$$

$$+ T_{p2} \left(d_{p2} - d_{pc} \right) + T_{s2} \left(d_{s2} - d_{pc} \right) \tag{13.2}$$

where d_{pc} locates the position of the plastic centroid of the cross-section.

Each of the internal forces can be calculated readily from the strain diagram. The magnitude of the compressive force in the concrete C_c is

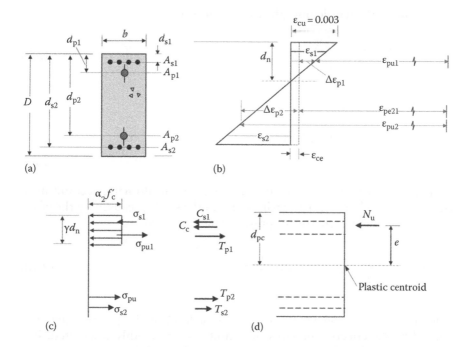

Figure 13.3 Ultimate stresses and strains on a rectangular cross-section in compression and uniaxial bending. (a) Cross section. (b) Strain. (c) Stresses and forces. (d) Stress resultant.

the volume of the rectangular stress block given by Equation 6.4 and reproduced here for ease of reference:

$$C_c = \alpha_2 f'_c \gamma d_n b \qquad (13.3)$$

The magnitude of the strain in the compressive non-prestressed steel A_{s1} is:

$$\varepsilon_{s1} = \frac{0.003(d_n - d_{s1})}{d_n} \qquad (13.4)$$

and the compressive force in A_{s1} is:

$$C_{s1} = A_{s1} E_s \varepsilon_{s1} \quad \text{if } \varepsilon_{s1} < \varepsilon_{sy} \ \left(= \frac{f_{sy}}{E_s} \right) \qquad (13.5)$$

$$= A_{s1} f_{sy} \quad \text{if } \varepsilon_{s1} \geq \varepsilon_{sy} \ \left(= \frac{f_{sy}}{E_s} \right) \qquad (13.6)$$

The strain in the tensile non-prestressed steel A_{s2} is:

$$\varepsilon_{s2} = \frac{0.003(d_{s2} - d_n)}{d_n} \tag{13.7}$$

and the force in A_{s2} is:

$$T_{s2} = A_{s2} E_s \varepsilon_{s2} \quad \text{if } \varepsilon_{s2} < \varepsilon_{sy} \tag{13.8}$$

$$= A_{s2} f_{sy} \quad \text{if } \varepsilon_{s2} \geq \varepsilon_{sy} \tag{13.9}$$

To determine the strain in the prestressing steel at ultimate, account must be taken of the large initial tensile strain in the steel ε_{pe} caused by the effective prestress. For each area of prestressing steel:

$$\varepsilon_{pe1} = \frac{P_{e.p1}}{A_{p1} E_p} \quad \text{and} \quad \varepsilon_{pe2} = \frac{P_{e.p2}}{A_{p2} E_p} \tag{13.10}$$

The strain in the concrete at the different level of the prestressing steel caused by the effective prestress (ε_{ce1} and ε_{ce2}) are readily calculated. It is here assumed that the prestressing forces in A_{p1} and A_{p2} are such that the effective prestress is axial, producing uniform compressive strain ε_{ce}, i.e. $\varepsilon_{ce} = \varepsilon_{ce1} = \varepsilon_{ce2}$. If this is the case, the resultant effective prestressing force $P_e(=P_{e.p1} + P_{e.p2})$ acts at the centroidal axis, and the magnitude of ε_{ce} is:

$$\varepsilon_{ce} = \varepsilon_{ce1} = \varepsilon_{ce2} = \frac{P_e}{(nA_{s1} + nA_{s2} + A_c)E_c} = \frac{P_e}{[(n-1)(A_{s1} + A_{s2}) + A_g]E_c} \tag{13.11}$$

where n is the modular ratio E_s/E_c; and A_c and A_g are the concrete area and the gross cross-sectional area, respectively.

The changes in strain in the bonded prestressing tendons due to the application of N_u at an eccentricity e may be obtained from the strain diagram of Figure 13.3b and are given by:

$$\Delta\varepsilon_{p1} = -\frac{0.003(d_n - d_{p1})}{d_n} + \varepsilon_{ce1} \tag{13.12}$$

and

$$\Delta\varepsilon_{p2} = \frac{0.003(d_{p2} - d_n)}{d_n} + \varepsilon_{ce2} \tag{13.13}$$

The final strain in each prestressing tendon is therefore:

$$\varepsilon_{pu1} = \varepsilon_{pe1} + \Delta\varepsilon_{p1} \tag{13.14}$$

and

$$\varepsilon_{pu2} = \varepsilon_{pe2} + \Delta\varepsilon_{p2} \tag{13.15}$$

The final stress in the prestressing tendons (σ_{pu1} and σ_{pu2}), may be obtained from a stress–strain curve for the prestressing steel, such as one of the curves shown in Figure 4.14. If the strain in the prestressing steel remains in the elastic range (and on the compressive side of the cross-section it does), then $\sigma_{pu} = \varepsilon_{pu}E_p$.

The forces in the tendons at ultimate are:

$$T_{p1} = \sigma_{pu1}A_{p1} \tag{13.16}$$

and

$$T_{p2} = \sigma_{pu2}A_{p2} \tag{13.17}$$

With the internal forces determined from Equations 13.3, 13.5, 13.6, 13.8, 13.9, 13.16 and 13.17, the ultimate compressive force N_u is obtained from Equation 13.1 and the eccentricity e is calculated using Equation 13.2. The resulting point (N_u, M [$=N_ue$]) represents the point on the strength interaction curve corresponding to the assumed strain distribution.

When the cross-section is subjected to pure compression (point 0 on the interaction curve), the eccentricity is zero and the strength (known as the *squash load*) is given by:

$$N_{u0} = C_{c0} + (A_{s1} + A_{s2})f_{sy} - \sigma_{pu1}A_{p1} - \sigma_{pu2}A_{p2} \tag{13.18}$$

where C_{c0} is the resultant compressive force carried by the concrete part of the cross-section in uniform compression. Because the total steel area on most column cross-sections is relatively small compared to the concrete area, it is usual to calculate C_{c0} based on the gross cross-sectional area. AS3600 2009 specifies that:

$$C_{c0} = \alpha_1 f'_c A_g \tag{13.19}$$

and

$$0.72 \le \alpha_1 = 1.0 - 0.003f'_c \le 0.85 \tag{13.20}$$

Example 13.1

Calculate the critical points on the strength interaction curve of the prestressed concrete column cross-section shown in Figure 13.4a. The cross-section is 600 mm wide and 800 mm deep. Steel quantities, prestressing details and material properties are as follows: $A_{s1} = A_{s2} = 2250$ mm^2, $A_{p1} = A_{p2} = 1000$ mm^2, $E_s = 200 \times 10^3$ MPa, $f_{sy} = 500$ MPa, $f_c' = 40$ MPa, $E_c = 32,000$ MPa, $n = E_s/E_c = 6.25$ and, from Equations 6.2 and 6.3, $\alpha_2 = 0.85$ and $\gamma = 0.77$. The properties of the prestressing steel are taken from the idealised stress–strain relationship shown in Figure 13.4b, where the initial elastic modulus $E_p = 195 \times 10^3$ MPa. The total effective prestress in each of the bonded prestressing tendons is $P_{e.p1} = P_{e.p2} = 1200$ kN and the effective strain in each tendon is obtained from Equation 13.10:

$$\varepsilon_{pe1} = \varepsilon_{pe2} = \frac{1,200 \times 10^3}{1,000 \times 195,000} = 0.00615$$

Because of the symmetry of the cross-section and the fact that the resultant effective prestress passes through the centroid of the section, the strain in the concrete caused by the effective prestress is uniform over the section and obtained from Equation 13.11:

$$\varepsilon_{ce} = \varepsilon_{ce1} = \varepsilon_{ce2}$$

$$= \frac{2 \times 1,200 \times 10^3}{[(6.25 - 1) \times (2,250 + 2,250) + (800 \times 600)] \times 32,000}$$

$$= 0.000149$$

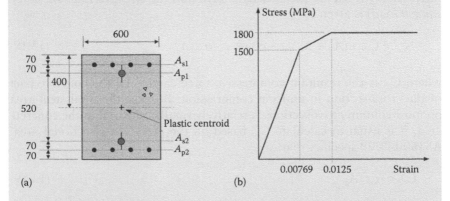

(a)

(b)

Figure 13.4 Cross-sectional details and idealized stress–strain relationship for tendons (Example 13.1). (a) Cross-section. (b) Stress versus strain for tendons.

Point 0: *Pure compression (e = 0)*
At the squash load, the strain on the cross-section ε'_{cu} is uniform with a magnitude of 0.0025 [1]. From Equation 13.20, $\alpha_1 = 0.85$ and the compressive force carried by the concrete in uniform compression is given by Equation 13.19:

$$C_{c0} = 0.85 \times 40 \times 800 \times 600 \times 10^{-3} = 16,320 \text{ kN}$$

The stress in the non-prestressed steel is $f_{sy} = 500$ MPa, and the stress in the prestressing steel is in the elastic range and given by:

$$\sigma_{pu1} \ (=\sigma_{pu2}) = E_p(\varepsilon_{pe1} - \varepsilon_{cu} + \varepsilon_{ce})$$

$$= 195 \times 10^3 \times (0.00615 - 0.0025 + 0.000149) = 741 \text{ MPa}$$

The strength of the cross-section in axial compression is given by Equation 13.18:

$$N_{u0} = 16,320 + (2,250 + 2,250) \times 500 \times 10^{-3} - (1,000 + 1,000) \times 741 \times 10^{-3}$$

$$= 17,090 \text{ kN}$$

Point 1: *Zero tension*
For the case of zero tension, $d_n = D = 800$ mm and the magnitude of ε_{cu} is 0.003 [1]. Equation 13.3 gives:

$$C_c = 0.85 \times 40 \times 0.77 \times 800 \times 600 \times 10^{-3} = 12,566 \text{ kN}$$

From Equation 13.4:

$$\varepsilon_{s1} = \frac{0.003(800 - 70)}{800} = 0.0028 > \varepsilon_{sy} \ (=0.0025)$$

The compressive non-prestressed steel has yielded and the compressive force in A_{s1} is given by Equation 13.6:

$$C_{s1} = 2250 \times 500 \times 10^{-3} = 1125 \text{ kN}$$

From Equation 13.7:

$$\varepsilon_{s2} = \frac{0.003 \times (730 - 800)}{800} = -0.000263$$

The strain in A_{s2} is compressive and from Equation 13.8:

$$T_{s2} = 2,250 \times 200,000 \times (-0.000263) \times 10^{-3}$$

$$= -118 \text{ kN (i.e. compressive)}$$

The change in strain at each level of prestressing steel is compressive and given by Equations 13.12 and 13.13:

$$\Delta\varepsilon_{p1} = -\frac{0.003 \times (800-140)}{800} + 0.000149 = -0.00233$$

$$\Delta\varepsilon_{p2} = \frac{0.003 \times (660-800)}{800} + 0.000149 = -0.000376$$

and the final strains in the prestressing tendons are obtained from Equations 13.14 and 13.15:

$$\varepsilon_{pu1} = 0.00615 - 0.00233 = 0.00382$$

$$\varepsilon_{pu2} = 0.00615 - 0.000376 = 0.00577$$

Both final strains are in the elastic range and the forces in the tendons are:

$$T_{p1} = \sigma_{pu1}A_{p1} = \varepsilon_{pu1}E_pA_{p1} = 0.00382 \times 195,000 \times 1,000 \times 10^{-3} = 745 \text{ kN}$$

$$T_{p2} = \sigma_{pu2}A_{p2} = \varepsilon_{pu2}E_pA_{p2} = 0.00577 \times 195,000 \times 1,000 \times 10^{-3} = 1,126 \text{ kN}$$

The resultant compressive force at ultimate is obtained from Equation 13.1:

$$N_u = 12,566 + 1,125 - 745 - 1,126 + 118 = 11,938 \text{ kN}$$

and the ultimate moment capacity for the case of zero tension is calculated using Equation 13.2:

$$M_u = \left[12,566 \times \left(400 - \frac{0.77 \times 800}{2} \right) \right.$$

$$+ 1,125 \times (400-70) - 745 \times (400-140)$$

$$\left. + 1,126 \times (660-400) - 118 \times (730-400) \right] \times 10^{-3}$$

$$= 1,587 \text{ kNm}$$

Therefore, the eccentricity corresponding to point 1 is:

$$e_1 = \frac{M_u}{N_u} = 132.9 \text{ mm}$$

Point 2: *The balanced failure point*

Point 2 corresponds to first yielding in the non-prestressed tensile steel, i.e. $\varepsilon_{s2} = 0.0025$, and therefore the force in the tensile non-prestressed steel is:

$$T_{s2} = 2250 \times 500 \times 10^{-3} = 1125 \text{ kN}$$

The depth to the neutral axis at point 2 is therefore:

$$d_n = \frac{\varepsilon_{cu}}{\varepsilon_{cu} + \varepsilon_{sy}} d_{s2} = \frac{0.003}{0.003 + 0.0025} \times 730 = 398 \text{ mm}$$

and the compressive force in the concrete (Equation 13.3):

$$C_c = 0.85 \times 40 \times 0.77 \times 398 \times 600 \times 10^{-3} = 6252 \text{ kN}$$

With $d_n = 398$ mm, the strain in the non-prestressed compressive reinforcement is:

$$\varepsilon_{s1} = \frac{0.003\,(398 - 70)}{398} = 0.00247 < \varepsilon_{sy}$$

and from Equation 13.5:

$$C_{s1} = 2,250 \times 200,000 \times 0.00247 \times 10^{-3} = 1,113 \text{ kN}$$

With the strain in the top prestressing steel still in the elastic range, the force T_{p1} is obtained from Equations 13.12 through 13.17:

$$T_{p1} = A_{p1}\left[\varepsilon_{pe1} - \frac{0.003(d_n - d_{p1})}{d_n} + \varepsilon_{ce}\right]$$

$$= 1,000 \times \left[0.00615 - \frac{0.003(398 - 140)}{398} + 0.000149\right] \times 195,000 \times 10^{-3}$$

$$= 849 \text{ kN}$$

Equations 13.12 through 13.15 give:

$$\varepsilon_{pu2} = \varepsilon_{pa2} + \frac{0.003(d_{p2} - d_n)}{d_n} + \varepsilon_{ce2} = 0.00615 + 0.00197 + 0.000149 = 0.00827$$

which is greater than the proportional limit in Figure 13.4b and, from that figure, we get:

$$\sigma_{pu2} = 1500 + 300 \times \frac{0.00827 - 0.00769}{0.0125 - 0.00769} = 1536 \text{ MPa}$$

and Equation 13.17 gives:

$$T_{p2} = 1536 \times 1000 \times 10^{-3} = 1536 \text{ kN}$$

The ultimate strength corresponding to the balanced point (point 2) is obtained from Equation 13.1:

$$N_{ub} = 6252 + 1113 - 849 - 1536 - 1125 = 3855 \text{ kN}$$

and the ultimate moment capacity at the balanced load point is calculated using Equation 13.2:

$$M_{ub} = \left[6252 \times \left(400 - \frac{0.77 \times 398}{2} \right) + 1113 \times (400 - 70) - 849 \times (400 - 140) \right.$$
$$\left. + 1536 \times (660 - 400) + 1125 \times (730 - 400) \right] \times 10^{-3}$$
$$= 2460 \text{ kNm}$$

The eccentricity corresponding to point 2 is:

$$e_2 = \frac{M_{ub}}{N_{ub}} = 638 \text{ mm}$$

Point 3: *Pure bending*
For equilibrium of the section in pure bending (i.e. bending without axial force), the magnitude of the resultant compression is equal to the magnitude of the resultant tension, i.e. $C = T$. A trial and error approach to determine the depth to the neutral axis indicates that:

$$d_n = 197.9 \text{ mm}$$

and the compressive force in the concrete is (Equation 13.3):

$$C_c = 0.85 \times 40 \times 0.77 \times 197.9 \times 600 \times 10^{-3} = 3109 \text{ kN}$$

From Equation 13.5:

$$C_{s1} = 2{,}250 \times 200{,}000 \times \frac{0.003(197.9 - 70)}{197.9} \times 10^{-3} = 873 \text{ kN}$$

Equations 13.12 through 13.17 give:

$$T_{p1} = 1{,}000 \times \left[0.00615 - \frac{0.003(197.9 - 140)}{197.9} + 0.000149 \right]$$
$$\times 195{,}000 \times 10^{-3} = 1{,}057 \text{ kN}$$

Equations 13.12 through 13.15 give:

$$\varepsilon_{pu2} = \varepsilon_{pe2} + \frac{0.003(d_{p2} - d_n)}{d_n} + \varepsilon_{ce2} = 0.00615 + 0.00700 + 0.000149 = 0.01330$$

which is greater 0.0125 in Figure 13.4b and, therefore σ_{pu2} = 1800 MPa and Equation 13.17 gives:

$$T_{p2} = 1800 \times 1000 \times 10^{-3} = 1800 \text{ kN}$$

The ultimate strength corresponding to point 2 is obtained from Equation 13.1:

$$N_u = 3109 + 873 - 1057 - 1800 - 1125 = 0$$

and the ultimate moment capacity is calculated using Equation 13.2:

$$M_{u0} = \begin{bmatrix} 3109 \times \left(400 - \frac{0.77 \times 197.9}{2} \right) + 873 \times (400 - 70) - 1057 \\ \times (400 - 140) + 1800 \times (660 - 400)1125 \times (730 - 400) \end{bmatrix} \times 10^{-3}$$

$$= 1859 \text{ kNm}$$

Point 4: Axial tension

The capacity of the section in tension is dependent only on the steel strength. Therefore, taking f_{sy} = 500 MPa and f_{pb} = 1800 MPa, as indicated in Figure 13.4b, the axial tensile strength is:

$$N_u = (A_{s1} + A_{s2})f_{sy} + (A_{p1} + A_{p2})f_{pb}$$

$$= [(2250 + 2250) \times 500 + (1000 + 1000) \times 1800] \times 10^{-3}$$

$$= 5850 \text{ kN}$$

Figure 13.5 shows the strength interaction curve for the cross-section. The design interaction curve (ϕN_u vs ϕM_u) is also shown in accordance with the provisions of AS3600-2009 [1], as discussed subsequently in Section 13.3.3. In addition, Figure 13.5 also illustrates the interaction curve for a cross-section with the same dimensions, material properties and steel quantities (both non-prestressed and prestressed), but without any effective prestress, i.e. $P_{e.p1} = P_{e.p2} = 0$. A comparison between the two curves indicates the effect of prestress. In this example, the prestressing steel induced an effective prestress of 5 MPa over the column cross-section. Evidently, the prestress

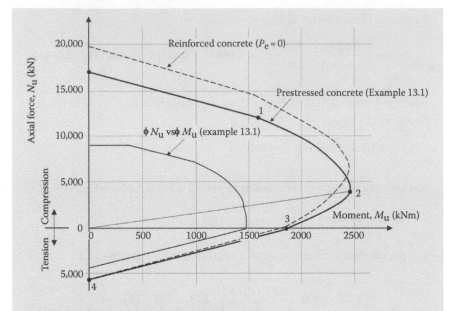

Figure 13.5 Strength interaction curve (Example 13.1).

reduces the axial load carrying capacity by about 13% (at point 0), but increases the bending strength of the cross-section in the primary tension region (i.e. between points 2 and 3).

13.3.3 Design interaction curves

For structural design in accordance with AS3600-2009 [1], the design actions N^* and M^* (obtained using the appropriate factored load combination for strength presented in Section 2.3.2) must lie on or inside the *design interaction curve*. This design curve is obtained by multiplying each point on the strength interaction curve by the strength reduction factor ϕ.

At the pure bending point, if the cross-section is under-reinforced (i.e. the tensile steel has yielded, as is the case in Example 13.1), the strength reduction factor $\phi = 0.8$ (for cross-sections containing Class N reinforcement and/or tendons). If the section is over-reinforced, the value of ϕ varies between 0.8 and 0.6 as described in row (b) of Table 2.3. For a cross-section containing any Class L reinforcement, ϕ should not exceed 0.64.

The strength reduction factor in the primary compression region where $N_u \geq N_{ub}$ (i.e. between points 0 and 2) is $\phi = 0.6$. When $N_u < N_{ub}$ (i.e. in the primary tension region between points 3 and 2), ϕ varies between 0.6 at the balanced point (point 2) to its value in pure bending, as described in Table 2.3.

A minimum eccentricity for the applied load of $0.05D$ is specified in AS3600-2009 [1], where D is the depth of the column in the plane of the bending moment. The AS3600-2009 [1] design interaction curve for the cross-section analysed in Example 13.1 is drawn in Figure 13.5.

13.3.4 Biaxial bending and compression

When a cross-section is subjected to axial compression and bending about both principal axes, such as the section shown in Figure 13.6a, the strength interaction diagram can be represented by the three-dimensional surface shown in Figure 13.6b. The shape of this surface may be defined by a set of contours obtained by taking horizontal slices through the surface. A typical contour is shown in Figure 13.6b. Each contour is associated with a particular axial force N. The equation of the contour represents the relationship between M_x and M_y at that particular value of axial force. In AS3600-2009 [1], the design expression given in Equation 13.21 is specified to model the shape of these contours. The form of Equation 13.21 was originally proposed by Bresler [2].

If the factored design actions N^*, M_x^* and M_y^* fall inside the *design interaction surface* (i.e. the strength interaction surface scaled down by the strength reduction factor ϕ), then the cross-section is adequate. According to AS3600-2009 [1], a cross-section subjected to biaxial bending should satisfy the following equation:

$$\left(\frac{M_x^*}{\phi M_{ux}} \right)^{\alpha_n} + \left(\frac{M_y^*}{\phi M_{uy}} \right)^{\alpha_n} \leq 1.0 \tag{13.21}$$

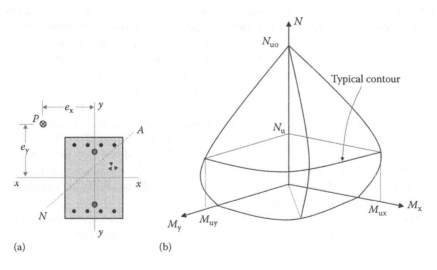

Figure 13.6 Biaxial bending and compression. (a) Cross-section. (b) Strength interaction surface.

where ϕM_{ux} and ϕM_{uy} are the design strength in bending calculated separately about the major and minor axis, respectively, under the design axial force N^*. The factored design moments M_x^* and M_y^* are magnified to account for slenderness, if applicable (see Section 13.4). The index α_n is a factor that depends on the axial force and defines the shape of the contour, and is given by:

$$\alpha_n = 0.7 + \frac{1.7N^*}{0.6N_{uo}} \quad \text{but within the limits } 1 \le \alpha_n \le 2 \qquad (13.22)$$

Biaxial bending is not a rare phenomenon. Most columns are subjected to simultaneous bending about both principal axes. AS3600-2009 [1] suggests that biaxial bending need not be considered where the eccentricity about one axes is less than the 0.1 times the column dimension in the direction of that eccentricity, or when the ratio of the eccentricities e_x/e_y falls outside the range 0.2–5.0. In each of the above situations, the code concedes that the cross-section can be designed for the axial force with each bending moment considered separately, that is in uniaxial bending and compression.

13.4 SLENDERNESS EFFECTS

13.4.1 Background

The strength of a short column is equivalent to the strength of the most heavily loaded cross-section and, for a given eccentricity, may be determined from the strength interaction curve (or surface). The strength of a long column (or slender column) depends not only on the strength of the cross-section, but also on the length of the member and its support conditions. A discussion of the behaviour of a slender pin-ended column was presented in Section 13.2 and the increase in secondary moments due to slenderness effects was illustrated in Figure 13.1. In general, as the length of a compression member increases, strength decreases.

To predict accurately the second-order effects in structures as they deform under load requires an iterative nonlinear computer analysis, and this generally involves considerable computational effort. For the design of concrete compression members, simplified procedures are available to account for slenderness effects and one such procedure is presented here. A more detailed study of geometric nonlinearity and instability in structures is outside the scope of this book.

The critical buckling load N_c of an axially loaded, perfectly straight, pin-ended, elastic column was determined by Euler and is given by:

$$N_c = \frac{\pi^2 EI}{L^2} \qquad (13.23)$$

where L is the length of the Euler column between the pinned ends. In practice, concrete columns are rarely, if ever, pinned at their ends. A degree of rotational restraint is usually provided at each end of a column by the supporting beams and slabs, or by a footing. In some columns, translation of one end of the column with respect to the other may also occur in addition to rotation. Some columns are completely unsupported at one end, such as a cantilevered column. The buckling load of these columns may differ considerably from that given by Equation 13.23.

In general, for design purposes, the critical buckling load of real columns is expressed in terms of the *effective length* $L_e = kL_u$. AS3600-2009 [1] defines L_u as the unsupported length of a column, taken as the distance between the faces of members providing lateral support to the column (and equals L for an idealised Euler column). Where column capitals or haunches are present, L_u is measured to the lowest extremity of the capital or haunch. The term k is an effective length factor that depends on the support conditions of the column. The critical load of a concrete column is therefore:

$$N_c = \frac{\pi^2 EI}{(kL_u)^2} \tag{13.24}$$

In structures that are laterally braced, the ends of the columns are not able to translate appreciably relative to each other, i.e. sidesway is prevented. Most concrete structures are braced, with stiff vertical elements such as shear walls, elevator shafts and stairwalls providing bracing for the more flexible columns. If the attached elements at each end of a braced column provide some form of rotational restraint, the critical buckling load will be greater than that of a pin-ended column (given in Equation 13.23), and therefore the effective length factor in Equation 13.24 is less than 1.0. Effective length factors specified in AS3600-2009 [1] for braced columns are shown in Figure 13.7a. The effective length of any column is the length associated with single curvature buckling, i.e. the distance between the points of inflection in the column, as shown in Figures 13.7a and 13.8a. For the column shown in Figure 13.7a(ii), the supports are neither pinned nor fixed. The effective length depends on the relative flexural stiffness of the column and the beams and other supporting elements at each end of the column, and may be calculated readily using end restraint coefficients and effective length factor charts contained in AS3600-2009 [1].

For columns in unbraced structures, where one end of the column can translate relative to the other (i.e. sidesway is not prevented), the effective length factor k is greater than 1.0, sometimes much greater, as shown in Figure 13.8b. The critical buckling load of an unbraced column is therefore significantly less than that of a braced column. Values of k specified in AS3600-2009 [1] for unbraced columns with various support conditions are shown in Figure 13.7b.

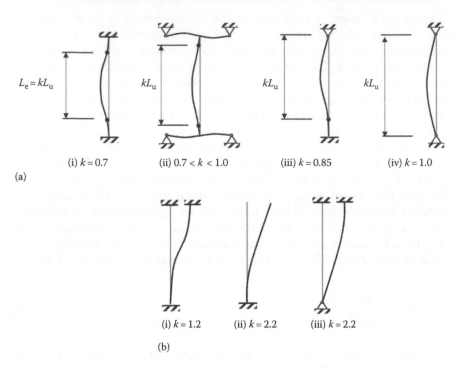

(a)

(i) $k = 0.7$ (ii) $0.7 < k < 1.0$ (iii) $k = 0.85$ (iv) $k = 1.0$

(b)

(i) $k = 1.2$ (ii) $k = 2.2$ (iii) $k = 2.2$

Figure 13.7 Effective length factors k [1]. (a) Braced columns. (b) Unbraced columns.

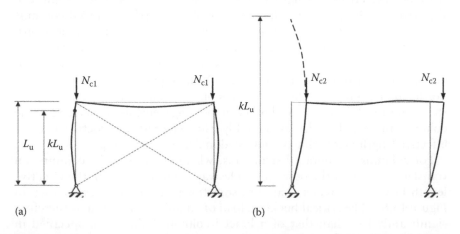

(a)   (b)

Figure 13.8 Effective lengths in a braced and an unbraced portal frame [1]. (a) Braced frame. (b) Unbraced frame.

A braced column is a compression member located within a storey of a building in which horizontal displacements do not significantly affect the moments in the column. AS3600-2009 [1] defines a braced column as a column in a structure where the lateral actions applied at the ends of the column are resisted by masonry infill panels, shear walls or lateral bracing. A compression member may be assumed to be braced, if it is located in a storey where the bracing elements (shear walls, masonry infills, bracing trusses and other types of bracing) have a total stiffness, resisting lateral movement of the storey, at least six times the sum of the stiffnesses of all the columns within the storey.

In the case of slender prestressed concrete columns, the question arises as to whether the longitudinal prestressing force P reduces the critical buckling load. In general, a concrete column prestressed with internally bonded strands or post-tensioned with tendons inside ducts within the member is no more prone to buckling than a reinforced concrete column of the same size and stiffness, and with the same support conditions. As a slender, prestressed concrete column displaces laterally, the tendons do not change position within the cross-section and the eccentricity of the line of action of the prestressing force does not change. The prestressing force cannot, therefore, generate secondary moments. However, if a member is externally prestressed, so that the line of action of the prestressing force remains constant, then prestress can induce secondary moments and hence affect the buckling load. Such a situation could exist, for example when a member is prestressed by jacking through an abutment.

13.4.2 Moment magnification method

In lieu of a detailed second-order analysis to determine the effects of short-term and time-dependent deformation on the magnitude of moment and forces in slender structures, AS3600-2009 [1] specifies a *moment magnifier method* to account for slenderness effects in columns. The idea behind the moment magnification method is based on the concept of using a factor to magnify the column moments to account for the change in geometry of the structure and the resulting secondary actions. The axial load and the magnified moment are then used in the design of the column cross-section. The effect of the secondary moments on the strength of a slender column is shown on the strength interaction curve in Figure 13.9. Line OA is the loading line corresponding to an initial eccentricity e on a particular cross-section. If the column is short, secondary moments are insignificant, the loading line is straight and the strength of the column corresponds to the axial force at A ($N_{u,short}$). If the column is slender, the secondary moments increase at a faster rate than the applied axial force and the loading line becomes curved, as shown. The strength of the slender column is the axial force corresponding to point B ($N_{u,slender}$), where the curved loading line

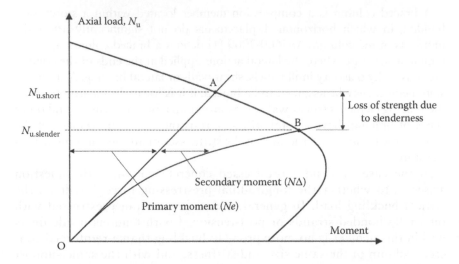

Figure 13.9 Strength interaction curve for a cross-section in a slender column.

meets the strength interaction curve. The loss of strength due to secondary moments is indicated in Figure 13.9.

The total moment at failure is the sum of the primary moment Ne and the secondary moment $N\Delta$ and may be expressed by a factor δ times the primary moment. That is:

$$\delta Ne = Ne + N\Delta = Ne\left(\frac{e + \Delta}{e}\right) \quad \text{and therefore} \quad \delta = \frac{e + \Delta}{e}$$

The factor δ may be used to magnify the primary moment to account for slenderness effects. This magnification factor depends on the ratio of the design axial force on the column N^* to the critical buckling load N_c, the ratio of the design moments at each end of the column M_1^*/M_2^* and the deflected shape of the column, which in turn depends on whether the column is braced or unbraced, and the rotational restraint at each end of the column.

To determine whether a particular column is slender, and therefore whether moment magnification is required, AS3600-2009 [1] specifies critical values of the slenderness ratio to mark the transition between short and slender columns. The slenderness ratio of a column is defined as the effective length of the column divided by its radius of gyration (taken about the axis of bending) $L_e/r = kL_u/r$. For a rectangular cross-section, the radius of gyration may be approximated as 0.3 times the overall column dimension in the direction in which stability is being considered or 0.25 times the diameter of a circular cross-section.

For a braced column, the column may be deemed to be short and the slenderness effect may be ignored when L_e/r satisfies the following condition:

$$L_e/r \le 25 \quad \text{or} \quad \alpha_c(38 - f_c'/15)(1 + M_1^*/M_2^*), \quad \text{whichever is the greater}$$

(13.25)

where:

$$\alpha_c = \sqrt{2.25 - 2.5 N^*/(0.6 N_{uo})} \quad \text{for} \quad \frac{N^*}{0.6 N_{uo}} \ge 0.15 \tag{13.26}$$

$$\alpha_c = \sqrt{1/\left[3.5 N^*/(0.6 N_{uo})\right]} \quad \text{for} \quad \frac{N^*}{0.6 N_{uo}} < 0.15 \tag{13.27}$$

and M_1^*/M_2^* is the ratio of the smaller to the larger of the two design bending moments at the ends of the column and may be taken as negative, when the column is bent in single curvature, and positive, when the column is bent in double curvature. AS3600-2009 [1] also states that when the absolute value of M_2^* does not exceed $0.05DN^*$, the ratio M_1^*/M_2^* may be taken as equal to –1.

For an unbraced column, the column may be deemed to be short and the slenderness effect ignored when:

$$\frac{L_e}{r} \le 22 \tag{13.28}$$

If the slenderness ratio exceeds 120, AS3600-2009 [1] states that the approximate moment magnifier method should not be used and that a second-order stability analysis should be undertaken. Studies have shown that for columns with slenderness ratios of up to 120, the moment magnifier method provides a conservative estimate of strength [3].

According to AS3600-2009 [1], the largest design moment (M_2^*) of a slender column is multiplied by a moment magnifier δ. For a braced column, the moment magnifier δ_b is given by:

$$\delta_b = \frac{k_m}{(1 - N^*/N_c)} \ge 1.0 \tag{13.29}$$

where k_m depends on the ratio of the end moments and whether the column is bent in single or double curvature. For members braced against sidesway and with no transverse loads applied to the column between the supports, k_m is given by:

$$k_m = (0.6 - 0.4 M_1^*/M_2^*) \ge 0.4 \tag{13.30}$$

while for all other cases $k_m = 1.0$.

The buckling load N_c is given by Equation 13.24 with the flexural rigidity EI at the ultimate limit state taken to be the ratio of moment to curvature at the point on the interaction curve corresponding to a depth to the neutral axis of $0.545d_o$ (i.e. the balanced failure point for a column containing non-prestressed tensile reinforcement with $f_{sy} = 500$ MPa). EI is estimated as follows:

$$EI = \frac{182d_o \phi M_{ub}}{1 + \beta_d} \qquad (13.31)$$

where $\phi = 0.6$; M_{ub} is the ultimate moment of the cross-section when the neutral axis parameter is $k_{uo} = 0.545$; the term β_d is a factor to account for the effects of creep and is given by $G/(G + Q)$, but taken as zero when $L_e/r \leq 40$ and $N^* \leq M^*/2D$; and G and Q are the factored design axial load components due to dead loads and live loads, respectively.

The moment magnifier for an unbraced column is taken as the larger value of δ_b (given by Equation 13.29) and δ_s calculated for each column in the storey using:

$$\delta_s = \frac{1}{(1 - \Sigma N^*/\Sigma N_c)} \qquad (13.32)$$

where the summations include all columns within the storey of a building and N_c is calculated using Equations 13.24 and 13.31 for each column.

For slender members in biaxial bending, the moment about each axis should be magnified using either Equation 13.29 or 13.32, as appropriate, with the restraint conditions applicable to each plane of bending.

EXAMPLE 13.2

Consider a 10 m long pin-ended column in a braced structure. The column cross-section is shown in Figure 13.4a, and the material properties and steel quantities are as outlined in Example 13.1. The strength interaction curve for the cross-section was calculated in Example 13.1 and illustrated in Figure 13.5. The column is laterally supported at close centres to prevent displacement perpendicular to the weak axis of the section but is unsupported between its ends in the direction perpendicular to the strong axis. The column is loaded by a compressive force N at a constant eccentricity e to produce compression and uniaxial bending about the strong axis. Establish two loading lines for the column corresponding to initial eccentricities of e = 100 mm and e = 400 mm and determine the strength of the slender column in each case, assuming that the ratio of factored dead load to total load is $\beta_d = G/(G + Q) = 0.7$.

Since the column is braced and pinned at each end, k = 1 (see Figure 13.7), and therefore $L_e = kL_u = 10$ m, with respect to the strong axis. The effective length about the weak axis is small due to the specified closely spaced lateral supports. For bending about the strong axis, the slenderness ratio is:

$$\frac{L_e}{r} = \frac{10,000}{0.3 \times 800} = 41.7$$

which is greater than the transition limit of 25 (specified in Equation 13.25) and, therefore, the column is slender. The member is subjected to single curvature bending with equal end moments, and therefore $M_1^*/M_2^* = -1$.

From Example 13.1, the bending strength determined at the balanced failure point (determined when $k_u = 0.545$ at point 2 on the interaction curve) is $M_{ub} = 2460$ kNm and from Equation 13.31:

$$EI = \frac{182 \times 730 \times 0.6 \times 2460 \times 10^6}{1+0.7} = 115.4 \times 10^{12} \text{ Nmm}^2$$

and from Equation 13.24:

$$N_c = \frac{\pi^2 \times 115.4 \times 10^{12}}{(10,000)^2} \times 10^{-3} = 11,390 \text{ kN}$$

The loading line for each initial eccentricity is obtained by calculating the magnification factor δ_b from Equation 13.29 for a series of values of axial force N^* and plotting the points ($N^*, \delta_b N^* e$) on a graph of axial force and moment.

Sample calculations are provided for the points on the loading lines corresponding to an axial force $N^* = 5000$ kN.

From Equation 13.30, for this column with $M_1^*/M_2^* = -1$, $k_m = 1.0$ and from Equation 13.29:

$$\delta_b = \frac{1.0}{(1 - 5,000/11,390)} = 1.783$$

When e = 100 mm, the magnified moment is:

$$\delta_b N^* e = 1.783 \times 5000 \times 100 \times 10^{-3} = 891 \text{ kNm}$$

and when e = 400 mm:

$\delta_b N^* e = 1.783 \times 5000 \times 400 \times 10^{-3} = 3566$ kNm

Other points on the loading lines are as follows:

N* (kN)	δ_b	e = 100 mm $\delta_b N^* e$ (kNm)	e = 400 mm $\delta_b N^* e$ (kNm)
2000	1.213	243	970
3000	1.358	407	1630
4000	1.541	616	2466
5000	1.783	891	3566
6000	2.113	1268	5071
7000	2.595	1817	—
8000	3.360	2690	—

The loading lines are plotted on Figure 13.10, together with the strength interaction curve reproduced from Figure 13.5 and the design interaction curve in accordance with AS3600-2009 [1]. The strength of each column is the axial load corresponding to the intersection of the loading line and the strength interaction curve. The maximum factored design load N* that can be applied to the slender column, and meet the strength design requirements of AS3600-2009 [1], is obtained from the point where the loading line crosses the design interaction curve. Note the significant reduction of strength in both columns due to slenderness.

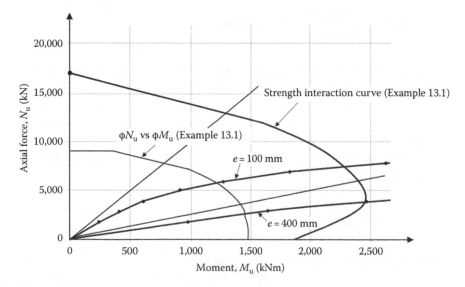

Figure 13.10 Loading lines and strength of the slender columns (Example 13.2).

Slenderness causes a far greater relative reduction in strength when the initial eccentricity is small in the primary compression region than when eccentricity is large and bending predominates. Note also that for very slender columns, the curved loading line crosses the strength interaction curve in the primary tension region, and this is the same region in which prestress provides additional strength (see Figure 13.5). There is some advantage in prestressing slender columns.

13.5 REINFORCEMENT REQUIREMENTS FOR COMPRESSION MEMBERS

AS3600-2009 [1] specifies design requirements related to the quantity and distribution of reinforcement in columns. The area of longitudinal reinforcement in a column A_s should not be less than $0.01A_g$ and should not be greater than $0.04A_g$. The upper limit may be exceeded if the amount and disposition of the reinforcement will not prevent the proper placement and compaction of the concrete at splice locations and at the junctions with other members. If a column has a larger area than that required for strength, A_s may be less than $0.01A_g$ provided $A_s f_{sy} > 0.15N^*$.

In addition to longitudinal reinforcement, transverse reinforcement in the form of closed ties or spirals (helical reinforcement) is required in columns. The behaviour of a column loaded to failure depends on the nature of the transverse reinforcement. If the column contains no transverse reinforcement, when the strength of the cross-section is reached, failure will be brittle and sudden. Transverse reinforcement imparts a measure of ductility to reinforced and prestressed concrete columns by providing restraint to the highly stressed longitudinal steel and by confining the inner core of compressive concrete. Ductility is a critical design requirement for columns in buildings located in earthquake prone regions, where the ability to absorb large amounts of energy without failure is needed. Spirally reinforced columns, in particular, exhibit considerable ductility at failure. In addition to imparting ductility, the transverse reinforcement also carries any diagonal tensile forces associate with shear and torsion, if these actions are carried in addition to axial compression and bending.

The detailing requirements for both the longitudinal and transverse reinforcement in columns are outline in Section 14.7.

13.6 TRANSMISSION OF AXIAL FORCE THROUGH A FLOOR SYSTEM

With the increasing use of high strength concrete in the columns of buildings, problems may arise when the high column load is required to pass through a lower strength concrete floor slab. Ospina and Alexander [4] showed that the strength and behaviour of column-slab joints are significantly affected by the load on the slab. The negative bending in a slab at a column location produces tension in the top of the slab. With compressive confinement only provided to the slab-column joint at the bottom of the slab, where compressive struts enter the joint.

According to AS3600-2009 [1], satisfactory transmission of the column forces through the joint will occur and the full column strength may be assumed in design provided the strength of the concrete in the slab (f'_{cs}) is greater than or equal to 75% of the strength of the concrete in the column (f'_{cc}) and provided the longitudinal reinforcement in the column is continuous through the joint. This allows for the concrete in columns to be one standard strength grade higher than that of the slab. For greater differences in column and slab strengths, calculations are required to determine whether additional longitudinal reinforcement is required through the joint, assuming an effective compressive strength of the concrete in the joint (f'_{ce}).

The strength of a column in the joint region is a function of the strengths of the column and slab concrete, the geometry of the connection and its location in the structure (i.e. a corner, edge or interior column).

For interior columns, restrained on four sides by beams of approximately equal depth or by a slab:

$$f'_{ce} = \left(1.33 - \frac{0.33}{h/D_c}\right)f'_{cs} + \frac{0.25}{h/D_c}f'_{cc} \tag{13.33}$$

except that f'_{ce} must be within the limits $\min(f'_{cc}, 1.33f'_{cs}) \le f'_{ce} \le \min(f'_{cc}, 2.5f'_{cs})$. In Equation 13.33, h is the overall depth of the joint and D_c is the smaller column cross-sectional dimension.

For edge columns restrained on two opposing sides by beams of approximately equal depth or by a slab:

$$f'_{ce} = \left(1.1 - \frac{0.3}{h/D_c}\right)f'_{cs} + \frac{0.2}{h/D_c}f'_{cc} \tag{13.34}$$

except that f'_{ce} must be within the limits $\min(f'_{cc}, 1.33f'_{cs}) \le f'_{ce} \le \min(f'_{cc}, 2.0f'_{cs})$.

For corner columns restrained on two adjacent sides by beams of approximately equal depth or by a slab:

$$f'_{ce} = 1.33f'_{cs} \le f'_{cc} \tag{13.35}$$

When f'_{ce} $(<f'_{cc})$ is determined, an increased area of longitudinal reinforcement may be placed through the joint to effectively replace the reduction in capacity from the influence of the weaker concrete. In this case, additional tie reinforcement should also be placed through the connection to confine

the weaker concrete and provide ductility to the joint that is under a high stress relative to its cylinder strength.

Another procedure sometimes used in construction is that of *puddling*, in which column strength concrete is placed within the slab-column connection and the surrounding region. The *puddled* column concrete should occupy the full slab thickness and extend beyond the face of the column by a distance greater than 600 mm and twice the depth of the slab or beam at the column face, whichever is greater. Special care has to be taken to avoid placement of the weaker slab concrete within the joint region. Proper vibration of the column and slab concretes is needed for optimal melding at the faces of the two materials. Since both column and slab concrete are to be cast simultaneously, with two different concrete strength grades on site during the one pour, a high level of on-site supervision and quality control is necessary.

13.7 TENSION MEMBERS

13.7.1 Advantages and applications

Prestressed concrete tension members are simple elements used in a wide variety of situations. They are frequently used as tie-backs in cantilevered construction, anchors for walls and footings, tie and chord members in trusses, hangers and stays in suspension bridges, walls of tanks and containment vessels and many other applications.

The use of reinforced concrete members in direct tension has obvious drawbacks. Cracking causes a large and sudden loss of stiffness, and crack control is difficult. Cracks occur over the entire cross-section and corrosion protection of the steel must be carefully considered, in addition to the aesthetic difficulties. By prestressing the concrete, however, a tension member is given strength and rigidity otherwise unobtainable from either the concrete or the steel acting alone. Provided that cracking does not occur in the concrete, the prestressing steel is protected from the environment and the tension member is suitable for its many uses. Compared with compression members, tension members usually have a high initial level of prestress.

The deformation of a prestressed concrete tension member can be carefully controlled. In situations where excessive elongation of a tension member may cause strength or serviceability problems, prestressed concrete is a design solution worth considering.

13.7.2 Behaviour

The analysis of a prestressed concrete direct tension member is straightforward. Both the prestressing force and the external tensile loads are generally concentric with the longitudinal axis of the member, and hence bending stresses are minimised.

Prior to cracking of the concrete, the prestressing steel and the concrete act in a composite manner and behaviour may be determined by considering a transformed cross-section. If required, a transformed section obtained using the effective modulus for concrete (Equation 4.16) may be used to include the time-dependent effects of creep and shrinkage.

Consider a tension member concentrically prestressed with an effective prestressing force P_e. The cross-section is symmetrically reinforced with an area of bonded prestressing steel A_p and non-prestressed steel A_s. The transformed area of the tie is therefore:

$$A = A_c + n_p A_p + n_s A_s = A_g + (n_p - 1)A_p + (n_s - 1)A_s \qquad (13.36)$$

where n_p and n_s are the modular ratios given by E_p/E_c and E_s/E_c, respectively. The uniform stress in the concrete σ due to the prestressing force and the applied external load N is:

$$\sigma = -\frac{P_e}{A_c} + \frac{N}{A} \qquad (13.37)$$

and the stress in the prestressing steel is:

$$\sigma_p = \frac{P_e}{A_p} + \frac{n_p N}{A} \qquad (13.38)$$

For most applications, it is necessary to ensure that cracking does not occur at service loads. To provide a suitable margin against cracking under the day to day loads, and to ensure that cracks resulting from an unexpected overload close completely when the overload is removed, it is common in design to insist that the concrete stress remains compressive under normal in-service conditions. By setting $\sigma = 0$ in Equation 13.37 and rearranging, an upper limit to the external tensile force is established, and is given by:

$$N \leq P_e \frac{A}{A_c} = P_e(1 + n_p p_p + n_s p_s) \qquad (13.39)$$

where $p_p = A_p/A_c$ and $p_s = A_s/A_c$.

When a tensile member is stressed beyond the service load range, cracking occurs when the concrete stress reaches the direct tensile strength, taken as $f'_{ct} = 0.36\sqrt{f'_c}$ in AS3600-2009 [1] (see Equation 4.4). If the tensile force at cracking is N_{cr}, then from Equation 13.37:

$$N_{cr} = \left(\frac{P_e}{A_c} + 0.36\sqrt{f'_c}\right) A \tag{13.40}$$

At a cracked section, when $N = N_{cr}$, the cross-section consists of only A_s and A_p, and the stress in non-prestressed steel is:

$$\sigma_s = \frac{N_{cr}}{A_s + (A_p E_p/E_s)} \tag{13.41}$$

and the stress in the prestressing steel is:

$$\sigma_p = \frac{N_{cr} - \sigma_s A_s}{A_p} = \frac{N_{cr}}{A_p}\left(\frac{1}{1 + (A_s E_s/A_p E_p)}\right) \tag{13.42}$$

Substituting Equation 13.40 into Equation 13.42 gives:

$$\sigma_p = \left(\frac{P_e}{A_c} + 0.36\sqrt{f'_c}\right)\frac{A}{A_p}\left(\frac{1}{1 + (A_s E_s/A_p E_p)}\right) \tag{13.43}$$

This is usually limited to a maximum of about $f_{pb}/1.2$ in order to obtain a minimum acceptable margin of safety between first cracking and ultimate strength. A conservative estimate of the minimum area of prestressing steel in a tension member can be obtained from Equation 13.43 by ignoring the area of non-prestressed reinforcement as follows:

$$A_p \geq \left(\frac{P_e}{A_c} + 0.36\sqrt{f'_c}\right)\frac{1.2A}{f_{pb}} \tag{13.44}$$

The ultimate strength of the member is equal to the tensile strength of the steel and is given by:

$$N_u = A_p f_{pb} + A_s f'_{sy} \tag{13.45}$$

and in design, the factored design tensile force must satisfy the design equation:

$$N^* \leq \phi N_u \tag{13.46}$$

where the strength reduction factor for direct tension is the same as for bending (ϕ = 0.8 in AS3600-2009 [1] when Class N non-prestressed reinforcement is used in addition to tendons).

The axial deformation of a prestressed tension member at service loads depends on the load history (i.e. the times at which the prestressing force(s) and the external loads are applied), and the deformation characteristics of the concrete. Stage stressing can be used to carefully control longitudinal deformation. The shortening of a tension member at any time t caused by an initial prestress P_i applied to the concrete at a particular time τ_0 and by shrinkage of the concrete (ε_{cs} assumed to begin at τ_0) may be approximated by:

$$\Delta_{Pi} = \frac{P_i L}{A_c E_{e.0}(1 + n_{ep.0} p_p + n_{es.0} p_s)} + \frac{\varepsilon_{cs} L}{(1 + n_{ep.0} p_p + n_{es.0} p_s)} \tag{13.47}$$

where L is the length of the member and $E_{e.0}$ is the effective modulus of the concrete obtained from Equation 4.16 using the creep coefficient associated with the age of the concrete at first loading (τ_0), $n_{ep.0} = E_p/E_{e.0}$ and $n_{es.0} = E_s/E_{e.0}$.

The elongation at any time caused by an external tensile force N applied at time τ_1 may be estimated by:

$$\Delta_N = \frac{NL}{A_c E_{e.1}(1 + n_{ep.1} p_p + n_{es.1} p_s)} \tag{13.48}$$

where $E_{e.1}$ is the effective modulus of the concrete using the creep coefficient associated with the age of the concrete when the force N is first applied (τ_1), $n_{ep.1} = E_p/E_{e.1}$ and $n_{es.1} = E_s/E_{e.1}$.

A detailed analysis of the time-dependent deformation of a tension member subjected to any load history can be made using the procedures outlined by Gilbert and Ranzi [5].

A satisfactory preliminary design of a tension member usually results if the prestressing force is initially selected so that, after losses, the effective prestress is between 10% and 20% higher than the maximum in-service tension. If the compressive stress in the concrete at transfer is limited to about $0.4f'_{cp}$, the minimum gross area of the cross-section A_g can be determined. The area of steel required to impart the necessary prestress is next calculated. The resulting member can then be checked for strength and serviceability, and details modified, if necessary.

EXAMPLE 13.3

Consider the vertical post-tensioned tension member acting as a tie-back for the cantilevered roof of a grandstand, shown in Figure 13.11. In the critical loading case, the tension member must transfer a design working dead load of 800 kN and live load of 200 kN to the footing, which is anchored to rock.

The material properties are: $f_c' = 40$ MPa; $E_c = 32{,}000$ MPa; $f_{cp}' = 30$ MPa; $E_{cp} = 27{,}500$ MPa; $f_{pb} = 1{,}870$ MPa; $E_p = 195{,}000$ MPa; and $n_p = 6.09$.

In accordance with the preceding discussion, an effective prestress that is 10% higher than the maximum applied tension is assumed:

$$P_e = 1.1 \times (800 + 200) = 1100 \text{ kN}$$

Owing to the small residual compression existing under sustained loads, the time-dependent loss of prestress is usually relatively small. In this short member, draw-in losses at transfer are likely to be significant. For the purposes of this example, the time-dependent losses are assumed to be 12% and the short-term losses are taken to be 15%.

The force immediately after transfer and the required jacking force are therefore:

$$P_i = \frac{1100}{0.88} = 1250 \text{ kN} \quad \text{and} \quad P_j = \frac{1250}{0.85} = 1470 \text{ kN}$$

If the maximum steel stress at jacking is $0.85f_{pb}$, then the area of prestressing steel is:

$$A_p \geq \frac{1470 \times 10^3}{0.85 \times 1870} = 925 \text{ mm}^2$$

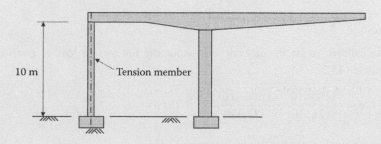

Figure 13.11 Tie-back member to be designed (Example 13.3).

Try ten 12.7 mm diameter strands (A_p = 986 mm²) post-tensioned within a 60 mm diameter duct located at the centroid of the cross-section.

The ultimate strength of the member is calculated using Equation 13.45 and using the strength reduction factor and load factors specified in AS3600-2009 [1] (see Chapter 2):

$$\phi N_u = \phi A_p f_{pb} = 0.8 \times 986 \times 1870 \times 10^{-3} = 1475 \text{ kN}$$

The design axial force is:

$$N^* = 1.2 \times 800 + 1.5 \times 200 = 1260 \text{ kN}$$

which is less than the design strength and is therefore satisfactory. If additional strength had been necessary, non-prestressed steel could be included to increase N_u to the required level.

If the concrete stress at transfer is limited to $0.4 f'_{cp}$ = 12 MPa, the area of concrete on the cross-section A_c must satisfy:

$$A_c \geq \frac{P_i}{0.4 f'_{cp}} = \frac{1,250 \times 10^3}{12} = 104,200 \text{ mm}^2$$

Try a 350 mm by 350 mm square cross-section with a centrally located 60 mm duct. Therefore, before the duct is grouted:

$$A_c = 350 \times 350 - (0.25 \times \pi \times 60^2) = 119,700 \text{ mm}^2$$

Under the effective prestress, after all losses and after the duct is fully grouted, the area of the transformed section is obtained using Equation 13.36:

$$A = 350 \times 350 + (6.09 - 1) \times 986 = 127,520 \text{ mm}^2$$

and

$$A_c = A_g - A_p = 121,510 \text{ mm}^2$$

The uniform stress in the concrete under the full service load is given by Equation 13.37:

$$\sigma = -\frac{1,100 \times 10^3}{121,510} + \frac{1,000 \times 10^3}{127,520} = -1.21 \text{ MPa}$$

and the steel stress is given by Equation 13.38:

$$\sigma_p = \frac{1,100 \times 10^3}{986} + \frac{6.09 \times 1,000 \times 10^3}{127,520} = 1,163 \text{ MPa}$$

Both stresses are satisfactory and cracking will not occur at service loads, even if the losses of prestress have been slightly underestimated.

The minimum area of steel required to ensure a factor of safety of 1.2 at cracking is checked using Equation 13.44:

$$A_p \geq \left(\frac{1,100 \times 10^3}{121,510} + 0.36\sqrt{40} \right) \frac{1.2 \times 127,520}{1,870} = 927 \text{ mm}^2$$

and the area of steel A_p = 986 mm² adopted here is just sufficient.

If the final creep coefficients associated with the age at transfer τ_o and the age when the external load is first applied τ_1 are $\varphi^*_{cc}(\tau_o)$ = 2.5 and $\varphi^*_{cc}(\tau_1)$ = 2.0, then Equation 4.16 gives the appropriate effective moduli:

$$E_{e.o} = \frac{27,500}{1+2.5} = 7,860 \text{ MPa} \quad \text{and} \quad E_{e.1} = \frac{32,000}{1+2.0} = 10,670 \text{ MPa}$$

If the final shrinkage strain is ε^*_{cs} = 650 × 10⁻⁶, the shortening of the member caused by prestress and shrinkage is obtained using Equation 13.47:

$$\Delta_{PI} = \frac{1,250 \times 10^3 \times 10,000}{121,510 \times 7,860 \times (1 + 24.81 \times 0.00811)} + \frac{650 \times 10^{-6} \times 10,000}{(1 + 24.81 \times 0.00811)}$$

$$= 16.31 \text{ mm}$$

and the elongation caused by N is given by Equation 13.48:

$$\Delta_N = \frac{1,000 \times 10^3 \times 10,000}{121,510 \times 10,670 \times (1 + 18.28 \times 0.00811)} = 6.72 \text{ mm}$$

The net effect is a shortening of the member by:

$$\Delta = 16.31 - 6.72 = 9.59 \text{ mm}$$

REFERENCES

1. AS3600-2009. (2009). *Australian Standard for Concrete Structures*. Standards Association of Australia, Sydney, New South Wales, Australia.
2. Bresler, B. (1960). Design criteria for reinforced concrete columns under axial load and biaxial bending. *ACI Journal*, 57, 481–490.
3. Gilbert, R.I. (1989). A procedure for the analysis of slender concrete columns under sustained eccentric loading. *Civil Engineering Transactions, Institution of Engineers, Australia*, CE31, 39–46.

4. Ospina, C.E. and Alexander, S.D.B. (1997). Transmission of high strength concrete column loads through concrete slabs. Structural Engineering Report No. 214, Department of Civil Engineering, University of Alberta, Canada.
5. Gilbert, R.I. and Ranzi, G. (2011). *Time-Dependent Behaviour of Concrete Structures*. London, U.K.: Spon Press.

Chapter 14

Detailing

Members and connections

14.1 INTRODUCTION

Detailing of a concrete structure is more than simply the preparation of working drawings that show the structural dimensions and the size and location of the reinforcing bars and tendons. It involves the communication of the engineer's ideas and specifications from the design office to the construction site and encompasses each aspect of the design process from preliminary analysis to final design. It involves the translation of a good structural design from the computer or calculation pad into the final structure. The most sophisticated or up-to-date methods of analysis and design are of little value if they remain in the calculations and do not find their way into the structure.

Detailing of the structural elements and the connections between them, perhaps more than any other single factor, decides the success or failure of a concrete structure. Good detailing ensures that the reinforcement and the concrete interact efficiently to provide satisfactory performance throughout the complete range of loading. Successful detailing requires experience, as well as a sound understanding of structural and material behaviour and an appreciation of appropriate construction practices and methodologies.

Too often detailing is the last thing considered by the engineer, or worse, not seriously considered at all. Yet it is critical if full strength and adequate ductility are to be achieved and if in-service performance is to be satisfactory. It should be remembered that reinforcement details that ensure satisfactory behaviour under service load conditions may not provide good collapse characteristics and, conversely, details that provide adequate strength and ductility do not necessarily ensure serviceability. Detailing must be considered when designing for both the ultimate load and the service load conditions.

In this chapter, guidelines for successful detailing of the structural elements and connections in prestressed concrete structures are outlined and, where necessary, the requirements of AS3600-2009 [1] introduced in previous chapters are revisited. The reasons for providing reinforcement and the sources of tension in concrete structures (some of which may not be immediately obvious) are also discussed. The importance of adequate anchorage for reinforcement is stressed and appropriate details are recommended. Many of the principles for successful detailing were presented by Park and Paulay [2] who, in turn, gained their inspiration from the pioneering work of Leonhardt and his colleagues [3–7]. Additional guidance for Australian practice is available in Reference [8].

14.2 PRINCIPLES OF DETAILING

14.2.1 When is steel reinforcement required?

The detailing requirements of a reinforcement bar or tendon depend on the reasons for its inclusion in the structure. It is only after the purpose of the steel bar or tendon is identified that the detailing requirements can be specified. Reinforcement is provided in concrete structures for a variety of reasons, including:

1. to carry the internal tensile forces that are determined from analysis, i.e. to provide tensile strength in regions of the structure where tensile strength is required and to impart ductility where ductility is required;
2. to control cracking (and in the case of prestressing tendons, to control or eliminate cracking), that is to ensure that the cracks caused by bending, axial tension, bursting and spalling stresses and shrinkage or temperature changes are serviceable;
3. to carry compressive forces in regions where the concrete alone is inadequate (e.g. in columns and in the compressive zone of heavily reinforced beams);
4. to provide restraint to bars in compression, that is to prevent lateral buckling of compressive reinforcing bars prior to reaching their full strength;
5. to provide confinement to the compressive concrete in the core of columns, in beams and within connections and other disturbed regions, thereby increasing both the strength and deformability of the concrete;
6. to provide protection against spalling, for example the *fire mesh* sometimes used in the protective concrete cover over fabricated steel sections;
7. to limit long-term deformation by providing restraint to creep and shrinkage of the concrete; and
8. to provide temporary support for other reinforcement during construction prior to and during the concreting operation.

A single reinforcing bar or tendon may be required for one or more of the aforementioned reasons. The ability of the bar or tendon to accomplish its task or tasks depends on how it is detailed and that is very much associated with the quality of its anchorage. When the location of a bar is determined, the question of how best to detail the bar can only be answered after its anchorage requirements are determined. Anchorage and stress development in conventional reinforcement are discussed in more detail in Section 14.3, and the anchorages of tendons were discussed in Chapter 8.

14.2.2 Objectives of detailing

When detailing reinforcement and tendons, the objectives are the same as the broad objectives in structural design and aim at:

1. achieving the required design strength at each cross-section in each member and at all connections;
2. preventing problems under the day-to-day in-service conditions that may impair the serviceability of the structure;
3. allowing ease of construction; and
4. maintaining economy.

To achieve these objectives, it is worthwhile to remember a number of general principles. When detailing for strength, the location and direction of all internal tensile forces should be determined by establishing the load path for the structure. Adequately anchored reinforcement or tendons should be used wherever a tensile force is required for equilibrium. The concrete should be assumed to carry no tension. The possible collapse mechanisms of the structure should be identified and adequate reinforcement should be specified to prevent premature collapse.

When detailing for serviceability, it is again essential that the location and direction of all internal tensile forces are determined. Often, this is not easy. The location of shrinkage and temperature-induced tension may not be easily recognised or intuitively obvious. By providing restraint, the bonded reinforcement itself can create tensile forces which may crack the surrounding concrete. When the sources of tension are identified and the magnitudes and locations of the tensile forces are determined, prestressing tendons can be used to reduce or eliminate the tension in the concrete or, alternatively, cracking can be permitted and adequate quantities of appropriately anchored reinforcement, with suitable bond characteristics, must be provided to maintain stiffness after cracking and to provide crack control.

Complex reinforcement details should be drawn to a suitably large scale to ensure that they are practical, i.e. to ensure that the reinforcement can be fixed and that the concrete can be placed and compacted adequately.

Some of the sources of tension in concrete structures, some obvious and some not so obvious, are discussed in the following section.

14.2.3 Sources of tension

The sources of tension in concrete structures are numerous and the following list is by no means exhaustive. However, the list does illustrate the wide variety of reasons why concrete structures crack and why successful detailing requires a sound understanding of structural behaviour. A significant amount of on-site experience also helps.

14.2.3.1 Tension caused by bending (and axial tension)

The primary roles of the longitudinal tensile reinforcement in a flexural member are to provide strength and ductility and to ensure crack control under service conditions.

14.2.3.2 Tension caused by load reversals

Temporary propping or bad handling during construction frequently causes tension in regions where tension may not normally be expected. In Figure 14.1a and b, two common instances of temporary propping are illustrated and these should be anticipated in design. Impact and rebound loading also causes tension. For example, considerable prestress is required in driven

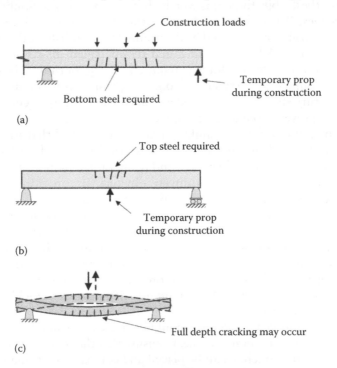

Figure 14.1 Examples where reversals of loads and internal actions may occur. (a) Cantilever. (b) Simple span. (c) Impact and rebound loading.

concrete piles to overcome the tension in the concrete caused by *rebound*. In slender members subjected to dynamic loads, such as precast stair treads, impact and rebound may cause tension on both sides of the member and result in full-depth cracking, as illustrated in Figure 14.1c.

14.2.3.3 Tension caused by shear and torsion

We have seen that the diagonal tension caused by the shear stresses causes inclined cracking in the webs of reinforced and prestressed concrete beams. Adequately anchored transverse reinforcement is required to carry this tension after inclined cracking has occurred. Some commonly used details for the anchorage of stirrups are examples of poor detailing and these are discussed in more detail in Section 14.6.2.

14.2.3.4 Tension near the supports of beams

The longitudinal tensile reinforcement required near the supports of beams is greater than indicated by the bending moment diagram. Consider the beam support shown in Figure 14.2a and the analogous truss of Figure 14.2b showing the flow of internal forces that transmit the applied loads into the support, i.e. the internal load path. The tensile force required to be carried by the longitudinal steel at the bottom of an inclined crack is equal to the compressive force at the top of the crack. This is clearly shown by the truss analogy which is a useful idealisation in any study of detailing and will be used in subsequent sections.

Shrinkage and thermal shortening also causes tension in beams with restrained ends. Considerable reinforcement may therefore be required throughout the beam length specifically to control the resulting direct tension cracking.

For the anchorage of positive moment reinforcement at a simple support of a beam, AS3600-2009 [1] requires that sufficient positive moment

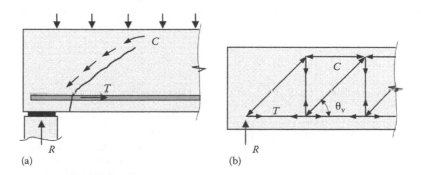

Figure 14.2 Tension at the discontinuous support of a beam. (a) Simple support. (b) Internal flow of forces – truss analogy.

reinforcement be anchored for a length (L_{st}) past the mid-point of the bearing such that the anchored reinforcement will be able to develop the tensile force resulting from the transfer of the design shear force (i.e. $V^*\cot \theta_v/\phi$) plus any tension that develops due to any other source (such as torsion and restraint to shrinkage or temperature change). V^* is the design shear force at a distance $d\cot \theta_v$ from the anchor point, θ_v is the truss angle (as shown in Figure 14.2b), ϕ is the strength reduction factor for shear (=0.7) and d is the effective depth of the longitudinal tensile reinforcement being anchored.

In addition, not less 50% of the tensile reinforcement required at mid-span is extended past the face of the support for a length of $12d_b$ or an equivalent anchorage, or not less than one-third of the tensile reinforcement required at mid-span extends past the face of the support for a length of $8d_b$ plus $D/2$, where d_b is the bar diameter and D is the overall depth of the concrete member.

For the anchorage of positive moment reinforcement at a continuous or flexurally restrained support of a beam, AS3600-2009 [1] requires that not less than one-quarter of the total positive moment reinforcement required at mid-span shall continue past the near face of the support.

14.2.3.5 Tension within the supports of beams or slabs

Shortening of beams and slabs occurs due to prestressing and due to shrinkage and drops in temperature, and this can cause tension and subsequent cracking in the supports if the longitudinal movement is restrained. For example, if adequate sliding joints are not introduced between a concrete slab and supporting masonry walls, shrinkage of the slab may cause considerable distress to the brickwork. Some illustrative examples are shown in Figure 14.3. This type of problem also frequently occurs in the supports of post-tensioned beams and slabs during the stressing operation if provision is not made for the elastic shortening of the beam or slab to be accommodated at the support. These sorts of problems are best avoided by the introduction of suitable movement joints.

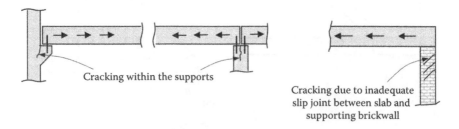

Figure 14.3 Cracking caused by tension within supports.

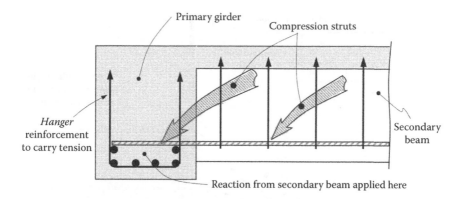

Figure 14.4 Primary girder supporting secondary beam.

14.2.3.6 Tension within connections

Tension occurs within connections where there is a sudden change in direction of the internal forces. The design of connections is covered in more detail in Section 14.8. One connection, in which significant tension is often overlooked by designers, is where a secondary beam is supported by a primary girder, as shown in Figure 14.4. Most of the reaction from the secondary beam flows into the bottom region of the primary girder via diagonal compression. It is essential that this force finds an effective support. Additional stirrups are required within the connection in the primary girder to transfer this *hanging* load from the bottom to the top of the girder (see also Section 14.6.3).

14.2.3.7 Tension at concentrated loads

The transverse tension caused by the dispersion of a concentrated load, such as exists behind the bearing plate in a post-tensioning anchorage (see Figure 14.5a), was discussed in Chapter 8. A similar situation exists at the anchorage of a reinforcing bar, such as the bend shown in Figure 14.5b or the hook shown in Figure 14.5c. A concentrated force is applied to the concrete where the bar changes direction as shown. As the concentrated force is dispersed, a transverse tension exists in the concrete that may cause splitting of the concrete in the plane of the hook or bend, particularly if the radius of curvature of the bend is small.

14.2.3.8 Tension caused by directional changes of internal forces

Wherever a reinforcement bar changes direction or a loaded concrete member is not straight, internal forces are generated in the surrounding concrete. Where these forces are compressive and concentrated over a small area, splitting may occur due to transverse tension, as discussed in the previous section.

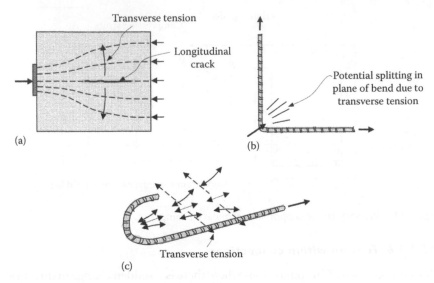

Figure 14.5 Tension due to concentrated loads. (a) Post-tensioned anchorage zone. (b) Bent bar. (c) Hooked anchorage.

Where the force is tensile, cracking may occur and additional reinforcement may be required. These internal tensile forces are often neglected when the structural member is being detailed and are a common cause of unsightly cracking, structural weakness and even premature failure.

Consider the haunched region of a beam shown in Figure 14.6a. In order to maintain equilibrium at the bend in the reinforcement, a tensile reactive

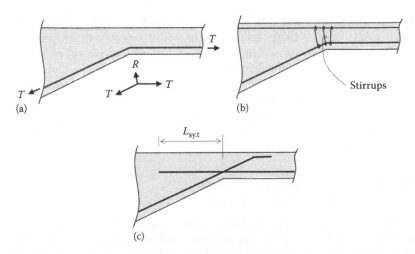

Figure 14.6 Haunched region of a beam. (a) Without transverse reinforcement. (b) With transverse reinforcement. (c) Lapped reinforcement arrangement.

force R is exerted on the concrete (as the bar tries to straighten). If the force R overcomes the tensile strength of the concrete, cracking along the bar will occur. The steel will straighten, resulting in the loss of concrete cover, structural integrity and strength.

Transverse reinforcement in the form of stirrups may be provided to carry the force R, thereby overcoming the problem, as shown in Figure 14.6b. However, a better solution may be to eliminate the problem altogether by anchoring each reinforcement bar with straight extensions so that no transverse force is generated (see Figure 14.6c).

The same principle applies when the internal compressive force changes direction, as illustrated in Figure 14.7. Adequately anchored transverse reinforcement must be provided to carry the resultant R after cracking.

In a curved member, such as that shown in Figure 14.8, the continuously changing direction of the internal compressive and tensile forces (caused by bending) creates a distributed transverse tensile force in the web. Stirrups at regular centres are required to carry these tensile forces. The transverse tension per unit length q_t produced by the tensile force in the longitudinal reinforcement is:

$$q_t = \frac{T}{r_m} = \frac{A_{st}f_{sy}}{r_m} \tag{14.1}$$

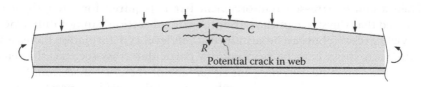

Figure 14.7 Transverse tension due to direction change of internal compression.

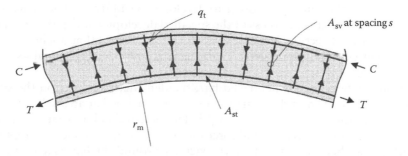

Figure 14.8 Transverse tension in a curved member.

and the required stirrup spacing is:

$$s = \frac{A_{sv} f_{sy.f}}{q_t} = \frac{A_{sv}}{A_{st}} \frac{f_{sy.f}}{f_{sy}} r_m \qquad (14.2)$$

where $f_{sy.f}$ and f_{sy} are the yield stresses of the stirrups and the longitudinal reinforcement; A_{sv} is the total area of stirrup legs within the length s; and A_{st} and r_m are the cross-sectional area and the radius of curvature of the curved longitudinal bars or tendons, respectively.

14.2.3.9 Other common sources of tension

Other sources of tension include: the internal restraint to shrinkage of the concrete by the bonded reinforcement or the external restraint to shrinkage by the supports of a structural member; restraint of deformation caused by temperature changes, including heat of hydration; restraint to load-independent movement caused by formwork or more permanent cladding; and settlement of supports.

14.3 ANCHORAGE OF DEFORMED BARS IN TENSION

14.3.1 Introductory remarks

When a non-prestressed reinforcement bar is required for strength, it is assumed that the stress in the bar at the critical section can not only reach the yield stress f_{sy}, but can be sustained at this level as deformation increases. If the yield stress is to be reached at a particular cross-section, the reinforcing bar must be *anchored* on either side of the critical section. Stress development can be obtained by embedment of the steel in concrete so that stress is transferred past the critical section by bond, or by some form of mechanical anchorage. Codes of practice specify a minimum length, called the development length $L_{sy.t}$ over which a straight bar in tension must be embedded in the concrete in order to develop its yield stress. The provision of anchorage lengths in excess of the specified development length for every bar at a critical section or peak stress location ensures that anchorage or bond failures do not occur before the design strength at the critical section is achieved.

At an anchorage of a deformed bar, the deformations bear on the surrounding concrete and the bearing forces F are inclined at an angle β to the bar axis as shown in Figure 14.9a [9]. The perpendicular components of the bearing forces exert a radial force on the surrounding concrete. Tepfers [10,11] described the concrete in the vicinity around the bar as acting like

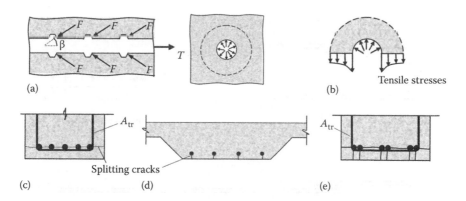

Figure 14.9 Splitting failures around developing bars. (a) Forces exerted on concrete by a deformed bar in tension. (b) Tensile stresses in concrete. (c) Horizontal splitting due to insufficient bar spacing. (d) Vertical splitting due to insufficient cover. (e) Splitting (bond) failure at a lapped splice.

a thick walled pipe as shown in Figure 14.9b and the radial forces exerted by the bar cause tensile stresses that may lead to splitting cracks radiating from the bar if the tensile strength of the concrete is exceeded. Bond failure may be initiated by these splitting cracks within the anchorage length of an anchored bar (Figure 14.9c and d) or within a lapped tension splice (Figure 14.9e).

Transverse reinforcement across the splitting planes (A_{tr} in Figure 14.9c and e) delays the propagation of splitting cracks and improves bond strength. Compressive pressure transverse to the plane of splitting delays the onset of cracking in the anchorage region, thereby improving bond strength.

For a reinforcing bar of diameter d_b, the design bond strength (i.e. the ultimate bond force over the development length) is $\phi \pi d_b L_{sy.t} f_b$ and this force must not be less than the design ultimate force in the bar $A_{st} f_{sy} = f_{sy} \pi d_b^2/4$. That is $\phi \pi d_b L_{sy.t} f_b \geq f_{sy} \pi d_b^2/4$ and therefore:

$$L_{sy.t} \geq \frac{d_b}{4} \frac{f_{sy}}{\phi f_b} \tag{14.3}$$

The average ultimate bond stress f_b in Equation 14.3 is proportional to the splitting strength of concrete and, in Australian practice, the appropriate strength reduction factor applied to f_b is $\phi = 0.6$.

Reinforcing bars in tension may be spliced together by welding or by a mechanical anchorage or by overlapping the bars by a specified length, $L_{sy.t.lap}$, as shown in Figure 14.10. In this latter anchorage, known as a lapped splice, each bar must be able to develop the yield stress within

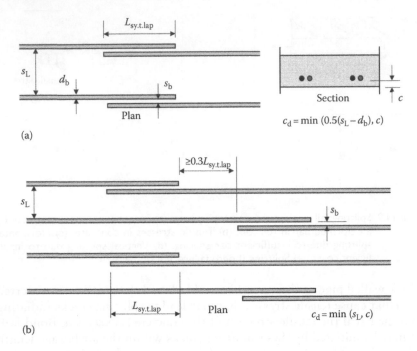

Figure 14.10 Contact and non-contact lapped splices. (a) 100% of bars spliced at the same location. (b) 50% of *bars* spliced at the same location (staggered splices).

the lap length $L_{sy.t.lap}$, and the design force in the bar on either side of the splice $(A_s f_{sy})$ must be safely carried across the splice without bond failure. Both contact splices $(s_b = 0)$ and non-contact lapped splices $(s_b > 0)$ are frequently used.

The mechanism of bond transfer at a lapped splice is quite different from that at a developing bar with no adjacent bar developing stress in close proximity, so in general where the bars at a lapped splice are required to develop the yield stress, the specified lap length is greater than the development length.

14.3.2 Development length for deformed bars in tension

14.3.2.1 Basic development length

According to AS3600-2009 [1], the development length $(L_{sy.t})$ required to develop the characteristic yield strength (f_{sy}) of a deformed bar in tension shall be taken as either the basic development length $(L_{sy.tb})$ calculated using Equation 14.4 or, where the beneficial effects of confinement by transverse reinforcement or transverse pressure are available, a refined development length calculated using Equation 14.6. The basic development length in tension of a deformed bar of diameter d_b (in mm) is given by:

$$L_{sy.tb} = \frac{0.5k_1k_3f_{sy}d_b}{k_2\sqrt{f_c'}} \geq 29k_1d_b \qquad (14.4)$$

where $k_1 = 1.3$ for a horizontal bar with more than 300 mm of concrete cast below the bar or $k_1 = 1.0$ for all other bars; $k_2 = (132 - d_b)/100$; $k_3 = 1.0 - 0.15(c_d - d_b)/d_b$ (within the range $0.7 \leq k_3 \leq 1.0$); and c_d is the thickness of the concrete annulus surrounding the bar (as shown in Figure 14.11) and is the smaller of the clear cover to the nearest concrete surface or half the clear spacing to the next parallel bar being anchored.

Because of the limited experimental data available for development lengths of deformed bars in high strength concrete, an upper limit of 65 MPa has been placed on the value of f_c' to be used in Equation 14.4. The minimum value of $L_{sy.tb}$ in Equation 14.4 depends on the yield stress of the steel and is taken to be $29k_1d_b$ when $f_{sy} = 500$ MPa.

The expression for the basic development length of a deformed bar in tension (Equation 14.4) is similar in form to Equation 14.2, with the average design ultimate bond stress ϕf_b given by:

$$\phi f_b = \frac{k_2\sqrt{f_c'}}{2k_1k_3} \qquad (14.5)$$

The average design ultimate bond stress ϕf_b is directly related to the tensile strength of concrete, which is taken to be proportional to $\sqrt{f_c'}$ in AS3600-2009 [1], and modified by coefficients of varying form and complexity to

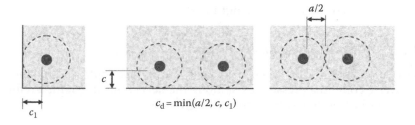

Figure 14.11 Concrete confinement dimension c_d.

account for the various factors that affect the bond strength, including bar location in the cross-section, bar diameter, bar spacing and concrete cover to the bar being developed. As the average ultimate bond stress is a property of the concrete and a brittle bond failure should be avoided, the appropriate magnitude of the strength reduction factor ϕ for inclusion in Equation 14.5 is 0.6. This value of ϕ is considered to be sufficient to accommodate the additional tensile force that may develop in the bar due to strain hardening and also to allow for considerable plastic deformation in the steel bar at the anchorage without bond failure and bar pull-out.

The factor k_1 in Equation 14.4 (and Equation 14.5) accounts for the position of the bar in the structure and increases the development length for bars with more than 300 mm of concrete cast below the bar (such as the top bars in a beam or thick slab). Such bars may be subjected to a reduction in bond strength due to cracking along the bar resulting from settlement of fresh concrete below the bar and an accumulation of bleed water. The factor applies only to *horizontal* bars in slabs, walls, beams and footings, and it does not apply to sloping or vertical bars, to fabric or to fitments. The factor k_2 accounts for the reduction in the average ultimate bond stress as the diameter of the reinforcing bar increases and varies linearly from $k_2 = 1.2$ when $d_b = 12$ mm to $k_2 = 0.92$ when $d_b = 40$ mm. The factor k_3 accounts for the confining effect of the concrete surrounding the bar and depends on the dimension c_d (illustrated in Figure 14.11): when c_d is less than or equal to the bar diameter $k_3 = 1.0$ and when c_d is greater than or equal to twice the bar diameter $k_3 = 0.7$. When c_d is between d_b and $2d_b$, the factor k_3 varies linearly between 1.0 and 0.7.

Due to reduced average ultimate bond stress, the development length for an epoxy-coated bar should be taken as 50% greater than for an uncoated bar. The tensile strength of lightweight concrete is significantly less than for normal weight concrete, and so the basic development length given by Equation 14.4 should be multiplied by 1.3 when lightweight concrete is used. Likewise the value should be increased by 30% when the structural element containing the deformed bar is built with slip forms (due to the possible adverse effects of slip form construction on concrete compaction and the steel–concrete bond strength).

14.3.2.2 Refined development length

In situations where the beneficial effects of transverse reinforcement and/or transverse confining pressure exist along the development length, a refined development length may be used and is given by:

$$L_{sy.t} = k_4 k_5 L_{sy.tb} \tag{14.6}$$

where:

$k_4 = 1.0 - K\lambda$ (but $0.7 \leq k_4 \leq 1.0$);

$k_5 = 1.0 - 0.04\rho_p$ (but $0.7 \leq k_5 \leq 1.0$);

$\lambda = (\Sigma A_{tr} - \Sigma A_{tr.min})/A_s$;

ΣA_{tr} is the cross-sectional area of the transverse reinforcement along the development length $L_{sy.t}$;

$\Sigma A_{tr.min}$ is the cross-sectional area of the minimum transverse reinforcement, which may be taken as $0.25A_s$ for beams and 0 for slabs;

A_s is the cross-sectional area of a single bar of diameter d_b being anchored;

K is a factor that accounts for the position of the bars being anchored with respect to the transverse reinforcement, with values given in Figure 14.12; and

ρ_p is the transverse compressive pressure (in MPa) at the ultimate limit state along the development length perpendicular to the plane of splitting.

The factor k_4 accounts for the presence of transverse reinforcement and is equal to 1.0 when there is no transverse reinforcement and may reduce to a minimum value of 0.7 depending on the amount and location of the transverse reinforcement.

The factor k_5 accounts for the increase in the average ultimate bond stress when transverse pressure (ρ_p in MPa) exists along the development length perpendicular to the plane of splitting. As ρ_p increases from 0 to 7.5 MPa, the factor k_5 decreases linearly from 1.0 to 0.7. When ρ_p exceeds 7.5 MPa, $k_5 = 0.7$. The average ultimate bond stress reduces when transverse tensile stress exists at the anchorage (i.e. when ρ_p is negative). Although this is not accounted for in AS3600-2009 [1], it is recommended here that, if transverse tensile stress exists, it should be considered in the determination of the development length by the inclusion of k_5 (=1.0 − 0.04ρ_p) greater than 1.0 in Equation 14.6.

In addition, the product $k_3k_4k_5$ should not be taken less than 0.7 [1].

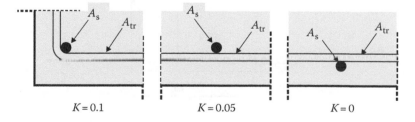

Figure 14.12 Values of K for anchored bars of area A_s [1].

14.3.2.3 Length required to develop a stress lower than the yield stress

When more reinforcement is provided than is necessary for strength at a particular location and the stress to be developed (σ_{st}) in a deformed bar is less than the yield stress (f_{sy}), the development length L_{st} may be reduced proportionally (i.e. $L_{st} = L_{sy.t}\sigma_{st}/f_{sy}$), with an absolute minimum value of $12d_b$. This reduction in development length is not to be applied to the calculation of lap splice lengths. Only full-strength lap splices are permitted by AS3600-2009 [1]. The minimum value of $12d_b$ can be reduced at the support of a slab, if at least 50% of the reinforcement is extended past the face of the support by $8d_b$.

14.3.2.4 Development length of a deformed bar in tension with a standard hook or cog

A bar ending in a standard hook or cog (shown in Figure 14.13) is fully anchored and able to develop the full yield stress at the point A in the figure (i.e. at a distance along the bar of $0.5L_{sy.t}$ measured from the outside of the hook or cog as shown). Standard hooks are defined as consisting of either a 180° bend or a 135° bend and are shown in Figure 14.13a and b. The minimum internal diameter of the bend d_{id} specified in AS3600-2009 [1] is

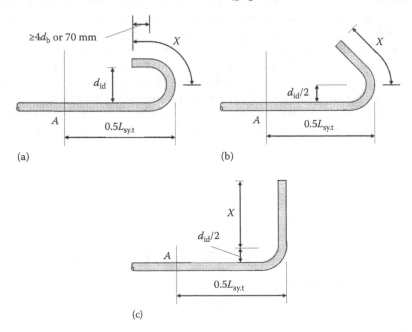

Figure 14.13 Development length of a deformed bar with a standard hook or cog. (a) Standard hook (180° bend). (b) Standard hook (135° bend). (c) Standard cog (90° bend).

Table 14.1 Minimum internal diameter of standard bends or hooks [1]

For fitments (1)	Reinforcement other than specified in Columns 3 and 4 (2)	Reinforcement where bend is to be straightened or rebent (3)	Reinforcement that is epoxy-coated or galvanised (4)
Class L bars – $3d_b$	All grades and sizes – $5d_b$	$d_b \le 16$ mm – $4d_b$	$d_b \le 16$ mm – $5d_b$
Class N (plain) – $3d_b$		$d_b = 20$ or 24 mm – $5d_b$	$d_b \ge 20$ mm – $8d_b$
Class N (deformed) – $4d_b$		$d_b \ge 28$ mm – $6d_b$	

given in Table 14.1 and depends on the grade of reinforcement, the purpose of the bar (main longitudinal reinforcement or fitment), whether the bar is to be bent and then straightened or rebent, the bar diameter, and whether the bar is epoxy-coated or galvanised. For the standard 180° hook, a minimum length of straight extension of the bar after the hook of $4d_b$ or 70 mm is required, whichever is the greater. The standard hook with a 135° bend has the same minimum length X as for the 180° hook with the same internal diameter (as shown in Figure 14.13). A standard cog consisting of a 90° bend with a minimum diameter of bend d_{id} is shown in Figure 14.13c and has the same minimum length X as for a 180° hook with the same internal diameter.

14.3.2.5 Development length of plain bars in tension

The average ultimate bond stress that can develop at the anchorage of a plain bar in tension is significantly smaller than that of a deformed bar and, as a consequence, AS3600-2009 [1] specifies that the development length in tension for a plain bar is 50% longer than for a deformed bar in the same location.

EXAMPLE 14.1: DEVELOPMENT LENGTH FOR A DEFORMED BAR IN TENSION

Calculate the development length required for the two terminated 28 mm diameter bottom bars in the beam shown in Figure 14.14. Take $f_{sy} = 500$ MPa, $f'_c = 32$ MPa, cover to the 28 mm bars = 40 mm and the clear spacing between the bottom bars $a = 60$ mm. The cross-sectional area of one N28 bar is $A_s = 620$ mm^2 and with N12 stirrups ($A_{tr} = 110$ mm^2) at 150 mm centres.

Considering the reinforcing bars in Figure 14.14:

For bottom bars: $k_1 = 1.0$
For 28 mm diameter bars: $k_2 = (132 - 28)/100 = 1.04$

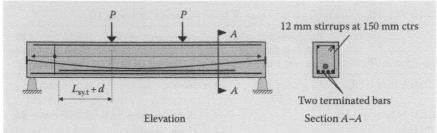

Figure 14.14 Development length of 28 mm diameter bottom bars.

With the concrete confinement dimension $c_d = a/2 = 30$ mm (see Figure 14.11), we have:

$$k_3 = 1.0 - 0.15(30 - 28)/28 = 0.99$$

From Equation 14.4, the basic development length is therefore:

$$L_{sy.tb} = \frac{0.5 \times 1.0 \times 0.99 \times 500 \times 28}{1.04\sqrt{32}} = 1178 \text{ mm} \quad (>29 \ k_1 d_b)$$

The minimum number of stirrups that can be located within the basic development length is 7. Therefore:

$$\Sigma A_{tr} = 7 \times 110 = 770 \text{ mm}^2$$

Taking $\Sigma A_{tr.min} = 0.25A_s = 155$ mm², the parameter λ is:

$$\lambda = (770 - 155)/620 = 0.99$$

From Figure 14.12, $K = 0.05$ (for each of the two interior bars) and therefore:

$$k_4 = 1.0 - 0.05 \times 0.99 = 0.95$$

It is assumed that in this location, the transverse pressure perpendicular to the anchored bar (ρ_p) is zero, and hence $k_5 = 1.0$.
From Equation 14.6:

$$L_{sy.t} = k_4 k_5 L_{sy.tb} = 0.95 \times 1.0 \times 1178 = 1120 \text{ mm}$$

Of course, the strength of the beam must be checked at the point where the two bars are terminated (at $L_{sy.t} + d$ from the constant peak moment region, as shown in Figure 14.14).

14.3.3 Lapped splices for bars in tension

Because of the particularly poor anchorage conditions that prevail at a lapped splice (at least two adjacent bars are anchored in close proximity), the lap length specified in AS3600-2009 [1] is larger than the development length ($L_{sy.t}$) for a single bar. Lapped splices should not be placed in critical regions unless absolutely necessary. At the ends of the splice where the bars terminate in a tension zone, a discontinuity exists and transverse cracks are usually initiated. These cracks may trigger the splitting cracks shown in Figure 14.9. Lapped splices should therefore be staggered where possible so that no free ends line up at the same section, unless the bars are further apart than $12d_b$ [2].

According to AS3600-2009 [1], in wide elements or members (e.g. flanges, band beams, slabs, walls and blade columns) where the bars being lapped are in the plane of the element or member, the tensile lap length ($L_{sy.t.lap}$) for either contact and non-contact splices is calculated from:

$$L_{sy.t.lap} = k_7 L_{sy.t} \quad (\geq 29 k_1 d_b) \tag{14.7}$$

where $L_{sy.t}$ is obtained from Equation 14.6, except that in the determination of $L_{sy.t}$ for use in Equation 14.7, the lower limit of $29\,k_1 d_b$ in Equation 14.4 does not apply; $k_7 = 1.25$. If the area of steel provided A_s is at least two times greater than the area of steel required and no more than one-half of the tensile reinforcement at the section is spliced, then k_7 can be reduced to 1.0.

In narrow elements or members (such as beam webs and columns), the tensile lap length ($L_{sy.t.lap}$) shall be not less than the larger of $k_7 L_{sy.t}$ and $L_{sy.t}$ + $1.5s_b$, where s_b is the clear distance between bars of the lapped splice as shown in Figure 14.10. However, if s_b does not exceed $3d_b$, then s_b may be assumed to equal zero for calculating $L_{sy.t.lap}$. For non-contact lapped splices, where the clear distance between the bars of the lapped splice s_b exceeds about $6d_b$, and the tensile forces either side of the splice are non-concurrent, the shear lag effect may require a longer lap length than $k_7 L_{sy.t}$ (i.e. $L_{sy.t} + 1.5s_b$).

14.4 ANCHORAGE OF DEFORMED BARS IN COMPRESSION

14.4.1 Introductory remarks

Unlike bars in tension, the anchorage of a bar in compression is provided by end bearing of the bar, as well as bond between the concrete and the steel bar along the development length. Consequently, the development length required to develop the yield stress in a deformed bar in compression is generally significantly less than that for a deformed bar in tension.

14.4.2 Development length of deformed bars in compression

As for deformed bars in tension, AS3600-2009 [1] specifies a two tiered approach for the development length of a deformed bar in compression. In any situation, a designer may adopt the simpler lower tier approach and specify the development length ($L_{sy.c}$) as the basic development length ($L_{sy.cb}$). Alternatively, in situations where the beneficial effects of transverse reinforcement and transverse confining pressure exist along the development length, the designer may opt for the refined upper tier approach.

14.4.2.1 Basic development length

The basic development length of a deformed bar in compression is given by:

$$L_{sy.cb} = \frac{0.22f_{sy}}{\sqrt{f'_c}}\,d_b \geq 0.0435f_{sy}d_b \quad \text{or 200 mm, whichever is greater} \quad (14.8)$$

When f'_c is greater or equal to than 25 MPa, the basic development length for a 500 grade deformed bar in compression ($f_{sy} = 500$ MPa) is $L_{sy.cb} = 22d_b$ (but for bar diameters of 9 mm or less, the basic development length is $L_{sy.cb} = 200$ mm). When $f'_c = 20$ MPa, the basic development length for 500 MPa grade bar is $L_{sy.cb} = 24.6d_b$.

14.4.2.2 Refined development length

In situations where the beneficial effects of transverse reinforcement and/or transverse confining pressure exist along the development length, a refined development length may be used and is given by:

$$L_{sy.c} = k_6L_{sy.cb} \quad (14.9)$$

The factor k_6 is equal to 0.75 when there are at least three transverse reinforcing bars or fitments located along the basic development length and outside the bar being anchored (i.e. located between the anchored bar and the nearest concrete surface) and the total area of transverse reinforcement within the basic development length (ΣA_{tr}) divided by the spacing of the transverse reinforcement s satisfies the relationship ($\Sigma A_{tr}/s$) $\geq$ ($A_s/600$). If this requirement is not satisfied, $k_6 = 1.0$.

For example if a 24 mm diameter bar ($A_s = 450$ mm^2) in a column with $f'_c = 40$ MPa is to be anchored and the column contains single R10 ties at 150 mm centres ($A_{tr} = 78.5$ mm^2), the basic development length is $L_{sy.cb} = 22 \times 24 = 528$ mm. At least three transverse bars are located within the basic development length and $\Sigma A_{tr}/s =$ three $\times$ 78.5/150 = 1.57 and this is greater than $A_s/600 = 450/600 = 0.75$. For these bars, the refined development length is $L_{sy.c} = k_6L_{sy.cb} = 0.75 \times 528 = 396$ mm.

14.4.2.3 Development length to develop a stress lower than the yield stress

When more reinforcement is provided than is necessary for strength at a particular location and the compressive stress to be developed (σ_{sc}) in a deformed bar is less than the yield stress (f_{sy}), the development length L_{sc} may be reduced proportionally to $L_{sc} = L_{sy.c}\sigma_{sc}/f_{sy}$, but not less than 200 mm.

14.4.2.4 Development length of a deformed bar in compression with a hook or cog

For a bar in compression ending in a hook or cog, the hook or cog is not considered to be effective in developing compressive stress and therefore should not be included in an assessment of anchorage. For example, where a bar in compression is bent for construction purposes, such as a 90° bend for a starter bar within a footing, the straight embedment into the footing must be not less than $L_{sy.c}$, as shown in Figure 14.15.

14.4.2.5 Development length of plain bars in compression

The average ultimate bond stress of a plain bar in compression is significantly smaller than that of a deformed bar. AS3600-2009 [1] requires that the development length for a plain bar in compression is 100% longer than for a deformed bar in the same location.

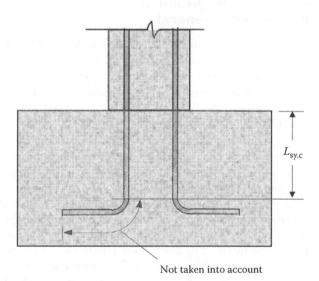

$L_{sy.c}$

Not taken into account

Figure 14.15 Development length of a hooked starter bar compression.

14.4.3 Lapped splices for bars in compression

The minimum length of a lapped splice for deformed bars in compression is $40d_b$ (but not less than 300 mm), except that, in a tied compression member, where at least three sets of fitments are present over the length of the lap and $A_{tr}/s \geq A_b/1000$, the minimum lap length is reduced to $32d_b$. In helically tied compression members, with n bars uniformly spaced around the helix, if at least three turns of helical reinforcement are present over the length of the lap and $A_{tr}/s \geq nA_b/6000$, the minimum lap length is $32d_b$.

14.5 STRESS DEVELOPMENT AND COUPLING OF TENDONS

A discussion of the anchorage of the tendons in a pretensioned member was presented in Section 8.2 and the minimum anchorage length specified in AS3600-2009 [1] of a pretensioned tendon from its end to the critical cross-section where the ultimate stress is required was given in Equation 8.3.

When tendons are coupled, AS3600-2009 [1] requires that the coupler is capable of developing at least 95% of the tendon characteristic breaking force and that the coupler is enclosed in grout-tight housings to facilitate grouting the duct (see Figure 3.11).

14.6 DETAILING OF BEAMS

14.6.1 Anchorage of longitudinal reinforcement: General

As mentioned earlier, a prerequisite for good detailing is favourable bond and anchorage conditions for each reinforcing bar and each tendon. The stress conditions surrounding a bar anchorage have considerable effect on the quality of bond. Where possible, bars should be anchored in regions where compressive stresses act in a transverse or normal direction to the bar. Bond strength increases considerably when normal pressure is present [12]. This increase is more pronounced for larger diameter bars. In Figure 14.16, it can be seen that the anchorage conditions for the bottom reinforcement at the support are more favourable than for the top reinforcement and consequently, a shorter anchorage length is required.

When bottom reinforcement is terminated away from the support, the diagonal compression in the web improves the anchorage, provided of course that there is sufficient web reinforcement to carry the diagonal force back to the top of the beam. The anchorage conditions of the terminating bars may be further improved by bending them into the web, as shown in Figure 14.17. This also reduces the possibility of premature shear failure at the discontinuity caused by the terminating flexural reinforcement.

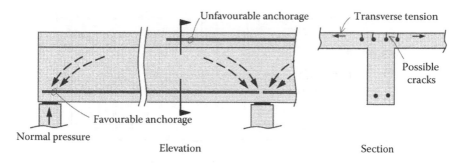

Figure 14.16 Anchorage of longitudinal reinforcement in a continuous beam.

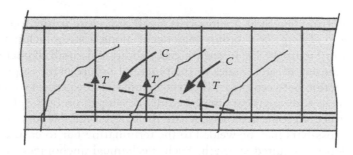

Figure 14.17 Anchorage of terminating bottom reinforcement.

The transverse tension that may cause splitting in the plane of a hooked anchorage (as illustrated in Figure 14.5) can be overcome at a beam support simply by tilting the hook (or better still, laying the hook in a near horizontal plane) to expose it to the normal reaction pressure at the support, as shown in Figure 14.18.

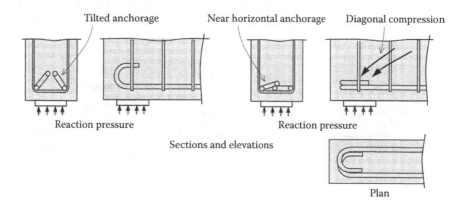

Figure 14.18 Hooked anchorages – preferred positions.

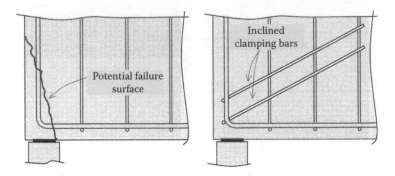

Figure 14.19 Detail when support length is short.

If the bearing length at a support is small and close to the free end of a member, a *sliding shear* failure may occur along a steep inclined crack, as shown in Figure 14.19. In such a case, additional small diameter bars may be required at right angles to the potential failure plane to provide a clamping action between the crack surfaces and thereby prevent sliding and failure. These additional bars must be fully developed on both sides of the failure surface. In some instances, where the length available for anchorage is small, cross-bars may be welded to the terminating bar to ensure that it can develop its required strength. Such mechanical anchorages are commonly used in precast elements and in regions of high concentrated loads such as corbels, brackets and other support points. Typical examples are illustrated in Figure 14.20.

For headed bars, such as shown in Figure 14.20b, AS3600-2009 [1] states that 'a head used to develop a deformed bar in tension shall consist of a nut or plate, having either a round, elliptical or rectangular shape, attached to the end(s) of the bar by welding, threading or swaging of suitable strength to avoid failure of the steel connection at ultimate load'.

AS3600-2009 [1] also requires that the net bearing area of the head should be at least four times the cross-sectional area of the bar, the cover

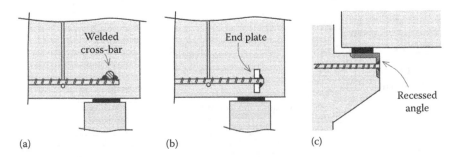

Figure 14.20 Mechanical anchorages. (a) Welded cross-bar. (b) Welded end plate (headed bar). (c) Recessed angle.

to the bar should be at least $2d_b$ and the clear spacing between bars being anchored should not be less than $4d_b$. Such anchorages should only be used for bar diameters less than or equal to 40 mm.

In addition, AS3600-2009 [1] states that if the cross-sectional area of the head of the headed reinforcement, or the area of the end plate for deformed bars mechanically anchored with an end plate (as shown in Figure 14.20b), is at least 10 times the cross-sectional area of the bar, the bar is considered to have a development length $(L_{sy.t})$ measured from the inside face of the head equal to $0.4L_{sy.t}$ of a bar of the same diameter.

In short span members, where load is carried to the support by arch action, it is essential that all bottom reinforcement bars (the tie of the arch) are fully developed at each support. To avoid bond failure in situations where the development length of each bottom bar is restricted, closely spaced transverse reinforcement in the form of stirrups can be used to bind the surfaces of both a potential horizontal splitting crack in the plane of the reinforcement and vertical splitting cracks developing from the anchored bar through the cover concrete (see Figure 14.21).

Apart from the reasons discussed earlier and illustrated in Figure 14.16, the anchorage conditions for top bars are always less favourable than for bottom bars because of increased sedimentation and poorer compaction of the concrete. Detailing for top bars therefore requires particular attention. If the top tensile reinforcement in a T-beam over an interior support is concentrated over the web in a multi-layered arrangement, a deterioration of bond strength may occur, resulting in increased crack widths and generally less favourable anchorage conditions.

It is better to place some of the top steel in the slab flange adjacent to the web. This improves crack control and provides better access to vibrators within the beam web. The measured crack widths in two-beams tested by Leonhardt et al. [7] are compared in Figure 14.22. In regions of high

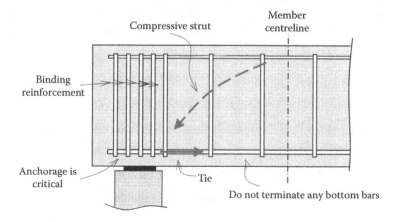

Figure 14.21 Anchorage in short span members.

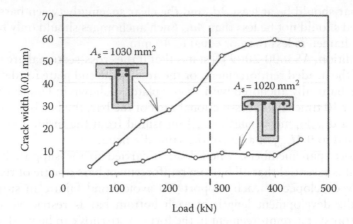

Figure 14.22 Comparison of crack widths in T-beams [6].

shear, sufficient longitudinal steel must be placed inside the stirrup cage to develop an efficient truss action.

14.6.2 Anchorage of stirrups

The flow of internal forces in a beam can be idealised as a parallel chord truss. This *truss analogy* was discussed in Section 7.2.3 and is illustrated in Figure 14.23. The compressive top chord and the diagonal web strut are the concrete portions of the truss, while the tensile bottom chord and vertical web ties must of course be steel reinforcement. The diagonal compression (in the concrete web strut) can only be resisted at the bottom of the beam at the intersection of the horizontal and vertical reinforcements, i.e. at the pin-joints of the analogous truss. It is evident that the tension in the vertical tie is constant over its entire height (i.e. from the pin joint at the bottom chord to the pin joint at the top chord). Therefore, adequate anchorage of the stirrups must be provided at every point along the vertical

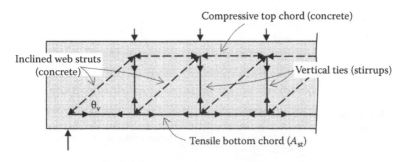

Figure 14.23 Truss analogy.

leg of the stirrup. After all, when calculating the shear strength provided by the stirrups, it is assumed that every vertical stirrup leg crossed by an inclined crack is at yield, irrespective of whether the inclined crack crosses the stirrup at its mid-depth or close to its top or bottom. The anchorage of the vertical leg of a stirrup may be achieved by a standard hook or cog (see Figure 14.13) or by welding of the fitment to the longitudinal bar or by a welded splice.

Stirrups should be anchored in the compression zone where anchorage conditions are most favourable. At ultimate loads, when diagonal cracks have developed, the compression zone may be relatively small. Stirrup hooks should therefore be as close to the compression edge as cover requirements allow. Stirrups depend on this transverse pressure for anchorage. It is common practice to locate the stirrup hooks near the top surface of a beam even in negative moment regions. When the top surface is in tension, the discontinuity created by a stirrup and its anchorage may act as a crack initiator. A primary crack therefore frequently occurs in the plane of the stirrup hook and anchorage is lost. As a consequence, in these regions, the beam may possess less than its required shear strength.

It is good practice to show the location of the stirrup hooks on the structural drawings and not to locate the hooks in regions where transverse cracking might compromise the anchorage of the stirrup.

Stirrup hooks should always be located around a larger diameter longitudinal bar that disperses the concentrated force at the anchorage and reduces the likelihood of splitting in the plane of the anchorage. Longitudinal bars are in fact required in each corner of the stirrup to distribute the concentrated force applied to the concrete at the corner. It is essential that the stirrup and stirrup hook fit snugly and are in contact with the longitudinal bars in each corner of the stirrup.

AS3600-2009 [1] requires that shear reinforcement, of area not less than that calculated as being necessary at any cross-section, must be provided for a distance (D) from that cross-section in the direction of decreasing shear. The first fitment at each end of a span should be located within 50 mm of the face of the support, and the shear reinforcement should extend as close to the compression face and the tension face of the member as cover requirements and the proximity of other reinforcement and tendons permit.

In Figure 14.24, some satisfactory and some unsatisfactory stirrup arrangements are shown. Stirrup hooks should be bent through an angle of at least 135°. A 90° bend (a cog) will become ineffective should the cover be lost, for any reason, and will not provide adequate anchorage. AS3600-2009 [1] states that fitment cogs are not to be used when the cog is located within 50 mm of any concrete surface.

In addition to carrying diagonal tension produced by shear, and controlling inclined web cracks, closed stirrups also provide increased ductility by confining the compressive concrete. The *open stirrups* shown in Figure 14.24b are commonly used, particularly in post-tensioned beams where the

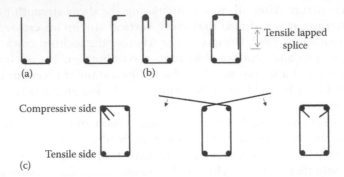

Figure 14.24 Stirrup shapes. (a) Incorrect. (b) Satisfactory in some situations. (c) Desirable.

opening at the top of the stirrup facilitates the placement and positioning of the post-tensioning duct along the member. This form of stirrup does not provide confinement for the concrete in the compression zone and is undesirable in heavily reinforced beams where confinement of the compressive concrete may be required to improve ductility of the member.

It is good practice to use adequately anchored stirrups (Figure 14.24c) even in areas of low shear, particularly when the longitudinal tensile steel quantities are relatively high and cross-section ductility is an issue. Although AS3600-2009 [1] suggests that no shear reinforcement is required in regions of low shear in shallow beams (where $D \leq 750$ mm), this is not recommended here.

In regions of high shear, it is desirable to use multi-leg stirrups when more than two longitudinal tensile bars are used. Park and Paulay [2] suggest that a *truss joint* should be formed at each longitudinal bar, i.e. the number of vertical stirrup legs should ideally equal the number of longitudinal bars and tendons. This is often not practical, but multi-leg stirrups should be used in members with wide webs to avoid the undesirable distribution of diagonal compression shown in Figure 14.25.

In Figure 14.25, the untied interior longitudinal bars with no nearby vertical stirrup leg cannot effectively resist diagonal compression and are therefore relatively inefficient in receiving bond forces.

Multi-leg stirrups are also far better for controlling the longitudinal splitting cracks (known as *dowel cracks*) that precipitate bond failure of the longitudinal bars in the shear span and are illustrated in Figure 14.26. The formation and propagation of this dowel crack usually triggers the sudden catastrophic shear failure that may occur in a heavily loaded shear span in a prestressed or reinforced concrete beam.

Closely spaced inclined stirrups (although in some instances not practical) are the most efficient form of shear reinforcement both in terms of strength and crack control. Vertical stirrups also perform well. Bent-up

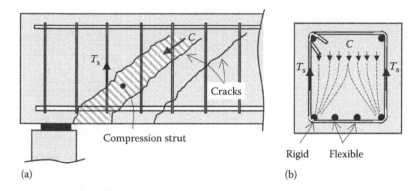

Figure 14.25 Undesirable distribution of diagonal compression due to wide stirrups [2]. (a) Elevation. (b) Cross-section.

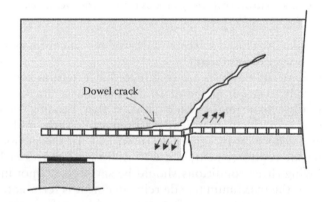

Figure 14.26 Shear failure triggered by bond failure of longitudinal bars.

longitudinal bars (once commonplace, now rarely used) are generally inefficient. Figure 14.27 shows the effect of various stirrup types on the control of inclined cracks in the web of beams measured by Leonhardt and Walther [7].

When a beam is subject to torsion, diagonal cracks exist on each face and the cracks spiral around the beam. Open stirrups of the type shown in Figure 14.24b are unsuitable. There is no point at which all stirrups can be effectively anchored, since the spiral cracking may occur in the plane of a stirrup hook. It is likely therefore that a number of stirrup anchorages are likely to be lost as the ultimate load is approached if conventional hooks are used. This can be accounted for in design by closing up the stirrup spacing somewhat, thereby allowing for some lost

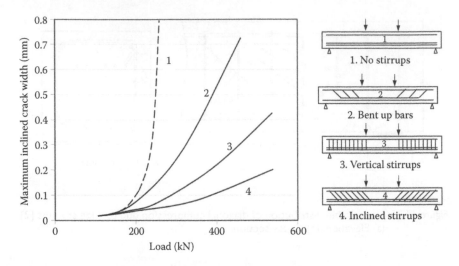

Figure 14.27   Crack control provided by various types of transverse reinforcement [2].

anchorages. Ideally, closed stirrups with welded anchorages should be used where torsion is significant.

Wherever longitudinal bars are terminated in a tension zone, primary cracks are likely to occur at the discontinuity. These cracks tend to be wider than adjacent primary cracks and, if they become inclined due to the presence of shear, may lead to premature shear failure (probably due to a reduction in aggregate interlock). In the vicinity of terminating tensile reinforcement, AS3600-2009 [1] requires at least one of the following three conditions should be satisfied: (1) not more than one-quarter of the maximum tensile reinforcement is terminated within any distance $2D$, (2) at the cut-off point, the design shear strength ϕV_u should not be less than $1.5V^*$ or (3) stirrups are provided to give an area of shear reinforcement of $A_{sv} + A_{sv.min}$ for a distance equal to the overall depth of the cross-section (D) along the terminated bar from the cut-off point, where $A_{sv.min}$ and A_{sv} are determined from Equations 7.7 and 7.20, respectively.

14.6.3 Detailing of support and loading points

When the support is at the soffit of a beam or slab, as shown in Figure 14.28a, the diagonal compression passes directly into the support as shown. However, when the support is at the top of the beam, as shown in Figure 14.28b, the diagonal compression must be carried back up to the support via an internal tie as shown. It is essential that adequately anchored

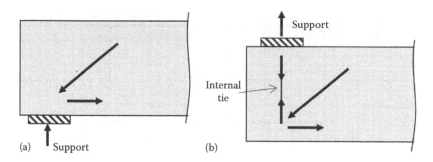

Figure 14.28 Support points. (a) Support at soffit. (b) Support at top of beam.

reinforcement be included to act as the tension tie and the reinforcement must pass into and be anchored within the support.

Consider the suspended slab supported from above by the upturned beam shown in Figure 14.29a. The horizontal component of the diagonal compression being delivered at the support of the slab must be resisted by the bottom slab steel.

The vertical component of the diagonal compression (i.e. the reaction from the slab) must be carried in tension up to the top of the upturned beam. This tension force must be carried across the unreinforced surface indicated in Figure 14.29a. The concrete on this surface may not be able to carry this tension and, if cracking occurs, premature and catastrophic failure could occur. The detail shown in Figure 14.29b overcomes the problem. The diagonal compression from the slab is now resisted by the stirrups in the upturned beam. No longer is there an unreinforced section of concrete required to carry tension. The vertical and horizontal members of the analogous truss have been effectively connected.

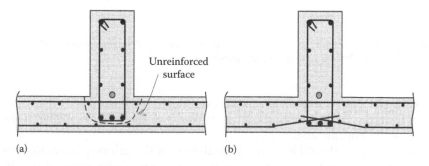

Figure 14.29 Suspended slab supported from above by upturned beam. (a) Incorrect detail. (b) Correct detail.

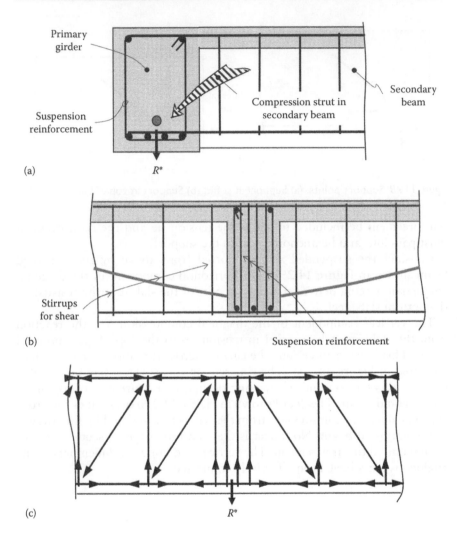

Figure 14.30 Beam-to-beam connection. (a) Section. (b) Primary grider – elevation. (c) Primary grider – truss anology.

Consider the beam to beam connection shown in Figure 14.30. The reaction from the secondary beam R^* is delivered to the primary girder at the level of the bottom steel.

This reaction should be carried by stirrups in the primary beam (*hanger* or *suspension reinforcement*) up to the top of the girder where it can be resolved into diagonal compression in a similar way to that of any other load applied to the top of the girder. For the reasons discussed in the previous paragraph, the bottom reinforcement in the secondary beam should always pass over the bottom reinforcement in the primary girder. The reinforcement details of the primary girder together with its truss analogy are shown in Figure 14.30b and c.

The *suspension* reinforcement is additional to the transverse reinforcement required for shear in the primary girder and must be located within the beam-girder connection. The area of additional suspension reinforcement A_{sr} required to carry the factored reaction R^* may be obtained from:

$$A_{sr} = \frac{R^*}{\phi f_{sy}}$$

where $\phi = 0.7$ (for strut and tie action) and f_{sy} is the characteristic yield stress of the hanger reinforcement.

When a load is applied to the underside of a reinforced concrete beam, some mechanical device must be used to transfer the *hanging* load to the top of the beam. Some typical devices are illustrated in Figure 14.31. When internal rods are used, plain round bars or bolts are suitable, since bond is not required to transfer the load to the top of the girder.

To form an internal hinge, particularly in precast construction, a *half joint* or *dapped-end joint*, as shown in Figure 14.32a, is frequently used. At such a connection, careful detailing is essential. Only half the beam depth is available and the internal forces are generally large. Figure 14.32b shows a strut-and-tie model from which an acceptable reinforcement layout can be determined. Figure 14.32c shows another commonly used strut-and-tie model and the corresponding reinforcement details.

The anchorage of all bars must be considered carefully. The bottom reinforcement at the right hand side of the joint in both details usually requires a hook or a 90° cog for adequate anchorage. The *suspension reinforcement* in Figure 14.32b must carry the full tension in the vertical tie and should be located as close to the connection as cover and spacing

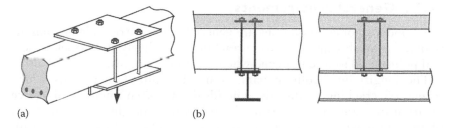

(a)   (b)

Figure 14.31 Mechanical means to support hanging loads. (a) External yoke. (b) Internal rods.

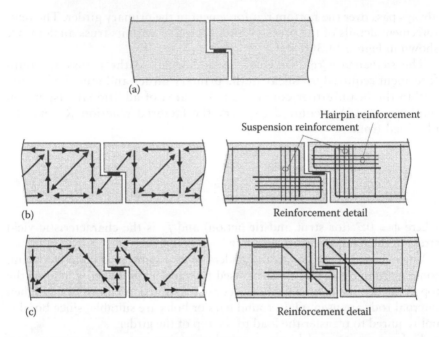

Figure 14.32 Half joint details. (a) Half joint. (b) Strut and tie model (No. 1). (c) Strut and tie model (No. 2).

requirements permit. The short cantilevered portions are designed as corbels (which are discussed in more detail in Section 14.8). Horizontal *hairpin reinforcement* should extend past the re-entrant corners where a potential crack may develop. It is usual for the area of this horizontal reinforcement to be taken as at least half the area of the suspension reinforcement.

14.7 DETAILING OF COLUMNS

14.7.1 General requirements

Bond and anchorage conditions are generally more favourable in columns than in beams, because transverse cracking is less likely. Nevertheless, several points need to be considered when detailing columns.

The longitudinal column bars should be spliced in regions where transverse cracking is unlikely. The ideal, but often impractical, splice location in many columns is at the mid-storey height, near the point of inflection where bending is small. In structures that may be subjected to earthquake loading, columns are often required to withstand large moments and possible plastic hinging at each end. Splices in these columns should always be near the mid-storey height, away from the peak moment region.

At a compressive splice, a large portion of the compression in the bar is transferred to the concrete by end bearing. In fact, before bond stresses can occur, the end bearing resistance of the bar must be overcome and some slip must occur. In tests reported by Leonhardt and Teichen [6], the concrete immediately under each of the spliced bars burst laterally before the ultimate load was reached. Additional transverse reinforcement at the ends of spliced bars is therefore important to provide confinement for the heavily stressed concrete. An arrangement suggested by Leonhardt and Teichen [6] is shown in Figure 14.33a.

If longitudinal bars are cranked to form an off-set, as shown in Figure 14.33b, additional transverse ties must be included to carry the resulting transverse tension R. According to AS3600-2009 [1], the slope of the inclined part of the bar in relation to the column axis should not exceed 1 in 6, the portions of the bar on either side of the offset should be parallel and adequate lateral support should be provided at the offset.

Where a single layer of reinforcement is used in a thin wall, a transverse tension splice, as shown in Figure 14.33c, should not be used. The internal couple resulting from the offset may lead to the cracking shown and this could precipitate premature failure, particularly if the wall is subjected to lateral loading. Lapped splices in thin walls should be in the plane of the wall.

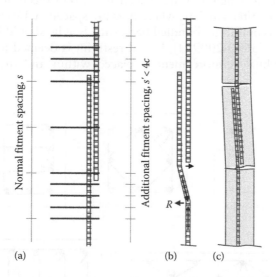

Figure 14.33 Lapped splices in columns and walls. (a) Additional fitments at compressive splice. (b) Tension at cranked bars. (c) Unsatisfactory tension splice in thin wall.

Lateral reinforcement (or fitments) in the form of closed ties or helices is required in columns for three reasons:

1. to provide restraint to the heavily stressed longitudinal reinforcement and thereby prevent outward buckling before the full strength of the bar is reached;
2. to provide confinement to the concrete core and thereby improve both the strength and ductility of the column. This confinement occurs at the points where the ties change direction around the longitudinal bars and results from the tension induced in the ties as the concrete core dilates under axial compression; and
3. to act as shear reinforcement when diagonal tension cracks are possible.

Whilst helical reinforcement is often used in piles and circular columns, closed ties are the most common form of lateral reinforcement used in rectangular columns. Typical ties arrangements are shown in Figure 14.34. Each main longitudinal bar is tied in two directions so as to effectively prevent outward buckling.

14.7.2 Requirements of AS3600-2009

Lateral restraint must be provided for: (1) each corner bar in a cross-section, (2) every other bar when bars are spaced at centres exceeding 150 mm and (3) at least every alternate bar, where bars are spaced at 150 mm or less.

For columns containing bundled bars, each bundle should be restrained. According to AS3600-2009 [1], lateral restraint is deemed to be provided if the longitudinal reinforcement is placed within and in contact with

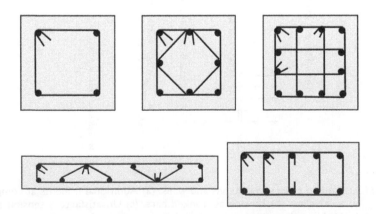

Figure 14.34 Typical tie arrangements in rectangular columns.

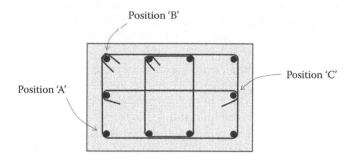

Figure 14.35 Lateral restraint to longitudinal bars.

a non-circular fitment and: (1) at a bend in the fitment, where the bend has an included angle of 135° or less (e.g. position A in Figure 14.35), or (2) between two 135° fitment hooks (e.g. position B in Figure 14.35) or (3) inside a single 135° fitment hook of a fitment that is approximately perpendicular to the column face (e.g. position C in Figure 14.35). Although in some circumstances, 90° fitment cogs are permitted by AS3600-2009 [1], their use is not recommended here. In the case of circular fitments or helical reinforcement, the longitudinal reinforcement inside the helix is deemed to be laterally restrained if the bars are equally spaced around the perimeter.

AS3600-2009 [1] specifies the following minimum bar diameter for fitments and helical reinforcement:

Longitudinal bar diameter (mm)	Minimum bar diameter for fitment and helix (mm)
Single bars up to 20	6
Single bars 24–28	10
Single bars 28–36	12
Single bar 40	16
Bundled bars	12

The spacing of fitments, or the pitch of a helix, should not exceed the smaller of D_c and $15d_b$ for single bars and $0.5D_c$ and $7.5d_b$ for bundled bars, where D_c is the smaller column cross-sectional dimension if the column is rectangular or the column diameter if the column is circular and d_b is the diameter of the smallest longitudinal bar.

One fitment, or the first turn of a helix, should be located not more than 50 mm vertically above the top of a footing, or the top of a slab in any storey. Another fitment, or the final turn of a helix, should be located not more than 50 mm vertically below the soffit of a slab, except that in a column with a capital, the fitment or turn of the helical reinforcement shall be located at a level at which the area of the cross section of the capital is not less than twice that of the column.

In situations where beams or brackets frame from four directions into a column and the column is restrained in all directions, the fitments or helical reinforcement may be terminated 50 mm below the highest soffit of the beams or brackets.

When splicing longitudinal reinforcement in columns, AS3600-2009 [1] has several additional requirements. At any splice, a tensile strength in each face of the column of not less than $0.25f_{sy}A_s$ shall be provided, where A_s is the cross-sectional area of longitudinal reinforcement in that face. In addition, at any splice in a column where tensile stress exists and where the tensile force in the longitudinal bars at any face of the column, due to strength design load effects, exceeds the minimum strength requirement (i.e. $0.25f_{sy}A_s$), the force in the bars shall be transmitted by either a welded or mechanical splice or a lap-splice in tension, as appropriate.

Where a splice is always in compression, the force in the longitudinal bar may be transmitted by end bearing. The mating ends of the bar must be square-cut and held in concentric contact by a sleeve. For such an end-bearing splice, an additional fitment should be placed above and below each sleeve. The bars should be rotated to achieve the maximum possible area of contact between the ends of the bars.

Where bending moments from a floor system are transferred to a column, lateral shear reinforcement, of area $A_{sv} \geq 0.35bs/f_{sy.f}$, should be provided through the connection unless restraint on all sides is provided by a floor system of approximately equal depth, where b is the smaller dimension of the column cross-section and s is stirrup spacing.

14.8 DETAILING OF BEAM-COLUMN CONNECTIONS

14.8.1 Introduction

Connections between structural members are often the weakest points in a structural system. Within connections, internal forces change direction abruptly and adequately anchored reinforcement must be inserted to carry all tension. Often the space available for anchorage of reinforcement is small and restricted, and due to extensive cracking within the connection at overloads, the anchorage conditions are particularly unfavourable.

Ideally, the strength of a connection should not govern the strength of the structure. Connections should therefore possess strength at least as great as the members they join. Connections should also perform satisfactorily at service loads (controlled cracks and small rotations) and be easy to construct. As will be seen, these requirements are often not easy to satisfy and detailing the reinforcement in connections requires careful attention.

The moments, shears and axial forces in a concrete structure are usually determined using elastic analyses. In such analyses, connections are usually assumed to be rigid, that is the ends of all members meeting at the connection are assumed to rotate by the same amount. In addition to being designed to carry the internal actions, the connections should respond with deformations at least similar to those assumed in the analysis. If the connection is too flexible, span moments will exceed those given by the analysis, causing increased deformations and, perhaps, premature failure in heavily reinforced members at overload where the ductility required for redistribution may not be available.

The difficulty in predicting the load-deformation characteristics of connections is a major obstacle to the use of collapse load analysis for reinforced and prestressed concrete frames. Of course, in any connection, ductility is essential and the use of low ductility steel should not be contemplated.

14.8.2 Knee connections (or two-member connections)

Two-member connections, such as those shown in Figure 14.36, are commonly used in concrete structures. It is often difficult to achieve 100% efficiency in such a connection (i.e. the strength of the connection is equal to the strength of the adjoining members), particularly when subjected to opening moments.

14.8.2.1 Opening moments

In Figure 14.37a, the flow of internal forces in a two-member connection subjected to an *opening moment* is illustrated and the crack pattern in such a connection as the applied loads are increased is shown in Figure 14.37b. For this connection to work efficiently, adequately anchored reinforcement must be included to carry the diagonal tension $\sqrt{2}T$ across the connection. Numerous tests [13,14] have shown that the reinforcement details illustrated in Figure 14.38a and b are quite unsatisfactory.

Unless secondary diagonal reinforcement is included to carry the tension within the connection (as shown in Figure 14.38c), the connection will fail at some small fraction of the strength of the adjoining members.

Provided the diagonal stirrups in Figure 14.38c are capable of developing the full diagonal tension at the ultimate limit state (i.e. provided the stirrups

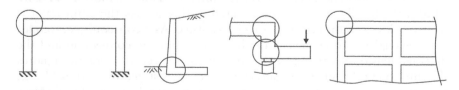

Figure 14.36 Two-member connections.

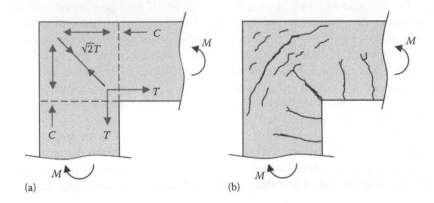

Figure 14.37 Knee connection in opening bending. (a) Internal forces. (b) Crack pattern.

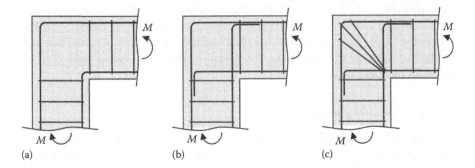

Figure 14.38 Unsatisfactory and potentially satisfactory reinforcement details. (a) Unsatisfactory. (b) Unsatisfactory. (c) Potentially satisfactory.

have an adequate cross-sectional area and provided they are adequately anchored), the detail shown can provide up to 100% efficiency, particularly in lightly reinforced beams (i.e. with $p = A_{st}/bd < 0.01$). In general, the more lightly reinforced the members, the more efficient is the connection. The diagonal stirrups must fit snugly around the longitudinal steel to effectively control the growth of cracks.

For lightly reinforced slabs and walls with extensive (wide) connections (such as the connection between the retaining wall and its footing in Figure 14.36), the tensile stress in the concrete acting across the connection is relatively low and the concrete area is extensive. As a consequence, secondary diagonal reinforcement is rarely used within the connection and the diagonal tension is carried by the concrete. However, this practice ignores the fundamental principle of successful detailing, stated in Section 14.2.2, namely to 'use adequately anchored reinforcement wherever a tensile force is required for equilibrium'.

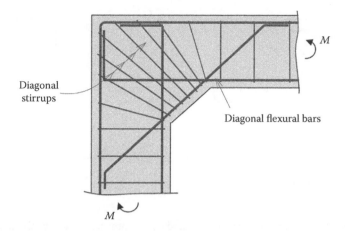

Figure 14.39 Suggested detail for large opening knee connections [2].

The area of the diagonal steel required in Figure 14.38c is easily obtained from the statics of Figure 14.37a and is given by:

$$A_{sv} = \frac{\sqrt{2}T}{\phi f_{sy.f}} = \frac{\sqrt{2}A_{st}f_{sy}}{\phi f_{sy.f}} \tag{14.10}$$

where $\phi = 0.7$; $f_{sy.f}$ and f_{sy} are the characteristic yield stresses of the diagonal and longitudinal tensile steel, respectively; and A_{st} is the area of the longitudinal tensile steel.

Park and Paulay [2] suggest that the detail shown in Figure 14.39 is suitable for relatively large knee connections. A haunch at the re-entrant corner will allow the inclusion of more diagonal flexural bars, reduce the magnitude of the internal tension and generally strengthen the connection.

14.8.2.2 Closing moments

The load path in two-member connections subjected to *closing moments* is shown in Figure 14.40a and the crack pattern caused by increasing the applied load is illustrated in Figure 14.40b. It is much easier to achieve full efficiency for a knee connection subjected to closing moments compared to one subjected to opening moments.

Typical reinforcement details for lightly reinforced wide members, such as slabs and walls, and for more heavily reinforced knee connections in closing bending are shown in Figure 14.41a and b, respectively. The outer tensile bars are easily developed, provided the radius of bend is large enough to avoid splitting failure. The inclusion of a larger diameter transverse bar tied inside the 90° bend in the top reinforcement (as shown in Figure 14.41)

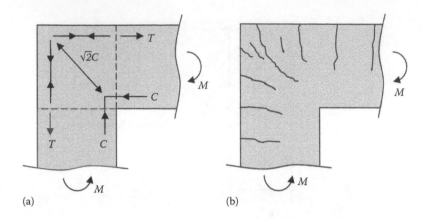

Figure 14.40 Knee connection in closing bending. (a) Internal forces. (b) Crack pattern.

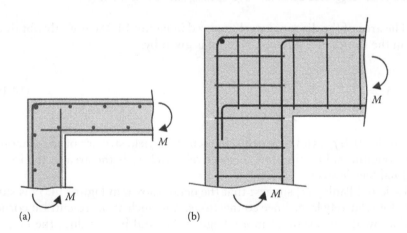

Figure 14.41 Reinforcement details – two-member connections in closing bending. (a) Wall or slab connection (when $p \le f_{ct.f}/f_{sy}$). (b) Beam to column knee connection.

will distribute the concentrated force applied to the concrete at the bend and avoid the development of splitting cracks. The main tensile bars should be continuous around the corner, and lapped splices within the connection should be avoided.

When transverse reinforcement is not included within the connection (Figure 14.41a), the amount of longitudinal tensile reinforcement should be limited in order to avoid the propagation of tensile cracks within the connection. Park and Paulay [2] suggest that $p \le f_{ct.f}/f_{sy}$, where $f_{ct.f}$ is the lower characteristic flexural tensile strength $(0.6\sqrt{f_c'})$ in MPa. For $f_c' = 25$ MPa and

$f_{sy} = 500$ MPa, the reinforcement ratio, therefore, should not exceed 0.006 in a wide closing connection without transverse reinforcement.

In isolated beam-column knee connections, in order to provide crack control, the vertical transverse reinforcement (stirrups) in the horizontal member should be extended into the connection, as should the horizontal transverse reinforcement in the vertical member. This reinforcement will also serve to confine the concrete within the connection and facilitate the development of the diagonal compressive strut between the inner corner and the bend in the longitudinal steel.

14.8.3 Exterior three-member connections

The flow of internal forces in a typical exterior beam-column connection is illustrated in Figure 14.42a and the crack pattern under increasing load is as shown in Figure 14.42b. The anchorage conditions of the longitudinal reinforcement in such a connection are particularly unfavourable. The top steel in the horizontal beam enters the connection in a region subjected to transverse tension. The surrounding concrete may be cracked, particularly in frames subjected to high lateral loads, and being in the top of the beam the concrete is subject to sedimentation. Bond along the bar from the face of the column to the beginning of the 90° bend should not be relied on for anchorage.

The outer column bars in Figure 14.42b at ultimate may be required to be at yield in tension below the connection and at yield, or close to it, in compression above the connection. Code provisions for anchorage may not be able to be met here. Tightly fitting and closely spaced column ties within the connection are required to control the splitting cracks caused by the usually very high bond stresses around these bars.

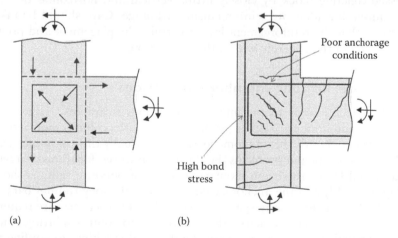

Figure 14.42 Three-member connection. (a) Internal forces. (b) Crack pattern.

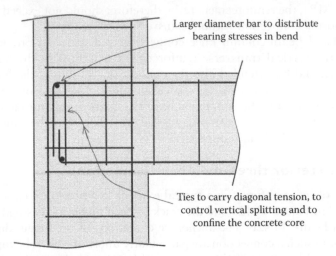

Figure 14.43 Reinforcement detail for a three-member connection.

A typical reinforcement detail is shown in Figure 14.43.

When an exterior three-member connection is subjected to reversals of load, such as may occur under seismic loading, the anchorage conditions in such a connection are at their worst.

Often mechanical anchorages are required for the longitudinal reinforcement in the beams and a large amount of transverse steel is required within the connection. Park and Paulay [2] point out that under conditions of alternating plasticity, it is unwise to assume that the concrete within the connection will contribute to the shear strength. Confinement of this highly stressed concrete block by closely fitting vertical and horizontal ties is a prerequisite for adequate reinforcement anchorage. Care should be taken to ensure that the reinforcement layout permits the placement and proper compaction of the concrete within the connection.

14.8.4 Interior four-member connections

Similar remarks apply to the interior connections in frames. If the beam moments equilibrate each other (or nearly so), i.e. the column moments are small compared to the beam moments (which is usually the case under gravity loads), no particular problems should arise. However, for frames carrying large lateral loads, the beam moments may be of opposite sign, as shown in Figure 14.44, and the anchorage conditions of the longitudinal bars are critical. Once again, closely spaced horizontal and vertical ties are required to carry diagonal tension across the connection, to control splitting cracks caused by high bond stresses, to control diagonal cracking, to confine the concrete core and, therefore, to improve the anchorage conditions generally within the connection.

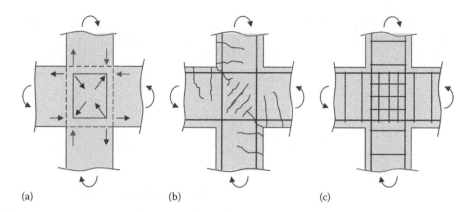

Figure 14.44 Interior four-member connection. (a) Internal forces. (b) Crack pattern.
(c) Reinforcement detail.

14.9 DETAILING OF CORBELS

14.9.1 Introduction

Corbels are short cantilevers that tend to act as simple trusses or deep beams, rather than flexural members. Corbels carry reactions or concentrated loads into columns or walls as shown in Figure 14.45a and have shear-span to depth ratios generally not greater than unity. From studies of stress trajectories in photoelastic models, Franz and Niedenhoff [15] verified the flow of internal forces shown. A typical reinforcement detail for a corbel is shown in Figure 14.45b. The horizontal reinforcement is required to carry the tension from under the bearing pad back into the column and must therefore be fully developed in the immediate vicinity of the bearing pad. Often the main tensile reinforcement is welded to a cross bar of at least the same diameter (as shown in Figures 14.45c and d) or directly to a steel bearing plate to ensure adequate anchorage. The welds should be designed to develop the yield strength of the primary tension reinforcement. A weld detail that has been used successfully in corbel tests is shown in Figure 14.45d [16]. Horizontal stirrups are usually included as shown in Figure 14.45b to control the inclined cracking which occurs from the top surface parallel to the compressive strut.

14.9.2 Design procedure

The strut and tie arrangement shown in Figure 14.46 suggests a simple design procedure for corbels. The required quantity of tensile reinforcement (A_s) is readily determined from equilibrium considerations using the strut-and-tie procedures of AS3600-2009 [1] (see Section 8.4).

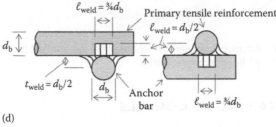

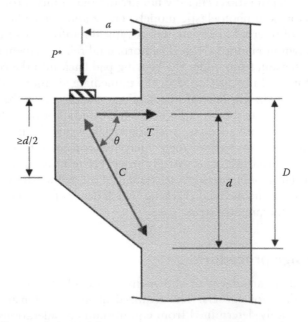

Figure 14.45 Reinforced concrete corbel details [16]. (a) Strut-and-tie action. (b) Reinforcement detail. (c) Welded reinforcement. (d) Satisfactory weld details.

Figure 14.46 Strut-and-tie model.

The tensile force T is given by:

$$T = \frac{P^*}{\tan\theta} = \phi A_s f_{sy}$$

and therefore:

$$A_s = \frac{P^*}{\phi f_{sy} \tan\theta} \tag{14.11}$$

The angle between the compression strut and the tie (θ) is determined from geometry and the capacity reduction factor for the tensile steel is $\phi = 0.8$.

If in addition to the vertical load P^*, a horizontal tensile force H^* is to be carried by the corbel, the area of tensile steel given by Equation 14.11 should be increased by $H^*/\phi f_{sy}$. In this case, it is advisable to mechanically connect (by welding) the main reinforcement to the bearing plate to ensure adequate anchorage.

The design strength of the concrete strut is given by Equation 8.16 as:

$$\phi_{st} C_u = \phi_{st}\beta_s 0.9 f_c' A_c$$

where $\phi_{st} = 0.6$; A_c is the smallest cross-sectional area of the strut measured normal to its line of action; and β_s is the strut efficiency factor given by Equation 8.17, i.e.:

$$\beta_s = \frac{1}{1.0 + 0.66\cot^2\theta} \quad \text{within the limits } (0.3 \le \beta_s \le 1.0)$$

The bursting reinforcement across the strut is calculated using Equations 8.18 through 8.23 and should be placed in the form of horizontal stirrups evenly spaced over the bursting length of the strut.

A good first estimate of the corbel dimensions can be made using the following:

$$P^*/b_w d \le 0.4\sqrt{f_c'} \tag{14.12}$$

In addition, ACI318-11M [16] suggests that corbels be proportioned such that:

$$a/d \le 1.0 \tag{14.13}$$

and

$$0.04f_c'/f_{sy} \le A_s/(b_w d) \le 0.2f_c'/f_{sy} \tag{14.14}$$

where a is defined in Figure 14.46.

EXAMPLE 14.2

The corbel of Figure 14.47a is to be designed to carry a dead load $G = 240$ kN and live load $Q = 100$ kN (i.e. a factored design load of $P^* = 1.2G + 1.5Q = 438$ kN) as shown. The column and the corbel width is $b_w = 300$ mm. The bearing plate is 250 mm × 300 mm in plan. The strut-and-tie model is shown in Figure 14.47b. The material properties are $f_c' = 32$ MPa, $f_{ct}' = 2.5$ MPa and $f_{sy} = 500$ MPa, and the clear concrete cover to the reinforcement is 30 mm.

Using Equation 14.12 for preliminary sizing, with $b_w = 300$ mm, the effective depth is selected to satisfy:

$$d \ge \frac{438 \times 10^3}{300 \times 0.4 \times \sqrt{32}} = 645 \text{ mm}$$

With $D = d$ + cover + 0.5 bar diameter and assuming 20 mm diameter bars, take $D = 700$ mm and therefore $d = 660$ mm.

From the geometry of the model shown in Figure 14.47b:

$$\tan\theta = \frac{d - 125\tan(90 - \theta)}{525} \text{ and solving gives } \theta = 45.7°$$

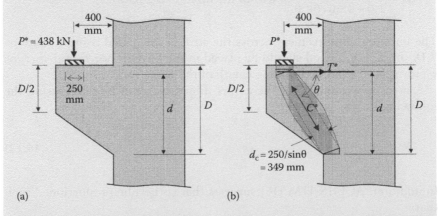

(a)   (b)

Figure 14.47 Corbel (Example 14.2). (a) Corbel dimensions. (b) Strut-and-tie model.

From Equation 14.11:

$$A_s = \frac{438 \times 10^3}{0.8 \times 500 \times \tan 45.7} = 1069 \text{ mm}^2$$

Try 4–20 mm diameter Grade N bars ($A_s = 1240$ mm²).

With $A_s/(b_w d) = 0.0063 = 0.098 f_c'/f_{sy}$, both Equations 14.13 and 14.14 are satisfied.

The strut efficiency factor given by Equation 8.17 is:

$$\beta_s = \frac{1}{1.0 + 0.66 \cot^2 \theta} = 0.614$$

and from Equation 8.16, with $A_c = b_w d_c = 300 \times 349 = 104{,}700$ mm², the design strength of the concrete strut is:

$$\phi_{st} C_u = 0.6 \times 0.614 \times 0.9 \times 32 \times 104{,}700 = 1{,}111 \text{ kN}$$

and this is greater than the design compressive force in the strut:

$$C^* = P^*/\sin \theta = 612 \text{ kN} \quad \therefore \text{ OK}$$

The length of the bursting zone l_b was defined in Figure 8.26c as the length of the strut (752 mm) minus d_c (349 mm), i.e. $l_b = 403$ mm. The compressive force in the strut at the serviceability and strength limit states are $(G + Q)/\sin \theta = 475$ kN and $(P^*)/\sin \theta = 612$ kN. From Equation 8.18, at the strength limit state, $\tan \alpha = 0.2$ and from Equation 8.15, $T_b^* = 612 \times 0.2 = 122$ kN. The minimum bursting reinforcement required for strength is obtained from Equation 8.22:

$$0.8 \times \Sigma A_{si} \times 500 \times \sin(45.7°) \geq 122 \times 10^3 \therefore \Sigma A_{si} \geq 427 \text{ mm}^2$$

At the serviceability limit state, $T_{b.s}^* = 475 \times 0.2 = 97.4$ kN. This is significantly less than the bursting force required to cause cracking given by Equation 8.20 (i.e. $T_{b.cr} = 0.7 \times 300 \times 403 \times 2.5 \times 10^{-3} = 212$ kN), so crack control at the serviceability limit state is unlikely to be a problem.

Use 6–12 mm bars (i.e. 3 horizontal 12 mm diameter ties) as bursting reinforcement.

Either of the reinforcement details shown in Figure 14.48 will prove to be acceptable.

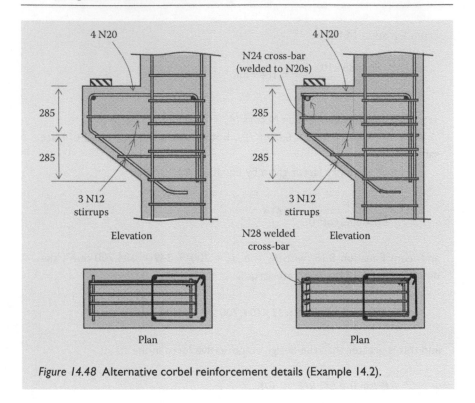

Figure 14.48 Alternative corbel reinforcement details (Example 14.2).

14.10 JOINTS IN STRUCTURES

14.10.1 Introduction

Joints are introduced into concrete structures for two main reasons:

1. as stopping places in the concreting operation. The location of these *construction joints* depends on the size and production capacity of the construction site and work force; and
2. to accommodate deformation (expansion, contraction, rotation, settlement) without local distress or loss of integrity of the structure. Such joints include *control joints* (contraction joints), *expansion joints*, *structural joints* (such as hinges, pin and roller joints), *shrinkage strips* and *isolation joints*. The location of these joints can usually be determined by consideration of the likely movements of the structure during its lifetime and the resulting effects on structural behaviour.

14.10.2 Construction joints

Construction joints are required in structures so that each concrete pour can be handled by the available workforce. Construction joints may be

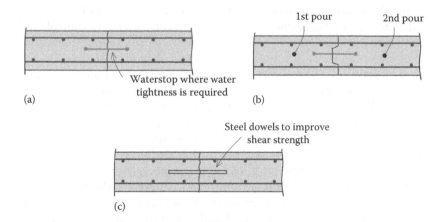

Figure 14.49 Construction joint details. (a) Butt joint. (b) Keyed joint. (c) Dowelled joint.

horizontal in slabs or vertical in long walls. Typical construction joint details for slabs and walls are shown in Figure 14.49.

All reinforcements should be continuous across the joint. The keyed or dowelled joints of Figure 14.49b and c are of questionable value, particularly in slabs with top and bottom reinforcement ratios exceeding about 0.0035. If the surface of the hardened concrete at the joint is properly prepared, friction and aggregate interlock on the concrete surface, together with the dowel action of the reinforcement, can provide shear strength at the joint as high as that of adjacent sections placed monolithically.

For a sound joint, after the first concrete pour, the exposed reinforcement should be cleaned and the aggregate of the hardened concrete exposed by wire brushing or water or sand blasting. The hardened concrete should be thoroughly wetted before the new concrete is poured. For a waterproof construction joint, a continuous plastic or rubber waterstop or waterbar is essential. Compaction of the concrete around the waterstop should be thorough and careful.

If possible, construction joints should coincide with other joints (e.g. expansion or contraction joints), so as to minimise the total number of joints in the structure. Construction joints should be disguised (or otherwise hidden) or incorporated as architectural features. It is therefore necessary that the number and location of joints be decided well before the concrete trucks arrive onto the construction site. Ideally, construction joints should not be placed in regions of high moment or shear, and should not occur where they could create stress concentrations.

14.10.3 Control joints (contraction joints)

In concrete that is not free to contract, restraint to shrinkage and temperature changes produces tension that can, and frequently does, cause excessive cracking. Restraint to shrinkage and temperature changes can be handled in two different ways. Sufficient reinforcement can be inserted to control cracking and cause a large number of very fine, serviceable cracks. Alternatively, control joints can be used in walls and slabs to concentrate the cracking into preformed grooves. The contraction of the concrete is taken up at the control joint and the restraint to shrinkage between cracks is largely removed.

A control joint is an intentionally introduced plane of weakness in a slab or wall. Control joints are spaced and positioned so that cracking and contraction takes place only on these preselected straight lines. The joints should be close enough together so that shrinkage- and temperature-induced tension in the concrete between the joints remains small.

Typical control joint details in reinforced concrete slabs and walls are shown in Figure 14.50. The joint must allow contraction of the concrete (i.e. it must be able to open longitudinally), but must resist relative transverse movement (i.e. it must allow for shear transfer, if required).

Control joints are formed by locally reducing the cross-sectional area of the slab or wall by about 25%. Form inserts can be used to form the joint (as illustrated in Figure 14.50b and c) or a saw cut in the fresh concrete on the surface of a slab can be used to create the weakened plane. To ensure relatively free contraction at the joint after a crack has formed, the reinforcement crossing the joint should be no more than $0.002A$, where A is the cross-sectional area of the slab or wall at the joint.

If significant shear is to be transferred across the joint, dowels can be used to improve shear strength, as shown in Figure 14.50d. The dowels should be debonded on one side of the joint to allow free contraction.

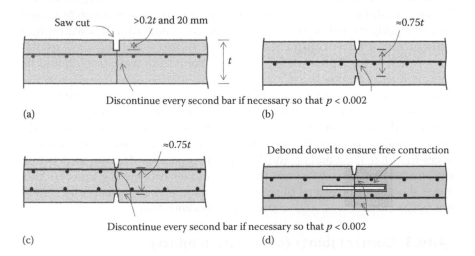

Figure 14.50 Control joint details. (a) Saw-cut joint in slab on ground. (b) Wall (t < 200 mm). (c) Wall (t ≥ 200 mm). (d) Dowelled joint.

Control joints should be located in regions of low moment, since the flexural strength of a slab or wall is usually quite low at a joint. Water tightness is always a potential problem at a control joint, and a continuous flexible (expandable) waterstop should be used if a waterproof joint is required.

The position and spacing of control joints depends on many factors including the shrinkage characteristics of the concrete, the curing and exposure conditions, the external restraint (due to supports, adjacent parts of the structure or friction with the ground) and the structural layout. No hard and fast rules can be made. However, as a general rule, a joint should be located wherever an abrupt change in dimension of the structure occurs (thickness, width or height).

In the design of walls and slabs, the designer in general has a choice between joints at close spacings (3–5 m – see Reference [17]) with small quantities of reinforcement (about $0.002A$) or joints at much wider centres and much larger quantities of reinforcement (at least $0.006A$). Reinforcement quantities of about $0.002A$ are unable to control shrinkage-induced cracking and therefore, joints at close centres are required to reduce restraint and accommodate contraction.

If small quantities of reinforcement are specified in reinforced concrete slabs with joints at close centres, joint spacing of as little as 3 m may be necessary in dry environments or with high shrinkage concretes. The first joint should be located no more than 3 m from a corner. The ratio of panel dimensions enclosed by joints in a wall or slab should be as close to 1.0 as possible, but not greater than 1.5.

Short walls restrained at their base by a more massive footing are particularly prone to shrinkage cracking (Figure 14.51a), as are cantilevered balcony slabs (Figure 14.51b).

Control joints at a spacing similar to the wall height (or the span of a cantilevered balcony) are required to accommodate all the concrete contractions unless relatively large quantities of reinforcement are specified.

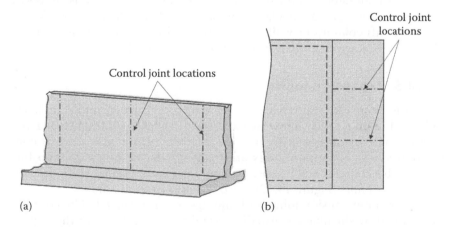

Figure 14.51 Control joint locations. (a) Wall elevation. (b) Balcony plan.

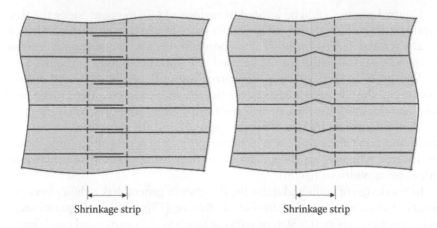

Shrinkage strip   Shrinkage strip

Figure 14.52 Alternative reinforcement details at shrinkage strips.

To prevent unsightly cracking at the exposed balcony edge at each joint location, additional reinforcement may be required along the balcony edge.

14.10.4 Shrinkage strips

Shrinkage strips serve the same purpose as control joints. A strip of about 1 m width across a building is left during concreting, thus allowing the concrete on either side of the strip to shrink freely. After several weeks, when a significant amount of the drying shrinkage has occurred, the strips are poured and continuity is established.

The reinforcement crossing a shrinkage strip is usually continuous, but is lapped or bent horizontally as shown in Figure 14.52 to allow unrestrained contraction on either side of the strip.

In long multi-storey framed structures without stiff columns or walls, shrinkage strips are often placed in slabs at about 40 m centre. When there are stiff columns or walls, strips are required at much closer centres. Vertical shrinkage strips may also be used in long walls.

14.10.5 Expansion joints

Expansion joints separate two adjacent parts of a structure into completely independent units. They allow for expansion of concrete during curing and due to temperature rises in-service and, by their nature, also serve as contraction joints. Expansion joints are frequently located on a column line with double columns and beams, as shown in Figure 14.53a. Half joints or dapped-end joints (Figure 14.53b) also act as expansion joints.

The use of expansion joints in buildings is controversial. The contraction caused by shrinkage is usually several times greater than the expansion caused by ambient temperature rises. Indeed, many large buildings

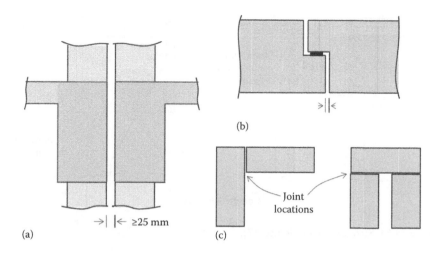

Figure 14.53 Expansion joint details. (a) Double column and beams. (b) Half joint. (c) Building plans – joint locations.

have been built successfully without expansion joints. Notwithstanding this, it is good practice to include expansion joints at abrupt changes in the plan dimensions of a building, as shown in Figure 14.53c, to avoid the stress concentrations and cracking that would otherwise occur at these locations.

It is important that movement joints in the concrete structure be accompanied by and be compatible with movement joints in the finishes, partitions and cladding attached to it. Movement in the concrete structure should not impose loads on the attached non-structural elements.

14.10.6 Structural joints

Structural joints allow free movement (translation and/or rotation) between two parts of a structure. The half joint of Figure 14.53b allows unrestrained translation along the axis of the member and unrestrained rotation and will serve the dual function of a structural hinge and an expansion/contraction joint.

When a hinge is required at the base of a column or when moment is not to be transferred from one element to another, the hinge joints shown in Figure 14.54 may be used. The joints must be able to transmit the imposed axial force and shear. It is essential that the strength and strain capacity of the concrete within the hinge is increased as high as possible using closely spaced confinement reinforcement, usually in the form of helical reinforcement.

When a concrete beam or slab is supported by a masonry wall, it is prudent to ensure that the concrete movements do not cause distress to the masonry.

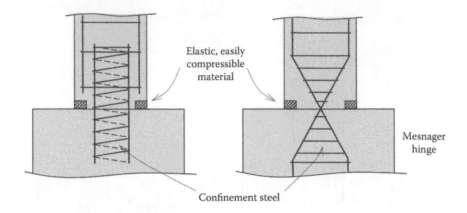

Figure 14.54 Structural hinge joints [17].

A sliding joint should always be used to break the bond between the concrete and the wall. Two layers of galvanised steel flashing (or equivalent) usually provide a satisfactory sliding joint between a concrete slab and a load-bearing masonry wall.

Isolation joints separate different parts of a structure and ensure that deformation in one part of the structure does not impose loads on the other. Differential shortening of adjacent columns in framed structures can induce significant moments and shears into floor slabs and beams. Isolation joints are frequently used to separate portions of a structure with significantly different sustained stress levels. For example tower columns in a multi-storey building may need to be separated from adjacent columns with low stress levels due to their long-term axial shortening. Structural members supporting vibrating machinery or pumps are usually isolated from the remainder of the structure. To avoid differential settlement problems, slabs on ground are isolated from the walls and columns passing through them. In structures subjected to seismic loads, parts with dissimilar mass and stiffness are separated so that they are able to oscillate without hammering against each other.

REFERENCES

1. AS3600-2009. (2009). *Australian Standard for Concrete Structures*. Standards Association of Australia, Sydney, New South Wales, Australia.
2. Park, R. and Paulay, T. (1975). The art of detailing (Chapter 13). In *Reinforced Concrete Structures*. New York: John Wiley and Sons, 769pp.
3. Leonhardt, F. (December 1965). Reducing the shear reinforcement in reinforced concrete beams and slabs. *Magazine of Concrete Research*, 17(53), 187–198.

4. Leonhardt, F. (1965). Über die Kunst des Bewehrens von Stahlbetontragwerken. *Beton-und Stahlbetonbau*, 60(8), 181–192; (9), 212–220.
5. Leonhardt, F. (1971). Das Bewehren von Stahlbetontragwerken. In *Beton-Kalender, Part II*. Berlin, Germany: Wilhelm Ernst & Sohn, pp. 303–398.
6. Leonhardt, F. and Teichen, K.T. (1972). *Druck-Stösse von Bewehrungstäben, Deutscher Ausschuss für Stahlbeton*, Bulletin No. 222. Berlin, Germany: Wilhelm Ernst & Sohn, pp. 1–53.
7. Leonhardt, F., Walther, R., and Dilger, W. (1964). *Schubversuche an Durchlaufträgern, Deutscher Ausschuss für Stahlbeton*, Heft 163. Berlin, Germany: Ernst & Sohn.
8. Concrete Institute of Australia. (2014). *Reinforcement Detailing Handbook*. Sydney, New South Wales, Australia: Concrete Institute of Australia.
9. Goto, Y. (1971). Cracks formed in concrete around deformed tension bars. *ACI Journal*, 68(4), 244–251.
10. Tepfers, R. (1979). Cracking of concrete cover along anchored deformed reinforcing bars. *Magazine of Concrete Research*, 31(106), 3–12.
11. Tepfers, R. (1982). Lapped tensile reinforcement splices. *Journal of the Structural Division, ASCE*, 108(1), 283–301.
12. Untrauer, R.E. and Henry, R.L. (1965). Influence of normal pressure on bond strength. *ACI Journal*, 62(5), 577–586.
13. Mayfield, B. et al. (May 1971). Corner joint details in structural lightweight concrete. *Journal ACI*, 68(5), 1971.
14. Swann, R.A. (November 1969). Flexural strength of corners of reinforced concrete portal frames. Technical Report TRA 434, Cement and Concrete Association, London, U.K.
15. Franz, G. and Niedenhoff, H. (1963). The reinforcement of brackets and short deep beams. Cement and Concrete Association Library Translation No. 114, London, U.K.
16. ACI 318-11M. (2011). Building code requirements for reinforced concrete. American Concrete Institute, Detroit, MI.
17. Fintel, M. (1974). *Handbook of Concrete Engineering*. New York: Van Nostrand Reinhold Company.

Index

ACI 209 (Committee), 74, 84
ACI 318M-11, 76, 109, 111, 219, 255, 265, 316, 352, 663
Actions
    characteristic (specified) actions, 30
    combinations for serviceability limit states, 30–31
    combinations for stability limit states, 30
    combinations for strength limit states, 29–30
    earth pressure, 26
    earthquake action, 26, 27
    expected actions, 31
    imposed action (live loads), 26–28
    liquid pressure 26
    permanent action (dead loads), 26
    snow action, 26, 27
    wind action, 26, 27
Aeberhard, H.U., 569
Age-adjusted effective modulus of concrete, 81
Age-adjusted effective modulus method, 142–144, 177
Age-adjusted modular ratio, 151
Age-adjusted transformed section, 151, 158
Ageing coefficient of concrete, 81
Aggregate interlock, 271
Alexande, S.D.B., 597, 606
Allowable stresses, see Stress limits
Alloy bar, 49, 51
Altouabat S.A., 106
Analogous truss, 290, 324, 344–346, 611, 632

Anchorage length, 313
Anchorage of deformed bars in compression, 625–628
Anchorage of deformed bars in tension, 616–625
Anchorage of longitudinal steel in beams, 628–629
Anchorage of stirrups, 271–273, 632–636
Anchorage set (slip) 186–187
Anchorage zones
    anchorage plates, 319
    bearing stresses, 319, 330–331
    bursting crack, 325
    bursting moment, 325
    methods of analysis, 322–331, 346–352
    post-tensioned members, 318–352
    pretensioned members, 314–318
    reinforcement requirements, 329–330
    single anchorage, 325–326, 331–334
    spalling moment, 326–327
    stress isobars, 320–323, 326, 329
    stress trajectories, 319–323, 347
    strut-and-tie analysis, 344–346
    strut-and-tie models, 324, 345
    symmetric prism, 327, 328, 337
    T-beam anchorage zone, 340–346
    transmission (transfer) length, 313–318
    transverse forces (bursting and splitting), 320

truss analogy, 324. 344–346
twin anchorages, 326–329,
    335–340
Angle of dispersion, 512
AS1012.9, 70
AS1012.10, 71
AS1012.11, 71
AS1012.13, 90
AS1012.16, 86
AS5100-2004, 137, 220
AS/NZS1170.0, 22, 26, 31, 44, 46
AS/NZS1170.1, 22, 26–29, 31
AS/NZS1170.2, 22, 26–28
AS/NZS1170.3, 22, 26, 29
AS/NZS1170.4, 22, 26
AS/NZS4671, 94–95
AS/NZS4672, 99–102
Attard, M.M., 73, 105

Balanced cantilever construction, 452
Balanced load, 9, 112, 485, 487, 494
Band-beam and slab systems, see Slabs
    (post-tensioned)
Bazant, Z.P., 81, 106, 220
Beam-type shear 299 (see Shear
    strength of beams)
Bearing failure, 319, 330–331
Beeby, A.W., 261, 265
Binding reinforcement, 631
Bischoff, P.H., 203, 220
Bond stress, factors affecting, 314–316
Bonded tendons, 4, 6, 59–60, 225–230
Branson, D.E., 47, 201, 220
Brooks, J.J., 90, 105, 106
Bresler, B., 587, 605
Bursting forces, 320
Bursting moment, 325

Cable profile and location, 7–9,
    122–123, 449–452, 471–474
Cantilever construction, 452
Capacity reduction factor, 31–34, 226,
    275, 280
Carreira, D.J., 81, 106
Carry-over factor, 463–464
Carry-over moment, 463
Chakrabarti, P., 569
Chu, K.-H. 81, 106
Circular prestressing, 61
Coefficient of thermal expansion for
    concrete, 92
Coefficient of thermal expansion for
    steel, 97

Collins, M.P., 73, 105, 349, 353
Column line tendons, 541–542
Column strip, 23, 519, 523, 541–543,
    545–547, 549–551
Combined load approach, 10–11, 14
Compatibility torsion, 288–289,
    291–292
Composite members
    advantages, 355–356
    behaviour, 356–358
    bond between mating surfaces, 356
    code provisions, 395–396
    determination of prestress, 361–363
    effective flange width, 358
    footbridge example, 369–392
    flexure-shear cracking, 400–401
    horizontal shear transfer, 393–395
    loading stages, 358–360
    short-term service-load analysis,
        364–366
    time-dependent service-load
        analysis, 366–369
    typical cross-sections, 357
    ultimate flexural strength, 392–393
    ultimate shear strength, 398–401
    web-shear cracking, 399–400
Compression members
    biaxial bending and compression,
        587–588
    braced and unbraced structures,
        589–590
    critical buckling load, 588–589
    design interaction curve, 586–587
    effective length, 589–590
    geometric non-linearity, 572
    moment magnifier method, 591–594
    primary and secondary moments,
        572–573, 592
    primary compressive failure, 575
    primary tensile failure, 575
    reinforcement requirements, 597,
        640–644
    slender columns, 572–573, 588–597
    slenderness effects, 588–594
    slenderness ratio, 592
    stocky columns, 593
    strength interaction curve, 573–586
    strength interaction surface, 587
    transmission of axial force through
        floor slab, 597–599
    types, 571
    ultimate strength analysis,
        575–579

Concordant tendons, 452,
    459–461, 494
Concrete components and properties
    admixtures, 66
    aggregates, 66
    biaxial strength envelope, 70
    cement replacements, 66
    coefficient of thermal expansion, 92
    composition, 66
    compressive strength
        (characteristic), 66–67, 70–71
    compressive strength (in-situ), 71
    confinement effects, 69
    creep strain, 77–82
    creep coefficient, 79–80, 85–88
    deformation of, 73–92
    density, 84
    elastic modulus, 76, 84–85
    hydration, 66
    instantaneous strain, 75–77
    normal class concrete, 70
    Poisson's ratio, 77
    shrinkage, 82–83, 88–92
    special class concrete, 70
    strain components, 73–75
    strength vs. time, 69
    stress limits, 109–110
    stress–strain curves, 68, 72–73
    tensile strength (direct, indirect and
        flexural), 69, 71–72
    water–cement ratio, 66
Concrete Institute of Australia, 663
Connections
    beam-column connection,
        644–651
    corbels, 651–654
    four member connection, 650–651
    knee-connection in closing bending,
        647–649
    knee-connection in opening
        bending, 645–647
    three member connection,
        649–650
Continuous members, see Statically
    indeterminate members
Control joints, 219, 656–660
Corbels, 651–654
Coupling of tendons, 57, 58, 628
Crack control, 215–219
    at openings and discontinuities, 219
    flexural cracks, 215–217
    restrained shrinkage cracks,
        217–219

Crack types
    direct tension (full depth), 165
    flexure-shear, 268, 276–277,
        400–401
    primary cracks, 165
    torsional cracks, 289–290
    web-shear, 268, 277–278, 399–400
Crack width limits, 39–40
Cracked section analysis, 164–172
Cracking moment, 165, 202
Creep of concrete
    creep coefficient, 79–80, 85–88
    creep function, 80
    creep-induced curvature, 208–209
    delayed elastic component, 78
    factors affecting, 77–78
    flow component, 78
    linear creep, 78
    losses, 190
    prediction of, 79–81
    recoverable and irrecoverable, 78
    specific creep, 79–80
    tensile, 81–82
    variation with time, 74
Critical buckling load, 588–589
Critical shear perimeter, 300–302
Cross, H., 462, 505
Cross-sectional analysis
    short-term, uncracked, 123–141
    short-term, cracked, 164–176
    time-dependent, uncracked,
        142–160
    time-dependent, cracked, 176–181
Cross-sections (sizing)
    initial trial dimensions, 407–410
    minimum dimensions based on
        flexural strength, 408–410
    minimum moment of inertia, 408
    minimum section modulus,
        115, 407
    types, 405–407
Crossing beam analogy, 536

Dapped-end joint, 639–640
Darwin, D., 105
Decompression moment, 222
Deflection
    allowable, 36–39
    approximate equations, 195–196
    calculation of, 193–215
    coefficients for slabs, 518, 536, 558
    creep, 116, 208–209
    double integration of curvature, 194

incremental, 37
instantaneous (short-term),
   201–207
limits, 36–39
long-term (time-dependent),
   207–215
problems, 36–39
shrinkage, 209–210
total, 37
Deformation of concrete, 73–92
Design action effects, 30, 31
Design procedures
   design for serviceability limit states,
      35–40, 107–219
   design for strength limit states,
      31–34, 221–265
   continuous beams, 484–504
   fully-prestressed beams (constant
      eccentricity), 427–437
   fully-prestressed beams (draped
      tendons), 410–427
   general requirements, 21–22,
      31–32, 35
   partially-prestressed beams,
      437–445
Design resistance effect, 31
Design strength, 31
Design strip, 23
Detailing, objectives, 609
Development length in
      compression, 626
   basic development length, 626
   refined development length, 626
Development length in tension, 616
   basic development length, 618
   refined development length, 620
Devine, P.J., 47
Diagonal compression struts, 270–271
Diagonal tension failure, 267
Dilger, W.H., 81, 105, 106, 663
Direct design method, 546–547
Distribution coefficient, 203
Distribution factor, 464
Disturbed region, 319
Dowel action, 271, 273
Drop panels, 508–509, 511–512
Ductility, 221–224, 226–227, 260–265
Durability
   design for, 40–42
   exposure classifications, 40–41
   minimum concrete cover, 43
   minimum curing requirements, 43
   minimum concrete strength, 43

Edge-supported two-way slabs, see
   Slabs
Effective area of a support, 301
Effective depth, 224
Effective length of columns, 589–590
Effective modulus of concrete, 80
Effective prestress, 107
Effective span, 38,
Effective width of flange, 255,
Elastic energy, 262
Elastic modulus of concrete, 76, 84–85
Elastic modulus of steel, 95, 100
Elastic stresses, calculation of, 9–16
Eldbadry, M., 105, 220
Equilibrium torsion, 288–289,
   292–299
Equivalent frame method, 544–546
Equivalent load method, 461–462,
   465–466, 472, 476, 481, 486,
   494
Eurocode, 2, 202, 220, 316, 352
Exposure classifications, 40–41
External prestressing, 62–63

Fan mechanisms, 565
Faulkes, K.A., 112, 219, 325, 353, 505
Favre, R., 105, 220
Finite element method, 517, 520
Fib Model Code 2010, 81, 106
Fintel, M., 663
Fire mesh, 608
Fire resistance
   design for, 43
   fire resistance period (FRP), 43
   minimum section dimensions,
      44–46
Fixed-end moments, 463–464
Flat-ducted tendons, 56, 508, 509, 514,
   532, 553, 555
Flat jacks, 6
Flat plates, see Slabs (post-tensioned)
Flat slabs, see Slabs (post-tensioned)
Flexibility coefficient, 32457, 460
Fixed-end moment, 463–464
Flexural behaviour
   general, 15–19
   overloads, 221–225
Flexural cracks, 165
Flexural strength theory, 224–240
   bonded tendons, 227–245
   design calculations, 247–254
   doubly reinforced sections, 233–
      240, 251–254

flanged sections, 254–260
idealized rectangular stress blocks, 224–227
singly-reinforced section, 229–233, 247–250
trial and error procedure, 229–233
ultimate moment, 224, 226, 229–240, 242–245, 247
unbonded tendons, 245–247
Flexure-shear cracking, 268, 276–277, 400–401
Force method, (flexibility method), 454–458
Foster, S.J., 320, 349, 353
Franz, G., 651, 663
Friction losses, 112, 183–185
Fully-prestressed concrete, 16, 18, 114, 116, 405, 407, 410–437

Gergely, P., 325, 353
Ghali, A., 74, 105, 220
Gilbert, R.I., 47, 74, 81, 84, 105, 106, 124, 142, 220, 352, 363, 403, 505, 518, 568
Goto, Y., 663
Guyon, Y., 320–322, 327, 328, 334, 343, 352

Hall, A.S., 300, 312
Helical reinforcement in columns, 642
Henry, R.L., 663
Headed bars, 630
Hicks, S.J., 49
Hilsdorf, H.K., 105, 106
Hinge rotation, 262–263
Hognestad, E., 311
Hooks and cogs, 622–623, 627
Hoyer, E., 314, 315, 352
Hoyer effect, 314–315
Hydration, 66
Hydraulic jacks, 54–56
Hyperstatic reactions, 452–454, 457–461, 465, 478

Inclined cracking, 267–268, 270–271, 273
Internal couple concept, 11–12, 14
Irwin, A.W., 27
Iyengar, K.T.S.R., 320, 353

Jennewein, M., 353
Jensen, J.J., 72, 105
Johansen, K.W., 561, 569

Joints in structures
construction joints, 656–657
control joints, 657–660
expansion joints, 660–661
shrinkage strips, 660
structural joints, 661–662
Jungwirth, D., 106

Korkosz, W., 352
Kupfer, H., 105

Lange, D.A., 106
Lapped splices in columns, 641
Lapped splices in compression, 628
Lapped splices in tension, 625
Leonhardt, F., 608, 631, 635, 641, 662, 663
Limit state design, 21–22
Lin, T. Y., 112, 219, 505
Linear transformation, 459–461, 470
Load patterns, 495
Load balancing, 9, 12–13, 120–122, 486
Load factors and combinations
for serviceability, 30–31
for stability, 30
for strength, 29–30
Loads, see Actions
Logan, D.R., 352
Loov, R. E., 101, 106, 241, 265
Losses of prestress, 65, 97, 101–104, 181–193
anchorage (slip), 186–87
creep, 190
elastic deformation, 182–183
friction, 183–185
immediate, 181–187
relaxation, 101–104, 190–191
shrinkage, 188–190
time-dependent, 181–182, 187–193

Magnel, G., 112, 219, 325, 353
Malik, A.R., 349, 353
Marshall, W.T., 318, 352
Marti, P., 346, 353
Martin, L.D., 352
Matt, P., 569
Mattock, A.H., 318, 352, 505
Mayer, H., 47
Mayfield, B., 663
Membrane action 520, 566
Mesnager hinge, 662

Metha, P.K., 65, 105
Mickleborough, N.C., 47, 518, 568
Middle strip, 23, 519, 541–543, 545,
    547, 549–551
Mitchell, D., 349, 353
Modular ratio, 131
Mohr's circle of stress, 268–269
Moment-area methods, 454, 470, 474
Moment distribution, 454, 461–465,
    467, 468
Moment-curvature relationship, 17,
    196–201
Moment magnifier method, 591–594
Moment of inertia
    average, 201, 202
    cracked section, 201
    effective, 201, 202
    uncracked section, 201
Montiero, P.J., 65, 105

Nawy, E.G., 569
Neville, A.M., 65, 74, 77, 81, 105
Niedenhoff, H., 651, 663
Nilson, A.H., 105, 505, 548, 551, 568
Nonlinear stress analysis, 34

Oh, B.H., 81, 106
One-way slabs, *see* Slabs
    (post-tensioned)
Ospina, C.E., 5, 97, 606
Ostergaard, L., 106
Over-reinforced beams, 223,
    260–261

Park, R., 608, 634, 647, 648, 650, 662
Partial safety factors, 31
Partially-prestressed concrete, 18, 39,
    116, 437–445
Paulay, T,. 608, 634, 647, 648,
    650, 662
Pauw, A., 76, 85, 105
Pecknold, D.A., 105
Permissible stresses, *see* Stress limits
Perry, C.J., 352
Plastic analysis, 25, 262–265, 561–566
Plastic energy, 262
Plastic hinge, (constant moment hinge),
    262–263, 561
Poisson's ratio for concrete, 77
Porasz, A., 73, 105
Post-tensioning
    anchorage wedge components, 56
    anchorage zones, 318–352

bonded *vs.* unbonded construction,
    59–60
coupling and intermediate
    anchorages, 58
dead end anchorages, 57
ducts, 53–54
friction, 183–185
grouting ducts, 54, 59
hydraulic jacks, 54–56
live end anchorages, 57–58
procedure, 5–6, 53–60
profiles, 7–9, 122–123, 449–452,
    471–474
stage stressing, 53
Post-Tensioning Institute, 515, 568
Precast elements, 356–358
Precast pretensioned trough girder, 369
Pressure line, 458–459
Prestressed concrete
    basic concepts, 1–19
    benefits, 1–2
    circular prestressing, 61
    external prestressing, 62–63
    introductory examples, 2–4, 13–16
    methods of prestressing, 4–6
    prestressing force, transverse
        component, 6–9
Prestressing steel
    cable curvature, 8
    cable layout, 7–9
    relaxation, 101–104
    strain components, 227–229
    types (wire, strand, bar), 49–51
Pretensioning
    anchorage zones, 314–318
    bed, 52
    multi-strand pretensioning, 51–52
    procedures, 4–5, 51–53
    single-strand pretensioning, 51–52
Primary direction (in a restrained
    slab), 217
Principal tensile stress, 268–269, 278
Principle of virtual work, 454–456,
    562–564
Profile of tendons, *see* Cable profile
    and locations
Punching shear strength, 299–311
    critical shear perimeter, 300–302
    design equation, 307
    edge column, 309–311
    effective area of support, 301
    forces on critical shear perimeter, 301
    interior column, 307–309

minimum moment transferred to
    column, 548
shear reinforcement details, 305
torsion strips, 303
with moment transfer, 303–307
with NO moment transfer, 302–303

Rangan, B.V., 300, 301, 304, 305, 312
Ranzi, G., 47, 74, 81, 105, 124, 142,
    220, 363, 403, 505, 602, 606
Raphael, J.M., 105
Reinforced concrete, 1–2
Reinforcement, see Steel reinforcement
Relaxation of steel, 101–104, 190–191
Ritter, W., 270, 311
Ritz, P., 569
Robustness
    design for, 43–46
    quantifying, 263–265
Rogowsky, D.M., 320, 353
Rose, D.R., 352
Rüsch, H., 47, 78, 105, 106
Russell, B.W., 352

Sargious, M., 320, 353
Schäfer, K., 353
Schlaich, J., 346, 353
Schlub, P., 569
Secondary direction (in a restrained
    slab), 217
Secondary moments and shears, 449,
    452, 453, 457–462, 465,
    467–471, 480–481, 491–492
Section modulus, minimum required,
    115, 407
Serviceability limit states
    design for, 35–40, 107–219
    load combinations, 30–31
Setunge, S., 73, 105
Sign convention, see Notation
Shear-compression failure, 268
Shear strength of beams
    anchorage of longitudinal
        reinforcement, 275–276
    anchorage of stirrups, 271–273,
        632–636
    concrete contribution, 273,
        276–278
    critical section, 279, 299
    design equation, 280–281
    design requirements, 279–280
    maximum and minimum
        strength, 275

maximum spacing of
    stirrups, 280
minimum transverse steel,
    275, 280
steel contribution, 273–274
ultimate strength, 273
web steel requirements, 279–280
Shrinkage of concrete
    chemical (autogenous) shrinkage, 82
    drying shrinkage, 82–83
    endogenous shrinkage, 83
    factors affecting, 82–83
    loss due to, 188–190
    plastic shrinkage, 82
    prediction of, 88–92
    shrinkage-induced curvature,
        209–210
Simpson's rule, 456
Singly-reinforced section, 229–233
Slabs (post-tensioned)
    balanced load stage, 513–515
    band-beam and slab systems, 508,
        515, 567–568
    bonded vs. unbonded tendons,
        508–509
    calculation of slab thickness,
        515–522
    cracking in slabs, 217–219,
        522–523
    edge-supported two-way slabs,
        507, 513, 515–518, 520,
        526–540
    effects of prestress, 510–513
    finite element modelling, 517, 520
    fire resistance, 43, 46, 515–516
    flat plate slabs, 507, 508, 513,
        516–520, 540–564
    flat slabs with drop panels, 23–24,
        507, 508, 513, 516–520,
        540–564
    frame analysis, 544–546
    one-way slabs, 507, 508, 515–519,
        525–526
    shear strength, see Punching shear
        strength
    span-to-depth ratios, 516–522
    yield line analysis of flat slabs,
        561–566
Slab system factor, 517, 519–520
Slate, F.O., 105
Slender columns, 572–573, 588–594
Slenderness ratio, 592
Smith, A.L., 47

Sozen, M.A., 325, 353
Spalling moment, 326–327
Span length, 24
Span-to-depth ratio
    edge-supported slabs, 516–522
    flat slabs, 516–522
    one-way slabs, 516–521
Spandrel beam, 288–289, 291, 300,
        301, 303, 305, 306
Spandrel strip, 326–327
Spiral reinforcement in columns, 642
Stage stressing, 53, 356, 514, 602
Stang, H., 106
Static moment, 542, 546–548
Statically indeterminate members
    advantages and disadvantages,
        447–449
    concordant tendons, 452,
        459–461, 494
    design of continuous beams,
        484–504
    design steps, 492–495
    effects of creep, 476–478
    equivalent load method, 461–462,
        465–466
    fixed-end moments, 463–464
    frames, 480–484
    hyperstatic reactions, 452–454,
        457–461, 465, 478
    linear transformation,
        459–461, 470
    moment distribution, 454, 461–465,
        467, 468
    moment redistribution at ultimate,
        490–491
    non-prismatic members, 475–476
    pressure line, 458, 459
    primary moments and shears,
        452–461, 475, 477, 480–481,
        485, 491, 494
    secondary effects at ultimate,
        491–492
    secondary moments and shears,
        449, 452, 453, 457–462, 465,
        467–471, 480–481, 491–492
    tendon profiles, 449–452, 471–474
    tertiary effects, 381
    virtual work, 453–457
Steel reinforcement, 92–97
    characteristic yield stress, 94–95
    coefficient of thermal expansion, 97
    elastic modulus, 95
    peak stress, 94–95

strength and ductility, 94–95
stress–strain relationships, 94–97
uniform elongation, 94–95
Steel used as tendons
    alloy bar, 49, 51
    creep coefficient, 102–103
    elastic modulus, 100
    relaxation, 101–104
    strand, 49–51
    stress–strain curves for tendons,
        98–101
    tensile strength, 98–100
    wire (cold drawn, stress-relieved),
        49–50
    yield stress, 98–100
Stewart, M.G., 73, 105
Stiffness coefficient, 463–464
Stirrups, 270–275, 280–281, 289–294,
        632–636
Strand, 49–51
Strength of concrete
    biaxial, 69–70
    characteristic compressive, 66–67
    cylinder vs. cube, 67
    gain with age, 69
    tensile, 69
    uniaxial, 66–68
Strength. see Flexural strength; Shear
        strength; Torsional strength
Strength limit states
    design for, 31–34
Strength reduction factor, see Capacity
        reduction factor
Stress analysis, 32–34
Stress block, rectangular, 224–227
Stress limits
    concrete, 109–110
    steel, 111
    satisfaction of, 112–120
Stress–strain relationships
    concrete, 75–76, 126, 142–144, 146
    prestressing steel, 98–99, 126, 146
    reinforcing bars, 95–97, 126, 146
Stress isobars, 320–323, 326, 329
Stress trajectories, 319–323, 347
Structural analysis, 24–26
Structural modelling, 22–24
Strut-and-tie analysis
    bursting reinforcement in struts,
        349–350
    concrete struts, 347–350
    design requirements, 34
    divergence angle, 348

nodes, 351–352
steel ties, 350
strut efficiency factor, 349
Strut-and-tie models, 324, 652
St. Venant's principle, 319
Superposition principle, 17, 121,
    176, 453
Support and loading points,
    636–638
Suspension reinforcement,
    638–639
Swann, R.A., 663
Symmetric prism, 327, 328, 337

Tasuji, M.E., 105
Teichen, K.T., 641, 663
Tellenbach, Ch., 569
Tendon profile, see Cable profile and
    location
Tension members
    advantages and applications, 599
    axial deformation, 602
    behaviour, 599–602
    design example, 603–605
Tension stiffening effect, 165, 166,
    199, 200
Tepfers, R., 616, 663
Theorem of complementary shear
    stress, 394
Thorenfeldt, E., 72, 105
Thornton, K., 505
Tie arrangements in columns,
    642–643
Tomaszewicz, A., 72, 105
Torsional constant, 292
Torsional cracking, 289–290, 292
Torsional strength
    additional longitudinal steel,
        291–292, 294
    before cracking, 292
    closed stirrups required, 293–294
    design equation, 294
    detailing of stirrups, 294–295
    minimum area of closed stirrups, 291
    truss analogy (3-D), 290
    web crushing, 294
Transfer length, 313–318
Transfer of prestress, 108–111
Transformed sections, 131–132, 177
Transmission length(see transfer
    length)
Transverse reinforcement in columns,
    642–644

Transverse forces imposed by tendons,
    6–9
Trost, H., 81, 106

Ultimate curvature, 226–227
Ultimate flexural strength, 221–265
    approximate code procedures, 241–7
    assumptions, 224
    doubly-reinforced cross-sections,
        233–240
    general, 23, 221–224
    idealised rectangular stress block,
        224–227
    singly-reinforced cross-sections,
        227–233
Ultimate load stage, 18, 221
Ultimate strength in shear and torsion,
    32, 267–312
Unbonded construction, 6, 59–60,
    245–247
Under-reinforced beams, 223
Untrauer, R.E., 663
Upper bound approach, 565

Variable angle truss model, 274
Vibration control, 39
Virtual force, 454–455
Virtual work, see Principle of virtual
    work
VSL prestressing, 512, 568
Volume integration, 456–457

Walters, D.B., 548, 551, 568
Walther, R., 635, 663
Warner, R.F., 112, 219, 325,
    353, 505
Web-crushing, 271, 275
Web-shear cracking, 268, 277–288,
    399–400
Weights of construction
    materials, 27
Wide beam method for slab deflection,
    548–551
Willford, M.R., 47
Wire mesh, welded, 92, 95
Wires, 49–50
Work products, external and internal,
    454–456, 562–564

Yield line theory, 561–566
Yield stress, 94–96, 98–100
Yogananda, C.V., 320, 353
Young, P., 47

Printed and bound by CPI Group (UK) Ltd, Croydon, CR0 4YY

23/10/2024

01777695-0005